Jeff
517
Davis, Calif. 95616

Linear Algebra with Applications
Including Linear Programming

The Appleton–Century Mathematics Series

Raymond W. Brink and John M. H. Olmsted, Editors

Linear Algebra with Applications
Including Linear Programming

HUGH G. CAMPBELL
*Virginia Polytechnic Institute
and State University*

New York

APPLETON–CENTURY–CROFTS
EDUCATIONAL DIVISION
MEREDITH CORPORATION

Copyright © 1971 by

MEREDITH CORPORATION

All rights reserved

This book, or parts thereof, must not be used or reproduced in any manner without written permission. For information address the publisher, Appleton–Century–Crofts, Educational Division, Meredith Corporation, 440 Park Avenue South, New York, N. Y. 10016.

781–2

Library of Congress Card Number: 72-133902

PRINTED IN THE UNITED STATES OF AMERICA

390-16712-6

*To my wife
Allen*

Preface

For many years calculus has served as the central freshman-sophomore mathematics course in college. One reason for this has been its frequent application in the engineering and physical sciences. Now, however, there is considerable evidence that the material called linear algebra rivals calculus in its usefulness in these disciplines;* moreover, linear algebra is fundamental in the rapidly increasing quantification that is taking place in the management and social sciences. Also, for the mathematics majors, experience has demonstrated that linear algebra serves as a satisfactory introductory course in the sequence of algebra courses that an undergraduate degree normally requires. It is not surprising, therefore, to find that many of the more recent freshman-sophomore calculus texts include several chapters pertaining to linear algebra. In fact, the Committee on the Undergraduate Program in Mathematics has recommended that a separate linear algebra course be taught in the sophomore year in the midst of three calculus courses.

This book is designed as a text for a linear algebra course at a level comparable to the beginning calculus sequence. Actually, such a linear algebra course may parallel or follow the calculus sequence, or interrupt it as recommended by the CUPM. In fact, by leaving out some of the applications that use calculus (designated with the symbol ★) this book could serve as a text for a linear algebra course that precedes Calculus.

Because many classes in elementary linear algebra will include non-mathematics majors, and because of the importance of motivation in the learning process, an unusual effort has been made to interest the reader by means of numerous and diverse examples that are given in subsections at the end of most sections. Although these subsections are labeled "Applications,"

---

* For instance, see Part IV of the book, *Applications of Undergraduate Mathematics in Engineering* by Ben Noble, The Macmillan Company, New York, 1967.

they are not designed as a comprehensive list of real world applications of linear algebra. Rather, they are meant to point out to the reader in a very limited way that the material he is studying is relevant to a broad spectrum of disciplines; consequently, they may motivate his investigation of other sources as well as his mastery of the linear algebra presented in this book. The so-called applied examples fall into four broad categories: (1) artificial or over-simplified models presented in an applied context, such models often being very useful in that they sometimes point the way toward genuine applications or are a first approximation of a real world situation; (2) realistic models that can be or are used currently in applications in the real world; (3) references to the latter type of application, and brief descriptions of them; and (4) mathematical applications. Many of the applications are documented to permit further investigation by the reader.

The theory is carefully presented: mathematical structure has been emphasized, with definitions, theorems, and proofs carefully set apart. Because of the elementary nature of the course, the proofs are presented with more than ordinary care and detail; many proofs are written (especially at first) in a formal statement-reason form.

At the end of some of the chapters a few special projects are suggested; the purpose of these projects is to encourage, and to give some direction to, individual study by the student. The text presupposes a knowledge of some trigonometry and analytic geometry, and various topics of calculus are assumed in some applications. These particular applications can be omitted, however, if the reader does not have an adequate background.

The text is adaptable to courses of different lengths and purposes. Chapters 1 through 9 form the core of the text; Chapters 10 and 11 are optional and are independent of each other. All of the applications are optional, and the format of the book allows the instructor to alter the emphasis of the course easily by omitting applications or proofs as desired. A desirable one-semester course would be the first nine chapters. Time restrictions, however, will probably impose some omissions. The following Sections may be omitted without destroying the continuity of the first nine chapters: 5.4, 6.5, 6.6, 7.4, 7.5, 7.6, 8.6, 8.7, and 8.8. The instructor can select the sections to be omitted to suit the needs of his particular class. A one-quarter course will probably require the omission of all of the sections listed above, plus a shift of emphasis away from some sections to allow more rapid coverage. A two-quarter course might well include a thorough coverage of the first nine chapters or sufficient omissions from those chapters to allow study of Chapter 10 or Chapter 11. A full year course, meeting three hours a week, will allow a comprehensive study of the whole book including the applications.

The guiding principle in the development of this textbook has been consideration for the student. Hopefully, the list of new vocabulary at the end of each chapter, the attempted motivation through the subsections on applications, the detailed proofs of theorems, the numerous exercises, and the many examples illustrating the concepts should all give evidence of the sincerity of the author in this regard.

The author is grateful to the many people who have assisted in the production of this book. Special appreciation is expressed to Dr. R. W. Brink and Professor R. E. Spencer for their many valuable suggestions; they deserve considerable credit for any success that this book may enjoy. Special appreciation also is expressed to my wife, Allen, who did most of the typing and who has been a constant source of encouragement.

H.G.C.

Contents

Preface   vii

1   BASIC DEFINITIONS

1.1	An Orientation	1
1.2	Matrices	3
1.3	Matrix Addition and Scalar Multiplication	10

2   SYSTEMS OF LINEAR EQUATIONS

2.1	An Introduction	16
2.2	Elementary Row Operations	24
2.3	The Gauss-Jordan Method	28

3   MATRIX MULTIPLICATION

3.1	An Introduction	40
3.2	The Inverse Matrix	52
3.3	Elementary Matrices	61
3.4	A Necessary and Sufficient Condition for the Existence of the Inverse	67

4   SPECIAL MATRICES

4.1	Symmetric and Skew-Symmetric Matrices	76
4.2	Hermitian and Skew-Hermitian Matrices	88
4.3	Coordinate Vectors	94

4.4	Orthogonal and Unitary Matrices	101
4.5	Diagonal and Triangular Matrices	104

5 DETERMINANTS

5.1	Definition of the Determinant of a Square C-Matrix	111
5.2	Cofactor Expansion	116
5.3	Properties of Determinants	121
5.4	Proofs of Determinant Theorems (optional)	127
5.5	Rank of a Matrix	131
5.6	Rank and Linear Systems	135
5.7	A Formula for A^{-1}	139

6 ALGEBRAIC SYSTEMS

6.1	Groups	149
6.2	Rings	154
6.3	Fields	157
6.4	Vector Spaces	161
6.5	Linear Algebras (optional)	166
6.6	Boolean Algebras (optional)	169

7 VECTOR SPACES

7.1	Linear Dependence and Independence	177
7.2	Basis of a Vector Space	185
7.3	Coordinate Vector Representation of Abstract Vectors	192
7.4	Inner Product Spaces (optional)	198
7.5	Length, Distance, and Angle (optional)	203
7.6	Gram-Schmidt Orthogonalization (optional)	210

8 LINEAR TRANSFORMATIONS

8.1	Definition and Examples	217
8.2	Matrix Representation of a Linear Transformation	226
8.3	Linear Operators	232
8.4	Change of Basis	239
8.5	The Effect of a Change of Basis on a Transformation Matrix	249
8.6	Linear Transformations of $V(F)$ into $U(F)$ (optional)	255
8.7	Change of Basis for $V(F) \xrightarrow{T} U(F)$ (optional)	259
8.8	Linear Algebra of Linear Transformations (optional)	263

9 CHARACTERISTIC VALUES AND VECTORS

9.1	Basic Definitions	269
9.2	Theorems	279
9.3	Diagonalization of Real Quadratic Forms	286

10 TRANSFORMATIONS OF MATRICES (optional)

10.1	Introduction	295
10.2	Similarity	297
10.3	Congruence	304
10.4	Orthogonal Transformations	314
10.5	Orthogonal Congruence	319
10.6	Definite and Semidefinite Forms	327

11 LINEAR PROGRAMMING (optional)

11.1	Introduction	335
11.2	Basic Definitions	342
11.3	A Fundamental Theorem	347
11.4	Introduction to the Simplex Method	351
11.5	Selection of the Vector to Leave the Basis	355
11.6	Selection of the Vector to Enter the Basis	363
11.7	Unbounded Feasible Set with No Optimal Solution	371
11.8	The Simplex Method and Tableau	375
11.9	Artificial Variables	382
11.10	The Revised Simplex Method	389

REFERENCES	395
APPENDICES	A1
ANSWERS TO SELECTED ODD-NUMBERED EXERCISES	A9
INDEX	A39

Linear Algebra with Applications
Including Linear Programming

1

BASIC DEFINITIONS

1.1 An Orientation

A fundamental concept of mathematics is that of a ***transformation*** of the elements of a given nonempty set A into uniquely determined elements of another (or the same) given nonempty set B. Such a transformation T frequently is called a ***function*** or a ***mapping*** from A into B and is designated $T: A \to B$. The set A is called the ***domain*** of the transformation, and the set B is called the ***codomain*** of the transformation.

[margin note: Thus a nonempty set maps into itself]

> **Example 1.** Consider the transformation, T, that transforms every real number, x, into the real number $2x + 1$. The effect of this transformation on elements of the domain can be designated by $x \overset{T}{\mapsto} (2x + 1)$ or by $x \overset{T}{\mapsto} T(x)$, where $T(x) = 2x + 1$; we say that $T(x)$ is the ***image*** of x. The transformation of $x = 1$ to $T(x) = 3$ by this mapping is shown in Figure 1.1.1. The codomain, as well as the domain, is the set of real numbers.

> **Definition 1.** *If for every element, x, of a set A there corresponds a unique element $T(x)$ of a set B, then the collection T of these correspondences is called a **transformation** (or **mapping** or **function**) from A into B. $T(x)$ is called the **image** of x under the transformation T. A transformation T from domain A into codomain B will be designated by $T: A \to B$ or by $A \overset{T}{\to} B$.*

[handwritten note: B may have elements other than the elements T(x) corresponding to elements x in A.]

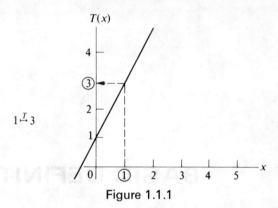

Figure 1.1.1

Example 2. Since for every element x of the set of integers $\mathscr{I}$ there corresponds a unique integer x^2, we say that the collection or set of such correspondences, $x \stackrel{T}{\mapsto} x^2$, is a transformation from the set of integers into the same set. A few of the correspondences are $1 \stackrel{T}{\mapsto} 1$, $-2 \stackrel{T}{\mapsto} 4$, $3 \stackrel{T}{\mapsto} 9$, $4 \stackrel{T}{\mapsto} 16$, $-5 \stackrel{T}{\mapsto} 25$, and $6 \stackrel{T}{\mapsto} 36$. It is customary to express these correspondences as ordered pairs $(1, 1)$, $(-2, 4)$, $(3, 9)$, $(4, 16)$, $(-5, 25)$, $(6, 36)$, etc. The transformation $T: \mathscr{I} \to \mathscr{I}$ can thus be expressed as a set of ordered pairs $T = \{(x, T(x)) \mid T(x) = x^2\}$.

The subset of the codomain consisting of those elements that are actually images of one or more elements of the domain is called *the **image set*** or ***range*** of the transformation. In Example 2, the domain is the set of integers $\mathscr{I}$, and the range is the set $\{0, 1, 4, 9, 16, \ldots\}$. If the range of a transformation T is equal to the codomain B, we can say that T maps the domain A *onto* B. Example 1 is an illustration of a transformation of the set of real numbers *onto* the set of real numbers. In Example 2, however, the transformation does not map the set of integers *onto* the set of integers. In both examples, T maps the domain *into* the codomain.

The primary purpose of this text is to study a special type of transformation known as a ***linear transformation***. Such a study is intimately associated with a study of matrices. We shall begin with the concept of a matrix as an element of an algebraic system and subject to certain operations; then, as we progress through the book, we shall repeatedly introduce and use the matrix as an instrument effecting linear transformations. Note that this dual role of matrices is similar to that of natural numbers: the number 2, for example, is an element subject to various operations, but the number 2 may also serve as an instrument effecting the transformation " double " (for example $x \stackrel{T}{\mapsto} 2x$).

A large and rapidly growing portion of applied mathematics stems from the study of linear transformations; one objective of this book is to point out a few of the many such applications, in the hope that the reader will be convinced of the relevance of the material he is studying.

EXERCISES

In each of the Exercises 1–6, for the given transformation T
(a) determine the domain and range of the transformation;
(b) determine the images of the elements of the domain;
(c) express the transformation as a set of ordered pairs;
(d) draw a diagram or a graph illustrating the transformation.

1. The transformation T maps a into p, b into q, and c into q, where $a, b, c, p,$ and q are distinct elements of a set.
2. The transformation T maps all odd positive integers into 0, and maps all even positive integers into 1.
3. The transformation T maps every real number x into the real number $x^3 + 1$.
4. The transformation T maps every nonnegative integer k into k.
5. The transformation
$$T = \{(k, T(k)) \mid T(k) = 2k + 1, k \text{ is a nonnegative integer}\}.$$
6. The transformation
$$T = \{(x, T(x)) \mid T(x) = 3x + 2, x \text{ is a real number}\}.$$
7. Let $A = \{9, 4\}$ and $B = \{3, 2, -3\}$. Why is the set of ordered pairs $\{(9, 3), (4, 2), (9, -3)\}$ not a transformation from A onto B?

1.2 Matrices

Undoubtedly the reader is familiar with rectangular arrays of data, such as the box score of a baseball game or a price chart for different sizes of a certain item. Such rectangular arrays when subject to certain operations are examples of matrices.

Definition 2. *Let m and n be positive integers; a rectangular array of entries arranged in m rows and n columns*

$$A = \begin{bmatrix} a_{11} & a_{12} & \cdots & a_{1n} \\ a_{21} & a_{22} & \cdots & a_{2n} \\ \cdots & \cdots & \cdots & \cdots \\ a_{m1} & a_{m2} & \cdots & a_{mn} \end{bmatrix},$$

*where the entries are elements of some algebraic system S, is called an **m by n matrix over S**.*[1]

---

[1] An alternate definition of a matrix is: *For specified positive integers m and n, an **m by n matrix over S** is a function whose domain is the set of ordered pairs of integers (i, j), where $1 \leq i \leq m$, and $1 \leq j \leq n$, and whose range is a subset of the codomain S; the rectangular array is a display of the elements of the range of the function.*

Until the study in Chapter 6 of certain algebraic systems, we shall assume that S is the complex number system. We call a matrix over the system of complex numbers a **C-matrix**. If, however, we further restrict the entries to the real numbers, then we say that the matrix is a **real matrix** or an **R-matrix**.

Example 1. The matrices

$$\begin{bmatrix} 2 & 3 \\ 4 & 0 \end{bmatrix}, \quad [\tfrac{1}{2} \quad 2 \quad 4], \quad \begin{bmatrix} \sqrt{3} & i & 0 \\ 1 & \tfrac{3}{2} & 2 \end{bmatrix}$$

are examples of C-matrices, or matrices over the system of complex numbers; furthermore, the first two matrices are R-matrices, or matrices over the system of real numbers.

In this book, when appropriate, the terminology "C-matrix" or "R-matrix" is used rather than the word "matrix"; this is done to impress upon the reader the role of the algebraic system to which the entries belong in the theory and applications of matrices. It also lays the groundwork for the discussion, in later chapters, of matrices over other systems.

If all of the entries of a matrix are zero, the matrix is called a **zero** or **null matrix** and is denoted by **0**; the bold print is used to distinguish a zero matrix from the number zero. The subscript i of the entry a_{ij} designates the *row* in which the entry appears, and the subscript j designates the *column* in which the entry appears; the double subscript ij is called the **address** of the entry. A matrix in which there are m rows and n columns of entries is said to have the **order** "**m by n**" (or **m** × **n**); the number of rows is always stated first. A 1 by n matrix, which consists of a single row, is called a **row matrix** or **row vector**, and an m by 1 matrix is called a **column matrix** or a **column vector**. A matrix is said to be a **square matrix** of **order n** if it has precisely n rows and n columns of entries. The **main diagonal** of a square matrix consists of the entries for which $i = j$ (that is, $a_{11}, a_{22}, \ldots, a_{nn}$). Frequently an m by n matrix A will be expressed as $A = [a_{ij}]_{(m,\,n)}$; this is referred to as the abbreviated notation, where i varies from 1 to m, and where j varies from 1 to n.

Example 2. The matrix

$$A = \begin{bmatrix} 4 & 2 & \sqrt{3} \\ 0 & 5 & 1 \\ 7 & \tfrac{3}{2} & 8 \end{bmatrix}$$

is a square R-matrix of order 3, where the entries 4, 5, 8 constitute the main diagonal. The entry a_{31} is 7 (third row, first column). Another matrix,

$$A = \begin{bmatrix} a_{11} & a_{12} & a_{13} \\ a_{21} & a_{22} & a_{23} \end{bmatrix},$$

can be written by using the abbreviated notation $A = [a_{ij}]_{(2,\,3)}$.

If all of the entries on the main diagonal of a square matrix are 1's and all other entries are 0's, the matrix is called an **identity matrix**, I_n, where n denotes the order.

Example 3.

$$I_3 = \begin{bmatrix} 1 & 0 & 0 \\ 0 & 1 & 0 \\ 0 & 0 & 1 \end{bmatrix}.$$

Definition 3. *If some rows or columns (or both) of a matrix A are deleted, the remaining array is called a **submatrix** of A. Also, it is customary to treat A as a submatrix of itself.*

Example 4. For

$$A = \begin{bmatrix} 3 & 2 & 1 \\ 4 & 6 & 9 \end{bmatrix},$$

we have the following submatrices:

$$\begin{bmatrix} 3 & 2 & 1 \\ 4 & 6 & 9 \end{bmatrix}, \begin{bmatrix} 3 & 2 \\ 4 & 6 \end{bmatrix}, \begin{bmatrix} 3 & 1 \\ 4 & 9 \end{bmatrix}, \begin{bmatrix} 2 & 1 \\ 6 & 9 \end{bmatrix}, \begin{bmatrix} 3 \\ 4 \end{bmatrix}, \begin{bmatrix} 2 \\ 6 \end{bmatrix}, \begin{bmatrix} 1 \\ 9 \end{bmatrix},$$

$$[3 \ 2 \ 1], \ [4 \ 6 \ 9], \ [3], \ [2], \ [1], \ [4], \ [6], \ [9],$$

$$[3 \ 2], \ [3 \ 1], \ [2 \ 1], \ [4 \ 6], \ [4 \ 9], \ [6 \ 9].$$

Sometimes it is convenient to divide a matrix into submatrices by means of broken lines. Thus we can write

$$A = \begin{bmatrix} A_{11} & A_{12} \\ A_{21} & A_{22} \end{bmatrix} = \begin{bmatrix} a_{11} & a_{12} & a_{13} \\ a_{21} & a_{22} & a_{23} \\ a_{31} & a_{32} & a_{33} \end{bmatrix}.$$

When this has been done we say that A has been **partitioned**.

Definition 4. *Two matrices of the same order, $A = [a_{ij}]_{(m,n)}$ and $B = [b_{ij}]_{(m,n)}$, are **equal** if and only if $a_{ij} = b_{ij}$ for all i and j.*

In other words, two matrices of the same order are equal if and only if their corresponding entries are equal.

Example 5.

(a) $A = \begin{bmatrix} 3 & \sqrt{4} & 2 \\ \frac{6}{2} & 1 & 0 \end{bmatrix}$, $B = \begin{bmatrix} 3 & 2 & 2 \\ 3 & 1 & 0 \end{bmatrix}$, $A = B$.

(b) $C = \begin{bmatrix} x & 3 \\ 4 & 1 \end{bmatrix}$, $D = \begin{bmatrix} 2 & 3 \\ 4 & 1 \end{bmatrix}$, $C = D$ only when $x = 2$.

(c) $E = \begin{bmatrix} 2 & 4 & 0 \\ 1 & 3 & 0 \end{bmatrix}$, $F = \begin{bmatrix} 2 & 4 \\ 1 & 3 \end{bmatrix}$, $E \neq F$.

APPLICATIONS

The applications given at the end of each section are certainly not meant to be an exhaustive listing of the applications of that particular section, nor is any claim made concerning the realism of many of the examples. Rather, these "evidences of usefulness" are supplied primarily in the interest of motivating the reader and with the hope that occasionally one will provide a spark that may arouse a much deeper, productive study in some areas of interest. All applications are optional in that they are not prerequisite to later material in the main part of the text.

The following examples illustrate a few of the many ways in which a matrix is useful.

Example 6. Consider the arrangement of four points (which may represent persons, nations, etc.) shown in Figure 1.2.1. This diagram can be translated into the language of matrix mathematics by

$$\begin{array}{c} \\ \#1 \\ \#2 \\ \#3 \\ \#4 \end{array} \begin{array}{cccc} \#1 & \#2 & \#3 & \#4 \end{array} \\ \begin{bmatrix} 0 & 1 & 0 & 1 \\ 1 & 0 & 1 & 1 \\ 0 & 1 & 0 & 0 \\ 1 & 1 & 0 & 0 \end{bmatrix},$$

where $a_{ij} = 1$ if the ith point is connected to the jth point, and $a_{ij} = 0$ if these points are not connected. We assume that no point is joined to itself. This representation fully describes whether or not there is a line segment connecting the points (or whether there is communication between persons, nations, or the like).

For large communication systems, diagrams like Figure 1.2.1 are very unwieldy, but by performing various operations on associated matrices one may reduce a large system to a simpler system, or make some other scientific analysis. An interesting variation of this application is given in Exercise 11 of this section.

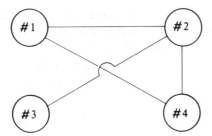

Figure 1.2.1

Example 7. Suppose that there exists a certain defined relation between persons, nations, numbers, or biological characteristics. Let us call this relation "dominance," and in Figure 1.2.2 suppose that an arrow from point i to point j

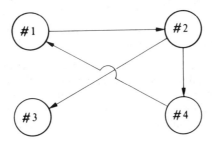

Figure 1.2.2

denotes the dominance of i over j. This dominance can be indicated by the following matrix

$$\begin{array}{c} \\ \#1 \\ \#2 \\ \#3 \\ \#4 \end{array} \begin{array}{c} \#1 \ \#2 \ \#3 \ \#4 \\ \begin{bmatrix} 0 & 1 & 0 & 0 \\ 0 & 0 & 1 & 1 \\ 0 & 0 & 0 & 0 \\ 1 & 0 & 0 & 0 \end{bmatrix} \end{array},$$

where $a_{ij} = 1$ if i dominates j, and $a_{ij} = 0$ if i does not dominate j. Obviously, no point dominates itself.

A variation of this example occurs when each point either dominates or is dominated by each of the other points. A very interesting use of this model was in a study of influence among the justices of the Michigan Supreme Court.[2] The use of the concept of dominance may pertain to biological characteristics, sociological behavior, and other areas, as well as to political influence.

[2] Ulmer, S., "Leadership in the Michigan Supreme Court" in *Judical Decision Making*, G. Shubert, ed., New York, The Free Press of Glencoe, 1963.

Example 8. Suppose that a certain company owns four factories which produce two products. The amount of product i produced by factory j can be represented by the entry a_{ij} in a 2 by 4 matrix such as

$$\begin{array}{c} \text{FACTORIES} \\ \begin{array}{cc} & \begin{array}{cccc} \#1 & \#2 & \#3 & \#4 \end{array} \\ \begin{array}{c} \text{PRODUCT } \#1 \\ \text{PRODUCT } \#2 \end{array} & \begin{bmatrix} 4 & 7 & 9 & 8 \\ 7 & 6 & 0 & 2 \end{bmatrix} \end{array} \end{array}.$$

Example 9. A standard problem in statistics is to study the effect that one variable has upon another in a given situation. The following matrix was taken from a statistical study of entering freshmen made by a certain University.[3]

	HIGH SCHOOL RANK	SAT VERBAL	SAT MATH	MATH ACHIEVEMENT
HIGH SCHOOL RANK	1.00	0.28	0.19	0.22
SAT VERBAL	0.28	1.00	0.36	0.37
SAT MATH	0.19	0.36	1.00	0.73
MATH ACHIEVEMENT	0.22	0.37	0.73	1.00

The four variables are the scores of three tests administered by the College Entrance Examination Board and a numerical representation of the rank of a freshman in his high school graduating class. Here a_{ij} represents the coefficient of correlation between the ith variable and the jth variable. A number near 1 shows a high degree of correlation, a number near 0 shows a low degree of correlation.

Example 10. Game theory is a relatively new area of mathematics and is of considerable interest in some disciplines, particularly the social sciences. The so-called payoff matrix is fundamental in the study of game theory. For example, consider the game of matching pennies, where Player A will win a penny from Player B if both players turn up a head on the coin or if both players turn up a tail on the coin; on the other hand, Player A will lose a penny to Player B if either player turns up a head and the other player turns up a tail. The payoff matrix considered from A's point of view is

$$P = \begin{array}{c} \overbrace{\begin{array}{cc} H & T \end{array}}^{B} \\ \begin{bmatrix} +1 & -1 \\ -1 & +1 \end{bmatrix} \begin{array}{c} H \\ T \end{array} \end{array} A.$$

The objective is to determine the optimal strategy of Player A if the game is played repeatedly. An introductory discussion of game theory may be found in [32] (Williams) and in Chapter 8 of [17] (Kemeny et al.); pages 222–233 of [22] (Nahikian) offer a brief but readable introduction to game theory. Linear algebra is very important in any serious study of game theory, and many of the concepts we develop in this book may be directly applied to such a study.

---

[3] Virginia Polytechnic Institute and State University Counseling Center, 1966.

EXERCISES

1. For each of the following matrices, give the order, and state whether or not the matrix is an R-matrix.

    (a) $\begin{bmatrix} 3 & 4 & i \\ 6 & 9 & 3 \end{bmatrix}$;   (b) $\begin{bmatrix} 2 \\ 1 \end{bmatrix}$;   (c) $[3 \quad i \quad 4]$;   (d) $[2+3+4]$.

2. Display the matrix $[a_{ij}]_{(2,3)}$ with entries $a_{13} = 3$, $a_{22} = 4$, $a_{11} = 5$, $a_{23} = 6$, $a_{12} = 7$, and $a_{21} = 8$.

3. If $[a_{ij}]_{(2,3)} = \begin{bmatrix} 6 & 9 & 4 \\ 3 & 2 & 1 \end{bmatrix}$, find: (a) a_{12};   (b) a_{23};   (c) a_{13}.

4. Write the matrices denoted by
    (a) $[a_{ij}]_{(3,1)}$;   (b) I_4;   (c) $[a_{ij}]_{(3,2)}$, where $a_{ij} = 0$ for all i and j.

5. Partition $\begin{bmatrix} 2 & 1+i & 4 \\ 2 & 1 & i \\ -3 & 0 & 2 \end{bmatrix}$ in such a way that one of the designated submatrices is a 2 by 2 R-matrix.

6. Write all of the four 3 by 3 submatrices of $\begin{bmatrix} 3 & 4 & 0 & 9 \\ 4 & 3 & 2 & 1 \\ 4 & 0 & 2 & 1 \end{bmatrix}$.

7. Partition $\begin{bmatrix} 2 & 3 & 4 & 1 & 0 \\ 9 & 3 & 2 & 0 & 1 \end{bmatrix}$ into 5 column vectors or column matrices.

8. What is the set of entries on the main diagonal of the 3 by 3 identity matrix?

9. Find, if possible, all values for each unknown that will make each of the following true.

    (a) $\begin{bmatrix} 2 & 4 \\ 5 & x+2 \end{bmatrix} = \begin{bmatrix} y & 4 \\ 5 & 7 \end{bmatrix}$;   (b) $\begin{bmatrix} x & 0 \\ 9 & y \end{bmatrix} = \begin{bmatrix} 2 & 0 \\ y & x \end{bmatrix}$;   (c) $\begin{bmatrix} x & y \\ y & x \end{bmatrix} = I_2$.

10. Let A be the 2 by 2 matrix of coefficients, and let B be the column matrix of the right members of the system

    $$\begin{cases} x_1 + 3x_2 = 6, \\ 2x_1 + 5x_2 = 4. \end{cases}$$

    Write the matrix $[A \mid B]$.

11. Suppose that in Example 6 of this section we think of the lines connecting the points as communication lines, and suppose that station #2 can speak to stations #1 and #4, but they cannot reciprocate. If we let arrows represent this one-way communication, the diagram in Example 6 is modified as shown in Figure 1.2.3. Express this communication system in matrix form.

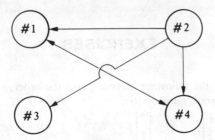

Figure 1.2.3

12. One variation of the dominance concept illustrated in Example 7 is that every object either dominates or is dominated by every other object. Suppose we have this situation among four objects. Let the matrix representation be

$$\begin{array}{c} \begin{array}{cccc} \#1 & \#2 & \#3 & \#4 \end{array} \\ \begin{array}{c} \#1 \\ \#2 \\ \#3 \\ \#4 \end{array} \begin{bmatrix} 0 & 0 & 1 & 1 \\ 1 & 0 & 1 & 0 \\ 0 & 0 & 0 & 0 \\ 0 & 1 & 1 & 0 \end{bmatrix} \end{array}$$

where $a_{ij} = 1$ if i dominates j, and $a_{ij} = 0$ if i does not dominate j. Draw a diagram representing this dominance.

1.3 Matrix Addition and Scalar Multiplication

We may add two matrices, when they are of the same order, by adding corresponding entries.

Definition 5. *Given two C-matrices $A = [a_{ij}]_{(m, n)}$ and $B = [b_{ij}]_{(m, n)}$; their sum is defined to be*

$$A + B = [a_{ij} + b_{ij}]_{(m, n)}.$$

*When two matrices are of the same order, they are said to be **conformable for addition**.*

Example 1.

$$\begin{bmatrix} 1 & 2 & -4 \\ 3 & 2 & 0 \end{bmatrix} + \begin{bmatrix} 7 & 0 & 3 \\ 6 & 5 & 8 \end{bmatrix} = \begin{bmatrix} (1+7) & (2+0) & (-4+3) \\ (3+6) & (2+5) & (0+8) \end{bmatrix} = \begin{bmatrix} 8 & 2 & -1 \\ 9 & 7 & 8 \end{bmatrix}.$$

Now that matrix addition has been defined, it is logical to inquire about its properties.

1.3 MATRIX ADDITION AND SCALAR MULTIPLICATION

Theorem 1. (*Commutative property*) If A and B are C-matrices and are conformable for addition, then

$$A + B = B + A.$$

The proof is left as Exercise 12.

Theorem 2. (*Associative property*) If A, B, and C are C-matrices and are of the same order, then

$$A + (B + C) = (A + B) + C.$$

Proof.

	STATEMENT	REASON
	$A + (B + C)$	
(1)	$= [a_{ij}]_{(m,n)} + ([b_{ij}]_{(m,n)} + [c_{ij}]_{(m,n)})$	(1) By change of notation.
(2)	$= [a_{ij}]_{(m,n)} + [b_{ij} + c_{ij}]_{(m,n)}$	(2) By definition of matrix addition.
(3)	$= [a_{ij} + (b_{ij} + c_{ij})]_{(m,n)}$	(3) By matrix addition.
(4)	$= [(a_{ij} + b_{ij}) + c_{ij}]_{(m,n)}$	(4) Addition of complex numbers is associative.
(5)	$= [a_{ij} + b_{ij}]_{(m,n)} + [c_{ij}]_{(m,n)}$	(5) By matrix addition.
(6)	$= ([a_{ij}]_{(m,n)} + [b_{ij}]_{(m,n)}) + [c_{ij}]_{(m,n)}$	(6) By matrix addition.
(7)	$= (A + B) + C$	(7) By change of notation. ☐

Theorem 3. (*Cancellation property*) If A, B, and C are C-matrices and are of the same order, then

$$A + B = A + C \Rightarrow B = C,$$

and

$$A + B = C + B \Rightarrow A = C,$$

where the symbol $\Rightarrow$ means "*implies that.*"

The proof is left as Exercise 13.

Consider the sum $A + A + A$. Can we say that this sum is equal to $3A$? An affirmative answer is justified if $3A$ is defined in the following manner:

Definition 6. *Given a C-matrix* $A = [a_{ij}]_{(m,n)}$ *and a complex number* c; *then* $cA = [ca_{ij}]_{(m,n)}$ *and* $Ac = [a_{ij}c]_{(m,n)}$. A corollary of this defn is that subtraction of matrices has now been defined (see next page)

When we are dealing with C-matrices it is customary to refer to a complex number as a *scalar*[4]; when we deal only with R-matrices, we call a real number a *scalar*. Thus, a matrix is multiplied by a scalar by multiplying

---

[4] A general definition of a scalar will be given in Section 6.4.

12 BASIC DEFINITIONS **1.3**

every entry of the matrix by that scalar. Making use of this concept, **subtraction** of C-matrices of the same order can be defined as

$$A - B = A + (-1)B.$$

Example 2.

$$2\begin{bmatrix} 3 & 1 \\ 4 & 3 \end{bmatrix} - \begin{bmatrix} 1 & 6 \\ 0 & 4 \end{bmatrix} = 2\begin{bmatrix} 3 & 1 \\ 4 & 3 \end{bmatrix} + (-1)\begin{bmatrix} 1 & 6 \\ 0 & 4 \end{bmatrix}$$

$$= \begin{bmatrix} 6 & 2 \\ 8 & 6 \end{bmatrix} + \begin{bmatrix} -1 & -6 \\ 0 & -4 \end{bmatrix} = \begin{bmatrix} 5 & -4 \\ 8 & 2 \end{bmatrix}.$$

Certain properties of scalar multiplication are listed in the following theorem.

Theorem 4. *If A and B are m by n C-matrices, and if c and d are scalars, then*

$$Ac = cA,$$
$$(cd)A = c(dA),$$
$$c(A + B) = cA + cB,$$
$$(c + d)A = cA + dA.$$

The proof is left as Exercise 14.

APPLICATIONS

Example 3. Assume that a manufacturer produces a certain alloy. The costs of purchasing and transporting the required amounts of three necessary raw materials, R, S, T, from two different locations, A, B, are given, respectively, by the following matrices:

$$A = \begin{bmatrix} 16 & 20 \\ 10 & 16 \\ 9 & 4 \end{bmatrix} \begin{matrix} \text{ORE } R \\ \text{ORE } S, \\ \text{ORE } T \end{matrix}$$

PURCHASE COST / TRANSPORTATION COST

$$B = \begin{bmatrix} 12 & 10 \\ 14 & 14 \\ 12 & 10 \end{bmatrix} \begin{matrix} \text{ORE } R \\ \text{ORE } S. \\ \text{ORE } T \end{matrix}$$

The matrix representing the total purchase and transportation costs of each type of ore from both locations is

$$A + B = \begin{bmatrix} 28 & 30 \\ 24 & 30 \\ 21 & 14 \end{bmatrix}.$$

1.3 MATRIX ADDITION AND SCALAR MULTIPLICATION

Example 4. An up-to-date inventory is a vital necessity in many businesses; if the process can be quantified, computer help becomes available for keeping a current record. Suppose that a district sales manager for a certain make of automobile wishes to keep a day-by-day inventory of the colors and models he has on hand. He requires that each retailer keep and report an inventory matrix of the following form, where A_k is the inventory of the kth dealer in the manager's territory:

$$A_k = \begin{bmatrix} & \text{TAN} & \text{RED} & \text{WHITE} & \text{GREEN} & \text{BLUE} & \\ a_{11} & a_{12} & a_{13} & a_{14} & a_{15} \end{bmatrix} \begin{matrix} \text{CONVERTIBLE} \\ \text{2-DOOR SEDAN} \\ \text{4-DOOR SEDAN} \end{matrix}$$

$$A_k = \begin{bmatrix} a_{11} & a_{12} & a_{13} & a_{14} & a_{15} \\ a_{21} & a_{22} & a_{23} & a_{24} & a_{25} \\ a_{31} & a_{32} & a_{33} & a_{34} & a_{35} \end{bmatrix} \begin{matrix} \text{CONVERTIBLE} \\ \text{2-DOOR SEDAN} \\ \text{4-DOOR SEDAN} \end{matrix}$$

If the sales manager has 99 retailers in his district, then the total inventory matrix T for this district is

$$A_1 + A_2 + \cdots + A_{99} = T.$$

This can be found very simply by programming a computer to do the necessary arithmetic calculations. Those items which are in total short supply can then be ordered.

The sales manager could calculate his inventory matrix T_2 after any given week by subtracting what was sold, S_2, that week from the total inventory matrix T_1 after the previous week; that is,

$$T_2 = T_1 - S_2 = T_1 + (-1)S_2.$$

However, in order to stay well ahead of the demand, suppose he orders and receives a shipment of new cars which is exactly double what he sold, S_1, the first week. Here we have

$$T_2 = T_1 - S_2 + 2S_1.$$

Example 5. In Chapter 9 we shall discuss certain important engineering problems which require consideration of the matrix $A - \lambda I_n$ or the matrix $\lambda I_n - A$, where A is an n by n R-matrix and λ is a real number.

Thus, if

$$A = \begin{bmatrix} 6 & 3 & 4 \\ 3 & 2 & 0 \\ 4 & 0 & 1 \end{bmatrix},$$

then

$$\lambda I_3 - A = \lambda \begin{bmatrix} 1 & 0 & 0 \\ 0 & 1 & 0 \\ 0 & 0 & 1 \end{bmatrix} + (-1) \begin{bmatrix} 6 & 3 & 4 \\ 3 & 2 & 0 \\ 4 & 0 & 1 \end{bmatrix}$$

$$= \begin{bmatrix} \lambda & 0 & 0 \\ 0 & \lambda & 0 \\ 0 & 0 & \lambda \end{bmatrix} + \begin{bmatrix} -6 & -3 & -4 \\ -3 & -2 & 0 \\ -4 & 0 & -1 \end{bmatrix}$$

$$= \begin{bmatrix} \lambda - 6 & -3 & -4 \\ -3 & \lambda - 2 & 0 \\ -4 & 0 & \lambda - 1 \end{bmatrix}.$$

EXERCISES

In Exercises 1–6, find the sum of each pair of matrices, if possible.

1. $\begin{bmatrix} 2 & 0 \\ 3 & 4 \end{bmatrix}, \begin{bmatrix} 4 & 1 \\ 2 & 1 \end{bmatrix}.$
2. $\begin{bmatrix} 2 \\ i \end{bmatrix}, \begin{bmatrix} 4 \\ 6 \end{bmatrix}.$

3. $[0 \quad 2 \quad 3], \begin{bmatrix} 2 \\ 4 \end{bmatrix}.$
4. $\begin{bmatrix} 2 & 4 \\ 6 & 3 \end{bmatrix}, I_2.$

5. $\begin{bmatrix} 6 & 3 & 2 \\ 4 & 9 & 2 \\ 2 & 2 & 3 \end{bmatrix}, \begin{bmatrix} 0 & 0 & 0 \\ 0 & 0 & 0 \\ 0 & 0 & 0 \end{bmatrix}.$
6. $\begin{bmatrix} 3 & 2 \\ 6 & -3 \\ 0 & 2 \end{bmatrix}, \begin{bmatrix} 6 & 3 \\ 4 & 9 \\ 5 & 2 \end{bmatrix}.$

7. Given $A = \begin{bmatrix} 2 & -1 \\ -3 & -4 \end{bmatrix}$ and $B = \begin{bmatrix} -2 & 0 \\ -1 & 3 \end{bmatrix}$, calculate:

  (a) $3A$; (b) $-2B$; (c) $-A$; (d) $A + 3B$; (e) $\frac{1}{2}B - 2A$.
  (f) Find C if $B + C = A$. (g) Find D if $A - 2D = 2B$.

8. Let $A = \begin{bmatrix} 5 & 10 & 20 \\ -65 & 15 & -10 \end{bmatrix}$. Find a matrix B which is a scalar multiple of A and which has 2 as its entry in the first row and second column.

9. Two matrices can be added only when they are of the same order. Is there any similar restriction on the multiplication of a matrix by a scalar?

10. Find $\lambda I_2 - A$ if $A = \begin{bmatrix} 3 & 2 \\ 1 & 6 \end{bmatrix}$. (See Example 5 of this section.)

11. If A is a 2 by 3 C-matrix and B is a 3 by 2 C-matrix, are A and B conformable for addition?

12. Prove Theorem 1.
13. Prove Theorem 3.
14. Prove Theorem 4.
15. If A, B, and C are R-matrices of the same order, why is it permissible to write $A + B + C$ without using any parentheses?
16. Prove that each of the following is true, giving the reason for each step. You may assume conformability and the theorems of this section. All matrices are C-matrices.

  (a) $A + (B + C) = (B + A) + C$; (b) $A + B = B + C \Rightarrow A = C$;
  (c) $A + 2(A + 3B) = 3A + 6B$.

NEW VOCABULARY

transformation 1.1
function 1.1
mapping 1.1
domain 1.1
codomain 1.1
image 1.1
image set 1.1
range 1.1
matrix over S 1.2
C-matrix 1.2
R-matrix 1.2
zero matrix 1.2
null matrix 1.2
address of an entry 1.2
order of a matrix 1.2

row matrix 1.2
row vector 1.2
column matrix 1.2
column vector 1.2
square matrix 1.2
main diagonal 1.2
identity matrix 1.2
submatrix 1.2
partitioned matrix 1.2
equal matrices 1.2
conformable for addition 1.3
matrix sum 1.3
scalar 1.3
scalar times a matrix 1.3
matrix subtraction 1.3

2

SYSTEMS OF LINEAR EQUATIONS

2.1 An Introduction

If a set of m equations in the n unknowns $x_1, x_2, \ldots, x_n$ is of the form

(2.1.1)
$$\begin{cases} a_{11}x_1 + a_{12}x_2 + \cdots + a_{1n}x_n = b_1, \\ a_{21}x_1 + a_{22}x_2 + \cdots + a_{2n}x_n = b_2, \\ \cdots\cdots\cdots\cdots\cdots\cdots\cdots\cdots\cdots\cdots\cdots\cdots\cdots \\ a_{m1}x_1 + a_{m2}x_2 + \cdots + a_{mn}x_n = b_m, \end{cases}$$

where the coefficients $a_{11}, a_{12}, \ldots, a_{mn}$, and the right-hand members $b_1, b_2, \ldots, b_m$, are elements of some number system S, then (2.1.1) is called a *system of linear equations over* S. A *solution* of a system of linear equations over S is an ordered set of elements of S, $(c_1, c_2, \ldots, c_n)$, such that m identities result when these n elements are substituted for the unknowns. In general, a system of linear equations over a number system S may have a unique solution, or more than one solution, or no solution; certainly one of these will be the case. Notice that in order to find a solution a number system S, such as the complex number system C or the real number system R, must be specified.

Systems of linear equations arise in a wide variety of disciplines, and it is important that practitioners in these disciplines have methods that can determine whether or not solutions to such equations exist and that they can find these solutions when they do exist. Because the digital computer is very

2.1 AN INTRODUCTION

useful in working with large systems of equations, any general method of solution is more useful if it is adaptable to computer capabilities.

Definition 1. *For the system*

$$\begin{cases} a_{11}x_1 + \cdots + a_{1n}x_n = b_1, \\ \phantom{a_{11}x_1 + \cdots + a_{1n}x_n = b_1,} \\ a_{m1}x_1 + \cdots + a_{mn}x_n = b_m, \end{cases}$$

the matrix

$$A = \begin{bmatrix} a_{11} & \cdots & a_{1n} \\ \vdots & & \vdots \\ a_{m1} & \cdots & a_{mn} \end{bmatrix}$$

is the **coefficient matrix** *of the system, and the matrix*

$$[A \mid B] = \begin{bmatrix} a_{11} & \cdots & a_{1n} & b_1 \\ \vdots & & \vdots & \vdots \\ a_{m1} & \cdots & a_{mn} & b_m \end{bmatrix}$$

is the **augmented matrix** *of the system.*

The following example illustrates the general method of solution that we shall develop in this chapter. Notice that this method is simply a variation of the old addition–subtraction method taught in a course in beginning algebra. In the second column of Example 1, we demonstrate a synthetic approach which illustrates that our manipulations are really performed only on the augmented matrix of our system.

Example 1. Find a solution (if one exists) of the given system over the real numbers. First, the augmented matrix of the given system is written; certain operations are then performed correspondingly on the system and on the matrix.

$$\begin{array}{l} R_1 \\ R_2 \\ R_3 \end{array} \begin{cases} x_1 + 3x_2 + 3x_3 = 2, \\ -2x_1 - 6x_2 - 2x_3 = 4, \\ x_2 + x_3 = 6. \end{cases} \begin{bmatrix} 1 & 3 & 3 & 2 \\ -2 & -6 & -2 & 4 \\ 0 & 1 & 1 & 6 \end{bmatrix}.$$

Add 2 times the terms of the first equation to the corresponding terms of the second equation (denoted by $2R_1 + R_2$, where R_i represents the ith row of the augmented matrix of the system).

$$\begin{cases} x_1 + 3x_2 + 3x_3 = 2, \\ 4x_3 = 8, \\ x_2 + x_3 = 6. \end{cases} \begin{bmatrix} 1 & 3 & 3 & 2 \\ 0 & 0 & 4 & 8 \\ 0 & 1 & 1 & 6 \end{bmatrix}.$$

Interchange the second and third equations (denoted by $R_2 \leftrightarrow R_3$).

$$\begin{cases} x_1 + 3x_2 + 3x_3 = 2, \\ x_2 + x_3 = 6, \\ 4x_3 = 8. \end{cases} \begin{bmatrix} 1 & 3 & 3 & 2 \\ 0 & 1 & 1 & 6 \\ 0 & 0 & 4 & 8 \end{bmatrix}.$$

Add -3 times the terms of the second equation to the corresponding terms of the first equation (denoted by $-3R_2 + R_1$).

$$\begin{cases} x_1 = -16, \\ x_2 + x_3 = 6, \\ 4x_3 = 8. \end{cases} \qquad \left[\begin{array}{ccc|c} 1 & 0 & 0 & -16 \\ 0 & 1 & 1 & 6 \\ 0 & 0 & 4 & 8 \end{array}\right].$$

Multiply the terms of the third equation by $\frac{1}{4}$ (denoted by $\frac{1}{4}R_3$).

$$\begin{cases} x_1 = -16, \\ x_2 + x_3 = 6, \\ x_3 = 2. \end{cases} \qquad \left[\begin{array}{ccc|c} 1 & 0 & 0 & -16 \\ 0 & 1 & 1 & 6 \\ 0 & 0 & 1 & 2 \end{array}\right].$$

Add -1 times the terms of the third equation to the terms of the second equation (denoted by $-R_3 + R_2$).

$$\begin{cases} x_1 = -16, \\ x_2 = 4, \\ x_3 = 2. \end{cases} \qquad \left[\begin{array}{ccc|c} 1 & 0 & 0 & -16 \\ 0 & 1 & 0 & 4 \\ 0 & 0 & 1 & 2 \end{array}\right].$$

The unique solution of the last system is $(-16, 4, 2)$. After Theorem 1 is stated and proved, it will follow that $(-16, 4, 2)$ is also the unique solution of the original system.

In Example 1 notice that at each stage of the method we had a system that was obtained from the preceding system by means of one of the following operations.

Definition 2. *The following operations performed on a given system of linear equations over C are known as* **elementary operations on equations***:*
  (1) *Interchange any two equations;*
  (2) *Multiply the terms of any equation by a nonzero complex number;*
  (3) *Add, to the terms of any equation, k times the corresponding terms of any other equation, where k belongs to C.*

As we pointed out earlier, some systems of linear equations have no solution, or have one solution, or have more than one solution; thus we say that a system has a set of solutions over C, where that set may be empty, have one element, or have more than one element. When two systems of linear equations over the same number system S have the same set of solutions, the systems are said to be **equivalent** systems. The following theorem answers the natural question of whether the set of solutions is invariant (unchanged) when elementary operations on equations are applied to a given system of linear equations.

Theorem 1. *If a system of linear equations over C is transformed into another linear system over C by means of one elementary operation on equations, then the two systems are equivalent.*

Proof. Let an arbitrary system of linear equations over C be represented by

(2.1.2)
$$\begin{cases} a_{11}x_1 + \cdots + a_{1n}x_n = b_1, \\ \vdots \qquad\qquad \vdots \qquad\qquad \vdots \\ a_{i1}x_1 + \cdots + a_{in}x_n = b_i, \\ \vdots \qquad\qquad \vdots \qquad\qquad \vdots \\ a_{j1}x_1 + \cdots + a_{jn}x_n = b_j, \\ \vdots \qquad\qquad \vdots \qquad\qquad \vdots \\ a_{m1}x_1 + \cdots + a_{mn}x_n = b_m, \end{cases}$$

where $m > 1$. The set of values

(2.1.3) $$x_1 = c_1, \ldots, x_n = c_n,$$

where c_i belongs to C, is a solution of (2.1.2) if and only if upon substitution of (2.1.3) into (2.1.2), m identities are obtained. We must show that the existence of these m identities is not altered by the three operations of Definition 2.

CASE I. The interchange of any two equations does not alter the existence of m identities when (2.1.3) is substituted into (2.1.2).

CASE II. The multiplication of an equation of (2.1.2) by a nonzero element k of C does not alter the existence of m identities when (2.1.3) is substituted into (2.1.2). Specifically, upon substitution of (2.1.3),

$$k(a_{i1}c_1 + \cdots + a_{in}c_n) = kb_i,$$

if and only if,

$$a_{i1}c_1 + \cdots + a_{in}c_n = b_i.$$

CASE III. Consider arbitrary and distinct ith and jth equations of (2.1.2). Add k times the terms of the jth equation to the corresponding terms of the ith equation. The resulting system can be expressed as

(2.1.4)
$$\begin{cases} a_{11}x_1 + \cdots + a_{1n}x_n = b_1, \\ \vdots \qquad\qquad \vdots \\ a_{i1}x_1 + \cdots + a_{in}x_n + k(a_{j1}x_1 + \cdots + a_{jn}x_n) = b_i + kb_j, \\ \vdots \qquad\qquad \vdots \\ a_{j1}x_1 + \cdots + a_{jn}x_n = b_j, \\ \vdots \qquad\qquad \vdots \\ a_{m1}x_1 + \cdots + a_{mn}x_n = b_m. \end{cases}$$

Notice that the ith equation is the only equation that is changed in the new system. Therefore, upon substitution of (2.1.3) into (2.1.4),

$$a_{i1}c_1 + \cdots + a_{in}c_n + k(a_{j1}c_1 + \cdots + a_{jn}c_n) = b_i + kb_j$$

and

$$a_{j1}c_1 + \cdots + a\ \ c = b_j$$

if and only if, upon substitution of (2.1.3) into (2.1.2),

$$a_{i1}c_1 + \cdots + a_{in}c_n = b_i$$

and

$$a_{j1}c_1 + \cdots + a_{jn}c_n = b_j. \quad \square$$

The importance of Theorem 1 should not be overlooked; repeated application of Theorem 1 assures us that the obvious solution of the last system in Example 1 is also the solution of the original system of Example 1. Moreover this will be true for any example in which a finite number of elementary operations on equations are used to transform the system. Theorem 1 is the key to the generalization of the method of solution used to solve the system in Example 1; this general method of solution is known as the Gauss-Jordan method and will be presented formally later in this chapter.

APPLICATIONS

Example 2. Suppose that a certain company produces three different products: R, S, and T. This output requires the services of two groups of workers; the members of one group are highly trained technicians, the members of the other group are unskilled laborers. Product R requires a day's work from each of 5 technicians and 5 laborers for each unit produced; product S needs 10 technicians and 10 laborers for each unit produced; and product T requires 2 technicians and 4 laborers for each unit produced. The company management wants to know how many units of each product to produce each day in order to keep each of its 100 technicians and 150 laborers employed. Let x_1, x_2, and x_3 stand for the number of units of R, S, and T, respectively, that should be manufactured each day to satisfy the stated conditions. Mathematically, the problem becomes

$$\begin{cases} 5x_1 + 10x_2 + 2x_3 = 100, \\ 5x_1 + 10x_2 + 4x_3 = 150. \end{cases}$$

There is no unique answer. Subtracting the members of the first equation from those of the second we find $x_3 = 25$. If this value is substituted into either equation, then reduction of the equation yields $x_1 = 10 - 2x_2$. The possible integral values of x_2 are then 0, 1, 2, 3, 4, and 5. Thus, with $x_3 = 25$, we have the following values for x_1 and x_2.

x_1	10	8	6	4	2	0
x_2	0	1	2	3	4	5

To choose among these possibilities the company must use other considerations. (Notice that in this problem each solution is a set of nonnegative integers rather than a set of real numbers; this explains why six solutions, rather than an infinite number of solutions, are obtained.)

Example 3. A system of linear equations in which there are more equations than unknowns can be obtained from the electrical circuit shown in Figure 2.1.1.

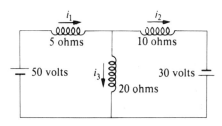

Figure 2.1.1

Let i_1, i_2, and i_3 represent the currents, measured in amperes, in their respective parts of the circuit. Using Kirchhoff's rule (named for Gustav Robert Kirchhoff, German, 1824–1887) for relating the voltages, resistances, and currents of a given circuit, we obtain the following system of linear equations

$$\begin{cases} i_1 - i_2 - i_3 = 0, \\ 5i_1 + 20i_3 = 50, \\ 10i_2 - 20i_3 = 30, \\ 5i_1 + 10i_2 = 80. \end{cases}$$

Example 4. The so-called Leontif input-output system of economy can be formed in the following way: Consider an economy with n industries that are dependent upon each other. Each of these industries is faced with demands for its products. These demands may come from the other industries, the general public, exports, government, etc. The demands from within the economy will be called internal, and those from outside the economy will be called external. We agree on units of measure such as dollar value, and then for a particular time interval we define the following symbols:

(1) Let x_j be the output or production of industry j.
(2) Let c_{ij} be the number of units of production of industry i used to produce one unit of production of industry j.
   Then $c_{ij}x_j$ represents the total amount of production from industry i used by industry j.
   Then $c_{i1}x_1 + \cdots + c_{in}x_n$ represents the total amount of production from industry i used by all industries in the economy. In other words this sum represents the internal consumption of the production of industry i.
(3) Let b_i be the demand on industry i in excess of that required by the other industries in the economy. In other words b_i is the external (external to the economy) demand on industry i.

Hence the total production of an industry will be divided between the internal demands and the external demands. For each of the n industries the last statement can be quantified by a single equation, and the production of the entire economy can be quantified by a system of equations:

(output) = (input) + (surplus),
$$\begin{cases} x_1 = (c_{11}x_1 + c_{12}x_2 + \cdots + c_{1n}x_n) + b_1, \\ x_2 = (c_{21}x_1 + c_{22}x_2 + \cdots + c_{2n}x_n) + b_2, \\ \vdots \\ x_n = (c_{n1}x_1 + c_{n2}x_2 + \cdots + c_{nn}x_n) + b_n. \end{cases}$$

These linear equations can be rearranged in standard form
$$\begin{cases} (1 - c_{11})x_1 - c_{12}x_2 - \cdots - c_{1n}x_n = b_1, \\ -c_{21}x_1 + (1 - c_{22})x_2 - \cdots - c_{2n}x_n = b_2, \\ \vdots \\ -c_{n1}x_1 - c_{n2}x_2 - \cdots + (1 - c_{nn})x_n = b_n. \end{cases}$$

A similar model can be developed to determine prices within a hypothetical economy.

EXERCISES

1. Write the augmented and coefficient matrices for:

(a) $\begin{cases} 3x_1 + x_2 + x_3 = 4, \\ 2x_1 + x_3 = 5, \\ x_1 + 2x_3 = 1. \end{cases}$
(b) $\begin{cases} 2x_1 + x_2 = 5, \\ x_1 - x_2 = 1, \\ x_1 + x_2 = 3. \end{cases}$

(c) $\begin{cases} x_1 + x_2 + x_3 + x_4 = 1, \\ 2x_1 + x_2 + x_4 = 2. \end{cases}$
(d) $\begin{cases} x_1 + x_2 + 3 = 0, \\ 2x_1 - x_2 + 6 = 0. \end{cases}$

In Exercises 2–9, use elementary operations on equations to find the solution of the given system over the real numbers. Write the entire system after each step, as illustrated in the first column of Example 1. Check your answer by substituting it into the original system to see if m identities result.

2. $\begin{cases} x_1 + x_2 = 3, \\ 2x_1 - x_2 = 4. \end{cases}$
3. $\begin{cases} x_1 + 2x_2 = 7, \\ -x_1 + 3x_2 = 13. \end{cases}$

4. $\begin{cases} x_1 + 2x_2 - x_3 = 6, \\ 2x_1 - x_2 + 3x_3 = -13, \\ 3x_1 - 2x_2 + 3x_3 = -16. \end{cases}$
5. $\begin{cases} x_1 - x_3 - 2 = 0, \\ x_2 + 3x_3 - 1 = 0, \\ x_1 - 2x_2 - 7 = 0. \end{cases}$

6. $\begin{cases} 6x_1 + 3x_2 + 2x_3 = 1, \\ 3x_1 - 4x_3 = 4, \\ 5x_1 - x_2 = 14. \end{cases}$
7. $\begin{cases} x_1 - x_2 = 1, \\ 2x_1 + x_2 = \frac{19}{2}, \\ 4x_1 - 2x_2 = 9. \end{cases}$

8. $\begin{cases} x_1 - x_2 - x_3 - x_4 = 5, \\ x_1 + 2x_2 + 3x_3 + x_4 = -2, \\ 2x_1 + 2x_3 + 3x_4 = 3, \\ 3x_1 + x_2 + 2x_4 = 1. \end{cases}$
9. $\begin{cases} x_1 + 2x_2 + x_3 = 2, \\ 2x_1 - 2x_3 + x_4 = 6, \\ 4x_2 + 3x_3 + 2x_4 = -1, \\ -x_1 + 6x_2 - x_3 - x_4 = 2. \end{cases}$

10. Solve Exercises 2 and 4 by performing the elementary operations on the augmented matrices, as illustrated in the second column of Example 1. Use the notation $(R_i \leftrightarrow R_j)$ or (kR_i) or $(kR_j + R_i)$ to specify the operations used.

11. Solve Exercises 3 and 5 by performing the elementary operations on the augmented matrices as illustrated in the second column of Example 1.

12. What justification can you give for declaring that the solution of the given system of Exercise 2 is the same as the solution of each new system that results upon application of an elementary operation on equations?

13. Graph the system of Exercise 3, and give a geometric interpretation of the solution.

14. Suppose that the system of Exercise 2 is defined to be a system over the set of integers. Does the system have a solution over the set of integers? State the reason for your answer.

15. Graph the system of Exercise 7, and give a geometric interpretation of the solution.

16. Are the systems of Exercises 2 and 3 equivalent? Give reason for your answer.

17. Does the system

$$\begin{cases} x_1 + 2x_2 = 4, \\ 2x_1 + 4x_2 = 2, \end{cases}$$

have a solution over C? Illustrate geometrically. Show how the method of Example 1 determines this.

18. Does the system

$$\begin{cases} x_1 + 2x_2 + x_3 = 8, \\ -2x_2 + 2x_3 = 2, \end{cases}$$

over C have a solution? Does the system have more than one solution? If more than one solution exists, find two solutions.

19. (a) State the converse of Theorem 1. (b) Is this converse valid?

20. Suppose that a system of two linear equations in two unknowns has the unique solution (3, 4). Is there any system of three linear equations in three unknowns that is equivalent to the given system? Is it possible for a system of three linear equations in two unknowns to be equivalent to the given system?

21. (a) If one system of linear equations is equivalent to a second system, is the second system equivalent to the first system?
(b) If one system of linear equations is equivalent to a second system and the second system is equivalent to a third system, is the first system equivalent to the third system?

22. If one system of linear equations can be transformed to a second system by elementary operations on equations, can the second system be transformed to the first system by elementary operations on equations? Why?

2.2 Elementary Row Operations

We have observed that a system of linear equations can be represented in terms of the augmented matrix of the system. We have also observed that the operations ordinarily performed on the equations of a given system can be performed on the corresponding rows of the associated matrix. The purpose of this section is to formalize these operations on a matrix. When performed on the rows of a matrix, these operations are called *elementary row operations*.

Definition 3. *The following operations performed on a C-matrix are called elementary row operations;*
 (1) *the interchange of any two rows;*
 (2) *the multiplication of the entries of any row by a nonzero complex number;*
 (3) *the addition to the entries of any row of k times the corresponding entries of any other row, where k belongs to C.*

Example 1. Consider the R-matrix $A = \begin{bmatrix} 5 & 2 \\ 4 & 6 \\ 3 & 1 \end{bmatrix}$.

(1) Interchange the second and third rows. The notation $(R_2 \leftrightarrow R_3)$ will be used to label this particular elementary operation. The resulting matrix B is

$$B = \begin{bmatrix} 5 & 2 \\ 3 & 1 \\ 4 & 6 \end{bmatrix}.$$

(2) Multiply each of the entries of the second row of B by 2. The notation $(2R_2)$ will be used to label this particular elementary operation. The resulting matrix C is

$$C = \begin{bmatrix} 5 & 2 \\ 6 & 2 \\ 4 & 6 \end{bmatrix}.$$

(3) Add -2 times the entries of the first row to the corresponding entries of the third row. The notation $(-2R_1 + R_3)$ will be used to designate this particular elementary operation. The resulting matrix D is

$$D = \begin{bmatrix} 5 & 2 \\ 6 & 2 \\ -6 & 2 \end{bmatrix}.$$

2.2 ELEMENTARY ROW OPERATIONS

Let M represent the set of all m by n C-matrices. Often it is useful to identify pairs of elements of this set that are related in a certain specified way. For example, we have already identified those pairs of m by n C-matrices having equal corresponding entries as "equal" matrices. After giving a formal definition of a *binary relation on a set* we will define another important binary relation on the set of m by n matrices M. More relations on M will follow later in the book.

Definition 4. *A **binary relation** on a set S is a subset R of the Cartesian product*[1] *$S \times S$.*

Example 2. Consider the set of all integers $\mathcal{I}$. "Divisor of" is the identification given to the binary relation on $\mathcal{I}$ consisting of the subset of elements of $\mathcal{I} \times \mathcal{I}$ in which the second element of the pair divided by the first element is an integer. Thus 2 "is a divisor of" 6, because $6 \div 2 = 3$ and 2, 6, and 3 are integers.

Definition 5. *If a C-matrix A can be transformed into a C-matrix B by one or more elementary row operations, we say that A is **row equivalent** to B, and this binary relation is denoted as $A \stackrel{row}{\sim} B$.*

Example 3. In Example 1 the work illustrates that $A \stackrel{row}{\sim} B$, $A \stackrel{row}{\sim} C$, $A \stackrel{row}{\sim} D$, $B \stackrel{row}{\sim} C$, $B \stackrel{row}{\sim} D$, and $C \stackrel{row}{\sim} D$.

We need one more definition before formally presenting the general method of solution of a system of linear equations.

Definition 6. *An **echelon matrix** is an m by n C-matrix having the following two properties.*
(1) *Each of the first k rows has some nonzero entries, where k is an integer such that $1 \leq k \leq m$. The entries are all zeros in the remaining $(m - k)$ rows.*
(2) *The first nonzero entry in each of the first k rows must be 1; moreover in each of rows $2, 3, \ldots, k$ (assuming $k \geq 2$) this 1 must appear in a column to the right of the column in which the first nonzero entry of the preceding row appeared.*

*If the matrix has the additional property (3), then it is a **reduced echelon matrix**.*
(3) *The first nonzero entry in each of the first k <u>rows</u> is the only nonzero entry in its column.*

---
[1] A discussion of a Cartesian product may be found in the Appendix.

Example 4. Although the following matrix is an echelon matrix, it is not a reduced echelon matrix, because the entries in bold print are not all zeros. The open parentheses denote that those entries are arbitrary complex numbers. Note that $k = 3$.

$$\begin{bmatrix} 0 & 1 & \mathbf{4} & (\,) & \mathbf{3} & (\,) \\ 0 & 0 & 1 & (\,) & \mathbf{8} & (\,) \\ 0 & 0 & 0 & 0 & 1 & (\,) \\ 0 & 0 & 0 & 0 & 0 & 0 \\ 0 & 0 & 0 & 0 & 0 & 0 \end{bmatrix}.$$

Example 5. The following matrix is a reduced echelon matrix.

$$\begin{bmatrix} 1 & (\,) & 0 & 0 & (\,) & (\,) \\ 0 & 0 & 1 & 0 & (\,) & (\,) \\ 0 & 0 & 0 & 1 & (\,) & (\,) \\ 0 & 0 & 0 & 0 & 0 & 0 \end{bmatrix}.$$

The open parentheses denote that those entries are arbitrary complex numbers.

Example 6. An identity matrix I_n is a reduced echelon matrix, but a null matrix is not an echelon matrix because of Part (1) of Definition 6.

If, by a sequence of elementary row operations, a given matrix is shown to be row equivalent to an echelon matrix, then that echelon matrix is frequently called an *echelon form* of the given matrix. An echelon form of a given matrix is not unique (Exercise 13).

APPLICATIONS

★ [2] **Example 7.** For those who have studied differential equations, an interesting example of the use of elementary row operations in the solution of certain nonlinear differential equations may be found on pages 245–250 of [23] (Noble). The treatment given in [23] is based on a paper by W. F. Ames [2], pages 214–218, and is concerned with the solution of nonlinear differential equations in certain chemical reactions.

Example 8. Most of the applications of elementary operations are indirect; as we shall see in the next section they will be used to solve systems of linear equations, which in turn can be directly applied. Elementary operations are also

---

[2] The symbol ★ will be used throughout this book to designate Examples and Exercises that require some knowledge of Calculus.

ELEMENTARY ROW OPERATIONS

used in later sections of this book to calculate the inverse matrix and the determinant of a matrix. Elementary row operations are fundamental to the simplex method of linear programming. Direct applications of the concepts mentioned above will be given as they are presented. One advantage in using elementary operations is that they are readily adaptable to computer use.

EXERCISES

1. Which of the following are echelon matrices, and which of those are reduced echelon matrices?

   (a) Any null matrix; (b) I_4; (c) $\begin{bmatrix} 1 & 2 \\ 0 & 2 \end{bmatrix}$; (d) $\begin{bmatrix} 1 & 4 & 0 & 0 \\ 0 & 1 & 1 & 0 \\ 0 & 0 & 1 & 2 \end{bmatrix}$;

   (e) $\begin{bmatrix} 1 & 0 & 0 & 0 \\ 0 & 0 & 0 & 0 \\ 0 & 0 & 0 & 1 \end{bmatrix}$; (f) $\begin{bmatrix} 0 & 1 \\ 0 & 0 \\ 0 & 0 \end{bmatrix}$; (g) $\begin{bmatrix} 1 & 2 \\ 0 & 1 \\ 0 & 0 \end{bmatrix}$; (h) $\begin{bmatrix} 1 & 2 & 0 \\ 0 & 0 & 1 \\ 0 & 0 & 0 \end{bmatrix}$;

   (i) $\begin{bmatrix} 1 & 0 & 0 \\ 0 & 1 & 0 \\ 0 & 1 & 0 \end{bmatrix}$.

2. If the augmented matrix of a given system of linear equations is row equivalent to the reduced echelon matrix $\begin{bmatrix} 1 & 0 & 0 & 4 \\ 0 & 1 & 0 & 3 \\ 0 & 0 & 1 & 2 \end{bmatrix}$, what is the solution of the original system?

3. Is $\begin{bmatrix} 2 & 1 & 4 \\ 0 & 1 & 3 \\ 2 & 0 & 1 \end{bmatrix}$ row equivalent to I_3? Why?

4. For each of the following pairs of matrices, if it is possible to do so, state a single elementary row operation that transforms the first matrix into the second matrix.

   (a) $\begin{bmatrix} 2 & 1 \\ 4 & 6 \end{bmatrix}, \begin{bmatrix} 2 & 1 \\ 0 & 4 \end{bmatrix}$; (b) $\begin{bmatrix} 2 & 4 & 6 \\ 1 & 2 & 4 \end{bmatrix}, \begin{bmatrix} 1 & 2 & 3 \\ 1 & 2 & 4 \end{bmatrix}$;

   (c) $\begin{bmatrix} 2 & 4 & 3 & 1 \\ 1 & 2 & 3 & 4 \\ 0 & 1 & 4 & 6 \end{bmatrix}, \begin{bmatrix} 1 & 2 & 3 & 4 \\ 0 & 1 & 4 & 6 \\ 2 & 4 & 3 & 1 \end{bmatrix}$.

In Exercises 5 and 6, using elementary row operations, write an echelon matrix to which the given matrix is row equivalent. Describe each step using the abbreviated notation shown in Example 1 of this section.

5. $\begin{bmatrix} 1 & 4 & 3 & 2 \\ 1 & 8 & 0 & 2 \\ 2 & 0 & 4 & 2 \end{bmatrix}$. 6. $\begin{bmatrix} 2 & 1 & 3 \\ 1 & 0 & -3 \\ 3 & 1 & 0 \end{bmatrix}$.

In Exercises 7 and 8, write the augmented matrix for the given system. Transform this augmented matrix into a row equivalent reduced echelon matrix. Then write the system for which this reduced echelon matrix is the augmented matrix. What is the solution over the real numbers of the resulting system? What is the solution over the real numbers of the original system?

7. $\begin{cases} 2x + y = 6, \\ 2x + 4y = 9. \end{cases}$

8. $\begin{cases} x + y + z = 4, \\ y + z = 6, \\ 3x - y - 2z = 9. \end{cases}$

9. (a) Show that if matrix B can be obtained from matrix A by one elementary row operation, then matrix A can be obtained from matrix B by one elementary row operation of the same type.
   (b) How does it follow from (a) that if $A \overset{\text{row}}{\sim} B$ then $B \overset{\text{row}}{\sim} A$?
   (c) Prove that if $A \overset{\text{row}}{\sim} B$ and $B \overset{\text{row}}{\sim} C$, then $A \overset{\text{row}}{\sim} C$.

10. If $A \overset{\text{row}}{\sim} B$, how does the order of A compare with the order of B?

11. Which of the following matrices are row equivalent?

$A = \begin{bmatrix} 2 & 1 \\ 4 & 3 \end{bmatrix}$; $B = \begin{bmatrix} 2 & 1 & 1 \\ 4 & 0 & 1 \end{bmatrix}$; $C = \begin{bmatrix} -2 & 1 & 0 \\ 4 & 3 & 2 \\ 0 & 5 & 2 \end{bmatrix}$;

$D = \begin{bmatrix} 1 & 0 & 0 \\ 0 & 4 & 0 \\ 2 & 0 & 1 \end{bmatrix}$; $E = \begin{bmatrix} 1 & 3 & 0 \\ 2 & 1 & 4 \end{bmatrix}$; $F = \begin{bmatrix} 4 & 1 \\ 3 & 2 \\ 0 & 1 \end{bmatrix}$; $G = \begin{bmatrix} 0 & 1 \\ 4 & 3 \end{bmatrix}$.

12. Change $\begin{bmatrix} 1 & 2 \\ 3 & 4 \end{bmatrix}$ to $\begin{bmatrix} 3 & 4 \\ 1 & 2 \end{bmatrix}$ in four steps, using only elementary row operations of types (2) and (3). (As this example suggests, it is possible to show that we can do the same things with elementary row operations (2) and (3) as we can with (1), (2), and (3).)

13. Find two different echelon forms for the matrix given in Exercise 5.

14. If A is an n by n reduced echelon C-matrix, and if the last row of A does not contain all zeros, then what can be said about A?

2.3 The Gauss-Jordan Method

A general method for finding the set of solutions of a system of linear equations is based upon a sequence of elementary row operations that transforms the augmented matrix of a given linear system into a reduced echelon matrix. The following theorem assures us that a C-matrix is always row equivalent to a reduced echelon matrix. The proof is particularly important because it gives a general statement of the primary form of the Gauss-Jordan method, which, in turn, can be programmed for a computer; in practice, certain modifications are usually made to reduce round-off error.

2.3 THE GAUSS-JORDAN METHOD

Theorem 2. *Any nonzero m by n C-matrix A is row equivalent to a reduced echelon matrix.*

Proof. PART I. Consider the first nonzero column of A, which we will designate as the kth column. If $a_{1k} \neq 0$, the following sequence of elementary row operations will force the first entry to be 1, and all other entries in the column to be 0:

$$\left(\frac{1}{a_{1k}} R_1\right), \quad (-a_{2k} R_1 + R_2), \quad (-a_{3k} R_1 + R_3), \quad \ldots, \quad (-a_{mk} R_1 + R_m).$$

If $a_{1k} = 0$, but some other entry in the kth column is nonzero, say, $a_{ik} \neq 0$, then the sequence of elementary operations listed above is preceded by the elementary operation $R_1 \leftrightarrow R_i$. In both of these cases the given matrix A is row equivalent to the resulting matrix, which will be denoted by B.

PART II. Consider the next column of B for which some entry other than the first is nonzero; call it the pth column. If $b_{2p} \neq 0$, the following sequence of elementary row operations will force the second entry of the column to be 1 and all other entries in the column to be 0:

$$\left(\frac{1}{b_{2p}} R_2\right), \quad (-b_{1p} R_2 + R_1), \quad (-b_{3p} R_2 + R_3), \quad \ldots, \quad (-b_{mp} R_2 + R_m).$$

If $b_{2p} = 0$ but some other entry in the second column *below* b_{2p} is nonzero, say, $b_{ip} \neq 0$ where $i > 2$, then the sequence of elementary operations listed above is preceded by the elementary operation $R_2 \leftrightarrow R_i$. In both of these cases the first $p - 1$ columns of B are unchanged, and the matrix B is row equivalent to the resulting matrix, which we will denote by C. Since $A \stackrel{\text{row}}{\sim} B$ and $B \stackrel{\text{row}}{\sim} C$, it follows that $A \stackrel{\text{row}}{\sim} C$ (by Exercise 9(c) of Section 2.2).

PART III. Repeat, for the rest of the columns, a procedure analogous to that listed in Part II. The given matrix A will be row equivalent to the resulting matrix N, because only elementary row operations were used to transform A to N. Moreover N must be a reduced echelon matrix, because each iteration forced the columns, one by one, into compliance with the demands of Definition 6 (definition of a reduced echelon matrix, page 25). □

The following examples illustrate (a) a general method of solution and (b) the three different types of linear systems (over the complex numbers) that can be encountered: (1) A system with a unique solution; (2) A system with an infinite number of solutions; (3) A system with no solution. In the first two cases we say that the system is **consistent**, and in the latter case that the system is **inconsistent**. The following examples also illustrate the importance of elementary row operations and matrix notation. The reader should observe that the method is systematic and readily adaptable to the computer.

Example 1. Consider the following system of linear equations over the real numbers:

$$\begin{cases} (1) & x_1 + x_2 + x_3 = 3, \\ (2) & \phantom{x_1 + {}} 2x_2 + x_3 = 2, \\ (3) & \phantom{x_1 + {}} x_2 + 2x_3 = 2. \end{cases}$$

A graph of this system is given in Figure 2.3.1. Notice that the graph consists of

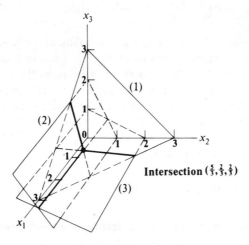

Figure 2.3.1

three planes intersecting in a single point. Algebraically, this single solution can be found as follows. Write the augmented matrix of the given system:

$$\begin{bmatrix} 1 & 1 & 1 & \vdots & 3 \\ 0 & 2 & 1 & \vdots & 2 \\ 0 & 1 & 2 & \vdots & 2 \end{bmatrix}.$$

Transform the augmented matrix to a reduced echelon matrix by the following elementary row operations. The operations performed in each step are listed below the relation symbol.

$$\begin{bmatrix} 1 & 1 & 1 & \vdots & 3 \\ 0 & 2 & 1 & \vdots & 2 \\ 0 & 1 & 2 & \vdots & 2 \end{bmatrix} \underset{\tfrac{1}{2}R_2}{\overset{\text{row}}{\sim}} \begin{bmatrix} 1 & 1 & 1 & \vdots & 3 \\ 0 & 1 & \tfrac{1}{2} & \vdots & 1 \\ 0 & 1 & 2 & \vdots & 2 \end{bmatrix}$$

$$\underset{\substack{-R_2+R_1 \\ -R_2+R_3}}{\overset{\text{row}}{\sim}} \begin{bmatrix} 1 & 0 & \tfrac{1}{2} & \vdots & 2 \\ 0 & 1 & \tfrac{1}{2} & \vdots & 1 \\ 0 & 0 & \tfrac{3}{2} & \vdots & 1 \end{bmatrix} \underset{\tfrac{2}{3}R_3}{\overset{\text{row}}{\sim}} \begin{bmatrix} 1 & 0 & \tfrac{1}{2} & \vdots & 2 \\ 0 & 1 & \tfrac{1}{2} & \vdots & 1 \\ 0 & 0 & 1 & \vdots & \tfrac{2}{3} \end{bmatrix} \underset{\substack{-\tfrac{1}{2}R_3+R_1 \\ -\tfrac{1}{2}R_3+R_2}}{\overset{\text{row}}{\sim}} \begin{bmatrix} 1 & 0 & 0 & \vdots & \tfrac{5}{3} \\ 0 & 1 & 0 & \vdots & \tfrac{2}{3} \\ 0 & 0 & 1 & \vdots & \tfrac{2}{3} \end{bmatrix}.$$

2.3 THE GAUSS-JORDAN METHOD

The final reduced echelon matrix is the augmented matrix of the system

$$\begin{cases} x_1 & = \tfrac{5}{3}, \\ x_2 & = \tfrac{2}{3}, \\ x_3 & = \tfrac{2}{3}. \end{cases}$$

The solution of the final system is obvious, and, since the original system is equivalent to the final system, it follows that $(\tfrac{5}{3}, \tfrac{2}{3}, \tfrac{2}{3})$ is a solution of the consistent original system.

Example 2. Consider the following system of linear equations over the real numbers:

$$\begin{cases} (1)\ x_1 + x_2 + x_3 = 3, \\ (2)\ x_1 - x_3 = 1, \\ (3)\ x_2 + 2x_3 = 2. \end{cases}$$

We will find that this system has an infinite number of solutions. The graph of this system, Figure 2.3.2, illustrates this fact by showing three planes that

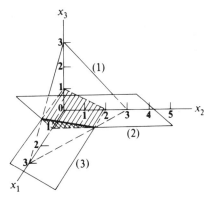

Figure 2.3.2

intersect along a line; the points on the line of intersection correspond to the solutions of the system. Algebraically, these solutions can be found as follows. The augmented matrix of the system is transformed to a reduced echelon matrix.

$$\begin{bmatrix} 1 & 1 & 1 & \vdots & 3 \\ 1 & 0 & -1 & \vdots & 1 \\ 0 & 1 & 2 & \vdots & 2 \end{bmatrix} \underset{-R_1+R_2}{\overset{\text{row}}{\sim}} \begin{bmatrix} 1 & 1 & 1 & \vdots & 3 \\ 0 & -1 & -2 & \vdots & -2 \\ 0 & 1 & 2 & \vdots & 2 \end{bmatrix}$$

$$\underset{-R_2}{\overset{\text{row}}{\sim}} \begin{bmatrix} 1 & 1 & 1 & \vdots & 3 \\ 0 & 1 & 2 & \vdots & 2 \\ 0 & 1 & 2 & \vdots & 2 \end{bmatrix} \underset{\substack{-R_2+R_1 \\ -R_2+R_3}}{\overset{\text{row}}{\sim}} \begin{bmatrix} 1 & 0 & -1 & \vdots & 1 \\ 0 & 1 & 2 & \vdots & 2 \\ 0 & 0 & 0 & \vdots & 0 \end{bmatrix}.$$

As before, the final reduced echelon matrix is the augmented matrix of a system

(2.3.1)
$$\begin{cases} x_1 \quad - \quad x_3 = 1, \\ \quad x_2 + 2x_3 = 2, \\ 0x_1 + 0x_2 + 0x_3 = 0, \end{cases}$$

that is equivalent to the given system. Obviously the last equation of (2.3.1) will produce an identity regardless of the values of the unknowns; hence (2.3.1) is equivalent to the system

(2.3.2)
$$\begin{cases} x_1 \quad - \quad x_3 = 1, \\ \quad x_2 + 2x_3 = 2. \end{cases}$$

here, 2
here, 2
here, n=3

Notice that in (2.3.2) there are more unknowns than equations. If there are r equations at this stage, then r unknowns (known as *fundamental variables*) are expressed in terms of the remaining $n - r$ unknowns, which are considered to be arbitrary parameters. System (2.3.2) is restated as

(2.3.3)
$$\begin{cases} x_1 = 1 + x_3, \\ x_2 = 2 - 2x_3. \end{cases}$$

As a check, (2.3.3) can be substituted into the original system to obtain three identities. System (2.3.3) is known as a *complete solution* of the given system. From a complete solution of a linear system of equations over the complex numbers, we may obtain what is called a *particular solution* if we substitute complex numbers for the parameters. For (2.3.3), x_3 is the only parameter, hence particular solutions may be obtained by substituting values for x_3.

If $x_3 = 2$, $\begin{cases} x_1 = \quad 3, \\ x_2 = -2, \\ x_3 = \quad 2. \end{cases}$ If $x_3 = 4$, $\begin{cases} x_1 = \quad 5, \\ x_2 = -6, \\ x_3 = \quad 4. \end{cases}$

If one or more of the parameters can take on an infinite number of values (as is the case when x_3 is any real number), then there will be an infinite number of solutions. Geometrically, the particular solutions $(3, -2, 2)$ and $(5, -6, 4)$ are two of an infinite number of points on the line of intersection shown in Figure 2.3.2; the complete solution is an algebraic representation of this line of intersection.

For systems having an infinite number of solutions the method of solution outlined so far can be extended to permit an unknown that is not originally a fundamental variable to become a fundamental variable. For example, if the reduced echelon form of the augmented matrix of Example 2,

$$\begin{bmatrix} 1 & 0 & -1 & | & 1 \\ 0 & 1 & 2 & | & 2 \\ 0 & 0 & 0 & | & 0 \end{bmatrix},$$

is transformed according to $-R_1$, and then $-2R_1 + R_2$, we obtain

$$\begin{bmatrix} -1 & 0 & 1 & | & -1 \\ 2 & 1 & 0 & | & 4 \\ 0 & 0 & 0 & | & 0 \end{bmatrix}.$$

2.3 THE GAUSS-JORDAN METHOD

Notice how columns 3 and 2 compare with the previous columns 1 and 2. From this matrix, another complete solution is seen to be

$$\begin{cases} x_3 = -1 + x_1, \\ x_2 = 4 - 2x_1. \end{cases}$$

Notice that x_3 and x_2 are now the fundamental variables instead of x_1 and x_2. The general method of solution outlined in the proof of Theorem 2 (with the variation just described permitted) is known as the **Gauss-Jordan** method of solution of a system of linear equations.

Example 3. Consider the following system of linear equations over the real numbers:

$$\begin{cases} (1) \quad x_1 = 1, \\ (2) \quad 2x_1 + x_2 + x_3 = 8, \\ (3) \phantom{2x_1+{}} \quad x_2 + x_3 = 4. \end{cases}$$

Upon analysis, a graph of this system (Figure 2.3.3) will illustrate that there are no points common to the three planes. Algebraically, this also can be determined by transforming the augmented matrix of the linear system to a reduced echelon form.

$$\begin{bmatrix} 1 & 0 & 0 & | & 1 \\ 2 & 1 & 1 & | & 8 \\ 0 & 1 & 1 & | & 4 \end{bmatrix} \underset{-2R_1+R_2}{\overset{\text{row}}{\sim}} \begin{bmatrix} 1 & 0 & 0 & | & 1 \\ 0 & 1 & 1 & | & 6 \\ 0 & 1 & 1 & | & 4 \end{bmatrix}$$

$$\underset{-R_2+R_3}{\overset{\text{row}}{\sim}} \begin{bmatrix} 1 & 0 & 0 & | & 1 \\ 0 & 1 & 1 & | & 6 \\ 0 & 0 & 0 & | & -2 \end{bmatrix} \underset{-\frac{1}{2}R_3}{\overset{\text{row}}{\sim}} \begin{bmatrix} 1 & 0 & 0 & | & 1 \\ 0 & 1 & 1 & | & 6 \\ 0 & 0 & 0 & | & 1 \end{bmatrix} \underset{-6R_3+R_2}{\overset{\text{row}}{\underset{-R_3+R_1}{\sim}}} \begin{bmatrix} 1 & 0 & 0 & | & 0 \\ 0 & 1 & 1 & | & 0 \\ 0 & 0 & 0 & | & 1 \end{bmatrix}.$$

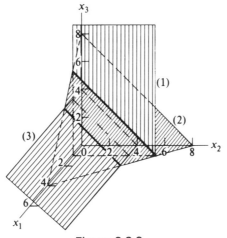

Figure 2.3.3

The corresponding system represented by the final reduced echelon matrix is

$$\begin{cases} x_1 = 0, \\ x_2 + x_3 = 0, \\ 0x_1 + 0x_2 + 0x_3 = 1. \end{cases}$$

The last system, however, obviously has no solution because there is no set of real numbers for x_1, x_2, and x_3 such that $0x_1 + 0x_2 + 0x_3 = 1$. Hence the original system is inconsistent.

A system of linear equations

$$\begin{cases} a_{11}x_1 + \cdots + a_{1n}x_n = b_1, \\ \cdots\cdots\cdots\cdots\cdots\cdots\cdots \\ a_{m1}x_1 + \cdots + a_{mn}x_n = b_m, \end{cases}$$

over C, in which $b_i = 0$ for all $i = 1, \ldots, m$, is known as a **homogeneous system**. If there exists at least one $b_i \neq 0$, then the system is **nonhomogeneous**.

APPLICATIONS

Example 4. One of the most important applications of the material in this section occurs in connection with the very useful characteristic value problems which will be studied in Chapter 9. There we seek solutions other than $(0, 0, \ldots, 0)$ to a homogeneous system of linear equations.

The reader should observe that because homogeneous systems are always consistent (because the *trivial solution* $(0, 0, \ldots, 0)$ is always a solution), then homogeneous systems possess nontrivial solutions only if there is more than one solution; the method developed in Chapter 9 and the Gauss-Jordan procedures provide such solutions.

Example 5. An interesting example from Radiative Heat Transfer that uses the material presented in this section may be found on pages 237–238 in [23] (Noble).

Example 6. Consider the system of linear equations

$$\begin{cases} p_{11}x_1 + p_{12}x_2 + p_{13}x_3 = \lambda x_1, \\ p_{21}x_1 + p_{22}x_2 + p_{23}x_3 = \lambda x_2, \\ p_{31}x_1 + p_{32}x_2 + p_{33}x_3 = \lambda x_3, \end{cases}$$

where λ is a constant, where p_{ij} is the probability that a purchaser who bought

Brand j last time will buy Brand i next time, and where x_j represents the number of buyers that bought Brand j the first time. Upon some reflection, one can see that if the p_{ij} are known and the x_j are unknown, then the solution of this linear system is equivalent to finding the number of buyers in each of the three groups that will result in the same proportion of purchases at the next purchase time; a proportional stability of purchases then can be anticipated, as long as the probabilitities do not change. To illustrate, suppose the probabilities are as shown in Figure 2.3.4, where an arrow extending from Brand j to Brand i represents p_{ij}.

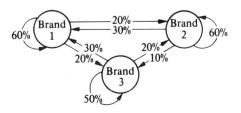

Figure 2.3.4

If we let $\lambda = 1$, then the resulting system of linear equations is

$$\begin{cases} 0.6x_1 + 0.3x_2 + 0.3x_3 = x_1, \\ 0.2x_1 + 0.6x_2 + 0.2x_3 = x_2, \\ 0.2x_1 + 0.1x_2 + 0.5x_3 = x_3, \end{cases}$$

or

$$\begin{cases} -0.4x_1 + 0.3x_2 + 0.3x_3 = 0, \\ 0.2x_1 - 0.4x_2 + 0.2x_3 = 0, \\ 0.2x_1 + 0.1x_2 - 0.5x_3 = 0. \end{cases}$$

This homogeneous system has an infinite number of solutions. By the Gauss-Jordan method we find a complete solution

$$\begin{cases} x_1 = \tfrac{9}{5}x_3, \\ x_2 = \tfrac{7}{5}x_3. \end{cases}$$

Example 7. The homogeneous system of equations in Example 6 could have been used to find stable proportions of three genetic characteristics of a population with respect to certain breeding probabilities, or to find stable proportions of three political affiliations of a population of voters with respect to certain probabilities of voting transition.

EXERCISES

In Exercises 1–13, apply the Gauss-Jordan method to the given system, and thereby determine either the unique solution or a complete solution and one particular solution, or show that the system is inconsistent.

1. $\begin{cases} 2x_1 + 6x_2 = 5, \\ 3x_1 + x_2 = 2. \end{cases}$

2. $\begin{cases} x_1 + x_2 - x_3 = 4, \\ x_1 - x_2 + x_3 = 2. \end{cases}$

3. $\begin{cases} 3x_1 + x_2 = 1, \\ 6x_1 + 2x_2 = 6. \end{cases}$

4. $\begin{cases} 2x_1 + x_2 - x_3 = -4, \\ x_1 - x_2 + 3x_3 = 3, \\ x_1 + 2x_2 - 4x_3 = 1. \end{cases}$

5. $\begin{cases} 2x_1 + x_2 - x_3 = 3, \\ x_1 + x_2 = 2, \\ x_1 - x_3 = 1. \end{cases}$

6. $\begin{cases} x_1 + 3x_2 - x_3 = 4, \\ x_1 + 2x_2 + x_3 = 2, \\ 3x_1 + 7x_2 + x_3 = 9. \end{cases}$

7. $\begin{cases} x_1 + 2x_2 + 3x_3 = 0, \\ 2x_1 - x_2 + x_3 = 0, \\ 2x_1 + 3x_2 + 4x_3 = 0. \end{cases}$

8. $\begin{cases} x_1 - 3x_2 + 2x_3 = 0, \\ -x_1 - 2x_2 + 2x_3 = 0, \\ -2x_1 + x_2 = 0. \end{cases}$

9. $\begin{cases} 2x_1 + x_2 = 1, \\ -4x_1 + 6x_2 = 6, \\ -2x_1 + 7x_2 = 7. \end{cases}$

10. $\begin{cases} x + y + 2z = 6, \\ x - y - 4z = -8, \\ 3x - 2y + 5z = 11, \\ 2x + 5y - 2z = 3. \end{cases}$

11. $\begin{cases} 3x_1 - x_2 + x_3 = 4, \\ x_1 - x_2 = 0, \\ 2x_1 + x_3 = 4, \\ 4x_1 - 2x_2 + x_3 = 4. \end{cases}$

12. $\begin{cases} x_1 - 2x_2 + x_3 - x_4 = 3, \\ 2x_1 - 3x_2 = 3, \\ x_1 - x_2 - x_3 + x_4 = 0. \end{cases}$

13. $\begin{cases} x_1 + x_3 = 2, \\ x_2 + x_3 = 6, \\ x_2 + x_4 = 0, \\ x_1 + x_2 + x_3 + x_4 = 2. \end{cases}$

14. Show both geometrically and algebraically that the system

$$\begin{cases} x_1 + x_2 = 2, \\ x_1 - x_2 = 0, \\ 2x_1 + x_2 = 2, \end{cases}$$

is inconsistent.

15. In the following consistent system can we solve for x_1 and x_2 in terms of x_3?

$$\begin{cases} x_1 + x_2 + x_3 = 4, \\ -x_1 - x_2 = 2, \\ 2x_1 + 2x_2 + 2x_3 = 8. \end{cases}$$

Which sets of unknowns will serve as fundamental variables?

2.3 THE GAUSS-JORDAN METHOD

16. Sketch the planes $x + 2y + 2z = 6$, $2x + y + z = 6$, and $3x + 2y + z = 9$. Estimate the point of intersection from your graph. Then, solve these equations by the Gauss-Jordan method to find the solution accurately.

17. Demonstrate graphically that the following system has an infinite number of solutions

$$\begin{cases} x \phantom{{}+y} + z = 1, \\ x + y + z = 3, \\ \phantom{x+{}} y \phantom{{}+z} = 2. \end{cases}$$

Then, using the Gauss-Jordan method, find a complete solution and three particular solutions. Be sure that the particular solutions agree with your graph.

18. In Exercises 1–13, which of the systems are homogeneous?

19. Following the Gauss-Jordan method outlined in the proof of Theorem 2, using the notation $kR_1 + R_2$, etc., whenever possible, state a general sequence of steps that will solve any linear system with two equations in two unknowns.

20. Two alloys P and Q are manufactured by a certain factory, which sells its products by the ton but purchases some of the ingredients in 100-pound units. For each ton of P produced, 4 units of metal A and 4 units of metal B are required. Each ton of Q requires 7 units of A and 3 units of B. If the factory can obtain, and wishes to use, exactly 60 units of metal A and 40 units of metal B per day, how many tons of each alloy can it produce in a day?

21. A trucking company owns three types of trucks, numbered 1, 2, and 3, which are equipped to haul three different types of machines per load, according to the following chart.

	Trucks		
	No. 1	No. 2	No. 3
Machine A	1	1	1
Machine B	0	1	2
Machine C	2	1	1

How many trucks of each type should be sent to haul exactly 12 of the type A machines, 10 of the type B machines, and 16 of the type C machines? Assume each truck is fully loaded.

22. To control a certain crop disease, it is determined that 6 units of chemical A, 10 units of chemical B, and 8 units of chemical C must be used. One barrel of commercial spray P contains 1, 3, and 4 units, respectively, of these chemicals; one barrel of commercial spray Q contains 3, 3, 3 units, respectively; and one barrel of commercial spray R contains 2 units of A and 5 units of B. How much of each type of spray should be used to spread the exact amounts of chemicals needed to control the disease?

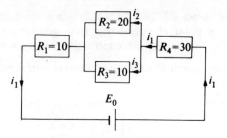

Figure 2.3.5

23. For the network shown in Figure 2.3.5, Kirchhoff's Laws yield the system of equations

$$\begin{cases} i_1 - i_2 - i_3 = 0, \\ 20i_2 - 10i_3 = 0, \\ 40i_1 + 20i_2 = E_0. \end{cases}$$

Solve for the currents i_1, i_2, i_3.

24. To control a certain crop disease, it is determined that 6 units of chemical A and 10 units of chemical B must be used. One barrel of commercial spray P contains 1 unit of A and 3 units of B; one barrel of spray Q contains 3 units of A and 3 units of B; and one barrel of spray R contains 2 units of A and 5 units of B. How much of each type of spray should be used to spread the exact amount of chemicals needed to control the disease? (See Exercise 22.)

25. (a) A trucking company owns three types of trucks which are equipped to haul two different types of machines per load according to the following chart.

	Trucks		
	No. 1	No. 2	No. 3
Machine A	1	1	1
Machine C	2	1	1

How many trucks of each type should be sent to haul exactly 12 of the type A machines and 16 of the type C machines? Assume each truck is fully loaded. (See Exercise 21.)

(b) Why is there not an infinite number of solutions to part (a)?

26. In a system of two linear equations in two unknowns over the real numbers, one equation is $x_1 - 2x_2 = 4$. By inspection, make up a possible second equation so that the system will (a) be inconsistent, (b) have an infinite number of solutions, and (c) have a unique solution.

27. In each of the following, the augmented matrix of a given system of linear equations over the real numbers has been transformed by elementary row operations to the reduced echelon matrix given. In each case, explain whether the original system was consistent and whether its solution was unique.

(a) $\begin{bmatrix} 1 & 0 & 0 & | & 5 \\ 0 & 1 & 0 & | & 4 \\ 0 & 0 & 1 & | & 3 \end{bmatrix}$; (b) $\begin{bmatrix} 1 & 0 & 2 & | & 5 \\ 0 & 1 & 3 & | & 4 \end{bmatrix}$; (c) $\begin{bmatrix} 1 & 0 & 2 & | & 5 \\ 0 & 1 & 3 & | & 4 \\ 0 & 0 & 0 & | & 0 \end{bmatrix}$;

(d) $\begin{bmatrix} 1 & 0 & 2 & | & 0 \\ 0 & 1 & 3 & | & 0 \\ 0 & 0 & 0 & | & 1 \end{bmatrix}$; (e) $\begin{bmatrix} 1 & 0 & 0 & | & 5 \\ 0 & 1 & 0 & | & 4 \\ 0 & 0 & 1 & | & 0 \end{bmatrix}$; (f) $[1 \; 2 \; | \; 5]$;

(g) $\begin{bmatrix} 1 & 0 & | & 5 \\ 0 & 1 & | & 4 \\ 0 & 0 & | & 0 \\ 0 & 0 & | & 0 \end{bmatrix}$; (h) $\begin{bmatrix} 1 & 0 & 1 & 2 & | & 5 \\ 0 & 1 & 3 & 1 & | & 4 \\ 0 & 0 & 0 & 0 & | & 0 \end{bmatrix}$.

28. In the proof of Theorem 2 it is important to consider the possibility that all entries of the first column are zero. Is this likely to happen when the matrix is the augmented matrix of a system of linear equations over the real numbers? Why?

NEW VOCABULARY

system of linear equations   2.1
solution of system of linear
  equations   2.1
coefficient matrix   2.1
augmented matrix   2.1
elementary operations on equations
  2.1
equivalent systems   2.1
elementary row operations   2.2
binary relation   2.2
row equivalent matrices   2.2
echelon matrix   2.2

reduced echelon matrix   2.2
echelon form   2.2
consistent system   2.3
inconsistent system   2.3
fundamental variables   2.3
complete solution   2.3
particular solution   2.3
Gauss-Jordan method of solution
  2.3
homogeneous system   2.3
nonhomogeneous system   2.3
trivial solution   2.3

3

MATRIX MULTIPLICATION

3.1 An Introduction

In Section 1.3 we learned how to add two matrices that were of the same order, and how to multiply any matrix by a scalar. Now we define a type of multiplication of two matrices that is sometimes called *Cayley multiplication* of matrices, since it was first defined by the English mathematician, Sir Arthur Cayley. We will be particularly interested in the use of this operation in a further study of systems of linear equations and elementary operations. Before giving a general definition of the Cayley product of two matrices, however, we consider the product of a row matrix by a column matrix,

Definition 1. *Let A be a 1 by p row C-matrix and let B be a p by 1 column C-matrix. The product $C = AB$ is a 1 by 1 C-matrix given by*

$$[a_{11} a_{12} \cdots a_{1p}] \begin{bmatrix} b_{11} \\ b_{21} \\ \vdots \\ b_{p1} \end{bmatrix} = [a_{11}b_{11} + a_{12}b_{21} + \cdots + a_{1p}b_{p1}]$$

$$= \left[\sum_{k=1}^{p} a_{1k} b_{k1} \right]_{(1,1)}.$$

3.1 AN INTRODUCTION

Example 1.

$$[4 \quad 2 \quad 7] \begin{bmatrix} 0 \\ 3 \\ 1 \end{bmatrix} = [(4 \cdot 0) + (2 \cdot 3) + (7 \cdot 1)] = [13].$$

Notice that the number of columns of A must equal the number of rows of B. One use of this operation is to express a linear equation as a matrix equation; that is,

$$[2 \quad 3] \begin{bmatrix} x \\ y \end{bmatrix} = [3]$$

means

$$[(2x + 3y)] = [3],$$

and, by the definition of matrix equality,

$$2x + 3y = 3.$$

Now, in order to express a system of equations such as

$$\begin{cases} 2x + 3y = 3, \\ x + 4y = 1, \end{cases}$$

as a matrix equation, we need the following definition in which we see that the general *Cayley product* is simply a succession of products of the type defined in Definition 1.

Definition 2. *Let A be an m by p C-matrix and B be a p by n C-matrix. The Cayley product $C = AB$ is an m by n C-matrix, where*

$$c_{ij} = a_{i1}b_{1j} + a_{i2}b_{2j} + \cdots + a_{ip}b_{pj},$$

or

$$c_{ij} = \sum_{k=1}^{p} a_{ik} b_{kj}.$$

There are various other ways of defining a product of two matrices, but the Cayley product is the most common. Hence, when we speak of the product of two matrices, we shall mean their Cayley product. The operation defined above can be illustrated in general by the following diagram.

$$\begin{bmatrix} a_{11} & a_{12} & \cdots & a_{1p} \\ a_{21} & a_{22} & \cdots & a_{2p} \\ \cdots & \cdots & \cdots & \cdots \\ a_{m1} & a_{m2} & \cdots & a_{mp} \end{bmatrix} \begin{bmatrix} b_{11} & b_{12} & \cdots & b_{1n} \\ b_{21} & b_{22} & \cdots & b_{2n} \\ \cdots & \cdots & \cdots & \cdots \\ b_{p1} & b_{p2} & \cdots & b_{pn} \end{bmatrix} = \begin{bmatrix} c_{11} & c_{12} & \cdots & c_{1n} \\ c_{21} & c_{22} & \cdots & c_{2n} \\ \cdots & \cdots & \cdots & \cdots \\ c_{m1} & c_{m2} & \cdots & c_{mn} \end{bmatrix},$$

where $c_{11} = a_{11}b_{11} + a_{12}b_{21} + \cdots + a_{1p}b_{p1}$.

$$\begin{bmatrix} a_{11} & \cdots & a_{1p} \\ \hdotsfor{3} \\ \boxed{a_{i1} \;\; \cdots \;\; a_{ip}} \\ \hdotsfor{3} \\ a_{m1} & \cdots & a_{mp} \end{bmatrix} \begin{bmatrix} b_{11} & \cdots & \boxed{b_{1j}} & \cdots & b_{1n} \\ \vdots & & \vdots & & \vdots \\ b_{p1} & \cdots & \boxed{b_{pj}} & \cdots & b_{pn} \end{bmatrix} = \begin{bmatrix} c_{11} & \cdots & & & c_{1n} \\ \cdot & & & & \cdot \\ \cdot & & \boxed{c_{ij}} & & \cdot \\ \cdot & & & & \cdot \\ c_{m1} & \cdots & & & c_{mn} \end{bmatrix},$$

where $c_{ij} = a_{i1}b_{1j} + a_{i2}b_{2j} + \cdots + a_{ip}b_{pj}$.

Example 2.

$$A = \begin{bmatrix} 3 & 0 \\ 4 & 1 \end{bmatrix}, \quad B = \begin{bmatrix} 2 & 6 & 0 \\ 0 & 3 & 5 \end{bmatrix}, \quad AB = \begin{bmatrix} 3 & 0 \\ 4 & 1 \end{bmatrix} \begin{bmatrix} 2 & 6 & 0 \\ 0 & 3 & 5 \end{bmatrix}.$$

From the definition, note that here $m = 2$, $p = 2$, and $n = 3$.

$$AB = \begin{bmatrix} (3 \cdot 2 + 0 \cdot 0) & (3 \cdot 6 + 0 \cdot 3) & (3 \cdot 0 + 0 \cdot 5) \\ (4 \cdot 2 + 1 \cdot 0) & (4 \cdot 6 + 1 \cdot 3) & (4 \cdot 0 + 1 \cdot 5) \end{bmatrix} = \begin{bmatrix} 6 & 18 & 0 \\ 8 & 27 & 5 \end{bmatrix}.$$

Notice that it is impossible to find BA.

In order to find the product of two matrices, the number of columns in the left matrix must equal the number of rows in the right matrix; when this is so, we say that the left matrix is *conformable for multiplication* to the right matrix.

In 1858 Cayley used the matrix product just defined as an abbreviation for expressing systems of linear equations; that is,

$$\begin{cases} a_{11}x_1 + a_{12}x_2 + \cdots + a_{1n}x_n = b_1, \\ a_{21}x_1 + a_{22}x_2 + \cdots + a_{2n}x_n = b_2, \\ \cdots\cdots\cdots\cdots\cdots\cdots\cdots\cdots\cdots\cdots \\ a_{m1}x_1 + a_{m2}x_2 + \cdots + a_{mn}x_n = b_m, \end{cases}$$

can be expressed as

$$AX = B,$$

where

$$A = \begin{bmatrix} a_{11} & a_{12} & \cdots & a_{1n} \\ a_{21} & a_{22} & \cdots & a_{2n} \\ \hdotsfor{4} \\ a_{m1} & a_{m2} & \cdots & a_{mn} \end{bmatrix}, \quad X = \begin{bmatrix} x_1 \\ x_2 \\ \vdots \\ x_n \end{bmatrix}, \quad B = \begin{bmatrix} b_1 \\ b_2 \\ \vdots \\ b_m \end{bmatrix}.$$

Example 3. The system of linear equations

$$\begin{cases} 3x_1 + 4x_2 - 5x_3 = 2, \\ x_1 + x_2 - x_3 = 1, \\ x_1 + x_3 = 2, \end{cases}$$

can be expressed as

$$AX = B,$$

where

$$A = \begin{bmatrix} 3 & 4 & -5 \\ 1 & 1 & -1 \\ 1 & 0 & 1 \end{bmatrix}, \quad X = \begin{bmatrix} x_1 \\ x_2 \\ x_3 \end{bmatrix}, \quad B = \begin{bmatrix} 2 \\ 1 \\ 2 \end{bmatrix}.$$

Example 4.

$$A = \begin{bmatrix} 2 & 1 \\ 3 & 2 \end{bmatrix}, \quad B = \begin{bmatrix} 0 & 4 \\ 1 & 3 \end{bmatrix}.$$

$$AB = \begin{bmatrix} 2 & 1 \\ 3 & 2 \end{bmatrix} \begin{bmatrix} 0 & 4 \\ 1 & 3 \end{bmatrix} = \begin{bmatrix} (2 \cdot 0 + 1 \cdot 1) & (2 \cdot 4 + 1 \cdot 3) \\ (3 \cdot 0 + 2 \cdot 1) & (3 \cdot 4 + 2 \cdot 3) \end{bmatrix} = \begin{bmatrix} 1 & 11 \\ 2 & 18 \end{bmatrix}.$$

$$BA = \begin{bmatrix} 0 & 4 \\ 1 & 3 \end{bmatrix} \begin{bmatrix} 2 & 1 \\ 3 & 2 \end{bmatrix} = \begin{bmatrix} (0 \cdot 2 + 4 \cdot 3) & (0 \cdot 1 + 4 \cdot 2) \\ (1 \cdot 2 + 3 \cdot 3) & (1 \cdot 1 + 3 \cdot 2) \end{bmatrix} = \begin{bmatrix} 12 & 8 \\ 11 & 7 \end{bmatrix}.$$

Notice that $AB \neq BA$; that is, matrix multiplication as defined in this section is *not* commutative. In the first part of this example we say that B was *premultiplied* by A and, in the second part, that B was *postmultiplied* by A.

In Section 1.2 the square matrix I_n, with 1's on the main diagonal and zeros elsewhere, was defined; this matrix is named the *identity matrix* because it serves as the identity element in the set of n by n C-matrices for Cayley multiplication. That is, I_n has the property that for any n by n C-matrix A,

$$AI_n = I_n A = A.$$

The summation notation is useful in expressing a matrix product, as illustrated in the following example.

Example 5. Express the product of two 2 by 2 C-matrices, using summation notation.

$$AB = \begin{bmatrix} a_{11} & a_{12} \\ a_{21} & a_{22} \end{bmatrix} \begin{bmatrix} b_{11} & b_{12} \\ b_{21} & b_{22} \end{bmatrix} = \begin{bmatrix} (a_{11}b_{11} + a_{12}b_{21}) & (a_{11}b_{12} + a_{12}b_{22}) \\ (a_{21}b_{11} + a_{22}b_{21}) & (a_{21}b_{12} + a_{22}b_{22}) \end{bmatrix}$$

$$= \begin{bmatrix} \sum_{k=1}^{2} a_{1k}b_{k1} & \sum_{k=1}^{2} a_{1k}b_{k2} \\ \sum_{k=1}^{2} a_{2k}b_{k1} & \sum_{k=1}^{2} a_{2k}b_{k2} \end{bmatrix} = \left[\sum_{k=1}^{2} a_{ik}b_{kj} \right]_{(2,2)}.$$

In general if A is an m by p C-matrix and B is a p by n C-matrix then

$$AB = \begin{bmatrix} \left(\sum_{k=1}^{p} a_{1k} b_{k1}\right) & \cdots & \left(\sum_{k=1}^{p} a_{1k} b_{kn}\right) \\ \vdots & & \vdots \\ \left(\sum_{k=1}^{p} a_{mk} b_{k1}\right) & \cdots & \left(\sum_{k=1}^{p} a_{mk} b_{kn}\right) \end{bmatrix}_{(m,n)} = \left[\sum_{k=1}^{p} a_{ik} b_{kj}\right]_{(m,n)}.$$

Assuming conformability, we may express the product $(AB)C$ by using repeated summation notation. Let

$$A = [a_{ij}]_{(m,p)}, \qquad B = [b_{ij}]_{(p,q)}, \qquad C = [c_{ij}]_{(q,n)};$$

then, if we let $AB = D$, we have

$$d_{ij} = \sum_{h=1}^{p} a_{ih} b_{hj},$$

where h is used as the index of summation rather than k. Hence

$$(AB)C = DC = \left[\sum_{k=1}^{q} d_{ik} c_{kj}\right]_{(m,n)} = \left[\sum_{k=1}^{q}\left(\sum_{h=1}^{p} a_{ih} b_{hk}\right) c_{kj}\right]_{(m,n)}.$$

Summation notation is used in the proofs of the next two theorems.

Theorem 1. *(Associative Property)* If $A = [a_{ij}]_{(m,p)}$, $B = [b_{ij}]_{(p,q)}$, and $C = [c_{ij}]_{(q,n)}$ are C-matrices, then $(AB)C = A(BC)$.

Proof.

	STATEMENT	REASON
(1)	$(AB)C = \left[\sum_{h=1}^{p} a_{ih} b_{hj}\right]_{(m,q)} [c_{ij}]_{(q,n)}$	(1) Definition of matrix multiplication.
(2)	$= \left[\sum_{k=1}^{q}\left(\sum_{h=1}^{p} a_{ih} b_{hk}\right) c_{kj}\right]_{(m,n)}$	(2) Same as (1).
(3)	$= \left[\sum_{h=1}^{p} a_{ih}\left(\sum_{k=1}^{q} b_{hk} c_{kj}\right)\right]_{(m,n)}$	(3) Property of summation notation (for complex numbers addition is commutative and associative, multiplication is associative, and multiplication is distributive with respect to addition).
(4)	$= [a_{ij}]_{(m,p)} \left[\sum_{k=1}^{q} b_{ik} c_{kj}\right]_{(p,n)}$	(4) Same as (1).
(5)	$= A(BC)$	(5) Same as (1). □

Theorem 2. *(Left Distributive Property)* If $A = [a_{ij}]_{(m,p)}$, $B = [b_{ij}]_{(p,n)}$, and $C = [c_{ij}]_{(p,n)}$ are C-matrices, then $A(B + C) = AB + AC$.

Proof.

STATEMENT	REASON
(1) $A(B + C) = [a_{ij}]_{(m,\,p)}[b_{ij} + c_{ij}]_{(p,\,n)}$	(1) Definition of matrix addition.
(2) $\quad = \left[\sum_{k=1}^{p} a_{ik}(b_{kj} + c_{kj})\right]_{(m,\,n)}$	(2) Definition of matrix multiplication.
(3) $\quad = \left[\sum_{k=1}^{p} (a_{ik}b_{kj} + a_{ik}c_{kj})\right]_{(m,\,n)}$	(3) Left distributive law for multiplication of complex numbers with respect to addition is valid.
(4) $\quad = \left[\sum_{k=1}^{p} a_{ik}b_{kj} + \sum_{k=1}^{p} a_{ik}c_{kj}\right]_{(m,\,n)}$	(4) Property of summation notation (addition of complex numbers is commutative and associative).
(5) $\quad = \left[\sum_{k=1}^{p} a_{ik}b_{kj}\right]_{(m,\,n)} + \left[\sum_{k=1}^{p} a_{ik}c_{kj}\right]_{(m,\,n)}$	(5) Same as (1).
(6) $\quad = AB + AC.$	(6) Same as (2). ◻

Theorem 3. *(Right Distributive Property) If $A = [a_{ij}]_{(m,\,p)}$, $B = [b_{ij}]_{(m,\,p)}$, and $C = [c_{ij}]_{(p,\,n)}$ are C-matrices, then*

$$(A + B)C = AC + BC.$$

The proof is left as Exercise 16.

Theorem 4. *If $A = [a_{ij}]_{(m,\,p)}$, and $B = [b_{ij}]_{(p,\,n)}$ are C-matrices, and if c is a complex number, then*

$$c(AB) = (cA)B = (Ac)B = A(cB).$$

The proof is left as Exercise 17.

Theorem 5. *If $A = [a_{ij}]_{(m,\,p)}$, and $B = [b_{ij}]_{(p,\,n)}$ are C-matrices, then*

[*the ith row of AB*] = [*the ith row of A*]B.

The proof is left as Exercise 18.

Positive integral powers of square matrices are defined similarly to those of complex numbers.

Definition 3. *If A is an n by n C-matrix, then*

$$A^1 = A,$$

and if k is a positive integer for which A^k has been defined, then

$$A^{k+1} = A^k A.$$

APPLICATIONS

Example 6. Consider a communications system in which some stations may not speak to other stations (these communication centers could represent women, nations, or electronic apparatus). If the ith station can speak to the jth station, then define $a_{ij} = 1$; otherwise, $a_{ij} = 0$. Suppose that we have four stations which communicate according to the following matrix. We assume that a station does not speak to itself directly, hence the entries on the main diagonal are 0.

$$A = \begin{matrix} & \text{RECEIVERS} \\ & \begin{matrix} \#1 & \#2 & \#3 & \#4 \end{matrix} \\ \begin{bmatrix} 0 & 0 & 0 & 1 \\ 1 & 0 & 0 & 1 \\ 1 & 0 & 0 & 0 \\ 0 & 1 & 1 & 0 \end{bmatrix} & \begin{matrix} \#1 \\ \#2 \\ \#3 \\ \#4 \end{matrix} \end{matrix} \quad \text{SPEAKERS.}$$

This matrix indicates that stations #2 and #4 are the only stations that reciprocate communication directly, whereas stations #2 and #3 do not communicate directly with each other at all.

This question can be raised: Which stations can communicate indirectly through the help of intermediaries? Let us look at the entry in the fourth row and first column of

$$A^2 = AA = \begin{bmatrix} 0 & 1 & 1 & 0 \\ 0 & 1 & 1 & 1 \\ 0 & 0 & 0 & 1 \\ 2 & 0 & 0 & 1 \end{bmatrix} \begin{matrix} \#1 \\ \#2 \\ \#3 \\ \#4 \end{matrix};$$

the entry b_{41} of the matrix $B = A^2$ equals

$$a_{41}a_{11} + a_{42}a_{21} + a_{43}a_{31} + a_{44}a_{41}$$

(these a_{ij} belong to the matrix A). Any one of these four products $a_{4k}a_{k1}$ (where $k = 1, 2, 3$, or 4) is equal to 1 only if both a_{4k} and a_{k1} are 1, that is, when station #4 can speak to station #k (where $k = 1, 2, 3$, or 4) and station #k can in turn speak to station #1. Here, we find $b_{41} = 2$; that is, there are two possible channels of communication by which station #4 can speak to station #1 through an intermediary (or by one relay).

The c_{ij} entry of $C = A^3$ tells how many [1] channels of communication are open between station #i and station #j using two relays, and so forth for A^n. Furthermore, $A + A^2$ gives the total number of channels of communication that are open between various stations for either zero or one relay; more generally, $A + A^2 + A^3 + \cdots + A^{k+1}$ gives the total number of channels of communication that are open between the various stations, with no more than k relays.

Example 7. Consider the following equations, involving products of matrices,

$$\sigma_x \sigma_y = -\sigma_y \sigma_x = i\sigma_z,$$
$$\sigma_y \sigma_z = -\sigma_z \sigma_y = i\sigma_x,$$
$$\sigma_z \sigma_x = -\sigma_x \sigma_z = i\sigma_y,$$

where

$$\sigma_x = \begin{bmatrix} 0 & 1 \\ 1 & 0 \end{bmatrix}, \quad \sigma_y = \begin{bmatrix} 0 & -i \\ i & 0 \end{bmatrix}, \text{ and } \sigma_z = \begin{bmatrix} 1 & 0 \\ 0 & -1 \end{bmatrix}.$$

These matrices are very important in atomic physics. The matrices σ_x, σ_y, and σ_z are known as the Pauli spin matrices (introduced by Wolfgang Pauli, Jr., Austrian and American, 1900–) and are used to describe the intrinsic angular momentum of an electron. Furthermore the reader can prove that

$$\sigma_x^2 = \sigma_y^2 = \sigma_z^2 = I_2.$$

Example 8. In Chapter 4 of [16] (Johnston et al.) the reader may find applications similar to Example 6. These applications are concerned with clique detection, sociometric relations, and intrabusiness communications, as well as with dominance. A relatively elementary and complete presentation, and a helpful bibliography (with notes), make [16] an excellent point of departure for readers interested in applications similar to those just mentioned. A very interesting application of clique detection, through the use of matrices, to an archeological problem is discussed on pages 401–404 of [18] (Kemeny et al.).

Example 9. Illustrations of the use of partitioned matrices and simple matrix operations in the Mathematics of Accounting may be found on pages 346–357 of [17] (Kemeny et al.). It is also both interesting and significant to note that Sir Arthur Cayley, the originator of the theory of matrices, wrote a book [5] on the double entry book-keeping system of accounting.

Example 10. A probability vector can be defined as a row matrix with non-negative entries whose sum is 1. A matrix that consists of rows of probability vectors is called a stochastic matrix. Consider a voting precinct in which the

---

[1] A word of caution is in order: for $C = A^3$, one communication route may be #$i \to$ #$k \to$ #$i \to$ #j which may not be a satisfactory situation for some applications. The problem of repeated segments of routes becomes greater as the exponent of A gets larger.

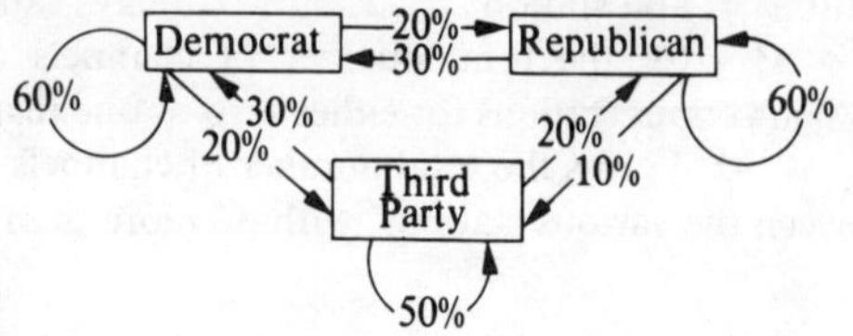

Figure 3.1.1

predicted voting probabilities of the people are as shown in Figure 3.1.1. For example, of the people who voted for the Republicans in the previous election, it is predicted that 60 percent will again vote Republican, while 30 percent will vote for the Democrats, and 10 percent will vote for a third party. This information can be stated as a stochastic matrix:

$$P = \begin{bmatrix} 0.60 & 0.20 & 0.20 \\ 0.30 & 0.60 & 0.10 \\ 0.30 & 0.20 & 0.50 \end{bmatrix} \begin{matrix} \text{From } D \\ \text{From } R \\ \text{From } T \end{matrix}$$

with columns labeled To D, To R, To T.

The matrix P represents a one-stage transition.

We calculate the square of P.

$$A = P^2 = \begin{bmatrix} 0.60 & 0.20 & 0.20 \\ 0.30 & 0.60 & 0.10 \\ 0.30 & 0.20 & 0.50 \end{bmatrix}^2 = \begin{bmatrix} 0.48 & 0.28 & 0.24 \\ 0.39 & 0.44 & 0.17 \\ 0.39 & 0.28 & 0.33 \end{bmatrix}.$$

Entry $a_{ij} = p_{i1}p_{1j} + p_{i2}p_{2j} + p_{i3}p_{3j}$ and therefore represents the probability of a two-stage transition from i to j; that is, 48 percent of the people who voted Democratic in the first election will be expected to vote Democratic in the third election, 28 percent of the people who voted Democratic in the first election will be expected to vote Republican in the third election, etc. Note that P^2 is also a stochastic matrix. Moreover, this method of stating predictions can be generalized to P^n, a stochastic matrix which represents the probabilities of an n-stage transition.

Example 11. In a manner similar to the last example, stochastic matrices and powers of stochastic matrices may be used to express transition probabilities from one state to another state, in a variety of ways. We list three illustrations: (1) The stochastic matrix of survival probabilities from one stage to another stage of a living organism's life span. (2) The stochastic matrix of customer purchasing probabilities of different brands of commodities on successive shopping trips (Kemeny et al [17], page 195). (3) The stochastic matrix of probabilities of a disease progressing from one state to another at different periods of diagnosis (Alling [1], pages 527–547).

★**Example 12.** For those readers who have studied differential equations, we point out that a single higher-order homogeneous linear differential equation with constant coefficients may be expressed as a single matrix equation. For instance, consider

(3.1.1) $$\frac{d^3x_1}{dt^3} + 4\frac{d^2x_1}{dt^2} + 5\frac{dx_1}{dt} - 7x_1 = 0,$$

which can be rewritten as

$$\frac{d^3x_1}{dt^3} = 7x_1 - 5\frac{dx_1}{dt} - 4\frac{d^2x_1}{dt^2}.$$

Then let

$$\frac{dx_1}{dt} = x_2, \quad \text{and} \quad \frac{dx_2}{dt} = x_3;$$

hence

$$\frac{d^2x_1}{dt^2} = \frac{dx_2}{dt} = x_3, \quad \text{and} \quad \frac{d^3x_1}{dt^3} = \frac{d^2x_2}{dt^2} = \frac{dx_3}{dt}.$$

The solution of equation (3.1.1) is the solution of the system

$$\begin{cases} \dfrac{dx_1}{dt} = x_2, \\[4pt] \dfrac{dx_2}{dt} = x_3, \\[4pt] \dfrac{dx_3}{dt} = 7x_1 - 5x_2 - 4x_3, \end{cases}$$

or the solution of

$$\begin{bmatrix} \dfrac{dx_1}{dt} \\[4pt] \dfrac{dx_2}{dt} \\[4pt] \dfrac{dx_3}{dt} \end{bmatrix} = \begin{bmatrix} 0 & 1 & 0 \\ 0 & 0 & 1 \\ 7 & -5 & -4 \end{bmatrix} \begin{bmatrix} x_1 \\ x_2 \\ x_3 \end{bmatrix}.$$

EXERCISES

1. Multiply the following, if possible:

   (a) $\begin{bmatrix} 2 & 1 \\ 3 & 4 \end{bmatrix} \begin{bmatrix} 0 & 1 \\ 2 & -1 \end{bmatrix}$;   (b) $\begin{bmatrix} 2 & 1 \\ 6 & 0 \end{bmatrix} \begin{bmatrix} -1 \\ 4 \end{bmatrix}$;

(c) $[2 \quad 1 \quad 0] \begin{bmatrix} 4 & 0 \\ 0 & 2 \\ -1 & 1 \end{bmatrix}$; (d) $\begin{bmatrix} 4 & 2 \\ 3 & 1 \end{bmatrix} \begin{bmatrix} 1 & 0 \\ 0 & 1 \end{bmatrix}$;

(e) $\begin{bmatrix} 9 & 6 & 2 \\ 4 & 3 & 1 \end{bmatrix} \begin{bmatrix} 2 & 4 \\ 0 & 2 \end{bmatrix}$; (f) $\begin{bmatrix} 2 \\ 3 \end{bmatrix} [3 \quad -1]$.

2. Let $A = [a_{ij}]_{(3,t)}$ and $B = [b_{ij}]_{(4,5)}$ be C-matrices.
   (a) Under what conditions does AB exist?
   (b) What is the order of AB?
   (c) Under what conditions, if any, does BA exist?

3. Let $A = [a_{ij}]_{(m,n)}$ and $B = [b_{ij}]_{(r,t)}$ be C-matrices.
   (a) Under what conditions does AB exist?
   (b) What is the order of AB?
   (c) Under what conditions does BA exist?
   (d) What is the order of BA?
   (e) Under what conditions will the order of AB be the same as that of BA?

4. Let $A = \begin{bmatrix} 2 & 0 \\ 3 & 1 \end{bmatrix}$, $B = \begin{bmatrix} 4 & -1 \\ 0 & 2 \end{bmatrix}$, $I = \begin{bmatrix} 1 & 0 \\ 0 & 1 \end{bmatrix}$, $\mathbf{0} = \begin{bmatrix} 0 & 0 \\ 0 & 0 \end{bmatrix}$.
   (a) Premultiply B by A; that is, find AB.
   (b) Postmultiply B by A; that is, find BA.
   (c) Find B^2. (d) Find B^3. (e) Find IB.
   (f) Find $\mathbf{0}B$. (g) Find I^3.

5. If $A = [a_{ij}]_{(m,p)}$ is a C-matrix and n is a positive integer, under what conditions does A^n exist?

6. Express the linear system
$$\begin{cases} x_1 + x_2 + x_3 = 4, \\ x_1 - x_2 + 2x_3 = 9, \\ 2x_1 + x_3 = 6, \end{cases}$$
as a single matrix equation, and define each matrix used.

7. Write out the system represented by
$$\begin{bmatrix} 2 & 0 \\ 1 & 3 \\ 4 & 2 \end{bmatrix} \begin{bmatrix} x_1 \\ x_2 \end{bmatrix} = \begin{bmatrix} 2 \\ 1 \\ 3 \end{bmatrix}.$$

8. If A and B are n by n C-matrices, does $(A+B)^2 = A^2 + 2AB + B^2$ in general? Why?

9. If A and B are n by n C-matrices, does $(AB)^2 = A^2B^2$ in general? Why?

10. If A and B are n by n C-matrices, does $A^2 - B^2 = (A-B)(A+B)$ in general? Why?

11. We have seen that AB does not always equal BA. In a case when AB does equal BA, we say that A and B *commute*.
    (a) Show that $A = \begin{bmatrix} 2 & 0 \\ 0 & 2 \end{bmatrix}$ and $B = \begin{bmatrix} 3 & 4 \\ 5 & 6 \end{bmatrix}$ commute.
    (b) Does every square matrix commute with itself?

3.1 AN INTRODUCTION

12. (*a*) Prove the following generalization of Theorem 2:
$$A(B + C + D) = AB + AC + AD.$$
(*b*) How can Theorem 2 be generalized further?

13. Can we say that matrix addition is right distributive with respect to matrix multiplication? Give an example.

14. Express $a_{11}b_{11} + a_{12}b_{21} + a_{13}b_{31}$ using summation notation and as the product of two matrices.

15. Express $\begin{bmatrix} a_{11} & a_{12} & a_{13} \\ a_{21} & a_{22} & a_{23} \end{bmatrix} \begin{bmatrix} b_{11} \\ b_{21} \\ b_{31} \end{bmatrix}$ using summation notation.

16. Prove Theorem 3.

17. Prove Theorem 4.

18. Prove Theorem 5.

19. Let the matrix $A = \begin{bmatrix} 2 & 1 \\ 4 & 3 \end{bmatrix}$ represent the number of gadgets R and S that factories P and Q can produce in a day, according to the table below.

	Factory P	Factory Q
Gadget R	2 per day	1 per day
Gadget S	4 per day	3 per day

Let $N = \begin{bmatrix} 5 \\ 6 \end{bmatrix}$ represent the number of days the two factories operate; that is, P operates 5 days per week and Q operates 6 days per week. Find AN, and state what it represents.

20. In Exercise 19 suppose that $N = \begin{bmatrix} x_1 \\ x_2 \end{bmatrix}$. What is the interpretation of
$$A \begin{bmatrix} x_1 \\ x_2 \end{bmatrix} = \begin{bmatrix} 9 \\ 8 \end{bmatrix}?$$

21. Suppose that four legislators influence each other according to Figure 3.1.2.

(*a*) Write the matrix that shows the number of ways in which any one legislator can influence another using at most one relay.

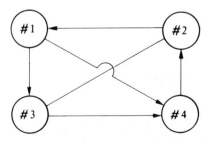

Figure 3.1.2

(b) Rank the legislators according to the total number of influence channels that each can exert using at most one relay.

22. In Example 7 of this section, justify each of the stated equations, including

$$\sigma_x^2 = \sigma_y^2 = \sigma_z^2 = I_2.$$

23. Draw a diagram illustrating the communication channels represented by matrix A in Example 6 of this section, and visualize the conclusions made concerning $A + A^2$.

24. In view of the fact that in general $AB \neq BA$, we can define two other types of multiplication. Prove that neither is associative, in contrast with Cayley multiplication which is associative.

Definition. *If A and B are two n by n C-matrices, their **Lie product** (or cross product) is*

$$A \times B = AB - BA,$$

*and their **Jordan product** is*

$$A * B = \frac{AB + BA}{2}.$$

*The Lie product is often called the **commutator** of two given matrices.*

25. Using Exercise 24 prove the following properties where A, B, and C are n by n C-matrices and c is a complex number.
    (a) $A \times B = -(B \times A)$.
    (b) $A \times A = 0$.
    (c) $A \times I_n = I_n \times A = 0$.
    (d) $A \times (B \times C) = B \times (A \times C) + C \times (B \times A)$.
    (e) $A \times (B + C) = (A \times B) + (A \times C)$.
    (f) $c(A \times B) = (cA) \times B = A \times (cB)$.
    (g) $A * B = B * A$.
    (h) $A * A = A^2$.
    (i) $A * I_n = A$.
    (j) $A * (B * C) = \frac{1}{4}(ABC + ACB + BCA + CBA)$.
    (k) $A * (B + C) = (A * B) + (A * C)$.
    (l) $c(A * B) = ((cA) * B) = (A * (cB))$.

26. If $C = [c_{ij}]_{(s,r)}$ and $B = [b_{ij}]_{(r,t)}$, write the Cayley product CB using summation notation. Then, assuming $D = [d_{ij}]_{(t,w)}$, write $(CB)D$ using summation notation.

3.2 The Inverse Matrix

Assume that a set S has an ***identity element*** for a given operation "$\circ$" (that is, there exists an element e of S such that for any element a of S, $a \circ e = e \circ a = a$). An ***inverse***, if one exists, of an element of S for an operation

3.2 THE INVERSE MATRIX

"∘", is an element of S such that when the two elements are combined by the operation in either order, the identity element of S results. For the system of rational numbers, $e = 0$ is the identity element for the operation of addition since $a + 0 = 0 + a = a$, and $e = 1$ is the identity element for the operation of multiplication since $a \cdot 1 = 1 \cdot a = a$. The inverse of a for addition is $(-a)$, since $a + (-a) = (-a) + a = 0$, and the inverse of a for multiplication (if $a \neq 0$) is $1/a$, since $a \cdot (1/a) = (1/a) \cdot a = 1$.

The reader will recall that the identity element for multiplication of the set of n by n C-matrices was I_n, because for an n by n C-matrix, A,

$$AI_n = I_n A = A.$$

The purpose of this section and of Section 3.4 is to discuss multiplicative inverses of n by n C-matrices. In addition to the theoretical importance of the inverse matrix, there are many practical applications.

Definition 4. *If, for a given n by n C-matrix A, there exists an n by n C-matrix designated A^{-1} such that*

$$AA^{-1} = A^{-1}A = I_n,$$

*then A^{-1} is an **inverse** of A with respect to matrix multiplication.*

Theorem 6. *If an n by n C-matrix has a multiplicative inverse then this inverse is unique.*

Proof. We will assume that there are two multiplicative inverses of A and then show that they are the same.
Assume: $AB = BA = I_n$ and $AC = CA = I_n$.
Statements: $B = BI_n = B(AC) = (BA)C = I_n C = C$.
The reasons for the statements are left as Exercise 14. □

It is common practice to refer to A^{-1} as *the inverse of A*, and it is understood that this means the inverse matrix with respect to Cayley multiplication. To distinguish between those matrices that do have inverses and those that do not have inverses, the following terminology is defined.

Definition 5. *An n by n C-matrix A is said to be **invertible** (or **nonsingular**) if A^{-1} exists, and **noninvertible** (or **singular**) if A does not have an inverse.*[2]

---

[2] Because it is impossible for a nonsquare matrix to have an inverse, we will adopt the convention that a noninvertible matrix is a *square* matrix that does not have an inverse. The reader should be aware, however, that some authors adopt the convention that any matrix (square or nonsquare) that does not have an inverse is noninvertible.

Example 1. Let $A = \begin{bmatrix} 3 & 5 \\ 1 & 2 \end{bmatrix}$ and $B = \begin{bmatrix} 2 & -5 \\ -1 & 3 \end{bmatrix}$.

The reader should show that $AB = BA = I_2$; therefore, $A^{-1} = B$ and $B^{-1} = A$. We say that both A and B are invertible or nonsingular. Moreover, by Theorem 6, A is the only inverse of B, and B is the only inverse of A. Now consider the matrix $\begin{bmatrix} 4 & 8 \\ 1 & 2 \end{bmatrix}$. By matrix multiplication we can show that $\begin{bmatrix} 4 & 8 \\ 1 & 2 \end{bmatrix} \begin{bmatrix} a & b \\ c & d \end{bmatrix} = \begin{bmatrix} 1 & 0 \\ 0 & 1 \end{bmatrix}$ leads to an inconsistent system

$$\begin{cases} 4a & + 8c & & & = 1, \\ & & 4b & + 8d & = 0, \\ a & + 2c & & & = 0, \\ & & b & + 2d & = 1; \end{cases}$$

therefore $\begin{bmatrix} 4 & 8 \\ 1 & 2 \end{bmatrix}$ is an example of a noninvertible matrix.

How do we determine when an n by n matrix is invertible, and how do we calculate A^{-1} for invertible A? In Section 3.4 we will justify the following answer to this important question. We will prove that A^{-1} exists if and only if A is row equivalent to I_n, that is, the reduced echelon form of A is I_n. We must be sure, however, that the number of nonzero rows of the reduced echelon form of a square C-matrix is the same, no matter what sequence of elementary row operations is used to obtain the reduced echelon form; the reader will be asked to prove this later in Exercise 20 of Section 5.5. In order to calculate A^{-1} we will show in Section 3.4 that if the matrix $[A \mid I_n]$ can be transformed by elementary row operations to the row equivalent form $[I_n \mid P]$, then P is the inverse of A.

Example 2. Find A^{-1} if $A = \begin{bmatrix} 1 & 2 & 1 \\ -1 & -1 & 1 \\ 0 & 1 & 3 \end{bmatrix}$.

$[A \mid I_3] = \begin{bmatrix} 1 & 2 & 1 & \mid & 1 & 0 & 0 \\ -1 & -1 & 1 & \mid & 0 & 1 & 0 \\ 0 & 1 & 3 & \mid & 0 & 0 & 1 \end{bmatrix} \underset{R_1 + R_2}{\overset{\text{row}}{\sim}} \begin{bmatrix} 1 & 2 & 1 & \mid & 1 & 0 & 0 \\ 0 & 1 & 2 & \mid & 1 & 1 & 0 \\ 0 & 1 & 3 & \mid & 0 & 0 & 1 \end{bmatrix}$

$\underset{\substack{-R_2 + R_3 \\ -2R_2 + R_1}}{\overset{\text{row}}{\sim}} \begin{bmatrix} 1 & 0 & -3 & \mid & -1 & -2 & 0 \\ 0 & 1 & 2 & \mid & 1 & 1 & 0 \\ 0 & 0 & 1 & \mid & -1 & -1 & 1 \end{bmatrix} \underset{\substack{-2R_3 + R_2 \\ 3R_3 + R_1}}{\overset{\text{row}}{\sim}} \begin{bmatrix} 1 & 0 & 0 & \mid & -4 & -5 & 3 \\ 0 & 1 & 0 & \mid & 3 & 3 & -2 \\ 0 & 0 & 1 & \mid & -1 & -1 & 1 \end{bmatrix}.$

Therefore $A^{-1} = \begin{bmatrix} -4 & -5 & 3 \\ 3 & 3 & -2 \\ -1 & -1 & 1 \end{bmatrix}$. Check: $AA^{-1} = A^{-1}A = I_3$.

Example 3. Now, consider $C = \begin{bmatrix} 1 & 2 \\ 3 & 6 \end{bmatrix}$.

$$[C \mid I_2] = \begin{bmatrix} 1 & 2 & \mid & 1 & 0 \\ 3 & 6 & \mid & 0 & 1 \end{bmatrix} \underset{-3R_1 + R_2}{\overset{\text{row}}{\sim}} \begin{bmatrix} 1 & 2 & \mid & 1 & 0 \\ 0 & 0 & \mid & -3 & 1 \end{bmatrix}.$$

Notice that C has been transformed to reduced echelon form and is not equivalent to I_2; hence C is noninvertible.

Properties that will be useful later in the book are stated in the following seven theorems.

Theorem 7. *If A and B are of the same order and are invertible, then $(AB)^{-1}$ exists. Moreover*

$$(AB)^{-1} = B^{-1}A^{-1}.$$

Proof. Since A and B are invertible, A^{-1} and B^{-1} exist. (Also they are unique).

	STATEMENTS	REASONS
(1)	$(B^{-1}A^{-1})AB = B^{-1}(A^{-1}A)B$	(These are left as Exercise 15.)
(2)	$= B^{-1}I_n B$	
(3)	$= B^{-1}B$	
(4)	$= I_n$.	
(5)	Likewise $AB(B^{-1}A^{-1}) = I_n$.	
(6)	Therefore $(AB)^{-1}$ exists and equals $B^{-1}A^{-1}$. □	

Theorem 8. *If $A_1, A_2, \ldots, A_p$ are of the same order and are invertible, then $(A_1 A_2 \cdots A_p)^{-1}$ exists. Moreover*

$$(A_1 A_2 \cdots A_p)^{-1} = A_p^{-1} \cdots A_2^{-1} A_1^{-1}.$$

The proof is left as Exercise 16.

Theorem 9. *If A is invertible, then A^{-1} is invertible, and, moreover, $(A^{-1})^{-1} = A$.*

The proof is left as Exercise 17.

There are some properties of scalar algebra that are not properties of matrix algebra. We have already seen that in general, $AB \neq BA$. Moreover, $AB = 0$ does not imply that $A = 0$ or $B = 0$; nor is the cancellation law for multiplication always valid (that is, neither $BA = CA$ nor $AB = AC$ implies that $B = C$).

Example 4. Consider

$$AB = \begin{bmatrix} -1 & 2 \\ -2 & 4 \end{bmatrix} \begin{bmatrix} 2 & 2 \\ 1 & 1 \end{bmatrix} = \begin{bmatrix} 0 & 0 \\ 0 & 0 \end{bmatrix}, \text{ and yet } A \neq \mathbf{0}, \text{ and } B \neq \mathbf{0}.$$

Now consider

$$CD = \begin{bmatrix} 2 & 0 \\ 0 & 0 \end{bmatrix} \begin{bmatrix} 4 & 0 \\ 2 & 1 \end{bmatrix} = \begin{bmatrix} 8 & 0 \\ 0 & 0 \end{bmatrix}, \quad \text{and} \quad CE = \begin{bmatrix} 2 & 0 \\ 0 & 0 \end{bmatrix} \begin{bmatrix} 4 & 0 \\ 6 & 7 \end{bmatrix} = \begin{bmatrix} 8 & 0 \\ 0 & 0 \end{bmatrix}.$$

Notice that $CD = CE$, but $D \neq E$.

Knowledge of the invertibility and noninvertibility of matrices allows us to make and prove some general statements concerning Example 4.

Theorem 10. *If $AB = \mathbf{0}$, and if A and B are square C-matrices of order n, then $A = \mathbf{0}$, or $B = \mathbf{0}$, or both A and B are noninvertible.*

Proof. Because A and B are square, two situations can exist: (1) At least one matrix is invertible; or (2) both are noninvertible.

STATEMENT	REASON
(1) Assume that A is invertible, hence A^{-1} exists.	(1) Definition 5. (Page 53.)
(2) $AB = \mathbf{0}$.	(2) Given.
(3) $A^{-1}(AB) = A^{-1}\mathbf{0}$.	(3) Premultiplication by A^{-1}.
(4) $I_n B = \mathbf{0}$.	(4) Associative Law and Definition 4 (page 53). Also, $A^{-1}\mathbf{0} = \mathbf{0}$ by Definition 2 (page 41).
(5) $B = \mathbf{0}$.	(5) Property of I_n.
(6) Likewise if B is invertible $A = \mathbf{0}$.	(6) Similar to steps (1) through (5).
(7) The only other possibility is that both A and B are noninvertible.	(7) By assumption that both A and B are square and the definition of a noninvertible matrix. $\square$

It should be noted that if A and B are not square it is also possible to have $AB = \mathbf{0}$ without having $A = \mathbf{0}$ or $B = \mathbf{0}$. For example,

$$\begin{bmatrix} 2 & 0 & 1 \\ 4 & 0 & 2 \end{bmatrix} \begin{bmatrix} 0 & -1 \\ 4 & 1 \\ 0 & 2 \end{bmatrix} = \begin{bmatrix} 0 & 0 \\ 0 & 0 \end{bmatrix}.$$

Theorem 11. *(Right Cancellation Property) If A, B, and C are n by n C-matrices, and if A is invertible, then*

$$BA = CA \Rightarrow B = C.$$

The proof is left as Exercise 5.

Theorem 12. *(Left Cancellation Property) If A, B, and C are n by n C-matrices, and if A is invertible, then*

$$AB = AC \Rightarrow B = C.$$

The proof is left as Exercise 6.

Theorem 13. *If A and B are n by n C-matrices, if $AB = I_n$, and if A or B is invertible, then $A = B^{-1}$ and $B = A^{-1}$.*

Proof. Let A be invertible. Reasons for the following statements are left as an exercise.

$B = I_n B = (A^{-1}A)B = A^{-1}(AB) = A^{-1}I_n = A^{-1}$, and $A = B^{-1}$. □

Negative integral powers as applied to an invertible matrix are defined in terms of the inverse matrix as follows: *$A^{-n} = (A^{-1})^n$ where n is a positive integer.* Negative integral powers are not defined for noninvertible matrices.

Our knowledge of the inverse can be applied to the solution of certain systems of n linear equations with n unknowns

$$AX = B.$$

If A^{-1} exists, premultiply both sides by A^{-1}. (The reader should observe why it is necessary to premultiply.)

$$A^{-1}(AX) = A^{-1}B,$$
$$(A^{-1}A)X = A^{-1}B,$$
$$I_n X = A^{-1}B,$$
$$X = A^{-1}B.$$

Thus $A^{-1}B$ is a unique column matrix that gives us the values of all the x_i's; we have proved the following theorem.

Theorem 14. *If A is invertible, then the linear system $AX = B$ has the unique solution $X = A^{-1}B$.*

Theorem 14 is especially useful in solving sets of systems where B varies but the coefficient matrix A remains unchanged, as illustrated in Examples 5 and 6.

APPLICATIONS

Example 5. The electrical system shown in Figure 3.2.1 gives rise to the following linear system:

$$\begin{cases} x_1 - x_2 - x_3 = 0, \\ 5x_1 + 20x_3 = 50, \\ 10x_2 - 20x_3 = 30, \end{cases}$$

or

$$AX = B,$$

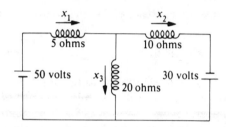

Figure 3.2.1

where X is the electric current vector, and where the last two components of B are the electromotive forces. Here,

$$A = \begin{bmatrix} 1 & -1 & -1 \\ 5 & 0 & 20 \\ 0 & 10 & -20 \end{bmatrix}, \text{ and we find } A^{-1} = \begin{bmatrix} \frac{4}{7} & \frac{3}{35} & \frac{4}{70} \\ -\frac{2}{7} & \frac{2}{35} & \frac{5}{70} \\ -\frac{1}{7} & \frac{1}{35} & -\frac{1}{70} \end{bmatrix}.$$

Since

$$B = \begin{bmatrix} 0 \\ 50 \\ 30 \end{bmatrix},$$

$$X = A^{-1}B = \begin{bmatrix} \frac{4}{7} & \frac{3}{35} & \frac{4}{70} \\ -\frac{2}{7} & \frac{2}{35} & \frac{5}{70} \\ -\frac{1}{7} & \frac{1}{35} & -\frac{1}{70} \end{bmatrix} \begin{bmatrix} 0 \\ 50 \\ 30 \end{bmatrix} = \begin{bmatrix} \frac{150}{35} + \frac{60}{35} \\ \frac{100}{35} + \frac{75}{35} \\ \frac{50}{35} - \frac{15}{35} \end{bmatrix} = \begin{bmatrix} 6 \\ 5 \\ 1 \end{bmatrix}.$$

If each of the applied electromotive forces of the circuit were changed to 70 volts, the solution would be

$$X = A^{-1} \begin{bmatrix} 0 \\ 70 \\ 70 \end{bmatrix} = \begin{bmatrix} 10 \\ 9 \\ 1 \end{bmatrix}.$$

3.2 THE INVERSE MATRIX

Example 6. The management of a certain company is faced with a decision. One of its factories uses two different machines M and N to manufacture two different products P and Q. Machine M can operate 12 hours per day, and machine N can operate 16 hours per day; the rest of the day is used for the maintenance of the machines. To produce one unit of product P, machine M must work for 2 hours in one day, and machine N must work for 1 hour in the same day. Each unit of product Q requires that machine M work 2 hours in one day, and that machine N work 3 hours in the same day. The management must determine the number of units of each product that the factory should make in a day in order to keep the machines working to capacity. They also want to know the effect on production if they buy more of each type of machine.

Solution. Let x_1 be the number of units of product P produced per day; let x_2 be the number of units of product Q produced per day. Machine M then spends $2x_1$ hours on product P and $2x_2$ hours on product Q. If machine M operates full time, we have the equation

$$2x_1 + 2x_2 = 12.$$

Similarly, if machine N operates full time, we get the equation

$$x_1 + 3x_2 = 16.$$

The two simultaneous equations can be expressed as the matrix equation

$$AX = B,$$

where

$$A = \begin{bmatrix} 2 & 2 \\ 1 & 3 \end{bmatrix} \quad \text{and} \quad B = \begin{bmatrix} 12 \\ 16 \end{bmatrix}.$$

Therefore,

$$X = A^{-1}B = A^{-1}\begin{bmatrix} 12 \\ 16 \end{bmatrix}.$$

By use of elementary row operations on $[A \mid I_2]$ we obtain $[I_2 \mid A^{-1}]$, from which we observe that

$$A^{-1} = \begin{bmatrix} \tfrac{3}{4} & -\tfrac{1}{2} \\ -\tfrac{1}{4} & \tfrac{1}{2} \end{bmatrix};$$

hence,

$$X = \begin{bmatrix} \tfrac{3}{4} & -\tfrac{1}{2} \\ -\tfrac{1}{4} & \tfrac{1}{2} \end{bmatrix}\begin{bmatrix} 12 \\ 16 \end{bmatrix} = \begin{bmatrix} 1 \\ 5 \end{bmatrix}.$$

Therefore, 1 unit of product P and 5 units of product Q should be produced per day to satisfy the initial stated conditions of the problem.

Now, if the mangement buys more machines, the resulting change in production can be determined by simply changing B and recalculating $A^{-1}B$. Notice that because A^{-1} does not change, it is relatively easy to find the new solution since the majority of the work—calculation of the inverse matrix—has already been done. Suppose that the management buys an extra machine M. This increases the total capacity of machines M to 24 hours; thus,

$$X = A^{-1}B = A^{-1}\begin{bmatrix} 24 \\ 16 \end{bmatrix} = \begin{bmatrix} \frac{3}{4} & -\frac{1}{2} \\ -\frac{1}{4} & \frac{1}{2} \end{bmatrix}\begin{bmatrix} 24 \\ 16 \end{bmatrix} = \begin{bmatrix} 10 \\ 2 \end{bmatrix}.$$

With a problem as simple as this, in practice one would probably use elementary operations on the original equations. The method outlined in this and the last example becomes of greater use in more difficult problems where B varies and A remains the same, in a system $AX = B$.

Example 7. A matrix approach to the representation of mineral chemical analysis is used by Perry in [27] and [28]; in particular, the Feldspars and Amphiboles are examined. The approach used makes use of systems of linear equations and the inverse matrix. Perry also presents computer programs in [26] which illustrate the use of matrix vocabulary and manipulation via the computer in geological research.

EXERCISES

1. Calculate the inverses of the following matrices, if possible. Use Theorem 13 to check your answer.

$$A = \begin{bmatrix} 2 & -1 \\ 4 & 3 \end{bmatrix}; \quad B = \begin{bmatrix} 4 & 2 \\ 2 & 1 \end{bmatrix}; \quad C = \begin{bmatrix} 2 & 0 & 3 \\ -1 & 0 & 2 \\ 0 & 1 & 1 \end{bmatrix}; \quad D = \begin{bmatrix} 3 & 2 & 1 \\ 2 & -1 & -1 \\ 1 & 4 & 0 \end{bmatrix}.$$

2. Calculate the inverses of the following matrices, if possible. Use Theorem 13 to check your answer.

$$A = \begin{bmatrix} 2 & 3 \\ -4 & 1 \end{bmatrix}; \quad B = \begin{bmatrix} 1 & 3 & 0 \\ 2 & 1 & 0 \\ 0 & 1 & -1 \end{bmatrix}; \quad C = \begin{bmatrix} 1 & 0 & 0 & 1 \\ 0 & 0 & 1 & 0 \\ 0 & 2 & 0 & 0 \\ 0 & 0 & 0 & 2 \end{bmatrix}.$$

3. If $A = \begin{bmatrix} 3 & 2 \\ 0 & 1 \end{bmatrix}$, find A^{-2} and A^{-3}.

4. State why it is impossible for a matrix that is not square to have an inverse.

5. Prove Theorem 11.

6. Prove Theorem 12.

7. If $A = \begin{bmatrix} 3 & 2 \\ 0 & 1 \end{bmatrix}$ and $B = \begin{bmatrix} 2 & 1 \\ 1 & 0 \end{bmatrix}$, verify Theorem 7. Also use A to verify Theorem 9.

In each of Exercises 8–11, solve the given system of equations by use of an inverse matrix.

8. $\begin{cases} x_1 + 3x_2 = 4, \\ 2x_1 - 2x_2 = 6. \end{cases}$

9. $\begin{cases} x_1 - 4x_2 = 2, \\ 2x_1 + x_2 = 1. \end{cases}$

10. $\begin{cases} x_1 + 2x_2 + x_3 = 0, \\ x_1 + x_3 = 1, \\ x_2 - x_3 = 3. \end{cases}$
11. $\begin{cases} 2x_1 + 5x_2 + 3x_3 = 1, \\ 3x_1 + x_2 + 2x_3 = 1, \\ x_1 + 2x_2 + x_3 = 0. \end{cases}$

12. Suppose that A and B are square C-matrices, that A is invertible, and that $AB = 0$. What can be said about B, and why?

13. Suppose that A and B are n by n non-null C-matrices and that $AB = 0$. Is $B \overset{\text{row}}{\sim} I_n$, and why?

14. Give the reasons for the statements in the proof of Theorem 6.

15. Give the reasons for the statements in the proof of Theorem 7.

16. Prove Theorem 8.

17. Prove Theorem 9.

18. Give the reasons for the statements in the proof of Theorem 13.

19. Prove or disprove the conjecture $(A + B)^{-1} = A^{-1} + B^{-1}$, if A and B are invertible and of the same order.

20. Rework Example 5 of this section with the applied electromotive force increased to 100 volts in each battery.

21. Rework Example 6 of this section assuming that the company has two of each type of machine.

22. Suppose that A, B, and C are square C-matrices of order n and that A is invertible. If $AB = CA$, does it follow that $B = C$? Why?

23. Assume that all matrices in this exercise are n by n invertible matrices. In each case, solve the matrix equation for D, and simplify as much as possible.
   (a) $ADB = C$;
   (b) $ACDB = C$;
   (c) $CADB = C$;
   (d) $(AB)^{-1}AD = I_n$;
   (e) $A(B + D) = (B^{-1})^{-1}$.

3.3 Elementary Matrices

In Section 2.2 we defined elementary row operations on a C-matrix A. *Elementary column operations* likewise may be defined by replacing the word "row" with the word "column" in Definition 3 (page 24) of Chapter 2; if two matrices A and B are related in such a way that B is obtained from A by means of elementary column operations, we say that A is *column equivalent* to B. This is denoted by $A \overset{\text{col}}{\sim} B$. We will denote the elementary column operations by $C_i \leftrightarrow C_j$, by kC_i where $k \neq 0$, and by $kC_i + C_j$. An *elementary operation* is any operation that is either an elementary row operation or an elementary column operation.

Definition 6. *If a C-matrix A is transformed to a C-matrix B by a finite number of elementary operations, then A is said to be equivalent to B. This is denoted by $A \sim B$.*

MATRIX MULTIPLICATION

Definition 7. *When a nonzero C-matrix has been reduced by elementary operations to one of the forms*

$$\begin{bmatrix} I_r & 0 \\ \hline 0 & 0 \end{bmatrix} \quad \text{or} \quad [I_r \mid 0] \quad \text{or} \quad \begin{bmatrix} I_r \\ \hline 0 \end{bmatrix} \quad \text{or} \quad I_r,$$

*we say that it has been reduced to **normal form** or **canonical form for equivalence**. The normal form of a zero matrix is that zero matrix.*

Example 1. We illustrate the last two definitions by transforming the matrix $\begin{bmatrix} 1 & 2 & 0 \\ 4 & 6 & 9 \\ 0 & -2 & 9 \end{bmatrix}$ into its equivalent normal form.

$$\begin{bmatrix} 1 & 2 & 0 \\ 4 & 6 & 9 \\ 0 & -2 & 9 \end{bmatrix} \xrightarrow{-4\tilde{R}_1 + R_2} \begin{bmatrix} 1 & 2 & 0 \\ 0 & -2 & 9 \\ 0 & -2 & 9 \end{bmatrix}$$

$$\xrightarrow{-2\tilde{C}_1 + C_2} \begin{bmatrix} 1 & 0 & 0 \\ 0 & -2 & 9 \\ 0 & -2 & 9 \end{bmatrix} \xrightarrow{-\frac{1}{2}\tilde{R}_2} \begin{bmatrix} 1 & 0 & 0 \\ 0 & 1 & -\frac{9}{2} \\ 0 & -2 & 9 \end{bmatrix}$$

$$\xrightarrow{2\tilde{R}_2 + R_3} \begin{bmatrix} 1 & 0 & 0 \\ 0 & 1 & -\frac{9}{2} \\ 0 & 0 & 0 \end{bmatrix} \xrightarrow{\frac{9}{2}\tilde{C}_2 + C_3} \begin{bmatrix} 1 & 0 & 0 \\ 0 & 1 & 0 \\ 0 & 0 & 0 \end{bmatrix} = \begin{bmatrix} I_2 & 0 \\ \hline 0 & 0 \end{bmatrix}$$

It can be proved that equivalence of matrices is both symmetric and transitive (see Exercises 17 and 20); that is, $A \sim B \Rightarrow B \sim A$, and $(A \sim B, B \sim C) \Rightarrow A \sim C$. After reading Example 1, the question naturally arises as to whether a given matrix is always equivalent to a unique normal form. The following theorem answers part of this question.

Theorem 15. *Any m by n nonzero C-matrix A can be reduced to normal form by elementary operations.*

Proof. PART I. Because A is nonzero, there exists some element a_{ij} of A which is nonzero. If necessary, the elementary operations $(R_i \leftrightarrow R_1)$ and $(C_j \leftrightarrow C_1)$ will move a_{ij} to the 1, 1 position. Hence we can assume without loss of generality that $a_{11} \neq 0$. Then the following sequence of elementary row and column operations

$$\left(\frac{1}{a_{11}} R_1\right), \quad (-a_{21} R_1 + R_2), \quad \ldots, \quad (-a_{m1} R_1 + R_m),$$
$$(-a_{12} C_1 + C_2), \quad \ldots, \quad (-a_{1n} C_1 + C_n),$$

will transform A to an equivalent matrix of the form

$$\begin{bmatrix} 1 & 0 \\ \hline 0 & B \end{bmatrix} \quad \text{if } m > 1 \text{ and } n > 1,$$

and B is $(m-1)$ by $(n-1)$. If either $m=1$, or $n=1$, or both $m=1$ and $n=1$, then A is equivalent to $[1 \ 0 \ \cdots \ 0]$ or $\begin{bmatrix} 1 \\ 0 \\ \vdots \\ 0 \end{bmatrix}$ or I_1, respectively, and the proof is complete.

PART II. If B is a null matrix, then the proof is complete; otherwise repeat PART I for the nonzero matrix B; if $m > 2$ and $n > 2$, then A will be equivalent to a matrix of the form

$$\left[\begin{array}{c|c} I_2 & 0 \\ \hline 0 & C \end{array}\right]$$

where C is $(m-2)$ by $(n-2)$. If either $m=2$ or $n=2$, or if both $m=2$ and $n=2$, then A is equivalent to $[I_2 \ \vdots \ 0]$ or $\begin{bmatrix} I_2 \\ \hdashline 0 \end{bmatrix}$ or I_2, respectively, and the proof is complete.

PART III. Continue this procedure until a null matrix appears in the lower right-hand corner or until all of the rows or columns are exhausted. In either event, the final matrix is in normal form, and, since only elementary row and column operations have been used, A is equivalent to the final matrix. □

Later in the book we shall prove (Exercise 21 in Section 5.5) that, for any C-matrix, the equivalent normal form is unique. The normal form then represents a class of matrices all of which are equivalent to it. The unique normal form of a matrix is frequently called the *canonical form* of the given matrix with respect to matrix equivalence, and the set of matrices equivalent to it is called a *canonical set under equivalence.*

Next, we begin to develop a procedure whereby elementary row and column operations are performed by means of matrix multiplication. For example, the operation $2R_1 + R_2$ on $\begin{bmatrix} 1 & 2 \\ -2 & 3 \end{bmatrix}$ can be accomplished by premultiplying by $\begin{bmatrix} 1 & 0 \\ 2 & 1 \end{bmatrix}$; that is

$$\begin{bmatrix} 1 & 0 \\ 2 & 1 \end{bmatrix} \begin{bmatrix} 1 & 2 \\ -2 & 3 \end{bmatrix} = \begin{bmatrix} 1 & 2 \\ 0 & 7 \end{bmatrix}.$$

Notice that we have used a matrix as an operator to transform another matrix according to $2R_1 + R_2$; this matrix operator is an example of an *elementary row transformation matrix.*

Definition 8. *An elementary row ⟨column⟩ transformation matrix is a matrix that can be obtained from an identity matrix I_n by a single elementary row ⟨column⟩ operation. A matrix that is either an elementary row transformation matrix or an elementary column transformation matrix is called an **elementary matrix**.*

From the preceding definition it is obvious that elementary matrices are of three types, of which the first type, $\begin{bmatrix} 1 & 0 & 0 \\ 0 & 0 & 1 \\ 0 & 1 & 0 \end{bmatrix}$, the second type, $\begin{bmatrix} 1 & 0 & 0 \\ 0 & k & 0 \\ 0 & 0 & 1 \end{bmatrix}$ where $k \neq 0$, and the third type, $\begin{bmatrix} 1 & 0 & 0 \\ 0 & 1 & 0 \\ 0 & k & 1 \end{bmatrix}$, are third-order examples. Each type corresponds to one of the three elementary operations. The following two theorems verify that a single elementary operation is accomplished by multiplying the matrix by a suitably chosen elementary matrix.

Theorem 16. *An elementary row operation is performed on a given C-matrix A by premultiplying A by an elementary row transformation matrix.*

Proof. Let A be a C-matrix of order m by n, where $m \geq 2$. The proof is divided into three parts—one part for each type of elementary row operation.

PART I. Let E represent an elementary row transformation matrix of the first type, in which the ith and jth rows $(i \neq j)$ of I_m have been interchanged. Let B be the result of premultiplying A by E. We will show that B is also the result of interchanging the ith and jth rows of A.

	STATEMENT	REASON
(1)	[jth row of B] = [jth row of EA]	(1) Given $B = EA$.
(2)	= [jth row of E]A	(2) Theorem 5, page 45.
(3)	= [ith row of I_m]A	(3) Definition of E.
(4)	= [ith row of $I_m A$]	(4) Theorem 5, page 45.
(5)	= [ith row of A].	(5) $I_m A = A$.
(6)	Similarly, we can show that [ith row of B] = [jth row of A].	(6) Similar to Statements (1)–(5).
(7)	Also [pth row of B] = [pth row of A] where $p \neq i, j$.	(7) Let k replace both i and j in Statements (1)–(5). If $m = 2$, Statement (7) is irrelevant.

PART II. This part of the proof is left as Exercise 15. (Let $m \geq 1$).

PART III. Let E be an elementary row transformation matrix of the third type, in which the jth row of E is the jth row of I_m, plus k times the ith row of I_m $(i \neq j)$. Let B be the result of premultiplying A by E. We will show that B is also the result of adding k times the ith row of A to the jth row of A.

3.3 ELEMENTARY MATRICES 65

	STATEMENT	REASON
(1)	[jth row of B] = [jth row of EA]	(The reasons are left as Exercise 16)
(2)	= [jth row of E]A	
(3)	= ([jth row of I_m] + k[ith row of I_m])A	
(4)	= [jth row of I_m]A + k[ith row of I_m]A	
(5)	= [jth row of $I_m A$] + k[ith row of $I_m A$]	
(6)	= [jth row of A] + k[ith row of A].	
(7)	Similarly we can show that if $p \neq j$, [pth row of B] = [pth row of A]. ☐	

Theorem 17. *An elementary column operation can be performed on a given C-matrix A by postmultiplying A by an elementary column transformation matrix.*

An easy proof of this theorem can be made after introducing the material found in Section 4.1, and for this reason the proof is left as an exercise for that section (Exercise 28).

Example 2. The first and second rows of a C-matrix $A = [a_{ij}]_{(3,2)}$ may be interchanged by premultiplying A by

$$E = \begin{bmatrix} 0 & 1 & 0 \\ 1 & 0 & 0 \\ 0 & 0 & 1 \end{bmatrix}.$$

Note that the desired elementary matrix E is found by simply performing the specified elementary operation on the identity matrix of the appropriate order.

$$EA = \begin{bmatrix} 0 & 1 & 0 \\ 1 & 0 & 0 \\ 0 & 0 & 1 \end{bmatrix} \begin{bmatrix} a_{11} & a_{12} \\ a_{21} & a_{22} \\ a_{31} & a_{32} \end{bmatrix} = \begin{bmatrix} a_{21} & a_{22} \\ a_{11} & a_{12} \\ a_{31} & a_{32} \end{bmatrix}.$$

The first and second columns are interchanged by postmultiplying A by $E = \begin{bmatrix} 0 & 1 \\ 1 & 0 \end{bmatrix}$.

$$AE = \begin{bmatrix} a_{11} & a_{12} \\ a_{21} & a_{22} \\ a_{31} & a_{32} \end{bmatrix} \begin{bmatrix} 0 & 1 \\ 1 & 0 \end{bmatrix} = \begin{bmatrix} a_{12} & a_{11} \\ a_{22} & a_{21} \\ a_{32} & a_{31} \end{bmatrix}.$$

Example 3. To multiply the second row of $A = [a_{ij}]_{(3,2)}$ by a nonzero scalar k, premultiply by $E = \begin{bmatrix} 1 & 0 & 0 \\ 0 & k & 0 \\ 0 & 0 & 1 \end{bmatrix}$.

$$EA = \begin{bmatrix} 1 & 0 & 0 \\ 0 & k & 0 \\ 0 & 0 & 1 \end{bmatrix} \begin{bmatrix} a_{11} & a_{12} \\ a_{21} & a_{22} \\ a_{31} & a_{32} \end{bmatrix} = \begin{bmatrix} a_{11} & a_{12} \\ ka_{21} & ka_{22} \\ a_{31} & a_{32} \end{bmatrix}.$$

To multiply the second column by k, postmultiply A by $E = \begin{bmatrix} 1 & 0 \\ 0 & k \end{bmatrix}$.

Example 4. To add k times the second row to the first row of $A = [a_{ij}]_{(3,2)}$, premultiply by $E = \begin{bmatrix} 1 & k & 0 \\ 0 & 1 & 0 \\ 0 & 0 & 1 \end{bmatrix}$.

$$EA = \begin{bmatrix} 1 & k & 0 \\ 0 & 1 & 0 \\ 0 & 0 & 1 \end{bmatrix} \begin{bmatrix} a_{11} & a_{12} \\ a_{21} & a_{22} \\ a_{31} & a_{32} \end{bmatrix} = \begin{bmatrix} (a_{11} + ka_{21}) & (a_{12} + ka_{22}) \\ a_{21} & a_{22} \\ a_{31} & a_{32} \end{bmatrix}.$$

To add k times the second column to the first column of A, postmultiply by $E = \begin{bmatrix} 1 & 0 \\ k & 1 \end{bmatrix}$.

EXERCISES

In Exercises 1–6, reduce each matrix to normal form.

1. $\begin{bmatrix} 3 & 2 & -1 \\ 7 & 8 & 0 \\ 4 & 6 & 1 \end{bmatrix}$.
2. $\begin{bmatrix} 0 & 1 & 3 & 0 \\ 0 & 4 & 0 & 2 \\ 1 & 0 & 3 & 0 \\ -1 & 1 & 0 & 0 \end{bmatrix}$.
3. $\begin{bmatrix} 3 & 1 & 4 & 6 & 2 \\ 4 & 1 & i & 9 & 6 \end{bmatrix}$.

4. $\begin{bmatrix} 3 & 2 & 7 \\ 4 & -3 & -2 \\ 0 & 1 & 2 \\ 6 & 1 & 8 \end{bmatrix}$.
5. $\begin{bmatrix} 4 & 9 & -3 & 1 \\ 6 & 9 & -4 & 0 \\ 2 & 9 & -2 & 2 \\ -2 & 0 & 1 & 1 \end{bmatrix}$.
6. $\begin{bmatrix} -1 & 0 & -1 & 3 & 2 \\ -1 & 4 & 0 & 0 & 1 \\ 0 & 0 & 0 & 4 & 2 \\ 1 & 0 & 1 & 1 & 0 \\ 0 & 4 & 1 & 1 & 1 \end{bmatrix}$.

In each of Exercises 7–12, write the elementary matrix E which performs the indicated elementary operation on $A = \begin{bmatrix} -3 & 2 & -1 \\ 4 & 0 & 1 \end{bmatrix}$. Then, multiply the matrices to perform the desired transformation.

7. Interchange the first and second rows.
8. Interchange the first and third columns.
9. Multiply the second row by 9.
10. Multiply the third column by 7.
11. Add 4 times the first row to the second row.
12. Add 5 times the third column to the first column.
13. Assume that $B = \begin{bmatrix} 3 & 2 \\ 1 & 4 \\ 6 & 0 \end{bmatrix}$ is obtained from a matrix A by adding 2 times

the second column of A to the first column of A. This elementary operation can be expressed as

$$AE = B, \quad \text{or} \quad AEE^{-1} = BE^{-1}, \quad \text{or} \quad A = BE^{-1}.$$

Find A by first finding E, then E^{-1}, and then multiply as shown above. Check by calculating AE.

14. Repeat Exercise 13 using $B = \begin{bmatrix} 3 & 2 \\ 4 & 1 \end{bmatrix}$.
15. Furnish Part II of the proof of Theorem 16.
16. Furnish reasons for Part III of the proof of Theorem 16.
17. (a) Prove that if $A \overset{\text{col}}{\sim} B$, then $B \overset{\text{col}}{\sim} A$.
    (b) Prove that if $A \sim B$, then $B \sim A$.
18. (a) List all possible normal forms of a 4 by 3 C-matrix.
    (b) List all possible normal forms of a 4 by 4 C-matrix.
19. Explain why any elementary row transformation matrix is also an elementary column transformation matrix, and vice versa.
20. (a) Prove that if $A \overset{\text{col}}{\sim} B$ and $B \overset{\text{col}}{\sim} C$, then $A \overset{\text{col}}{\sim} C$.
    (b) Prove that if $A \sim B$ and $B \sim C$, then $A \sim C$.
21. How can you determine whether two given m by n C-matrices are equivalent? (*Hint:* Use Theorem 15, Exercise 17, and Exercise 20.)

3.4 A Necessary and Sufficient Condition for the Existence of the Inverse

The purpose of this section is to develop further properties of elementary matrices and invertible matrices; at the same time, we will justify necessary and sufficient conditions for the existence of the inverse of a square C-matrix and the method used in Section 3.2 for its calculation.

Theorem 18. *An elementary row transformation matrix is invertible, and its inverse is an elementary row transformation matrix of the same type.* i.e., which contains the same elementary row operation.

Outline of proof. PART I. Show that an elementary row transformation matrix of the first type is its own inverse. This seems plausible, because the inverse operation of interchanging two rows is to interchange the same two rows.

PART II. Show that an elementary row transformation matrix which multiplies a row by $1/k$ is the inverse of an elementary row transformation matrix which multiplies the same row by k.

PART III. Show that an elementary row transformation matrix which adds $-k$ times the ith row to the jth row is the inverse of the elementary row transformation matrix which adds k times the ith row to the jth row. □

Example 1. We illustrate Part III of the preceding proof for a 3 by 3 matrix. Let E be the elementary row transformation matrix which adds k times the third row to the second row.

$$E = \begin{bmatrix} 1 & 0 & 0 \\ 0 & 1 & k \\ 0 & 0 & 1 \end{bmatrix}.$$

We show that

$$E^{-1} = \begin{bmatrix} 1 & 0 & 0 \\ 0 & 1 & -k \\ 0 & 0 & 1 \end{bmatrix},$$

because

$$\begin{bmatrix} 1 & 0 & 0 \\ 0 & 1 & k \\ 0 & 0 & 1 \end{bmatrix} \begin{bmatrix} 1 & 0 & 0 \\ 0 & 1 & -k \\ 0 & 0 & 1 \end{bmatrix} = \begin{bmatrix} 1 & 0 & 0 \\ 0 & 1 & -k \\ 0 & 0 & 1 \end{bmatrix} \begin{bmatrix} 1 & 0 & 0 \\ 0 & 1 & k \\ 0 & 0 & 1 \end{bmatrix} = \begin{bmatrix} 1 & 0 & 0 \\ 0 & 1 & 0 \\ 0 & 0 & 1 \end{bmatrix}.$$

Theorem 19. *An elementary column transformation matrix is invertible, and its inverse is an elementary column transformation matrix of the same type.*

The proof is left as Exercise 11.

Theorem 20. *Let A and B be m by n C-matrices. Then $A \stackrel{row}{\sim} B$ if and only if there exists an invertible matrix P such that $B = PA$, where P is a product of elementary row transformation matrices.*

Outline of Proof. PART I. If $A \stackrel{row}{\sim} B$, then there exists a sequence of p elementary row operations which transforms A into B. By Theorem 16 (page 64) there exist corresponding elementary row transformation matrices $E_1, E_2, \ldots, E_p$ such that, by repeated use of the associative law, we obtain

$$(E_p \cdots E_2 E_1)A = B$$

or

$$PA = B \quad \text{where} \quad P = E_p \cdots E_2 E_1.$$

Also, by Theorems 18 and 7, elementary row transformation matrices are invertible, and products of invertible matrices are invertible; hence P is invertible.

PART II. Conversely, if there exist elementary row transformation matrices $E_1, E_2, \ldots, E_p$ such that $B = (E_p \cdots E_2 E_1)A$, then by repeated use of the associative law and Theorem 16, $A \stackrel{row}{\sim} B$. Note that associated with each of the elementary row transformation matrices is an elementary row

3.4 A CONDITION FOR THE EXISTENCE OF THE INVERSE

operation; when an elementary row operation is applied to a matrix, the result is row equivalent to the original matrix. ☐

Theorem 21. *Let A and B be m by n C-matrices. Then $A \stackrel{col}{\sim} B$ if and only if there exists an invertible matrix Q such that $B = AQ$, where Q is a product of elementary column transformation matrices.*

The proof is left as Exercise 12.

We are now in a position to justify the procedure that we used in Section 3.2 to calculate the inverse of an invertible matrix. Actually, the calculation procedure is a by-product of the proof of a necessary and sufficient condition for a square matrix to have an inverse.

Theorem 22. *An n by n C-matrix A is invertible if and only if $A \stackrel{row}{\sim} I_n$.*

Proof. PART I. $(A \stackrel{row}{\sim} I_n) \Rightarrow (A \text{ is invertible})$.

STATEMENT	REASON
(1) $A \stackrel{row}{\sim} I_n$ implies that there exists an n by n invertible product P of elementary row transformation matrices such that $PA = I_n$.	(1) Theorem 20, page 68.
(2) $P = A^{-1}$ and hence A is invertible.	(2) Theorem 13, page 57.

PART II. $(A \text{ is invertible}) \Rightarrow (A \stackrel{row}{\sim} I_n)$.

STATEMENT	REASON
(1) Any nonzero C-matrix is row equivalent to a reduced echelon matrix; hence we can say $A \stackrel{row}{\sim} B$ where B is a reduced echelon matrix.	(1) Theorem 2 of Chapter 2, page 29. Why can we assume that A is nonzero?
(2) Therefore there exists an invertible matrix P, which is the product of elementary row transformation matrices, such that $PA = B$.	(2) Theorem 20, page 68.
(3) B is invertible.	(3) B is the product of two invertible matrices.
(4) Assume $B \neq I_n$. Then the last row of B consists entirely of zeros.	(4) Follows from definition of reduced echelon matrix and fact that B is square. *(since A is $n \times n$)*

(5) The last row of BB^{-1} consists entirely of zeros.

(5) By assumption of (4) and by definition of matrix multiplication. Statement (3) guarantees the existence of B^{-1}.

(6) But $BB^{-1} = I_n$, therefore last row of BB^{-1} does not consist entirely of zeros.

(6) By Statement (3), and the definitions of B^{-1} and I_n.

(7) $B = I_n$.

(7) The assumption that $B \neq I_n$ in (4) led to a contradiction.

(8) $A \overset{\text{row}}{\sim} I_n$.

(8) Statements (1) and (7), and Theorem 20. □

From Statements (1) and (2) of Part I of the proof of Theorem 22, we observe that A^{-1} is equal to a product of the elementary row transformation matrices used to transform A into I_n.

Corollary. *If $[A \vdots I_n] \overset{\text{row}}{\sim} [I_n \vdots P]$ for a given C-matrix A, then P is the inverse of A.*

The proof is left as Exercise 13.

Observe that this corollary justifies the method of calculation of A^{-1} illustrated in Section 3.2.

Theorem 23. *Any n by n invertible C-matrix A can be written as a product of elementary row transformation matrices.*

Proof.

STATEMENT

(1) (A is invertible) $\Rightarrow (A \overset{\text{row}}{\sim} I_n)$.
(2) There exist elementary row transformation matrices $E_1, \ldots, E_p$ such that $(E_p \cdots E_1)A = I_n$.
(3) $A = (E_p \cdots E_1)^{-1} I_n$.
(4) $A = E_1^{-1} \cdots E_p^{-1}$.
(5) A is a product of elementary row transformation matrices.

REASON

(1) Theorem 22, page 69.
(2) Theorem 20, page 68.
(3) (This is left as Exercise 18).
(4) Theorem 8, page 55. Property of I_n.
(5) Theorem 18, page 67. □

3.4 A CONDITION FOR THE EXISTENCE OF THE INVERSE

Theorem 24. *Any invertible C-matrix can be written as a product of elementary column transformation matrices.*

The proof is left as Exercise 14.

Theorems 20, 21, 23, and 24 can be used to prove the following theorem.

Theorem 25. *Let A and B be m by n C-matrices. $A \sim B$ if and only if there exist invertible matrices P and Q such that*

$$PAQ = B.$$

The proof is left as Exercise 15.

APPLICATIONS

Example 2. Consider a certain corporation that has three fields of operation; it mines coal, produces gasoline, and generates electricity. Each of these activities makes use of varying amounts of the three products. Suppose that in order to produce one unit of coal the corporation consumes

> 0 units of coal,
> 1 unit of gasoline,
> 1 unit of electricity;

to produce one unit of gasoline the corporation consumes

> 0 units of coal,
> $\frac{1}{5}$ unit of gasoline,
> $\frac{2}{5}$ unit of electricity;

and to produce one unit of electricity the corporation consumes

> $\frac{1}{5}$ unit of coal,
> $\frac{2}{5}$ unit of gasoline,
> $\frac{1}{5}$ unit of electricity.

These three columns form what is known as a consumption matrix:

$$C = \begin{bmatrix} \text{COAL} & \text{GAS} & \text{ELECTRICITY} \\ 0 & 0 & \frac{1}{5} \\ 1 & \frac{1}{5} & \frac{2}{5} \\ 1 & \frac{2}{5} & \frac{1}{5} \end{bmatrix} \begin{array}{l} \text{COAL CONSUMED} \\ \text{GAS CONSUMED} \\ \text{ELECTRICITY CONSUMED.} \end{array}$$

For a given time interval, let x_1 be the number of units of coal produced, let x_2 be the number of units of gas produced, and let x_3 be the number of units of electricity produced. Thus the column vector

$$X = \begin{bmatrix} x_1 \\ x_2 \\ x_3 \end{bmatrix}$$

shows the production, and the product CX shows the internal consumption necessary for this desired level of production. Naturally, the corporation wishes to produce more than its internal needs require. Suppose that it has an order for 100 units of each product. How much of each product should the corporation produce in order to meet this demand? The solution is found by considering the nonnegative solution of the matrix equation $X - CX = D$, if such a solution exists, where $D = \begin{bmatrix} 100 \\ 100 \\ 100 \end{bmatrix}$. This equation can be rewritten as

$$(I_3 - C)X = D \quad \text{or} \quad X = (I_3 - C)^{-1} D$$

if $(I_3 - C)^{-1}$ exists. For our data, this becomes

$$X = \left(\begin{bmatrix} 1 & 0 & 0 \\ 0 & 1 & 0 \\ 0 & 0 & 1 \end{bmatrix} - \begin{bmatrix} 0 & 0 & \frac{1}{5} \\ 1 & \frac{1}{5} & \frac{2}{5} \\ 1 & \frac{2}{5} & \frac{1}{5} \end{bmatrix} \right)^{-1} \begin{bmatrix} 100 \\ 100 \\ 100 \end{bmatrix} = \begin{bmatrix} 300 \\ 1000 \\ 1000 \end{bmatrix}.$$

Once a corporation's internal consumption is established, then $(I_n - C)^{-1}$ (if it exists) can be calculated once and for all; with each new demand column vector, the problem then amounts to evaluating a simple product of matrices.

Example 3. In Chapter 6 of [16] (Johnston et al.) the reader may find several variations of the model presented in Example 2; these variations are applied to problems in Business and Economics. Particular attention is called to the Model for Allocation of Service Charges in Accounting that is begun on page 262 of [16]. Students of Business and Economics are encouraged to at least glance at Chapter 6 of [16] in order to assure them of the relevance of the material they are studying in this book. The bibliography and bibliographical notes on pages 272–273 of [16] may also be helpful.

EXERCISES

1. Let the operations $R_1 + R_2$ and $3R_2$ be applied in that order to the matrix $\begin{bmatrix} 1 & 4 \\ 2 & 6 \end{bmatrix}$ by means of premultiplying by two elementary row transformation matrices. Illustrate Theorem 20.

2. Use what you learned in the proof of Theorem 18 to write the inverse of each of the following matrices:

3.4 A CONDITION FOR THE EXISTENCE OF THE INVERSE

(a) $\begin{bmatrix} 0 & 1 & 0 & 0 \\ 1 & 0 & 0 & 0 \\ 0 & 0 & 1 & 0 \\ 0 & 0 & 0 & 1 \end{bmatrix}$; (b) $\begin{bmatrix} 1 & 0 & 0 \\ 0 & 1 & 0 \\ 0 & 0 & 7 \end{bmatrix}$; (c) $\begin{bmatrix} 1 & 0 & 0 \\ 5 & 1 & 0 \\ 0 & 0 & 1 \end{bmatrix}$.

3. The elementary row operations $R_1 \leftrightarrow R_2$ and $3R_1 + R_3$ are performed in that order on a 3 by 3 C-matrix A and the result is denoted by B. Illustrate Theorem 20.

4. The elementary column operations $4C_1$ and $2C_2 + C_3$ are performed in that order on a 2 by 3 C-matrix A and the result is denoted by B. Illustrate Theorem 21.

5. Is the matrix $\begin{bmatrix} 3 & 0 & 1 \\ 3 & 2 & 0 \\ 0 & 2 & -1 \end{bmatrix}$ invertible? Justify your answer.

6. Use the Corollary of Theorem 22 to evaluate

$$\begin{bmatrix} 3 & 4 \\ 6 & 1 \end{bmatrix}^{-1}.$$

7. Illustrate Theorem 23 with

$$A = \begin{bmatrix} 1 & 2 \\ 0 & 2 \end{bmatrix}.$$

8. Illustrate Theorem 24 with

$$A = \begin{bmatrix} 1 & 0 & 0 \\ 0 & 3 & 0 \\ 0 & 2 & 1 \end{bmatrix}.$$

9. Suppose that the same elementary row transformations were performed on two C-matrices

$$A = [a_{ij}]_{(3,3)} \quad \text{and} \quad K = \begin{bmatrix} k & 0 & 0 \\ 0 & k & 0 \\ 0 & 0 & k \end{bmatrix}$$

in such a way that they became

$$I_3 = \begin{bmatrix} 1 & 0 & 0 \\ 0 & 1 & 0 \\ 0 & 0 & 1 \end{bmatrix} \quad \text{and} \quad B = [b_{ij}]_{(3,3)},$$

respectively. What is the relationship between A and B?

10. Provide a detailed proof of Theorem 18.
11. Prove Theorem 19.
12. Prove Theorem 21.
13. Prove the Corollary to Theorem 22. *Hint*: Read the sentence immediately before Corollary.
14. Prove Theorem 24.

15. Prove Theorem 25. *Hint:* Read the sentence immediately before Theorem 25.
16. (a) State a theorem analogous to Theorem 22 using column equivalence rather than row equivalence.
    (b) Prove your statement in part (a).
17. (a) State a Corollary analogous to the Corollary of Theorem 22 using column equivalence rather than row equivalence.
    (b) Illustrate part (a) with $A = \begin{bmatrix} 1 & 2 \\ 1 & 3 \end{bmatrix}$.
18. Give the reason for Statement (3) in the proof of Theorem 23.
19. (a) In Example 2 of this section what is the internal consumption?
    (b) Verify that $X = \begin{bmatrix} 300 \\ 1000 \\ 1000 \end{bmatrix}$ is the correct answer in Example 2.
20. Compare Example 2 of this section with Example 4 of Section 2.1.

NEW VOCABULARY

Cayley product 3.1
conformable for multiplication 3.1
premultiplied 3.1
postmultiplied 3.1
positive integral powers of square matrices 3.1
Pauli spin matrices 3.1
Lie product (cross product) 3.1
Jordan product 3.1
commutator 3.1
inverse of a matrix 3.2
identity element 3.2
inverse element 3.2
invertible matrix 3.2
nonsingular matrix 3.2
noninvertible matrix 3.2

singular matrix 3.2
negative integral powers of invertible matrices 3.2
elementary column operations 3.3
column equivalent matrices 3.3
equivalent matrices 3.3
elementary operations 3.3
normal form for equivalence 3.3
canonical form for equivalence 3.3
canonical set under equivalence 3.3
elementary row transformation matrix 3.3
elementary column transformation matrix 3.3
elementary matrix 3.3

SPECIAL PROJECTS

I. On pages 98–99 of Eves, Howard, *Elementary Matrix Theory*, Allyn and Bacon, Boston, 1966, the author defines a left inverse matrix of an m by n matrix and a right inverse of an m by n matrix; he then states some theorems pertaining to these definitions. Write a paper in which the material referred to is explained and illustrated. Also prove some of the given theorems in the paper.

II. An elementary discussion of so-called "Generalized Inverse Matrices" may be found on pages 144–162 of Searle, S. R., *Matrix Algebra for the Biological Sciences*, John Wiley & Sons, New York, 1966, or on pages 8, 28–33, 146–147 of Ijiri, Yuji, *Management Goals and Accounting for Control*, Rand McNally & Co., Chicago, 1965. Both of these discussions are based on the paper, Penrose, R. A., A Generalized Inverse for Matrices, *Proceedings of the Cambridge Philosophical Society* 51, 1955, pages 406–413. From a study of one or more references, write a short paper on a generalized inverse matrix.

4

SPECIAL MATRICES

4.1 Symmetric and Skew-Symmetric Matrices

A very useful transformation over the set of all C-matrices consists of interchanging the rows and columns of each matrix of the set.

Definition 1. *Let A be a C-matrix. The **transpose** of A is the matrix which is formed by interchanging the rows and columns of A; that is, the ith row of A becomes the ith column of the transpose of A for all i. The transpose of A is denoted by A^T.*

Definition 1 states that if $A = [a_{ij}]_{(m,n)}$ then $A^T = [a_{ji}]_{(n,m)}$. Notice that if A is an m by n matrix then A^T is n by m, and if A is square then A^T is also square.

Example 1. Given $A = \begin{bmatrix} 3 & 4 & 6 \\ 2 & 1 & 5 \end{bmatrix}$, and $B = [1 \quad 3 \quad 4]$.

Then $A^T = \begin{bmatrix} 3 & 2 \\ 4 & 1 \\ 6 & 5 \end{bmatrix}$, and $B^T = \begin{bmatrix} 1 \\ 3 \\ 4 \end{bmatrix}$.

4.1 SYMMETRIC AND SKEW-SYMMETRIC MATRICES

Now that a new transformation has been defined, we can investigate its properties and applications. A few of the properties will be stated now; others will follow throughout the book.

Theorem 1. *Let A and B be C-matrices, let k be a complex number, and let the orders of A and B be such that conformability requirements are met. Then*
  (i) $(A^T)^T = A$,
  (ii) $(A + B)^T = A^T + B^T$,
  (iii) $(AB)^T = B^T A^T$,
  (iv) $(kA)^T = kA^T$.

The proofs of parts (i), (ii), and (iv) are left as Exercise 13.

Proof of part (iii). Let $A = [a_{ij}]_{(m, p)}$ and $B = [b_{ij}]_{(p, n)}$.

	STATEMENT	REASON
(1)	$B^T A^T = [b_{ji}]_{(n,p)}[a_{ji}]_{(p,m)}$	(1) Definition 1.
(2)	$= \left[\sum_{k=1}^{p} b_{ki} a_{jk}\right]_{(n, m)}$	(2) By definition of matrix multiplication.
(3)	$= \left[\sum_{k=1}^{p} a_{jk} b_{ki}\right]_{(n, m)}$	(3) Multiplication of complex numbers is commutative.
(4)	$= \left[\sum_{k=1}^{p} a_{ik} b_{kj}\right]_{(m, n)}^{T}$	(4) Definition 1.
(5)	$= (AB)^T.$	(5) By definition of matrix multiplication. ∎

Example 2.

(i) $\left(\begin{bmatrix} 0 & 1 \\ 4 & 3 \\ 2 & 2 \end{bmatrix}^T\right)^T = \begin{bmatrix} 0 & 4 & 2 \\ 1 & 3 & 2 \end{bmatrix}^T = \begin{bmatrix} 0 & 1 \\ 4 & 3 \\ 2 & 2 \end{bmatrix}.$

(ii) $\left(\begin{bmatrix} 2 & 4 \\ 3 & 2 \end{bmatrix} + \begin{bmatrix} 1 & 5 \\ 3 & 2 \end{bmatrix}\right)^T = \begin{bmatrix} 3 & 9 \\ 6 & 4 \end{bmatrix}^T = \begin{bmatrix} 3 & 6 \\ 9 & 4 \end{bmatrix},$

and

$$\begin{bmatrix} 2 & 4 \\ 3 & 2 \end{bmatrix}^T + \begin{bmatrix} 1 & 5 \\ 3 & 2 \end{bmatrix}^T = \begin{bmatrix} 2 & 3 \\ 4 & 2 \end{bmatrix} + \begin{bmatrix} 1 & 3 \\ 5 & 2 \end{bmatrix} = \begin{bmatrix} 3 & 6 \\ 9 & 4 \end{bmatrix}.$$

$$\left(\begin{bmatrix} 3 & 4 \\ 5 & 0 \\ 2 & 1 \end{bmatrix}\begin{bmatrix} 2 & 1 \\ 4 & 6 \end{bmatrix}\right)^T = \begin{bmatrix} 22 & 27 \\ 10 & 5 \\ 8 & 8 \end{bmatrix}^T = \begin{bmatrix} 22 & 10 & 8 \\ 27 & 5 & 8 \end{bmatrix},$$

SPECIAL MATRICES

and

$$\begin{bmatrix} 2 & 1 \\ 4 & 6 \end{bmatrix}^T \begin{bmatrix} 3 & 4 \\ 5 & 0 \\ 2 & 1 \end{bmatrix}^T = \begin{bmatrix} 2 & 4 \\ 1 & 6 \end{bmatrix} \begin{bmatrix} 3 & 5 & 2 \\ 4 & 0 & 1 \end{bmatrix} = \begin{bmatrix} 22 & 10 & 8 \\ 27 & 5 & 8 \end{bmatrix}.$$

Know what a(constant)x(matrix) is.

(iv) $\left(k \begin{bmatrix} 3 & 4 & 2 \\ 1 & 0 & 2 \end{bmatrix} \right)^T = \begin{bmatrix} 3k & 4k & 2k \\ k & 0 & 2k \end{bmatrix}^T = \begin{bmatrix} 3k & k \\ 4k & 0 \\ 2k & 2k \end{bmatrix},$

and

$$k \begin{bmatrix} 3 & 4 & 2 \\ 1 & 0 & 2 \end{bmatrix}^T = k \begin{bmatrix} 3 & 1 \\ 4 & 0 \\ 2 & 2 \end{bmatrix} = \begin{bmatrix} 3k & k \\ 4k & 0 \\ 2k & 2k \end{bmatrix}.$$

Theorem 2. *If A is an invertible C-matrix, then*

$$(A^{-1})^T = (A^T)^{-1}.$$

The proof is left as Exercise 7.

Example 3.

$$\left(\begin{bmatrix} 3 & 5 \\ 1 & 2 \end{bmatrix}^{-1} \right)^T = \begin{bmatrix} 2 & -5 \\ -1 & 3 \end{bmatrix}^T = \begin{bmatrix} 2 & -1 \\ -5 & 3 \end{bmatrix},$$

and

$$\left(\begin{bmatrix} 3 & 5 \\ 1 & 2 \end{bmatrix}^T \right)^{-1} = \begin{bmatrix} 3 & 1 \\ 5 & 2 \end{bmatrix}^{-1} = \begin{bmatrix} 2 & -1 \\ -5 & 3 \end{bmatrix}.$$

By using the transpose, we can define several special types of matrices.

Definition 2. *A C-matrix A is said to be **symmetric** if $A = A^T$.*

Example 4. The matrix $A = \begin{bmatrix} 3 & 1 & 4 \\ 1 & 2 & 0 \\ 4 & 0 & 5 \end{bmatrix}$ is symmetric, because $A = A^T$. The matrix $B = \begin{bmatrix} 3 & 2 & 4 \\ 0 & 2 & 5 \\ 1 & 0 & 4 \end{bmatrix}$ is not symmetric, because $B \neq B^T$. The name symmetric matrix is appropriate because of the physical symmetry with respect to the main diagonal, as illustrated in Figure 4.1.1.

Notice that a matrix must be square if there is to be any hope of symmetry.

4.1 SYMMETRIC AND SKEW-SYMMETRIC MATRICES

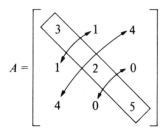

Figure 4.1.1

Definition 3. *A C-matrix A is said to be **skew-symmetric** if $A = -A^T$.*

Example 5. The matrix $A = \begin{bmatrix} 0 & 0 & 2 \\ 0 & 0 & -4 \\ -2 & 4 & 0 \end{bmatrix}$ is skew-symmetric, because $A = -A^T$. The matrix $B = \begin{bmatrix} 1 & 0 & 2 \\ 0 & 2 & -4 \\ -2 & 4 & 3 \end{bmatrix}$ is not skew-symmetric, because $B \neq -B^T$. Notice that all skew-symmetric C-matrices must be square and must have only zeros on the main diagonal (in Exercise 8 the reader is asked to prove this property). Also notice that if a matrix is skew-symmetric, two entries that are symmetrically placed with respect to the main diagonal across from each other are negatives of each other, as shown in Figure 4.1.2.

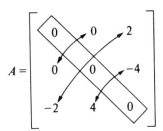

Figure 4.1.2

A few of the properties of symmetric and skew-symmetric matrices are listed in the next three theorems. Other properties are stated in the Exercises and in later sections of this book.

Theorem 3. *If A is a symmetric C-matrix, then*
  (i) kA is symmetric, where k is any complex number;
  (ii) $AA^T = A^TA$;
  (iii) A^2 is symmetric.

The proofs of parts (i) and (ii) are left as Exercise 14.

Proof of part (iii): We must show that $(A^2)^T = A^2$.

	STATEMENT		REASON
(1)	$(A^2)^T = (AA)^T$	(1)	By definition of A^2.
(2)	$= A^T A^T$	(2)	By Theorem 1 (*iii*).
(3)	$= AA$	(3)	Given that A is symmetric.
(4)	$= A^2$.	(4)	By definition of A^2. □

Theorem 4. *If A is a skew-symmetric C-matrix, then*
    (i) kA *is skew-symmetric, where k is any complex number;*
    (ii) $AA^T = A^T A$;
    (iii) A^2 *is symmetric.*

The proof is left as Exercise 15.

Theorem 5. *A square C-matrix A can be expressed as the sum of a symmetric and a skew-symmetric matrix. Moreover, this expression is unique.*

Outline of Proof. Construct a symmetric matrix S and a skew-symmetric matrix K as follows:

$$S = \frac{A + A^T}{2}, \quad K = \frac{A - A^T}{2}; \quad \text{then } A = S + K.$$

(It is proved in Exercises 9 and 10 that $A + A^T$ is symmetric and that $A - A^T$ is skew-symmetric for square A.) To prove uniqueness, let S_1 be symmetric, let K_1 be skew-symmetric, and assume that $A = S_1 + K_1$. We can show that $S_1 = S$ and $K_1 = K$ as follows:

$$(A = S_1 + K_1) \Rightarrow (A^T = S_1^T + K_1^T = S_1 - K_1)$$

$$\Rightarrow \left(\frac{A + A^T}{2} = S_1 \quad \text{and} \quad \frac{A - A^T}{2} = K_1 \right)$$

$$\Rightarrow (S_1 = S \quad \text{and} \quad K_1 = K). \quad □$$

APPLICATIONS

Example 6. In many engineering and social science problems, quadratic expressions of the form

$$f(x, y, z) = x^2 + 4xy + 2y^2 + 6yz + 4z^2 + xz$$

4.1 SYMMETRIC AND SKEW-SYMMETRIC MATRICES

appear. It is possible to express $f(x, y, z)$ using matrix notation as follows:

$$f(x, y, z) = X^T A X,$$

where

$$A = \begin{bmatrix} 1 & 2 & \tfrac{1}{2} \\ 2 & 2 & 3 \\ \tfrac{1}{2} & 3 & 4 \end{bmatrix} \quad \text{and} \quad X = \begin{bmatrix} x \\ y \\ z \end{bmatrix}.$$

The reader should verify this multiplication. Notice that the transpose of X was used and that A is a symmetric matrix which is constructed by letting the entries along the main diagonal represent the coefficients of the squared terms of $f(x, y, z)$; the other entries are one-half of the coefficients of the corresponding product terms. For those who are faced with problems of this general type, it is useful to investigate the properties of symmetric matrices.

Example 7. Background for this example may be found in Chapter 5 of [14], a book by Y. Ijiri which contains a host of applications of linear algebra in accounting. Consider the accounting problem[1] of computing the closing balance for a firm which has the following beginning balance sheet:

Beginning Balance Sheet

Debit Balances		Credit Balances	
Cash	$30	Accounts Payable	$15
Inventories	20	Equity	35
	$50		$50

The firm is in the merchandising business; it buys on credit and makes only cash sales. In this particular example assume that only the accounts listed above are in use. First, arrange the beginning balances, starting with asset accounts, into the column matrix U:

$$U = \begin{bmatrix} u_1 \\ u_2 \\ u_3 \\ u_4 \end{bmatrix} = \begin{bmatrix} 30 \\ 20 \\ -15 \\ -35 \end{bmatrix}.$$

Assume that single transactions have been recorded during a certain time period. The dollar representation of these transactions is next arranged into a "spread sheet matrix" W with a row for each account and a column for each

---

[1] The basic model used in this example was contributed by Veikko Jaaskelainen, Visiting Professor of Business, Virginia Polytechnic Institute and State University, 1968–69.

account. Rows represent accounts to be debited, and columns represent accounts to be credited, with transactions of the period. Arranging the transactions,

Purchases of goods on credit	$w_{23} = \$40$
Cash payments to suppliers	$w_{31} = 35$
Fixed costs paid in cash	$w_{41} = 15$
Sales	
Contribution margin, cash sales	$w_{14} = 25$
Cost of goods sold	$w_{12} = 30$
Goods returned to suppliers	$w_{32} = 5$

into a spread sheet matrix W yields

$$W = \begin{bmatrix} 0 & 30 & 0 & 25 \\ 0 & 0 & 40 & 0 \\ 35 & 5 & 0 & 0 \\ 15 & 0 & 0 & 0 \end{bmatrix}.$$

The first transaction is w_{23} because, by accounting convention, item 2 (inventory) is debited and item 3 (accounts payable) is credited. The main diagonal of W is filled with zeros because we do not recognize a debit and a credit to the same account as representing a genuine transaction. By transposing matrix W, we form matrix W^T. Then the change in U, called ΔU, is equal to the expression $(W - W^T)e$, where e is a column vector which has all its elements equal to 1 and which has as many elements as there are accounts in the system. (Notice that $W - W^T$ is skew-symmetric.)

$$\Delta U = (W - W^T)e = \left(\begin{bmatrix} 0 & 30 & 0 & 25 \\ 0 & 0 & 40 & 0 \\ 35 & 5 & 0 & 0 \\ 15 & 0 & 0 & 0 \end{bmatrix} - \begin{bmatrix} 0 & 0 & 35 & 15 \\ 30 & 0 & 5 & 0 \\ 0 & 40 & 0 & 0 \\ 25 & 0 & 0 & 0 \end{bmatrix} \right) \begin{bmatrix} 1 \\ 1 \\ 1 \\ 1 \end{bmatrix}$$

$$= \begin{bmatrix} 0 & 30 & -35 & 10 \\ -30 & 0 & 35 & 0 \\ 35 & -35 & 0 & 0 \\ -10 & 0 & 0 & 0 \end{bmatrix} \begin{bmatrix} 1 \\ 1 \\ 1 \\ 1 \end{bmatrix} = \begin{bmatrix} 5 \\ 5 \\ 0 \\ -10 \end{bmatrix}.$$

This enables us to compute the closing balance U' as follows

$$U' = U + \Delta U = \begin{bmatrix} 30 \\ 20 \\ -15 \\ -35 \end{bmatrix} + \begin{bmatrix} 5 \\ 5 \\ 0 \\ -10 \end{bmatrix} = \begin{bmatrix} 35 \\ 25 \\ -15 \\ -45 \end{bmatrix}.$$

The closing balance can be presented as

Closing Balance Sheet

Debit Balances		Credit Balances	
Cash	$35	Accounts Payable	$15
Inventories	25	Equity	45
	$60		$60

4.1 SYMMETRIC AND SKEW-SYMMETRIC MATRICES

Example 8. Suppose an appellate court, consisting of four judges, ruled on six appeals that were brought before it during a given time interval. On a motion to favor each appeal the judges' responses are *yes*, or *no*, or *abstain*. The responses given by the four judges on the six cases can be characterized by three matrices. Suppose that the *yes* matrix is

$$\begin{array}{c} \text{APPEALS} \\ \#1 \; \#2 \; \#3 \; \#4 \; \#5 \; \#6 \end{array}$$

$$R_1 = \begin{bmatrix} 1 & 0 & 0 & 1 & 1 & 0 \\ 1 & 0 & 0 & 1 & 0 & 0 \\ 0 & 0 & 1 & 0 & 0 & 1 \\ 1 & 1 & 1 & 0 & 1 & 0 \end{bmatrix} \begin{array}{c} \#1 \\ \#2 \\ \#3 \\ \#4 \end{array} \text{JUDGES}$$

where a 1 in the ij position means that Judge i voted *yes* on Appeal j. A zero in the ij position indicates that Judge i did *not* vote *yes* on Appeal j. In other words, $r_{ij} = 1$ if Judge i voted *yes* on Appeal j, and $r_{ij} = 0$ otherwise. Suppose that the *no* matrix is

$$\begin{array}{c} \text{APPEALS} \\ \#1 \; \#2 \; \#3 \; \#4 \; \#5 \; \#6 \end{array}$$

$$R_2 = \begin{bmatrix} 0 & 1 & 1 & 0 & 0 & 1 \\ 0 & 1 & 1 & 0 & 0 & 1 \\ 1 & 0 & 0 & 1 & 0 & 0 \\ 0 & 0 & 0 & 1 & 0 & 1 \end{bmatrix} \begin{array}{c} \#1 \\ \#2 \\ \#3 \\ \#4 \end{array} \text{JUDGES}$$

where $r_{ij} = 1$ if Judge i voted *no* on Appeal j, and $r_{ij} = 0$ otherwise. Finally suppose that the *abstain* matrix is

$$\begin{array}{c} \text{APPEALS} \\ \#1 \; \#2 \; \#3 \; \#4 \; \#5 \; \#6 \end{array}$$

$$R_3 = \begin{bmatrix} 0 & 0 & 0 & 0 & 0 & 0 \\ 0 & 0 & 0 & 0 & 1 & 0 \\ 0 & 1 & 0 & 0 & 1 & 0 \\ 0 & 0 & 0 & 0 & 0 & 0 \end{bmatrix} \begin{array}{c} \#1 \\ \#2 \\ \#3 \\ \#4 \end{array} \text{JUDGES}$$

where $r_{ij} = 1$ if Judge i abstained on Appeal j, and $r_{ij} = 0$ otherwise. Now form the partitioned matrix $R = \begin{bmatrix} R_1 \\ \hline R_2 \\ \hline R_3 \end{bmatrix}$, and then the product

$$A = RR^T = \begin{bmatrix} R_1 \\ \hline R_2 \\ \hline R_3 \end{bmatrix} [R_1^T \mid R_2^T \mid R_3^T] = \begin{bmatrix} R_1 R_1^T & R_1 R_2^T & R_1 R_3^T \\ R_2 R_1^T & R_2 R_2^T & R_2 R_3^T \\ R_3 R_1^T & R_3 R_2^T & R_3 R_3^T \end{bmatrix}$$

$$A = \begin{bmatrix}
3 & 2 & 0 & 2 & 0 & 0 & \underline{2} & 1 & 0 & 1 & 1 & 0 \\
2 & 2 & 0 & 1 & 0 & 0 & 2 & 1 & 0 & 0 & 0 & 0 \\
0 & 0 & 2 & 1 & 2 & 2 & 0 & 1 & 0 & 0 & 0 & 0 \\
2 & 1 & 1 & 4 & 2 & 2 & 1 & 0 & 0 & 1 & 2 & 0 \\
\hline
0 & 0 & 2 & 2 & 3 & 3 & 0 & 1 & 0 & 0 & 1 & 0 \\
0 & 0 & 2 & 2 & 3 & 3 & 0 & 1 & 0 & 0 & 1 & 0 \\
\underline{2} & 2 & 0 & 1 & 0 & 0 & 2 & 1 & 0 & 0 & 0 & 0 \\
1 & 1 & 1 & 0 & 1 & 1 & 1 & 2 & 0 & 0 & 0 & 0 \\
\hline
0 & 0 & 0 & 0 & 0 & 0 & 0 & 0 & 0 & 0 & 0 & 0 \\
1 & 0 & 0 & 1 & 0 & 0 & 0 & 0 & 0 & 1 & 1 & 0 \\
1 & 0 & 0 & 2 & 1 & 1 & 0 & 0 & 0 & 1 & 2 & 0 \\
0 & 0 & 0 & 0 & 0 & 0 & 0 & 0 & 0 & 0 & 0 & 0
\end{bmatrix}$$

Columns grouped as Responses: Yes (Judges #1–#4), No (Judges #1–#4), Abstain (Judges #1–#4). Rows grouped correspondingly as Responses: Yes (Judges #1–#4), No (Judges #1–#4), Abstain (Judges #1–#4).

It can be shown that the ijth entry of the submatrix $R_h R_k^T$ of A represents the number of appeals that caused Response h by Judge i and Response k by Judge j. For instance, from matrix A we can determine at a glance that Judge 1 voted *yes* at the same time that Judge 3 voted *no* exactly two times. (The entries that determine this fact are underlined; there are two such entries because A is a symmetric matrix.) Also the ijth entry of the sum $R_1 R_1^T + R_2 R_2^T + R_3 R_3^T$, which in this problem is

$$\begin{bmatrix} 6 & 5 & 0 & 3 \\ 5 & 6 & 1 & 2 \\ 0 & 1 & 6 & 2 \\ 3 & 2 & 2 & 6 \end{bmatrix},$$

represents the number of times that Judges i and j agree; for instance Judges 1 and 2 agree 5 times whereas Judges 1 and 3 never agree. Additional results can be found by forming a matrix for each judge or for each appeal rather than for each response; these matrices can be manipulated in ways similar to those of this example. Although the matrix method illustrated in this example derives results that can be obtained by other methods, this matrix method has the advantage of displaying and manipulating vast amounts of data in a systematic way. Of course, an added advantage is that computer programs needed to perform the necessary matrix operations are already available.

This example will be continued in Example 8 in the applications of Section 7.5.

4.1 SYMMETRIC AND SKEW-SYMMETRIC MATRICES

Example 9. The matrix method of the last example applies to any problem in which m objects yield p responses when subjected to n stimuli; moreover, the responses need *not* be mutually exclusive as they were in the last example. Also it can be observed that objects, stimuli, and responses need only to be listed, numbered, and counted. No other quantification is necessary. This matrix method might apply to such diverse problems as: (1) studies of responses of animals or plants when subjected to certain diseases, medication, weather, or food, etc.; (2) studies of responses of legislatures or legislators, or nations of the UN, when subjected to votes on certain issues; (3) studies of responses of certain business or economic indicators when subjected to various national or international events; (4) psychological studies of responses of individuals or groups of individuals when subjected to certain emotional stresses or physical stresses or mental testing.

Example 10. Consider the number of roads connecting the four cities shown in Figure 4.1.3. (Of course, many other interpretations besides roads may be

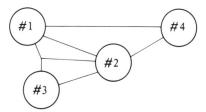

Figure 4.1.3

given to the diagram, as, for instance, communication lines or mutual influence among people or nations.) This information can be represented by a matrix A, where

$$a_{ij} = a_{ji} = \text{(number of roads connecting the } i\text{th and } j\text{th cities without passing through another city),}$$

$$A = \begin{matrix} & \begin{matrix} \#1 & \#2 & \#3 & \#4 \end{matrix} \\ & \begin{bmatrix} 0 & 2 & 1 & 1 \\ 2 & 0 & 2 & 1 \\ 1 & 2 & 0 & 0 \\ 1 & 1 & 0 & 0 \end{bmatrix} & \begin{matrix} \#1 \\ \#2 \\ \#3 \\ \#4 \end{matrix} \end{matrix}$$

A is a symmetric matrix. It can be shown by a method similar to that of Example 6 of Section 3.1 that A^2 represents the number of ways to travel between any two cities by passing through exactly one city; opposite directions on the same road are considered different. A^3 represents the number of ways to travel between any two cities by passing through exactly two cities. It is left as an exercise to show that if A is symmetric then A^n, where n is a positive integer, is symmetric. Also $A + A^2$ (which represents the total number of ways to travel between two cities with at most one intermediate city) is symmetric.

EXERCISES

1. Determine which of the following are symmetric or skew-symmetric; if any one is neither, state why.

    (a) $\begin{bmatrix} 1 & 2 & 5 \\ 2 & 2 & -1 \\ 5 & -1 & 3 \end{bmatrix}$; (b) $\begin{bmatrix} 2 & 3 \\ 3 & 4 \\ 0 & 0 \end{bmatrix}$;

    (c) $\begin{bmatrix} 0 & 0 & 2 \\ 0 & 0 & -1 \\ -2 & 1 & 0 \end{bmatrix}$; (d) $\begin{bmatrix} 1 & -2 & 3 \\ 2 & 1 & 4 \\ -3 & -4 & 1 \end{bmatrix}$.

2. When is A conformable to A^T for addition? For multiplication?

3. Using $A = \begin{bmatrix} 3 & 2 \\ 4 & 0 \end{bmatrix}$ and $B = \begin{bmatrix} 0 & 1 & 2 \\ 1 & 2 & 0 \end{bmatrix}$, verify parts (i) and (iii) of Theorem 1.

4. Does $(AB)^T = A^T B^T$? Why?

5. Consider the conjecture $(A - B)^T = A^T - B^T$. If the conjecture is valid, then prove it, using parts (ii) and (iv) of Theorem 1. If the conjecture is not valid, then give a counterexample.

6. Simplify $(A^T B^T + 3C)^T$. Give the reason for each step.

7. Prove Theorem 2. (Answer provided.)

8. (a) Prove that each entry on the main diagonal of a skew-symmetric C-matrix must be 0.
   (b) What can be said about the entries on the main diagonal if the matrix is symmetric rather than skew-symmetric?

9. Prove that if A is a square C-matrix, then $A + A^T$ is symmetric. (Answer provided.)

10. Prove: If A is a square C-matrix, then $A - A^T$ is skew-symmetric.

11. Given: A and B are symmetric n by n C-matrices.
    (a) Prove that $A + B$ is symmetric;
    (b) Prove that AB need not be symmetric.

12. (a) Show that the converse of Exercise 11(a) is not true.
    (b) Prove: If A and B are symmetric C-matrices, and if $AB = BA$, then AB is symmetric.

13. (a) Prove part (i) of Theorem 1;
    (b) Prove part (ii) of Theorem 1;
    (c) Prove part (iv) of Theorem 1.

14. (a) Prove part (i) of Theorem 3;
    (b) Prove part (ii) of Theorem 3.

4.1 SYMMETRIC AND SKEW-SYMMETRIC MATRICES

15. (a) Prove part (i) of Theorem 4;
(b) Prove part (ii) of Theorem 4;
(c) Prove part (iii) of Theorem 4.

16. Assuming conformability for multiplication, prove that

$$(AB \cdots M)^T = M^T \cdots B^T A^T.$$

17. Prove: If A is a square C-matrix, and if n is a positive integer, then

$$(A^T)^n = (A^n)^T.$$

18. Prove that if A is a symmetric C-matrix, and if n is a positive integer, then A^n is a symmetric C-matrix.

19. Prove: If A is skew-symmetric, and if n is a positive integer, then A^n is symmetric if n is even and skew-symmetric if n is odd.

20. Prove: If A is an m by n C-matrix, then AA^T and A^TA are symmetric.

21. Express the following matrix as the sum of a skew-symmetric matrix and a symmetric matrix:

$$\begin{bmatrix} 3 & 3 & -1 \\ 0 & 3 & -2 \\ -1 & 2 & 2 \end{bmatrix}.$$

22. Express the quadratic expression

$$f(x, y, z) = x^2 - 4xy + y^2 + 6yz + z^2$$

in (symmetric) matrix form.

23. Consider a small telephone system between three towns in which there are 5 lines between town #2 and town #3, 3 lines between town #1 and town #3 and 4 lines between town #1 and town #2. Use matrices to determine the total number of different communication channels between each pair of the towns; a channel between two towns may or may not pass through a third town.

24. (a) Is the transpose of an echelon matrix necessarily an echelon matrix?
(b) Is the transpose of a reduced echelon matrix necessarily an echelon matrix?

25. (a) If $[a_{ij}]_{(4, 4)}$ is symmetric and $a_{23} = 7$, what is a_{32}?
(b) If $[a_{ij}]_{(4, 4)}$ is skew-symmetric and $a_{23} = 7$, what is a_{32}?

26. Verify Theorem 2 for $A = \begin{bmatrix} 1 & 2 \\ 1 & 4 \end{bmatrix}$.

27. (a) Prove that if A is invertible and symmetric then A^{-1} is symmetric.
(b) Prove that if A is invertible and symmetric then A^{-2} is symmetric.

28. Prove Theorem 17 of Section 3.3. (Answer provided in booklet of Answers to Selected Even Numbered Exercises.)

4.2 Hermitian and Skew-Hermitian Matrices

Basic to the theory of C-matrices are the special matrices defined in this section.

Definition 4. Let $A = [a_{ij}]_{(m,n)}$ be a C-matrix. The matrix with entries that are complex conjugates of the corresponding entries of A is called the **conjugate** of A and is denoted as $\bar{A} = [\bar{a}_{ij}]_{(m,n)}$. The transpose of $\bar{A}$ is called the **tranjugate** of A and is denoted by A^*.

Example 1. If

$$A = [a_{ij}]_{(2,2)} = \begin{bmatrix} 2+i & 4 \\ 3-i & -i \end{bmatrix},$$

then

$$\bar{A} = [\bar{a}_{ij}]_{(2,2)} = \begin{bmatrix} 2-i & 4 \\ 3+i & i \end{bmatrix},$$

and

$$A^* = (\bar{A})^T = \begin{bmatrix} 2-i & 3+i \\ 4 & i \end{bmatrix}.$$

Theorem 6. If A and B are C-matrices, and if conformability is assumed, then

(i) $\overline{(\bar{A})} = A$;
(ii) $\overline{A + B} = \bar{A} + \bar{B}$;
(iii) $\overline{AB} = \bar{A}\bar{B}$;
(iv) $\overline{cA} = \bar{c}\bar{A}$, where c is a complex number;
(v) $\overline{(A^T)} = (\bar{A})^T$.

The proof is left as Exercise 14.

If we use Theorem 6, it is easy to prove the following theorem, which is analogous to Theorem 1 of Section 4.1.

Theorem 7. If A and B are C-matrices, and if conformability is assumed, then

(i) $(A^*)^* = A$;
(ii) $(A + B)^* = A^* + B^*$;

4.2 HERMITIAN AND SKEW-HERMITIAN MATRICES

(iii) $(AB)^* = B^*A^*$;
(iv) $(cA)^* = \bar{c}A^*$, where c is a complex number.

The proof is left as Exercise 15.

In view of the importance of matrices with the property that $A = A^T$, it seems reasonable to investigate matrices that have the property that $B = B^*$.

Definition 5. *A C-matrix A is said to be **Hermitian**[2] if $A = A^*$.*

Example 2. $B = \begin{bmatrix} 3 & i & 2-i \\ -i & 4 & 2 \\ 2+i & 2 & 0 \end{bmatrix}$ is Hermitian, because

$$B^* = [\bar{b}_{ij}]^T_{(3,3)} = \begin{bmatrix} 3 & -i & 2+i \\ i & 4 & 2 \\ 2-i & 2 & 0 \end{bmatrix}^T = B,$$

but matrix A of Example 1 is not Hermitian. Notice that a Hermitian matrix must be square and must have real entries on the main diagonal; in Exercise 5 the reader is asked to prove this property. Also notice that entries that are symmetrically placed across the main diagonal must be complex conjugates, as shown in Figure 4.2.1.

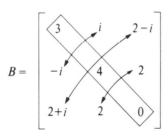

Figure 4.2.1

Definition 6. *A C-matrix A is said to be **skew-Hermitian** if $A = -A^*$.*

Example 3. Let $B = [b_{ij}]_{(2,2)} = \begin{bmatrix} i & 2+i \\ -2+i & 0 \end{bmatrix}$.

Then $\bar{B} = [\bar{b}_{ij}]_{(2,2)} = \begin{bmatrix} -i & 2-i \\ -2-i & 0 \end{bmatrix}$; $-B^* = -(\bar{B})^T = \begin{bmatrix} i & 2+i \\ -2+i & 0 \end{bmatrix} = B$;

therefore B is skew-Hermitian. Notice that a skew-Hermitian C-matrix must be square and must have either pure imaginary numbers or zeros on the main diagonal; in Exercise 6 the reader is asked to prove this property. Also notice

---
[2] Named after the French mathematician Charles Hermite (1822–1901).

that entries across the main diagonal from each other must be negative complex conjugates of each other, as shown in Figure 4.2.2.

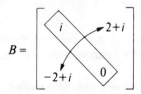

Figure 4.2.2

As might be expected, many of the properties of Hermitian and skew-Hermitian matrices are analogous to those properties developed in the last section.

Theorem 8. *If A is a Hermitian matrix, then*
   (i) iA and $-iA$ are skew-Hermitian;
   (ii) $AA^* = A^*A$;
   (iii) A^2 is Hermitian;
   (iv) A can be expressed as $A = S + iK$ where S is real and symmetric and where K is real and skew-symmetric. Moreover this expression is unique.

Proof of part (i). We must show that $(iA)^* = -(iA)$ and that $(-iA)^* = iA$.

	STATEMENT		REASON
(1)	$(iA)^* = \bar{i}A^*$	(1)	Theorem 7 Part (iv).
(2)	$= -iA$.	(2)	$\bar{i} = -i$ and A is Hermitian.
(3)	Also, $(-iA)^* = \overline{(-i)}A^*$	(3)	Theorem 7 Part (iv).
(4)	$= iA$.	(4)	$\overline{(-i)} = i$ and A is Hermitian.

The proofs of parts (ii) and (iii) are left as Exercise 16.

Outline of proof of part (iv). From Exercise 9 of Section 4.1, $A + A^T$ is symmetric. Furthermore, if A is Hermitian, then $A = (\bar{A})^T$ or $A^T = \bar{A}$. Hence $A + A^T = A + \bar{A}$, which must be real as well as symmetric. Likewise, by Exercise 10 from Section 4.1, $A - A^T$ is skew-symmetric and, because A is Hermitian, $A - A^T = A - \bar{A}$. But $A - \bar{A}$ has entries which are pure imaginary or zero, and hence $(A - \bar{A})/i$ is real as well as skew-symmetric. Therefore if $S = (A + \bar{A})/2$ and if $K = (A - \bar{A})/2i$, then $A = S + iK$. The uniqueness of this expression can be proved in a manner similar to that used in the proof of Theorem 5. □

Example 4. As an illustration of parts (i) and (iv) of Theorem 8, let A be the Hermitian matrix

$$A = \begin{bmatrix} 2 & 1 & 1+i \\ 1 & 0 & i \\ 1-i & -i & 1 \end{bmatrix}.$$

4.2 HERMITIAN AND SKEW-HERMITIAN MATRICES

Then

$$iA = \begin{bmatrix} 2i & i & -1+i \\ i & 0 & -1 \\ 1+i & 1 & i \end{bmatrix}, \text{ and } -iA = \begin{bmatrix} -2i & -i & 1-i \\ -i & 0 & 1 \\ -1-i & -1 & -i \end{bmatrix},$$

which are both skew-Hermitian. To illustrate part (iv), consider the same Hermitian matrix A and its conjugate

$$\bar{A} = \begin{bmatrix} 2 & 1 & 1-i \\ 1 & 0 & -i \\ 1+i & i & 1 \end{bmatrix}.$$

We find

$$A + \bar{A} = \begin{bmatrix} 2 & 1 & 1+i \\ 1 & 0 & i \\ 1-i & -i & 1 \end{bmatrix} + \begin{bmatrix} 2 & 1 & 1-i \\ 1 & 0 & -i \\ 1+i & i & 1 \end{bmatrix} = \begin{bmatrix} 4 & 2 & 2 \\ 2 & 0 & 0 \\ 2 & 0 & 2 \end{bmatrix},$$

and

$$A - \bar{A} = \begin{bmatrix} 0 & 0 & 2i \\ 0 & 0 & 2i \\ -2i & -2i & 0 \end{bmatrix}.$$

Therefore

$$S = \tfrac{1}{2}(A + \bar{A}) = \begin{bmatrix} 2 & 1 & 1 \\ 1 & 0 & 0 \\ 1 & 0 & 1 \end{bmatrix},$$

which is real and symmetric, and

$$K = \frac{1}{2i}(A - \bar{A}) = \begin{bmatrix} 0 & 0 & 1 \\ 0 & 0 & 1 \\ -1 & -1 & 0 \end{bmatrix},$$

which is real and skew-symmetric. Obviously $A = S + iK$.

Theorem 9. *If A is a skew-Hermitian matrix, then*
  (i) *iA and $-iA$ are Hermitian;*
  (ii) *$AA^* = A^*A$;*
  (iii) *A^2 is Hermitian;*
  (iv) *A can be expressed as $A = K + iS$, where S is real and symmetric, and where K is real and skew-symmetric. Moreover this expression is unique.*

The proof, which is similar to the proof of Theorem 8, is left as Exercise 17.

Theorem 10. *A square C-matrix can be expressed as the sum of a Hermitian matrix and a skew-Hermitian matrix. Moreover this expression is unique.*

The proof, which is similar to the proof of Theorem 5, is left as Exercise 18. The next theorem easily follows from Theorems 10 and 9(i).

Theorem 11. *Every square C-matrix M can be expressed as $M = A + iB$ where A and B are Hermitian matrices. Moreover this expression is unique.*

The proof is left as Exercise 19.

Other properties of Hermitian and skew-Hermitian matrices are stated in the Exercises.

APPLICATIONS

Example 5. In the theory of relativity, the C-matrix

$$\begin{bmatrix} 1 & 0 & 0 & 0 \\ 0 & 1 & 0 & 0 \\ 0 & 0 & \gamma & \dfrac{iv\gamma}{c} \\ 0 & 0 & \dfrac{-iv\gamma}{c} & \gamma \end{bmatrix}$$

is a transformation matrix from a stationary system to a system moving with a velocity v with respect to the stationary system. Here c represents the velocity of light, and $\gamma = 1/\sqrt{1-(v^2/c^2)}$. Notice that the matrix is Hermitian.

Example 6. The Pauli spin matrices σ_x, σ_y, σ_z (defined in Example 7 of Section 3.1) are Hermitian. It can be shown that the product of any two of the Pauli spin matrices is skew-Hermitian.

EXERCISES

1. Which of the following matrices are Hermitian? Which are skew-Hermitian? Which are neither, and why?

   (a) $\begin{bmatrix} i & 2-i \\ 0 & 3i \\ 2 & 4 \end{bmatrix}$; (b) $\begin{bmatrix} 2 & 2+i \\ i & 2i \end{bmatrix}$; (c) $\begin{bmatrix} 0 & 3+i \\ 3-i & 2 \end{bmatrix}$;

   (d) $\begin{bmatrix} 2 & 3 \\ 3 & 1 \end{bmatrix}$; (e) $\begin{bmatrix} 0 & i & 2i \\ i & 0 & -4 \\ 2i & 4 & 0 \end{bmatrix}$.

2. With $A = \begin{bmatrix} i & 2+i \\ 2 & -2i \end{bmatrix}$ and $B = \begin{bmatrix} 0 & i \\ 1+i & 2 \end{bmatrix}$, and with the complex number $c = 1+i$, illustrate
   (a) Theorem 6; (b) Theorem 7.

4.2 HERMITIAN AND SKEW-HERMITIAN MATRICES

3. Is a symmetric R-matrix Hermitian? Why?
4. Is a skew-symmetric R-matrix skew-Hermitian? Why?
5. Is a Hermitian matrix always square with real entries on the main diagonal? Why?
6. Is a skew-Hermitian matrix always square with either pure imaginary numbers or zeros on the main diagonal? Why?
7. Use the matrix $\begin{bmatrix} i & 2+i \\ 2 & -2i \end{bmatrix}$ to illustrate
   (a) Theorem 10;   (b) Theorem 11.
8. Use the Hermitian matrix $\begin{bmatrix} 2 & 2-i \\ 2+i & 3 \end{bmatrix}$ to illustrate Theorem 8.
9. Use the skew-Hermitian matrix $\begin{bmatrix} i & 1+i \\ -1+i & 0 \end{bmatrix}$ to illustrate Theorem 9.
10. Prove: If A and B are Hermitian m by m matrices, then $A + B$ is Hermitian.
11. Prove: If A and B are Hermitian m by m matrices, and if $AB = BA$, then AB is Hermitian.
12. Prove: If A is a square C-matrix, then $A + A^*$ is Hermitian.
13. Prove: If A is a square C-matrix, then $A - A^*$ is skew-Hermitian.
14. For Theorem 6: (a) prove part (i);   (b) prove part (ii);
                  (c) prove part (iii);   (d) prove part (iv);
                  (e) prove part (v).
15. For Theorem 7: (a) prove part (i);   (b) prove part (ii);
                  (c) prove part (iii);   (d) prove part (iv).
16. For Theorem 8: (a) prove part (ii);   (b) prove part (iii).
17. For Theorem 9: (a) prove part (i);   (b) prove part (ii);
                  (c) prove part (iii);   (d) prove part (iv).
18. Prove Theorem 10. (*Hint:* See outline of proof of Theorem 5, and use Exercises 12 and 13.)
19. Prove Theorem 11 in Statement–Reason form. (*Hint:* Use Theorems 10 and 9(*i*).)
20. Assuming conformability, prove
$$(AB \cdots M)^* = M^* \cdots B^* A^*.$$
21. Prove: If A is a square C-matrix and n is a positive integer, then
$$(A^*)^n = (A^n)^*.$$
22. Prove: If A is Hermitian, and if n is a positive integer, then A^n is Hermitian.
23. Prove: If A is skew-Hermitian, and if n is a positive integer, then A^n is Hermitian if n is even, and skew-Hermitian if n is odd.
24. Verify the last statement in Example 6.
25. Prove that if A is invertible then $(A^{-1})^* = (A^*)^{-1}$.

4.3 Coordinate Vectors

In Section 1.2, a 1 by n C-matrix

$$[a_{11} a_{12} \cdots a_{1n}]$$

was called a row matrix or a row vector, and an m by 1 C-matrix

$$\begin{bmatrix} a_{11} \\ a_{21} \\ \vdots \\ a_{m1} \end{bmatrix}$$

was called a column matrix or a column vector. In both cases there is an ordered set of entries; if no attention is paid to how the ordered set is displayed except that the order of the entries is maintained, then the set is called simply a *coordinate vector*.

Definition 7. *An ordered set of entries, say $(a_1, a_2, \ldots, a_n)$, from an algebraic system S is a* **coordinate vector***; the entries of a coordinate vector are known as* **components** *or* **coordinates***. The number of components is the* **dimension** *of the coordinate vector.*

Until Chapter 6, as we mentioned previously, S is assumed to be the complex number system. If, however, we further restrict the components of a coordinate vector to be real numbers, then we have a **real coordinate vector**. If all of the components of a coordinate vector are zero, then the vector is called a **zero coordinate vector**; if at least one component is nonzero, then the vector is a **nonzero coordinate vector**. In the past the reader has probably referred to a real coordinate vector as simply a "vector." We prefer to reserve the word "vector" for a more abstract concept to be developed later.

It is reemphasized that a coordinate vector can be represented by an n-tuple or by either a row or column matrix, or in any ordered fashion.

Example 1. The set of coordinates $(1, 4, 3)$ of a point P in ordinary three-space with reference to the Cartesian coordinate system (see Figure 4.3.1) is an example of a three-dimensional coordinate vector; this real nonzero coordinate vector may be expressed in matrix form as

$$[1 \quad 4 \quad 3] \quad \text{or} \quad \begin{bmatrix} 1 \\ 4 \\ 3 \end{bmatrix}.$$

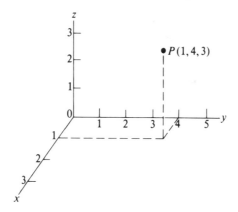

Figure 4.3.1

Example 2. The set of closing prices of the Dow-Jones industrials in alphabetical order is an example of a coordinate vector with real components. These components may be arranged as a row or column matrix.

As might be expected from the matrix representation, two coordinate vectors of the same dimension are **equal** if each component of one vector is equal to the corresponding component of the other vector. Also, two coordinate vectors of the same dimension may be added by adding corresponding components, and the product of a complex number times a coordinate vector is obtained by multiplying each component by the complex number. A type of product of two coordinate vectors is defined next.

Definition 8. Let α and β be real coordinate vectors with components $a_1, a_2, \ldots, a_n$ and $b_1, b_2, \ldots, b_n$, respectively. The **standard inner product** or **dot product**, $\alpha \cdot \beta$, is the real number

$$\alpha \cdot \beta = a_1 b_1 + a_2 b_2 + \cdots + a_n b_n.$$

Example 3. Let $\alpha = [3 \quad 2 \quad 0 \quad 4]$ and $\beta = [4 \quad 2 \quad 9 \quad 6]$. Then

$$\alpha \cdot \beta = 3 \cdot 4 + 2 \cdot 2 + 0 \cdot 9 + 4 \cdot 6 = 40.$$

Of course, $\alpha \cdot \beta$ is not defined unless α and β have the same number of components.

If the coordinate vectors are not restricted to be real, then Definition 8 may be regarded as a special case of the following definition.

Definition 9. Let α and β be coordinate vectors with complex components $a_1, a_2, \ldots, a_n$ and $b_1, b_2, \ldots, b_n$ respectively. The **standard inner product** or **dot product**, $\alpha \cdot \beta$, is the complex number

$$a_1 \bar{b}_1 + a_2 \bar{b}_2 + \cdots + a_n \bar{b}_n.\text{[3]}$$

Notice that in general the standard inner product operation is not commutative since

$$\beta \cdot \alpha = b_1 \bar{a}_1 + b_2 \bar{a}_2 + \cdots + b_n \bar{a}_n = \overline{\alpha \cdot \beta} \neq \alpha \cdot \beta.$$

Example 4. Let $A = [i \quad 4 \quad 3+i]$ and $B = [1 \quad 2-i \quad i]$.

$$A \cdot B = (i)(1) + (4)(2+i) + (3+i)(-i) = 9 + 2i.$$

The reader should find $B \cdot A$ and observe that $A \cdot B \neq B \cdot A$. Also notice that $A \cdot A$ is a nonnegative real number even though A is not a real vector; the reader can easily show that the dot product of any complex coordinate vector with itself is a nonnegative real number.

Definition 10. The **magnitude** (or **length**) of a coordinate vector α, with complex components, is denoted by $\|\alpha\|$ and is equal to the real number $\|\alpha\| = \sqrt{\alpha \cdot \alpha}$. The coordinate vector α is **normal** if $\|\alpha\| = 1$, and in this case we say that α is a **unit coordinate vector**.

Example 5. Consider the real coordinate vector $\alpha = (3, 4)$, which may be interpreted geometrically as a point in the plane shown in Figure 4.3.2. Then the length of α is $\|\alpha\| = \sqrt{\alpha \cdot \alpha} = \sqrt{3 \cdot 3 + 4 \cdot 4} = \sqrt{25} = 5$.

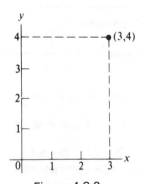

Figure 4.3.2

---

[3] In some books $\alpha \cdot \beta$ is defined to be the conjugate, $\bar{a}_1 b_1 + \bar{a}_2 b_2 + \cdots + \bar{a}_n b_n$.

Geometrically, the number 5 represents the distance from the origin to the point (3, 4). In this example $\|\alpha\| \neq 1$, and therefore α is not normal.

Example 6. For the coordinate vector $A = [6 \quad 2 + i]$,

$$\|A\| = \sqrt{A \cdot A} = \sqrt{a_1 \bar{a}_1 + a_2 \bar{a}_2}$$
$$= \sqrt{(6)(6) + (2 + i)(2 - i)}$$
$$= \sqrt{36 + (4 - i^2)}$$
$$= \sqrt{36 + (4 + 1)}$$
$$= \sqrt{41}.$$

In this example $\|A\| \neq 1$, and therefore A is not normal.

When a nonzero coordinate vector α is multiplied by $1/\|\alpha\|$, the resulting coordinate vector is obviously normal, and we frequently say that α has been *normalized*, or that α has been reduced to a unit coordinate vector.

Definition 11. *Two nonzero coordinate vectors α and β with complex components are* **orthogonal** *if $\alpha \cdot \beta = 0$. A given set $\{\alpha_1, \alpha_2, \cdots, \alpha_n\}$ of nonzero coordinate vectors of the same dimension is an* **orthogonal set** *if $\alpha_i \cdot \alpha_j = 0$ for all i and j, where $i \neq j$.*

Definition 12. *Let $n > 1$; if a set of n normal coordinate vectors is orthogonal, then the set is said to be an* **orthonormal set**.

Example 7. The set of real coordinate vectors

$$B_1 = \begin{bmatrix} 1 \\ 0 \\ 0 \end{bmatrix}, \quad B_2 = \begin{bmatrix} 0 \\ 1 \\ 0 \end{bmatrix}, \quad B_3 = \begin{bmatrix} 0 \\ 0 \\ 1 \end{bmatrix}$$

is an orthogonal set because $B_i \cdot B_j = 0$ for all i and j, where $i \neq j$. Moreover notice that $\|B_i\| = 1$ for all i, hence the set is orthonormal. Geometrically, we may think of each vector as representing an arrow that is one unit long (see Figure 4.3.3) emanating from the origin and terminating at the respective points (1, 0, 0), (0, 1, 0), and (0, 0, 1), where each pair of these arrows is perpendicular. The set of the real coordinate vectors $\begin{bmatrix} 1 \\ -1 \\ 0 \end{bmatrix}$, $\begin{bmatrix} 1 \\ 1 \\ 2 \end{bmatrix}$, and $\begin{bmatrix} 1 \\ 1 \\ -1 \end{bmatrix}$ is orthogonal but not orthonormal (see Figure 4.3.4).

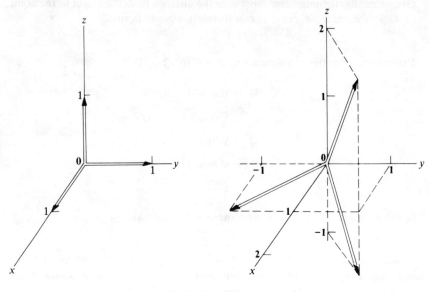

Figure 4.3.3   Figure 4.3.4

APPLICATIONS

Example 8. We live in a three-dimensional world, and therefore many applied problems employ only three-dimensional vectors. However, n-dimensional vectors are used more today than in the past; computers have played a big part in this development. Suppose an investment company decides to sell four different types of stocks. In one transaction 200 shares of stock A, 300 shares of stock B, 100 shares of stock C, and 200 shares of stock D are sold. The selling prices per share are \$20, \$30, \$50, and \$10, respectively. Let the total quantity of stocks sold be represented by the vector $\alpha = (200, 300, 100, 200)$, and let the selling prices be denoted by the vector $\beta = (20, 30, 50, 10)$. The total receipt from the stock sale is, then,

$$\alpha \cdot \beta = (200, 300, 100, 200) \cdot (20, 30, 50, 10)$$
$$= 20{,}000 \text{ dollars.}$$

Example 9. Assume that four items, R, S, T, and U, are to be distributed by three arbitrators, A, B, and C. (Items R, S, T, and U may be thought of as four allocations of a property being divided by a board of arbitrators, or appropriations for four items in a budget, or allocations of four chemicals for the control of a disease, etc., whereas A, B, and C are the agencies that must take a certain position on the distribution of R, S, T, and U.) Let the coordinate vectors

$\alpha = (a_1, a_2, a_3, a_4)$, $\beta = (b_1, b_2, b_3, b_4)$, and $\gamma = (c_1, c_2, c_3, c_4)$ represent the initial positions of the three arbitrators, and let $\pi = (p_1, p_2, p_3, p_4)$ represent their final compromise position. The three coordinate vectors $\pi - \alpha$, $\pi - \beta$, and $\pi - \gamma$ show both the magnitude and direction of change in position of the three arbitrators. The magnitudes $\|\pi - \alpha\|$, $\|\pi - \beta\|$, $\|\pi - \gamma\|$ may be used to rank the degree of movement by the arbitrators and thereby, over a period of time, to study such things as leadership, influence, middle of the road positions, etc.

Example 10. Suppose we construct a simple model economy in which there are three industries—the crude oil industry, the refining industry which produces gasoline, and the utility industry which supplies electricity. Then suppose that there are six types of consumers—the general public, the government, the export firms, and the three industries themselves. The industries and the consumers both exercise certain consumptive demands on each of the industries. For instance, suppose that the crude oil industry needs to use 4 units of gasoline and 2 units of electricity; therefore the demand vector for the crude oil industry is expressed as $\alpha_c = (0, 4, 2)$. Likewise, we specify the other demand vectors. In each case we list the goods demanded, in the form

$$\alpha = (\text{crude oil, gasoline, electricity}).$$

The demand vectors are

crude oil industry: $\alpha_c = (0, 4, 2)$,
refining industry: $\alpha_r = (8, 0, 6)$,
utility industry: $\alpha_u = (1, 6, 0)$,
general public: $\alpha_1 = (1, 9, 5)$,
government: $\alpha_2 = (8, 8, 8)$,
export firms: $\alpha_3 = (7, 2, 0)$.

The total demand on all the industries is then

$$\alpha_{\text{total}} = \alpha_c + \alpha_r + \alpha_u + \alpha_1 + \alpha_2 + \alpha_3 = (25, 29, 21).$$

Now suppose that the price of crude oil is $4 per unit, the price of gasoline $3 per unit, and the price of electricity $2 per unit. This can be expressed as a vector $\beta = (4, 3, 2)$. Assuming that the industries produce exactly what is demanded of them, the income of the crude oil industry is 25 units times $4, which equals $100. The crude oil industry had to have gasoline and electricity to operate; therefore, its costs were

$$\alpha_c \cdot \beta = (0, 4, 2) \cdot (4, 3, 2) = \$16.$$

Hence, the gross profit in the crude oil industry is

$$\$100 - \$16 = \$84.$$

Finding the gross profits (or losses) of the other industries is left as an exercise.

Example 11. One primary and often advantageous use of the column vector or column matrix notation is for expressing a system of linear equations

$$\begin{cases} a_{11}x_1 + a_{12}x_2 + \cdots + a_{1n}x_n = b_1, \\ \vdots \quad\quad \vdots \quad\quad\quad\quad \vdots \quad\quad \vdots \\ a_{m1}x_1 + a_{m2}x_2 + \cdots + a_{mn}x_n = b_m, \end{cases}$$

as

$$\begin{bmatrix} a_{11} \\ \vdots \\ a_{m1} \end{bmatrix} x_1 + \begin{bmatrix} a_{12} \\ \vdots \\ a_{m2} \end{bmatrix} x_2 + \cdots + \begin{bmatrix} a_{1n} \\ \vdots \\ a_{mn} \end{bmatrix} x_n = \begin{bmatrix} b_1 \\ \vdots \\ b_m \end{bmatrix}.$$

The latter notation is particularly prevalent in the literature of linear programming.

EXERCISES

1. Find the standard inner products (or dot products) of the coordinate vectors given below, if possible. If $\alpha \cdot \beta \neq \beta \cdot \alpha$ or $A \cdot B \neq B \cdot A$, state the reason.

   (a) $\alpha = (3, 2, 4)$, $\quad \beta = (6, 0, 1)$;

   (b) $A = \begin{bmatrix} \frac{1}{5} \\ -\frac{2}{5} \end{bmatrix}$, $\quad B = \begin{bmatrix} 0 \\ 3 \end{bmatrix}$;

   (c) $A = \begin{bmatrix} i \\ 3 \end{bmatrix}$, $\quad B = \begin{bmatrix} 2-i \\ 4 \\ i \end{bmatrix}$;

   (d) $A = [6 \ 9 \ 4 \ 3]$, $\quad B = \begin{bmatrix} \frac{4}{7} \\ 0 \\ \frac{4}{7} - \frac{3}{7}i \\ \frac{1}{4}\sqrt{6} \end{bmatrix}$;

   (e) $A = \begin{bmatrix} 6i \\ 4 \end{bmatrix}$, $\quad B = [1 \ \ 0]$.

2. Find the magnitude of each coordinate vector labeled A in Exercise 1. Which, if any, are normal?
3. Find the magnitude of each coordinate vector labeled B in Exercise 1. Which, if any, are normal?
4. Normalize the coordinate vector $(-5, 12)$ and give a geometric interpretation on a two-dimensional graph.
5. Normalize the coordinate vector $(12, 1, 0)$ and give a geometric interpretation on a three-dimensional graph.
6. Which of the following sets are orthogonal?

   (a) $\left\{ \begin{bmatrix} \frac{1}{14}\sqrt{14} \\ \frac{3}{14}\sqrt{14} \\ \frac{2}{14}\sqrt{14} \end{bmatrix}, \begin{bmatrix} 0 \\ 1 \\ 0 \end{bmatrix}, \begin{bmatrix} \frac{2}{5}\sqrt{5} \\ 0 \\ \frac{1}{5}\sqrt{5} \end{bmatrix} \right\};$

(b) $\left\{ \begin{bmatrix} 0 \\ i \\ 0 \end{bmatrix}, \begin{bmatrix} -i \\ 0 \\ 0 \end{bmatrix}, \begin{bmatrix} 0 \\ 0 \\ 1 \end{bmatrix} \right\};$

(c) $\{[2+i \quad 1], \quad [i \quad 1+i]\};$

(d) $\{[\frac{1}{5}\sqrt{5} \quad \frac{2}{5}\sqrt{5}i], \quad [\frac{1}{5}\sqrt{5}i \quad \frac{2}{5}\sqrt{5}]\};$

(e) $\left\{ \begin{bmatrix} 1 \\ -3 \end{bmatrix}, \begin{bmatrix} 3 \\ 1 \end{bmatrix} \right\};$

(f) $\{[\frac{1}{2}\sqrt{2} \quad \frac{1}{2}\sqrt{2} \quad 0], \quad [-\frac{1}{2}\sqrt{2} \quad \frac{1}{2}\sqrt{2} \quad 0], \quad [0 \quad 0 \quad 1]\}.$

7. Which of the sets of Exercise 6 are orthonormal?
8. (a) Find $\alpha + \beta$ or $A + B$, if possible, in each part of Exercise 1.
   (b) Find 5α or $5A$ in each part of Exercise 1.
9. Find $A \cdot B$, if possible: $A = [3 \quad 5 \quad 2 \quad 0]$, $B = [1 \quad -2 \quad 0]$. Find $\|A\|$ and $\|B\|$.
10. Under what conditions does $A \cdot B = B \cdot A$?
11. Why is Definition 8 a special case of Definition 9?
12. Determine whether each of the following is a unit vector. If it is not a unit vector, find a multiple of it that is normal.
    (a) $A = [\frac{1}{6}\sqrt{6} \quad -\frac{2}{6}\sqrt{6} \quad \frac{1}{6}\sqrt{6}];$ (b) $B = [1 \quad 1 \quad 1].$
13. Find a nonzero vector orthogonal to (5, 10) and show both vectors on the same graph. What do you notice about your graph?

4.4 Orthogonal and Unitary Matrices

We have previously investigated some of the properties of the matrices A for which $A = A^T$ and of the matrices A for which $A = A^*$. Now we investigate the properties of matrices for which $A^{-1} = A^T$ and for which $A^{-1} = A^*$.

Definition 13. *An R-matrix A is **orthogonal** if $A^{-1} = A^T$.*

To be orthogonal, of course, a matrix must be square and invertible. An orthogonal matrix should not be confused with an orthogonal set of vectors, although we shall see later (Theorem 13) that there is a connection between these concepts, as the names may suggest.

Example 1. Consider the matrix $A = \begin{bmatrix} \frac{1}{5}\sqrt{5} & -\frac{2}{5}\sqrt{5} \\ \frac{2}{5}\sqrt{5} & \frac{1}{5}\sqrt{5} \end{bmatrix}.$

$[A \mid I_2] = \begin{bmatrix} \frac{1}{5}\sqrt{5} & -\frac{2}{5}\sqrt{5} & 1 & 0 \\ \frac{2}{5}\sqrt{5} & \frac{1}{5}\sqrt{5} & 0 & 1 \end{bmatrix} \underset{\sim}{\text{row}} \begin{bmatrix} 1 & 0 & \frac{1}{5}\sqrt{5} & \frac{2}{5}\sqrt{5} \\ 0 & 1 & -\frac{2}{5}\sqrt{5} & \frac{1}{5}\sqrt{5} \end{bmatrix} = [I_2 \mid A^{-1}].$

Therefore $A^{-1} = \begin{bmatrix} \frac{1}{5}\sqrt{5} & \frac{2}{5}\sqrt{5} \\ -\frac{2}{5}\sqrt{5} & \frac{1}{5}\sqrt{5} \end{bmatrix}$, which equals A^T, and hence A is orthogonal.

Theorem 12. *An n by n R-matrix A is orthogonal if and only if $A^T A = I_n$.*

Proof. PART I: $(A^{-1} = A^T) \Rightarrow (A^T A = I_n)$; the proof of this is left as Exercise 9.

PART II: $(A^T A = I_n) \Rightarrow (A^{-1} = A^T)$. It will be shown on page 147 after a study of determinants that if $AB = I_n$, where A and B are square, then both A and B are invertible. Therefore since $A^T A = I_n$, $A^T = A^{-1}$ by Theorem 13, page 57, of Chapter 3. □

Theorem 13. *An n by n R-matrix A $(n > 1)$ is orthogonal if and only if the columns ⟨rows⟩ of A form an orthonormal set.*

Proof. The reasons for the following statements are left as Exercise 10(a). Let A_i represent the ith column of A.

STATEMENTS: $(A^{-1} = A^T) \Leftrightarrow (A^T A = I_n)$

$$\Leftrightarrow A_i \cdot A_j = \begin{cases} 1 \text{ if } i = j \\ 0 \text{ if } i \neq j \end{cases}$$

$\Leftrightarrow \{A_1, A_2, \ldots, A_n\}$ is an orthonormal set.

A similar proof if A has an orthonormal set of *rows* is left as Exercise 10(b). □

Definition 14. *An n by n C-matrix A is **unitary** if $A^{-1} = A^*$.*

Notice that if A is real, then $A^* = A^T$, and hence an R-matrix is unitary if and only if it is orthogonal.

Example 2. Consider the matrix $A = \begin{bmatrix} \frac{1}{10}\sqrt{10}\,i & -\frac{3}{10}\sqrt{10} \\ \frac{3}{10}\sqrt{10} & -\frac{1}{10}\sqrt{10}\,i \end{bmatrix}$.

$$[A \mid I_2] = \begin{bmatrix} \frac{1}{10}\sqrt{10}\,i & -\frac{3}{10}\sqrt{10} & \mid & 1 & 0 \\ \frac{3}{10}\sqrt{10} & -\frac{1}{10}\sqrt{10}\,i & \mid & 0 & 1 \end{bmatrix}$$

$$\underset{\sim}{\text{row}} \begin{bmatrix} 1 & 0 & \mid & -\frac{1}{10}\sqrt{10}\,i & \frac{3}{10}\sqrt{10} \\ 0 & 1 & \mid & -\frac{3}{10}\sqrt{10} & \frac{1}{10}\sqrt{10}\,i \end{bmatrix} = [I_2 \mid A^{-1}].$$

Therefore $A^{-1} = \begin{bmatrix} -\frac{1}{10}\sqrt{10}\,i & \frac{3}{10}\sqrt{10} \\ -\frac{3}{10}\sqrt{10} & \frac{1}{10}\sqrt{10}\,i \end{bmatrix}$ which is equal to $(\bar{A})^T$; hence, $A^{-1} = A^*$, and A is unitary. Observe, however, that $A^{-1} \neq A^T$.

4.4 ORTHOGONAL AND UNITARY MATRICES

Theorem 14. *An n by n C-matrix A is unitary if and only if $A^*A = I_n$.*

The proof is left as Exercise 11.

Theorem 15. *An n by n C-matrix A is unitary if and only if the columns ⟨rows⟩ of A form an orthonormal set.*

The proof is left as Exercise 12.

Additional properties of orthogonal and unitary matrices are given in the Exercises.

EXERCISES

1. Determine whether the following matrices are orthogonal, by using the definition.

    (a) $\begin{bmatrix} -\frac{1}{10}\sqrt{10} & \frac{3}{10}\sqrt{10} \\ -\frac{3}{10}\sqrt{10} & -\frac{1}{10}\sqrt{10} \end{bmatrix}$; (b) $\begin{bmatrix} 1 & 2 \\ 0 & 3 \end{bmatrix}$; (c) $\begin{bmatrix} \frac{1}{2}\sqrt{2} & 0 & -\frac{1}{2}\sqrt{2} \\ 0 & 1 & 0 \\ \frac{1}{2}\sqrt{2} & 0 & \frac{1}{2}\sqrt{2} \end{bmatrix}$.

2. Determine whether the matrices of Exercise 1 are orthogonal, by using Theorem 12.

3. Determine whether the matrices of Exercise 1 are orthogonal, by using Theorem 13.

4. Determine whether the following matrices are unitary by using the definition.

    (a) $\begin{bmatrix} \frac{1}{5}\sqrt{5} & -\frac{1}{5}\sqrt{5}i \\ \frac{2}{5}\sqrt{5}i & -\frac{2}{5}\sqrt{5} \end{bmatrix}$; (b) $\begin{bmatrix} 1 & i \\ 0 & i \end{bmatrix}$;

    (c) $\begin{bmatrix} 1 & 0 & 0 \\ 0 & 0 & -i \\ 0 & i & 0 \end{bmatrix}$; (d) $\begin{bmatrix} 1 & 0 & 0 \\ 0 & \frac{1}{2}\sqrt{2}i & \frac{1}{2}\sqrt{2}i \\ 0 & -\frac{1}{2}\sqrt{2}i & -\frac{1}{2}\sqrt{2}i \end{bmatrix}$.

5. Determine whether the matrices of Exercise 4 are unitary, by using Theorem 14.

6. Determine whether the matrices of Exercise 4 are unitary, by using Theorem 15.

7. Determine which of the matrices of Exercise 1 are unitary, by using the answer to Exercise 1. Justify your conclusions.

8. Do the columns of $A = \begin{bmatrix} 1 & -1 \\ -1 & -1 \end{bmatrix}$ form an orthogonal set of vectors? Is A an orthogonal matrix?

9. Give Part I of the proof of Theorem 12.
10. (a) Give reasons for the statements of the proof of Theorem 13.
    (b) Prove: An n by n R-matrix A is orthogonal if and only if the rows of A form an orthonormal set.
11. Prove Theorem 14. (*Hint:* See proof of Theorem 12.)
12. Prove Theorem 15. (*Hint:* See proof of Theorem 13.)
13. Prove: The product of two orthogonal matrices of the same order is orthogonal.
14. Prove: The product of two unitary matrices of the same order is unitary.
15. Prove: The transpose of an orthogonal matrix is orthogonal.
16. Prove: The transpose of a unitary matrix is unitary.
17. Prove: the conjugate of a unitary matrix is unitary.
18. Prove: The tranjugate of a unitary matrix is unitary.
19. Prove: The inverse of an orthogonal matrix is orthogonal.
20. Prove: The inverse of a unitary matrix is unitary.
21. (a) Find all real values of x for which $\begin{bmatrix} x & 0 \\ 0 & x \end{bmatrix}$ is an orthogonal matrix.
    (b) Find all real values of x for which $\begin{bmatrix} -x & 0 \\ 0 & x \end{bmatrix}$ is an orthogonal matrix.
22. Is an orthogonal matrix necessarily symmetric?
23. Prove: The rows of an n by n R-matrix A form an orthonormal set if and only if the columns of A form an orthonormal set. (*Hint:* Use Theorem 13.)
24. State and prove a corollary of Theorem 15 that is analogous to the statement in Exercise 23.

4.5 Diagonal and Triangular Matrices

If all the entries both above and below the main diagonal are zero, then a square C-matrix is called a *diagonal* matrix.

Diagonal matrices are very important in the theory of matrices and in many applications.

Definition 15. *Let A be a square C-matrix. If $a_{ij} = 0$ for all $i \neq j$, then A is a diagonal matrix.*

Example 1. The following C-matrices are diagonal.

$$A = \begin{bmatrix} 3i & 0 & 0 \\ 0 & 1 & 0 \\ 0 & 0 & 2 \end{bmatrix}; \quad B = \begin{bmatrix} 3 & 0 & 0 \\ 0 & 3 & 0 \\ 0 & 0 & 3 \end{bmatrix}; \quad 0_3 = \begin{bmatrix} 0 & 0 & 0 \\ 0 & 0 & 0 \\ 0 & 0 & 0 \end{bmatrix}.$$

4.5 DIAGONAL AND TRIANGULAR MATRICES

Definition 16. *A diagonal matrix in which all of the entries on the main diagonal are equal is a **scalar matrix**.*

Both B and 0_3 in Example 1 are scalar matrices.

Definition 17. *If a square C-matrix A can be partitioned in such a way,*

$$A = \begin{bmatrix} A_{11} & A_{12} & \cdots & A_{1n} \\ \vdots & \vdots & & \vdots \\ A_{n1} & A_{n2} & \cdots & A_{nn} \end{bmatrix},$$

*that when $i = j$ all A_{ij} are square, and when $i \neq j$ all A_{ij} are zero matrices, then A is a **diagonal block matrix**.*

Example 2. The following matrices are diagonal block matrices:

$$A = \begin{bmatrix} 3 & 2 & 0 & 0 & 0 \\ 0 & 4 & 0 & 0 & 0 \\ 0 & 0 & 3 & 4 & 1 \\ 0 & 0 & 2 & 1 & 0 \\ 0 & 0 & 2 & 4 & 0 \end{bmatrix}; \quad B = \begin{bmatrix} 3 & 0 & 0 & 0 \\ 0 & 3 & 0 & 0 \\ 0 & 0 & 2 & 1 \\ 0 & 0 & 4 & 0 \end{bmatrix}.$$

Diagonal block matrices can be obtained by the matrix operation defined next.

Definition 18. *If A and B are square C-matrices, then*

$$A \oplus B = \begin{bmatrix} A & 0 \\ 0 & B \end{bmatrix}$$

*is the **direct sum** of A and B.*

Theorem 16. *If A, B, and C are square C-matrices, then*

$$(A \oplus B) \oplus C = A \oplus (B \oplus C);$$

furthermore, if A and B are invertible, then

$$(A \oplus B)^{-1} = A^{-1} \oplus B^{-1}.$$

The proof is left as Exercise 14.

Example 3.

$$\begin{bmatrix} 3 & 5 & | & 0 \\ 1 & 2 & | & 0 \\ \hline 0 & 0 & | & 6 \end{bmatrix}^{-1} = \left(\begin{bmatrix} 3 & 5 \\ 1 & 2 \end{bmatrix} \oplus [6] \right)^{-1} = \begin{bmatrix} 3 & 5 \\ 1 & 2 \end{bmatrix}^{-1} \oplus [6]^{-1} = \begin{bmatrix} 2 & -5 \\ -1 & 3 \end{bmatrix} \oplus \begin{bmatrix} \frac{1}{6} \end{bmatrix}$$

$$= \begin{bmatrix} 2 & -5 & 0 \\ -1 & 3 & 0 \\ 0 & 0 & \frac{1}{6} \end{bmatrix}.$$

See Exercise 18 for other properties of the direct sum.

Theorem 16 allows us to generalize Example 3 and to calculate the inverse of a diagonal block matrix by

$$\begin{bmatrix} A_{11} & & 0 \\ & \ddots & \\ 0 & & A_{nn} \end{bmatrix}^{-1} = \begin{bmatrix} A_{11}^{-1} & & 0 \\ & \ddots & \\ 0 & & A_{nn}^{-1} \end{bmatrix},$$

provided all A_{ii} are invertible.

A more general class of matrices, which includes diagonal matrices as a special case, is that of triangular matrices.

Definition 19. *Let A be a square C-matrix. If $a_{ij} = 0$ for all $i > j$, then A is an **upper triangular matrix**. If $a_{ij} = 0$ for all $i < j$, then A is a **lower triangular matrix**. Any upper or lower triangular matrix is a **triangular matrix**.*

Example 4. The following matrices are upper triangular.

$$A = \begin{bmatrix} 6 & 3 & 4 \\ 0 & 1 & 6 \\ 0 & 0 & 2 \end{bmatrix}; \quad B = \begin{bmatrix} 0 & 2 & 1 \\ 0 & 0 & 3 \\ 0 & 0 & 0 \end{bmatrix}; \quad C = \begin{bmatrix} 0 & 0 & 2 \\ 0 & 0 & 0 \\ 0 & 0 & 0 \end{bmatrix}.$$

The following matrices are lower triangular:

$$D = \begin{bmatrix} 6 & 0 & 0 \\ 3 & 9 & 0 \\ 2 & 2 & 5 \end{bmatrix}; \quad E = \begin{bmatrix} 0 & 0 & 0 \\ 0 & 0 & 0 \\ 1 & 2 & 0 \end{bmatrix}; \quad F = \begin{bmatrix} 0 & 0 & 0 \\ 0 & 0 & 0 \\ 0 & 2 & 0 \end{bmatrix}.$$

Notice that the entries on the main diagonal are immaterial to considerations of triangularity.

All diagonal matrices are both upper and lower triangular.

Theorem 17. *The sum and product of two upper ⟨lower⟩ triangular matrices of the same order are upper ⟨lower⟩ triangular.*

The proof is left as Exercise 15.

4.5 DIAGONAL AND TRIANGULAR MATRICES

Definition 20. *If a square C-matrix A can be partitioned in such a way,*

$$A = \begin{bmatrix} A_{11} & A_{12} & \cdots & A_{1n} \\ \vdots & \vdots & & \vdots \\ A_{n1} & A_{n2} & \cdots & A_{nn} \end{bmatrix},$$

*that all A_{ii} are square, and if, for $i > j$, all A_{ij} are zero, then A is an **upper triangular block matrix**. If all A_{ii} are square, and if for $i < j$ all A_{ij} are zero, then A is a **lower triangular block matrix**. Any upper or lower triangular block matrix is a **triangular block matrix**.*

Example 5. The following matrices are triangular block matrices, as the partitioning indicates.

$$A = \begin{bmatrix} 1 & 2 & 6 & 1 & 1 \\ 3 & 2 & 3 & 2 & 0 \\ 0 & 0 & 2 & 4 & 4 \\ 0 & 0 & 1 & 6 & 2 \\ 0 & 0 & 0 & 0 & 3 \end{bmatrix}$$

is an upper triangular block matrix, and

$$B = \begin{bmatrix} 2 & 1 & 3 & 0 & 0 & 0 \\ 4 & 2 & 1 & 0 & 0 & 0 \\ 6 & 0 & 9 & 0 & 0 & 0 \\ 2 & 1 & 3 & 2 & 0 & 0 \\ 1 & 6 & 9 & 2 & 3 & 4 \\ 3 & 2 & 4 & 4 & 6 & 2 \end{bmatrix}$$

is a lower triangular block matrix. All diagonal block matrices are both upper triangular and lower triangular block matrices.

Theorem 18. *If an invertible upper triangular block C-matrix A is of the form*

$$A = \begin{bmatrix} A_{11} & A_{12} \\ 0 & A_{22} \end{bmatrix},$$

where A_{11} and A_{22} are invertible, then

$$A^{-1} = \begin{bmatrix} A_{11}^{-1} & (-A_{11}^{-1} A_{12} A_{22}^{-1}) \\ 0 & A_{22}^{-1} \end{bmatrix}.$$

Outline of proof. Show that $AA^{-1} = A^{-1}A = I_p$, where A is a p by p C-matrix. ☐

It will be left as an exercise to state a corresponding theorem for lower triangular block matrices.

For certain invertible matrices, computation of the inverse is much less tedious by methods of this section than by the method learned in Chapter 3. (See matrix D of Exercise 16.)

APPLICATIONS

Example 6. Later in this book we will develop some transformations of matrices that will map a certain matrix into a diagonal matrix or into a triangular matrix. The ability to accomplish such transformations, when possible, is very important in applied work. Example 11 of Section 5.3 and the applied examples in Sections 8.5, 9.3, and 10.2 will illustrate the latter statement.

Example 7. One interesting illustration of the use of triangular matrices may be found on pages 268–273 of [17] (Kemeny et al.). The illustration deals with the manufacture of a finished product, called an assembly, and of components, called parts and subassemblies. Powers and inverses of triangular matrices are used.

EXERCISES

1. Determine which of the following matrices are upper triangular.

$$A = \begin{bmatrix} 4 & 6 & 3 \\ 0 & 1 & 0 \\ 0 & 0 & 1 \end{bmatrix}, \quad B = \begin{bmatrix} 0 & 0 & 0 \\ 0 & 0 & 0 \\ 3 & 4 & 0 \end{bmatrix}, \quad C = \begin{bmatrix} 0 & 0 \\ 4 & 0 \\ 6 & 9 \end{bmatrix},$$

$$D = \begin{bmatrix} 4 & 0 & 0 \\ 6 & 0 & 0 \\ 9 & 4 & 2 \end{bmatrix}, \quad E = \begin{bmatrix} 0 & 0 & 0 \\ 0 & 0 & 0 \\ 0 & 0 & 1 \end{bmatrix}, \quad F = \begin{bmatrix} 2 & 0 & 0 \\ 0 & 2 & 0 \\ 0 & 0 & 2 \end{bmatrix}.$$

2. In Exercise 1 determine which of the matrices are lower triangular.
3. In Exercise 1 determine which of the matrices are diagonal, and also determine which are scalar matrices.
4. Determine which of the following matrices are upper triangular block matrices.

$$A = \begin{bmatrix} 1 & 2 & 0 & 0 \\ 3 & 4 & 0 & 0 \\ 0 & 0 & 6 & 4 \\ 0 & 0 & 3 & 2 \end{bmatrix}, \quad B = \begin{bmatrix} 1 & 0 & 0 & 0 \\ 2 & 3 & 0 & 0 \\ 4 & 5 & 6 & 0 \\ 0 & 9 & 4 & 0 \end{bmatrix}, \quad C = \begin{bmatrix} 1 & 2 & 3 & 1 & 6 & 2 \\ 4 & 5 & 6 & 2 & 0 & 0 \\ 9 & 2 & 1 & 3 & 0 & 0 \\ 0 & 0 & 0 & 3 & 0 & 0 \\ 0 & 0 & 0 & 0 & 2 & 0 \\ 0 & 0 & 0 & 0 & 1 & 1 \end{bmatrix},$$

$$D = \begin{bmatrix} 1 & 2 & 3 & 0 & 0 \\ 3 & 4 & 6 & 0 & 0 \\ 0 & 0 & 0 & 6 & 2 \\ 0 & 0 & 0 & 3 & 9 \end{bmatrix}, \quad E = \begin{bmatrix} 0 & 0 & 0 & 0 \\ 0 & 0 & 0 & 0 \\ 0 & 0 & 0 & 0 \\ 0 & 0 & 0 & 1 \end{bmatrix}.$$

4.5 DIAGONAL AND TRIANGULAR MATRICES

5. In Exercise 4 determine which of the matrices are lower triangular block matrices.
6. In Exercise 4 determine which of the matrices are diagonal block matrices.
7. Tell whether each of the following is true or false. If false, give a counter-example.
   (a) The transpose of a diagonal matrix is a diagonal matrix.
   (b) Every diagonal matrix is also a scalar matrix.
   (c) Every scalar matrix is a square matrix.
   (d) Every scalar matrix is Hermitian.
   (e) Every diagonal matrix is a triangular matrix.
   (f) Every scalar matrix is an echelon matrix.
   (g) The product of two diagonal matrices of the same order is a diagonal matrix.
   (h) Every diagonal block matrix is an upper triangular matrix.
   (i) Every upper triangular block matrix is an upper triangular matrix.
8. Let D be a diagonal matrix. Is D invertible, and if so what is D^{-1}?
9. If D is a diagonal matrix, if A is an arbitrary m by n C-matrix, and if DA is defined, what is the relationship between A and DA?
10. If D is a diagonal matrix, if A is an arbitrary m by n C-matrix, and if AD is defined, what is the relationship between A and AD?
11. What is the transpose of an upper triangular matrix?
12. Find the inverse of the direct sum of $A = \begin{bmatrix} 1 & 2 \\ 3 & 5 \end{bmatrix}$ and $B = \begin{bmatrix} 1 & 2 \\ 0 & 1 \end{bmatrix}$.
13. Illustrate Theorem 17 if $A = \begin{bmatrix} 0 & 0 & 0 \\ 3 & 2 & 0 \\ 1 & 1 & 1 \end{bmatrix}$ and $B = \begin{bmatrix} 2 & 0 & 0 \\ 3 & 1 & 0 \\ 6 & 0 & 4 \end{bmatrix}$.
14. Prove Theorem 16.
15. Prove Theorem 17.
16. Determine the inverses of the following matrices, given $\begin{bmatrix} 1 & 1 \\ 3 & 4 \end{bmatrix}^{-1} = \begin{bmatrix} 4 & -1 \\ -3 & 1 \end{bmatrix}$:

$$A = \begin{bmatrix} 1 & 1 & 0 \\ 3 & 4 & 0 \\ \hline 0 & 0 & 4 \end{bmatrix}, \quad B = \begin{bmatrix} 1 & 1 & 2 \\ 3 & 4 & 1 \\ \hline 0 & 0 & 3 \end{bmatrix},$$

$$C = \begin{bmatrix} 1 & 1 & 0 & 0 & 0 & 0 \\ 3 & 4 & 0 & 0 & 0 & 0 \\ \hline 0 & 0 & 1 & 1 & 0 & 0 \\ 0 & 0 & 3 & 4 & 0 & 0 \\ \hline 0 & 0 & 0 & 0 & 6 & 0 \\ \hline 0 & 0 & 0 & 0 & 0 & 9 \end{bmatrix}, \quad D = \begin{bmatrix} 3 & 0 & 0 & 1 & 1 \\ 0 & 2 & 0 & 1 & 1 \\ 0 & 0 & 1 & 1 & 1 \\ \hline 0 & 0 & 0 & 1 & 1 \\ 0 & 0 & 0 & 3 & 4 \end{bmatrix}.$$

17. State and prove a theorem corresponding to Theorem 18 for lower triangular block C-matrices.
18. Prove the following properties of the direct sum for square C-matrices A, B, C, and D:
    (a) If k is a complex number, $k(A \oplus B) = kA \oplus kB$.
    (b) Assuming conformability for addition,
    $$(A \oplus B) + (C \oplus D) = (A + C) \oplus (B + D).$$
    (c) Assuming conformability for multiplication
    $$(A \oplus B)(C \oplus D) = AC \oplus BD.$$
    (d) $(A \oplus B)^T = A^T \oplus B^T$.
19. Derive the p by p matrix A^{-1} in Theorem 18 by letting $A^{-1} = B = \begin{bmatrix} B_{11} & B_{12} \\ \hline B_{21} & B_{22} \end{bmatrix}$ and by using the fact that $AB = I_p$. Also verify that $AA^{-1} = A^{-1}A = I_p$.
20. By inspection, write the inverse of matrix A in Example 1.

NEW VOCABULARY

transpose of a matrix   4.1
symmetric matrix   4.1
skew-symmetric matrix   4.1
conjugate of a matrix   4.2
tranjugate of a matrix   4.2
Hermitian matrix   4.2
skew-Hermitian matrix   4.2
coordinate vector   4.3
components of a coordinate vector   4.3
coordinates of a coordinate vector   4.3
dimension of a coordinate vector   4.3
real coordinate vector   4.3
zero coordinate vector   4.3
nonzero coordinate vector   4.3
standard inner product of coordinate vectors   4.3
dot product of coordinate vectors   4.3

magnitude of a coordinate vector   4.3
length of a coordinate vector   4.3
unit coordinate vector   4.3
normal coordinate vector   4.3
orthogonal coordinate vectors   4.3
orthogonal set   4.3
orthonormal set   4.3
orthogonal matrix   4.4
unitary matrix   4.4
diagonal matrix   4.5
scalar matrix   4.5
diagonal block matrix   4.5
direct sum of matrices   4.5
upper triangular matrix   4.5
lower triangular matrix   4.5
triangular matrix   4.5
upper triangular block matrix   4.5
lower triangular block matrix   4.5
triangular block matrix   4.5

5

DETERMINANTS

5.1 Definition of the Determinant of a Square *C*-Matrix

It has been observed that the definition of a matrix A does *not* assign a numerical value to A. *The first purpose of this chapter is to define a function or mapping over the set of square C-matrices in such a way that to every square matrix of complex numbers there corresponds a single complex number. This function or mapping is called the determinant function.* Its value is denoted by det A or by $|A|$ and is called the determinant of A.

The origin of determinants was closely associated with techniques for solving systems of linear equations. There is evidence that the Chinese, before the time of Christ, made use of bamboo rods to develop methods of solving simultaneous equations. Their methods indicate that they had an idea similar to what we now call the *expansion* of a determinant. The idea of determinants did not begin to take definite form, however, until about 1683 in Japan and about 1693 in Germany. Seki Kowa, the Japanese mathematician, and G. W. Leibniz, the famous German, were independently responsible for this initial formulation. The works of these men were amplified by G. Cramer of Switzerland in 1750. The gist of these early developments for a simple special case was essentially this: A system of linear equations over the complex numbers

$$\begin{cases} ax + by = c, \\ a'x + b'y = c', \end{cases}$$

can be solved for x by the formula

$$x = \frac{cb' - bc'}{ab' - ba'},$$

when the denominator is not 0. Using present-day terminology, we say that the numerator and denominator in this formula are expansions of second-order determinants, written

$$x = \frac{\begin{vmatrix} c & b \\ c' & b' \end{vmatrix}}{\begin{vmatrix} a & b \\ a' & b' \end{vmatrix}}.$$

Thus, $\begin{vmatrix} c & b \\ c' & b' \end{vmatrix}$ means $(cb' - bc')$ and $\begin{vmatrix} a & b \\ a' & b' \end{vmatrix}$ means $(ab' - ba')$. For example, $\begin{vmatrix} 2 & 4 \\ 9 & 3 \end{vmatrix} = 2 \cdot 3 - 4 \cdot 9 = 6 - 36 = -30$. This method of solving for x will be amplified in Section 5.2.

In 1771, the French mathematician Vandermonde recognized determinants as independent functions, and in 1812 Cauchy, another French mathematician, gave the function its present name, "determinant." After Cauchy had begun what is known as the theory of determinants, the subject attracted widespread attention for about a hundred years. By this time the determinant was considered an established tool of the trade. It is interesting to note that Arthur Cayley, the inventor of the matrix, was one of the greatest contributors to determinant theory. He developed the notation presently in use—a square array between two vertical bars. The historical relationship between a matrix and a determinant may be seen in Cayley's own words. In 1894, in reply to a question about what led him to matrices, he wrote, "I certainly did not get the notion of a matrix in any way through quaternions: it was either directly from that of a determinant or as a convenient mode of expression of the equations

$$\begin{cases} x' = ax + by, \\ y' = cx + dy. \end{cases}"\ [1]$$

Before proceeding to the definition of the determinant of an n by n C-matrix, consider the column numbers $1, 2, \ldots, n$ of such a matrix. Suppose that these consecutive integers are not arranged in natural order, and further suppose that we wish to place them in natural order by making interchanges

---

[1] C. G. Knott, *Life and Scientific Work of Peter Guthrie Tait*, New York, G. P. Putnam Sons, 1911, page 164.

5.1 DETERMINANT OF A SQUARE C-MATRIX

of one integer with another; it can be proved[2] that although the number of interchanges used to restore the natural order is not unique, the oddness or evenness of the number of interchanges is uniquely determined by the sequence to be rearranged.

Example 1. Determine the oddness or evenness of any number of interchanges necessary to reduce the sequence 2, 4, 3, 1 to natural order, 1, 2, 3, 4. One interchange (of 2 and 1) changes 2, 4, 3, 1 to 1, 4, 3, 2. A second interchange (of 4 and 2) changes 1, 4, 3, 2 to 1, 2, 3, 4. If we proceeded in a different manner we might require more interchanges, but in any event we would need an even number of interchanges.

Definition 1. *Let A be the square C-matrix $[a_{ij}]_{(n,n)}$. The **determinant** of A is the sum of all terms of the form $(-1)^t a_{1j_1} a_{2j_2} \cdots a_{nj_n}$, where the column subscripts assume all possible arrangements in which each column is represented exactly once in each term of the sum, and the exponent, t, is any number of interchanges necessary to bring the column subscripts into natural order (that is, $1, 2, 3, \ldots, n$). The determinant of A is denoted by $\det A$ or $|A|$.*

Example 2. By definition,

$$\begin{vmatrix} a_{11} & a_{12} \\ a_{21} & a_{22} \end{vmatrix} = \det \begin{bmatrix} a_{11} & a_{12} \\ a_{21} & a_{22} \end{bmatrix} = (-1)^0 a_{11} a_{22} + (-1)^1 a_{12} a_{21} = a_{11} a_{22} - a_{12} a_{21}.$$

The right-hand side is called the *expansion* of the determinant of a 2 by 2 matrix.

Note: The row subscripts maintain natural order, and the column subscripts assume all possible arrangements.

Example 3. By definition,

$$\det \begin{bmatrix} a_{11} & a_{12} & a_{13} \\ a_{21} & a_{22} & a_{23} \\ a_{31} & a_{32} & a_{33} \end{bmatrix} = (-1)^0 a_{11} a_{22} a_{33} + (-1)^1 a_{11} a_{23} a_{32}$$
$$+ (-1)^1 a_{12} a_{21} a_{33} + (-1)^2 a_{12} a_{23} a_{31}$$
$$+ (-1)^2 a_{13} a_{21} a_{32} + (-1)^1 a_{13} a_{22} a_{31}.$$

Example 4. By definition,

$$\det \begin{bmatrix} 2 & 1 & 3 \\ 4 & 8 & 6 \\ 0 & 7 & 5 \end{bmatrix} = (-1)^0 2 \cdot 8 \cdot 5 + (-1)^1 2 \cdot 6 \cdot 7 + (-1)^1 1 \cdot 4 \cdot 5 + (-1)^2 1 \cdot 6 \cdot 0$$
$$+ (-1)^2 3 \cdot 4 \cdot 7 + (-1)^1 3 \cdot 8 \cdot 0$$
$$= 80 - 84 - 20 + 0 + 84 - 0 = 60.$$

[2] For a proof see Lancaster, Peter, *Theory of Matrices*, New York, N.Y., Academic Press, 1969, pages 30–31.

It is useful at times to use the notation

$$\sum_{(j)} (-1)^t a_{1j_1} a_{2j_2} \cdots a_{nj_n}$$

to indicate the summation of the $n!$ terms specified in the definition of the determinant of a matrix. As an illustration, the expansion of $\det[a_{ij}]_{(3,3)}$ of Example 3 can be expressed as

$$\det A = \sum_{(j)} (-1)^t a_{1j_1} a_{2j_2} a_{3j_3}.$$

The $j_1, \ldots, j_n$ represent the arrangement of column subscripts in an arbitrary term of the expansion, and t represents, as in the definition, the number of interchanges needed to bring the column subscripts $j_1, \ldots, j_n$ of the arbitrary term into natural order. As illustrated in Example 3 the value of t varies from term to term, and, as we stated earlier, t is not unique for a specific term; we restate for emphasis, however, that the oddness or evenness of t for a specific term is unique, and hence the sign of the arbitrary term is unique.

It can be verified[3] that $\det A$ is also equal to

$$\sum_{(i)} (-1)^t a_{i_1 1} a_{i_2 2} \cdots a_{i_n n},$$

where, now, the row subscripts take on all specified arrangements and the column subscripts are in natural order. When square A is of order n, we say that $\det A$ is of **order n**.

APPLICATIONS

Example 5. One very important elementary application of determinants is simplifying notation. Consider the calculation of the area of the triangle shown in Figure 5.1.1. To do this, one first calculates the areas of the three trapezoids shown in Figure 5.1.1. From these results one can verify that the area of the triangle is

$$\text{Area} = \tfrac{1}{2}(y_1 + y_3)(x_3 - x_1) + \tfrac{1}{2}(y_3 + y_2)(x_2 - x_3) - \tfrac{1}{2}(y_1 + y_2)(x_2 - x_1).$$

It can be shown (Exercise 9) that the expansion of this formula is the same as the expansion of the determinant

$$\frac{1}{2} \begin{vmatrix} x_1 & y_1 & 1 \\ x_2 & y_2 & 1 \\ x_3 & y_3 & 1 \end{vmatrix}.$$

This is just one of many examples of the use of determinant notation to express complicated formulas.

---

[3] This will be verified in the first part of the proof of Theorem 3 on page 129.

5.1 DETERMINANT OF A SQUARE C-MATRIX

Example 6. A unit cell of a crystal, as defined by a certain set of basis vectors

$$\alpha = (a_1, a_2, a_3),$$
$$\beta = (b_1, b_2, b_3),$$
$$\gamma = (c_1, c_2, c_3),$$

forms a parallelepiped whose volume is the absolute value of

$$\det \begin{bmatrix} a_1 & a_2 & a_3 \\ b_1 & b_2 & b_3 \\ c_1 & c_2 & c_3 \end{bmatrix}.$$

Verification of this fact may be found in many books on elementary vector analysis. Further use of determinants in connection with finding volumes of unit cells of crystals when bases are changed will be discussed in Example 6 of Section 8.4.

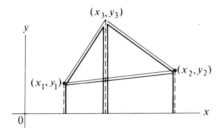

Figure 5.1.1

EXERCISES

1. Evaluate by the definition:

   (a) $\begin{vmatrix} 2 & 3 \\ -6 & 1 \end{vmatrix}$;
   (b) $\det \begin{bmatrix} 0 & -2 \\ -1 & 4 \end{bmatrix}$;
   (c) $\begin{vmatrix} 1 & -i \\ i & 2i \end{vmatrix}$;

   (d) $\det \begin{bmatrix} 1 & t \\ 4 & t \end{bmatrix}$;
   (e) $\begin{vmatrix} 3 & 2 & 0 \\ 0 & 1 & 4 \\ 3 & 3 & 4 \end{vmatrix}$;
   (f) $\det \begin{bmatrix} 6 & 0 & 1 \\ 0 & 0 & 3 \\ 2 & 1 & 0 \end{bmatrix}$.

2. Write a second-order matrix whose determinant is equal to $cb - ad$.

3. How many terms are there in the expansion of a fourth-order determinant? (*Hint:* A knowledge of permutations will be helpful.) How many terms are there in the expansion of an nth-order determinant?

4. Write any two of the terms of the expansion of a fourth-order determinant.

5. Is t odd or even in the term $(-1)^t a_{14} a_{22} a_{31} a_{45} a_{53}$ of the expansion of a fifth-order determinant? Is t unique?

6. (a) Evaluate by definition: $\det[5]$.
   (b) Evaluate by definition: $\det[a]$.

7. If $A = [a_{ij}]_{(3,3)}$, find det A using det $A = \sum_{(i)} (-1)^t a_{i_1 1} a_{i_2 2} a_{i_3 3}$. Verify that your result agrees with that for Example 3 on page 113.

8. The six terms of the expansion of the determinant of a 3 by 3 matrix can be obtained by a method illustrated in Figure 5.1.2. Use this method to evaluate the determinant of the matrix given in Example 4. The reader should be cautioned, however, that a similar diagram for a determinant of higher order $(n \geq 4)$ does not give the $n!$ terms of that expansion.

9. In Example 5 of this section, verify that the formula for the area is equal to the expansion of the given determinant.

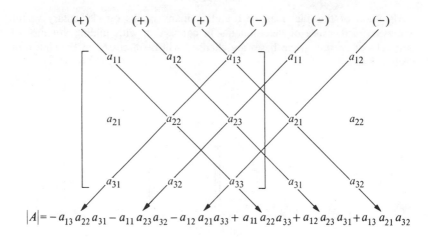

Figure 5.1.2

5.2 Cofactor Expansion

In order to establish another general procedure for evaluating a determinant, we first make the following definitions.

Definition 2. *The **minor** of an entry a_{ij} of a square C-matrix A $(n \geq 2)$ is the determinant of the submatrix of A obtained by deleting the ith row and jth column.*[4]

Example 1. The minor of a_{32} in $[a_{ij}]_{(3,3)}$ is

$$\det \begin{bmatrix} a_{11} & a_{12} & a_{13} \\ a_{21} & a_{22} & a_{23} \\ a_{31} & a_{32} & a_{33} \end{bmatrix} = \det \begin{bmatrix} a_{11} & a_{13} \\ a_{21} & a_{23} \end{bmatrix} = a_{11} a_{23} - a_{13} a_{21}.$$

---

[4] In some books a minor is defined to be the submatrix rather than the determinant of the submatrix.

5.2 COFACTOR EXPANSION

Definition 3. The **cofactor** of an entry a_{ij} of a square C-matrix A ($n \geq 2$) is the product of $(-1)^{i+j}$ and the minor of a_{ij}. This cofactor is denoted by A_{ij}.

Example 2. For a third-order C-matrix A:

$$A_{31} = (-1)^{3+1} \det \begin{bmatrix} a_{12} & a_{13} \\ a_{22} & a_{23} \end{bmatrix} = a_{12}a_{23} - a_{13}a_{22},$$

$$A_{32} = (-1)^{3+2} \det \begin{bmatrix} a_{11} & a_{13} \\ a_{21} & a_{23} \end{bmatrix} = (-1)(a_{11}a_{23} - a_{13}a_{21}),$$

$$A_{33} = (-1)^{3+3} \det \begin{bmatrix} a_{11} & a_{12} \\ a_{21} & a_{22} \end{bmatrix} = a_{11}a_{22} - a_{12}a_{21}.$$

Example 3. In Example 3 of Section 5.1, suppose that we rearrange the products in such a way that the entries of the third row appear first.

$$|A| = a_{33}(a_{11}a_{22}) - a_{32}(a_{11}a_{23}) - a_{33}(a_{12}a_{21})$$
$$+ a_{31}(a_{12}a_{23}) + a_{32}(a_{13}a_{21}) - a_{31}(a_{13}a_{22}).$$

Then, factoring out these entries of the third row, we obtain

$$|A| = a_{31}(a_{12}a_{23} - a_{13}a_{22}) + a_{32}(a_{13}a_{21} - a_{11}a_{23}) + a_{33}(a_{11}a_{22} - a_{12}a_{21}).$$

By the preceding example this can be expressed as

$$|A| = a_{31}A_{31} + a_{32}A_{32} + a_{33}A_{33}.$$

The final expression in Example 3 is called the **cofactor expansion** (or simply the *expansion*) about the third row. A generalization of Example 3 leads us to the following theorem.

Theorem 1. If $A = [a_{ij}]_{(n, n)}$ is a C-matrix ($n \geq 2$), then

$$\det A = a_{i1}A_{i1} + a_{i2}A_{i2} + \cdots + a_{in}A_{in},$$

and

$$\det A = a_{1j}A_{1j} + a_{2j}A_{2j} + \cdots + a_{nj}A_{nj}.$$

If we write the cofactor expansion about the other two rows of $A = [a_{ij}]_{(3, 3)}$ as we did in Example 3, then we obtain a justification of the first part of the theorem when $n = 3$. For the general case, however, matters become considerably more involved. The student is referred to Section 5.4 for a complete proof of Theorem 1. This theorem is extremely useful in the calculation of higher-order determinants, as the following examples illustrate.

Example 4. To evaluate $|A| = \det A = \det \begin{bmatrix} 1 & 6 & 2 \\ 3 & 2 & 0 \\ 4 & 6 & 4 \end{bmatrix} = \begin{vmatrix} 1 & 6 & 2 \\ 3 & 2 & 0 \\ 4 & 6 & 4 \end{vmatrix}$, expand $|A|$ about the second row. We get

$$|A| = (3)(-1)^{2+1}\begin{vmatrix} 6 & 2 \\ 6 & 4 \end{vmatrix} + 2(-1)^{2+2}\begin{vmatrix} 1 & 2 \\ 4 & 4 \end{vmatrix} + 0$$
$$= (3)(-1)(12) + (2)(+1)(-4) = -36 - 8 = -44.$$

Example 5. Evaluate
$$|A| = \begin{vmatrix} 3 & 4 & 6 & 1 \\ 0 & 1 & 0 & 3 \\ 0 & 1 & 0 & 4 \\ 1 & -2 & 1 & 3 \end{vmatrix}.$$

If we expand $|A|$ about the third column, we obtain

$$6(-1)^{1+3}\begin{vmatrix} 0 & 1 & 3 \\ 0 & 1 & 4 \\ 1 & -2 & 3 \end{vmatrix} + 0 + 0 + (1)(-1)^{4+3}\begin{vmatrix} 3 & 4 & 1 \\ 0 & 1 & 3 \\ 0 & 1 & 4 \end{vmatrix}$$

$$= 6\begin{vmatrix} 0 & 1 & 3 \\ 0 & 1 & 4 \\ 1 & -2 & 3 \end{vmatrix} - \begin{vmatrix} 3 & 4 & 1 \\ 0 & 1 & 3 \\ 0 & 1 & 4 \end{vmatrix}.$$

Both of these determinants can be expanded about any row or column; we choose the first columns in both because of the zero entries.

$$|A| = 6\left\{ 0 + 0 + (1)(-1)^{3+1}\begin{vmatrix} 1 & 3 \\ 1 & 4 \end{vmatrix}\right\} - \left\{(3)(-1)^{1+1}\begin{vmatrix} 1 & 3 \\ 1 & 4 \end{vmatrix} + 0 + 0\right\}$$
$$= 6\{1\} - \{3\} = 3.$$

APPLICATIONS

Determinants are useful in the study of systems of linear equations. One way, but probably *not* the best way, to solve n linear equations in n unknowns is by means of what is called Cramer's Rule (after the Swiss mathematician G. Cramer, 1704–1752). Consider the following system over the complex number system.

$$\begin{cases} a_{11}x_1 + a_{12}x_2 + \cdots + a_{1n}x_n = b_1, \\ \cdots\cdots\cdots\cdots\cdots\cdots\cdots\cdots\cdots\cdots\cdots\cdots \\ a_{n1}x_1 + a_{n2}x_2 + \cdots + a_{nn}x_n = b_n, \end{cases}$$

or
$$AX = B.$$

5.2 COFACTOR EXPANSION

Matrix A is an n by n C-matrix. Let $({}^jA)$ denote the matrix obtained from A by replacing the jth column of A by the vector B.

Cramer's Rule: If $\det A \neq 0$, then the system $AX = B$ of n linear equations in n unknowns over the complex numbers has exactly one solution; this solution is

$$x_j = \frac{\det({}^jA)}{\det A}, \quad j = 1, 2, \ldots, n.$$

The proof of this assertion is left as Exercise 18 in Section 5.7 after adjoint matrices have been studied.

Example 6. Solve by Cramer's rule

$$\begin{cases} 2x_1 + x_2 + x_3 = 0, \\ x_1 - x_2 + 5x_3 = 0, \\ x_2 - x_3 = 4. \end{cases}$$

By cofactor expansion:

$$x_1 = \frac{\begin{vmatrix} 0 & 1 & 1 \\ 0 & -1 & 5 \\ 4 & 1 & -1 \end{vmatrix}}{\begin{vmatrix} 2 & 1 & 1 \\ 1 & -1 & 5 \\ 0 & 1 & -1 \end{vmatrix}} = \frac{4\begin{vmatrix} 1 & 1 \\ -1 & 5 \end{vmatrix}}{-6} = \frac{24}{-6} = -4.$$

$$x_2 = \frac{\begin{vmatrix} 2 & 0 & 1 \\ 1 & 0 & 5 \\ 0 & 4 & -1 \end{vmatrix}}{\det A} = \frac{-4\begin{vmatrix} 2 & 1 \\ 1 & 5 \end{vmatrix}}{-6} = \frac{-36}{-6} = 6.$$

$$x_3 = \frac{\begin{vmatrix} 2 & 1 & 0 \\ 1 & -1 & 0 \\ 0 & 1 & 4 \end{vmatrix}}{\det A} = \frac{4\begin{vmatrix} 2 & 1 \\ 1 & -1 \end{vmatrix}}{-6} = \frac{-12}{-6} = 2.$$

Upon substitution, these values are seen to satisfy the given equations.

EXERCISES

1. Given $A = \begin{bmatrix} 2 & 3 & 0 \\ 3 & 2 & 2 \\ -1 & 4 & 0 \end{bmatrix}$.

(a) Expand $|A|$ about the first column.
(b) Expand $|A|$ about the third row.

(c) Expand $|A|$ about the third column.
(d) What is the cofactor of the entry in the third row and second column?
(e) What is the minor of the entry in the first row and second column?

2. Evaluate $|A|$ by row or column expansion. (*Hint:* Make good use of the zeros.)

$$A = \begin{bmatrix} 3 & 0 & 2 & 0 \\ 3 & 0 & 0 & 2 \\ 0 & 2 & 3 & 0 \\ 2 & 0 & 1 & 3 \end{bmatrix}.$$

3. The value of the minor of the entry in the 13th row and 11th column of a 22nd-order C-matrix is found to be 4. What is the cofactor of this entry? Why?

4. The cofactor A_{ij} of a_{ij} was defined as the product of $(-1)^{i+j}$ and the minor of a_{ij}. Write a matrix $B = [b_{ij}]_{(3, 3)}$ of cofactors where $b_{ij} = A_{ij}$ if

$$A = \begin{bmatrix} a_{11} & a_{12} & a_{13} \\ a_{21} & a_{22} & a_{23} \\ a_{31} & a_{32} & a_{33} \end{bmatrix}.$$

5. (a) Evaluate $\begin{vmatrix} 1 & -1 & 2 \\ 4 & 0 & 5 \\ -3 & -3 & 3 \end{vmatrix}$; (b) Evaluate $\begin{vmatrix} 0 & 2 & 5 & 1 \\ 1 & -1 & 0 & 1 \\ 2 & 0 & 2 & -1 \\ 3 & 2 & 1 & 0 \end{vmatrix}.$

6. (a) Verify: $\begin{vmatrix} 0 & -1 & 1 & 1 \\ 0 & 2 & -1 & -1 \\ 2 & -3 & 2 & 4 \\ 2 & -2 & 1 & 3 \end{vmatrix} = 0;$

(b) Verify: $\begin{vmatrix} 1 & 0 & 2 & -1 \\ 3 & -2 & 6 & 4 \\ 5 & 4 & 3 & 0 \\ 2 & 2 & -5 & 6 \end{vmatrix} = -132.$

7. In each of the following systems, find y by Cramer's rule, and then find the other unknowns by substitution.

(a) $\begin{cases} x + y = 5, \\ 2x - y = 7. \end{cases}$ (b) $\begin{cases} x + y + z = 0, \\ - y + 2z = 4, \\ 2x + 2y = 1. \end{cases}$

8. In each of the following systems, find all of the unknowns by Cramer's rule.

(a) $\begin{cases} 2x - 2y = 5, \\ x - 4y = 7. \end{cases}$ (b) $\begin{cases} x + y + z = 2, \\ 2x + 3y + 4z = 3, \\ x - 2y - z = 1. \end{cases}$

9. Follow the pattern of Example 3, page 117, to show that if $A = [a_{ij}]_{(3,3)}$, then $\det A = a_{12} A_{12} + a_{22} A_{22} + a_{32} A_{32}$.

5.3 Properties of Determinants

For future reference some of the properties of determinants will be stated now. Many of these properties will prove to be very useful in later portions of this book. All matrices mentioned in this section must be considered square. Also, the word "*line*" is used to signify either row or column. Proofs of the following theorems are given in the next section.

Theorem 2. *If a matrix B is formed from an n by n C-matrix A by the interchange of two parallel lines (rows or columns), then $|A| = -|B|$.*

Example 1. Let $A = \begin{bmatrix} 2 & 1 & 3 \\ 1 & 1 & 0 \\ 4 & 1 & 3 \end{bmatrix}$, and let $B = \begin{bmatrix} 3 & 1 & 2 \\ 0 & 1 & 1 \\ 3 & 1 & 4 \end{bmatrix}$ where the first and third columns have been interchanged. By Theorem 2, $|A| = -|B|$. On expanding $|A|$ and $|B|$, we obtain $|A| = -6$ and $|B| = 6$.

Corollary. *Let E be an elementary matrix such that premultiplication of A by E effects the interchange of any two parallel rows of A, then $|EA| = -|A|$ and $|E| = -1$.*

Theorem 3. *The determinants of an n by n C-matrix and its transpose are equal; that is, $|A| = |A^T|$.*

Example 2. Let $A = \begin{bmatrix} 1 & 2 & 3 \\ 0 & 2 & 1 \\ 0 & 2 & 4 \end{bmatrix}$, and hence $A^T = \begin{bmatrix} 1 & 0 & 0 \\ 2 & 2 & 2 \\ 3 & 1 & 4 \end{bmatrix}$.

By Theorem 3, $|A| = |A^T|$.

Theorem 4. *If all of the entries of any line of an n by n C-matrix A are zero, then $|A| = 0$.*

Example 3. Because all of the entries in the third row are zero, we know that
$|A| = \begin{vmatrix} 0 & 1 & 4 \\ 2 & 1 & 6 \\ 0 & 0 & 0 \end{vmatrix} = 0$, by Theorem 4.

Theorem 5. *If A is an n by n C-matrix, and if $c_1, c_2, \ldots, c_n$ are complex numbers, then the expressions*

$$c_1 A_{i1} + c_2 A_{i2} + \cdots + c_n A_{in}$$

and

$$c_1 A_{1j} + c_2 A_{2j} + \cdots + c_n A_{nj}$$

are equal to the determinants of matrices which are the same as A except that the entries of the ith row and the jth column, respectively, have been replaced by $c_1, c_2, \ldots, c_n$.

Example 4. Expanding the following determinant about the first column, we obtain

$$\det A = \begin{vmatrix} 2 & 6 & 1 \\ 3 & 3 & 2 \\ 4 & 4 & 1 \end{vmatrix} = 2 \begin{vmatrix} 3 & 2 \\ 4 & 1 \end{vmatrix} - 3 \begin{vmatrix} 6 & 1 \\ 4 & 1 \end{vmatrix} + 4 \begin{vmatrix} 6 & 1 \\ 3 & 2 \end{vmatrix}.$$

However, if the entries of the first column are replaced by the complex numbers c_1, c_2, c_3, we have

$$\begin{vmatrix} c_1 & 6 & 1 \\ c_2 & 3 & 2 \\ c_3 & 4 & 1 \end{vmatrix} = c_1 \begin{vmatrix} 3 & 2 \\ 4 & 1 \end{vmatrix} - c_2 \begin{vmatrix} 6 & 1 \\ 4 & 1 \end{vmatrix} + c_3 \begin{vmatrix} 6 & 1 \\ 3 & 2 \end{vmatrix} = c_1 A_{11} + c_2 A_{21} + c_3 A_{31}.$$

Theorem 6. *The determinant of an n by n C-matrix with two identical parallel lines is zero.*

Example 5. Because the first two columns are identical, Theorem 6 assures us that $\begin{vmatrix} 2 & 2 & 1 \\ 3 & 3 & 1 \\ 4 & 4 & 0 \end{vmatrix} = 0.$

Theorem 7. *The sum of the products of the entries of one line of an n by n C-matrix A by the cofactors of the corresponding entries of a different parallel line of A is zero; that is,*

row expansion $\quad a_{i1} A_{k1} + a_{i2} A_{k2} + \cdots + a_{in} A_{kn} = 0, \quad \text{if } i \neq k,$

column expansion $\quad a_{1j} A_{1k} + a_{2j} A_{2k} + \cdots + a_{nj} A_{nk} = 0, \quad \text{if } j \neq k.$

Example 6. If $|A|$ is expanded about the first column,

$$|A| = \begin{vmatrix} 2 & 4 & 1 \\ 3 & 4 & 6 \\ 4 & 0 & 1 \end{vmatrix} = 2 \begin{vmatrix} 4 & 6 \\ 0 & 1 \end{vmatrix} - 3 \begin{vmatrix} 4 & 1 \\ 0 & 1 \end{vmatrix} + 4 \begin{vmatrix} 4 & 1 \\ 4 & 6 \end{vmatrix}.$$

However, if we replace the elements of the first column in the expansion by the elements of any other column, say, the second, the value is zero; that is,

$$4 \begin{vmatrix} 4 & 6 \\ 0 & 1 \end{vmatrix} - 4 \begin{vmatrix} 4 & 1 \\ 0 & 1 \end{vmatrix} + 0 \begin{vmatrix} 4 & 1 \\ 4 & 6 \end{vmatrix} = 0.$$

minus and plus signs are unchanged

5.3 PROPERTIES OF DETERMINANTS

Theorem 8. *If A is an n by n C-matrix, then the value of the determinant, $|A|$, is multiplied by a complex number c when every entry of one line of A is multiplied by c.*

Notice that this is different from the multiplication of a matrix by a complex number.

Example 7.

(a) $c|B| = c\begin{vmatrix} b_{11} & b_{12} & b_{13} \\ b_{21} & b_{22} & b_{23} \\ b_{31} & b_{32} & b_{33} \end{vmatrix} = \begin{vmatrix} cb_{11} & cb_{12} & cb_{13} \\ b_{21} & b_{22} & b_{23} \\ b_{31} & b_{32} & b_{33} \end{vmatrix} = \begin{vmatrix} b_{11} & cb_{12} & b_{13} \\ b_{21} & cb_{22} & b_{23} \\ b_{31} & cb_{32} & b_{33} \end{vmatrix}$.

(b) If $A = \begin{bmatrix} 2 & 1 \\ 2 & 3 \end{bmatrix}$, then $2|A| = \begin{vmatrix} 2 & 1 \\ 4 & 6 \end{vmatrix}$.

However, $|2A| = \begin{vmatrix} 4 & 2 \\ 4 & 6 \end{vmatrix} = 2^2|A| = 4|A|$.

Corollary. *Let E be an elementary matrix such that premultiplication of A by E effects the multiplication of a row of A by a nonzero complex number c, then $|EA| = c|A|$ and $|E| = c$.*

Theorem 9. *If a matrix B is obtained from an n by n C-matrix A by adding to each entry of a line of A a constant complex multiple of the corresponding entry of a parallel line, then $|B| = |A|$.*

Example 8. Let

$$A = \begin{bmatrix} 1 & 0 & 1 \\ 0 & 4 & 6 \\ -2 & 2 & 1 \end{bmatrix}.$$

If we multiply the first row of A by 2 and add to the third row (abbreviated $2R_1 + R_3$), we obtain a matrix whose determinant is equal to det A. That is,

$$\begin{vmatrix} 1 & 0 & 1 \\ 0 & 4 & 6 \\ -2 & 2 & 1 \end{vmatrix}_{2R_1+R_3} = \begin{vmatrix} 1 & 0 & 1 \\ 0 & 4 & 6 \\ 0 & 2 & 3 \end{vmatrix}.$$

Notice that the only row that changed was the third row, the recipient of the operation. The reader should also observe that an operation such as $2R_1 + 3R_2$ is not considered in Theorem 9.

Theorem 9 will prove to be of great help in evaluating higher-order determinants. For instance, in Example 8, above, if we expand the last determinant about the first column we see that the original third-order determinant is reduced to a constant times a second-order determinant.

Example 9. Evaluate
$$\begin{vmatrix} 198 & 0 & 99 & 99 \\ 1 & 1 & -2 & 0 \\ 1 & 2 & 1 & 2 \\ 1 & -3 & 6 & 1 \end{vmatrix}.$$

By Theorems 8, 9, and 1 we have

$$99\begin{vmatrix} 2 & 0 & 1 & 1 \\ 1 & 1 & -2 & 0 \\ 1 & 2 & 1 & 2 \\ 1 & -3 & 6 & 1 \end{vmatrix} \underset{-2R_2+R_3}{=} 99\begin{vmatrix} 2 & 0 & 1 & 1 \\ 1 & 1 & -2 & 0 \\ -1 & 0 & 5 & 2 \\ 1 & -3 & 6 & 1 \end{vmatrix}$$

$$\underset{3R_2+R_4}{=} 99\begin{vmatrix} 2 & 0 & 1 & 1 \\ 1 & 1 & -2 & 0 \\ -1 & 0 & 5 & 2 \\ 4 & 0 & 0 & 1 \end{vmatrix} \underset{\text{(Expand on } C_2\text{)}}{=} (99)(-1)^{2+2}(1)\begin{vmatrix} 2 & 1 & 1 \\ -1 & 5 & 2 \\ 4 & 0 & 1 \end{vmatrix}$$

$$\underset{-4C_3+C_1}{=} 99\begin{vmatrix} -2 & 1 & 1 \\ -9 & 5 & 2 \\ 0 & 0 & 1 \end{vmatrix} \underset{\text{(Expand on } R_3\text{)}}{=} 99(-1)^{3+3}(1)\begin{vmatrix} -2 & 1 \\ -9 & 5 \end{vmatrix} = -99.$$

Corollary. *Let E be an elementary matrix such that premultiplication of A by E effects the addition of a complex multiple of one row of A to another row of A, then $|EA| = |A|$ and $|E| = 1$.*

Theorem 10. *The determinant of the product of an elementary matrix E and a square C-matrix A of the same order is equal to the product of the determinants of the two matrices; that is $|EA| = |E||A|$ and $|AE| = |A||E|$.*

Example 10. Consider the elementary matrix $E = \begin{bmatrix} 1 & 2 \\ 0 & 1 \end{bmatrix}$ whose premultiplication of $A = \begin{bmatrix} 3 & 0 \\ 2 & 1 \end{bmatrix}$ effects the addition of twice the second row of A to the first row of A. By Theorem 10, $|EA| = |E| \, |A| = 1 \cdot 3 = 3$. This can be verified by

$$|EA| = \det\left(\begin{bmatrix} 1 & 2 \\ 0 & 1 \end{bmatrix}\begin{bmatrix} 3 & 0 \\ 2 & 1 \end{bmatrix}\right) = \det\begin{bmatrix} 7 & 2 \\ 2 & 1 \end{bmatrix} = 3.$$

APPLICATIONS

Example 11. One efficient way of calculating with a computer the determinant of a matrix is to use Theorem 9 repeatedly to triangularize the given matrix; then the determinant of the resulting triangular matrix is equal to the product of the entries on the main diagonal, as the following numerical problem illustrates.

5.3 PROPERTIES OF DETERMINANTS

$$\det \begin{bmatrix} 1 & 6 & 9 & 3 \\ 0 & 2 & 2 & 3 \\ 0 & 4 & 5 & 7 \\ 0 & 0 & 6 & 9 \end{bmatrix} \underset{-2R_2+R_3}{=} \det \begin{bmatrix} 1 & 6 & 9 & 3 \\ 0 & 2 & 2 & 3 \\ 0 & 0 & 1 & 1 \\ 0 & 0 & 6 & 9 \end{bmatrix} \underset{-6R_3+R_4}{=} \det \begin{bmatrix} 1 & 6 & 9 & 3 \\ 0 & 2 & 2 & 3 \\ 0 & 0 & 1 & 1 \\ 0 & 0 & 0 & 3 \end{bmatrix}.$$

The last matrix is triangular, and, by repeated use of the cofactor expansion about the first column of successive cofactors, it is easily seen that the value of the determinant is the product of the entries on the main diagonal.

$$\det \begin{bmatrix} 1 & 6 & 9 & 3 \\ 0 & 2 & 2 & 3 \\ 0 & 0 & 1 & 1 \\ 0 & 0 & 0 & 3 \end{bmatrix} = (1) \det \begin{bmatrix} 2 & 2 & 3 \\ 0 & 1 & 1 \\ 0 & 0 & 3 \end{bmatrix} = (1)(2) \det \begin{bmatrix} 1 & 1 \\ 0 & 3 \end{bmatrix} = (1)(2)(1)(3) = 6.$$

EXERCISES

1. Tell whether the following statements are true or false. Give your reasons. Do *not* expand the given determinants.

   (a) $\begin{vmatrix} 1 & 2 & 3 & 4 \\ 5 & 6 & 7 & 8 \\ 8 & 7 & 6 & 5 \\ 4 & 3 & 2 & 1 \end{vmatrix} = \begin{vmatrix} 2 & 1 & 4 & 3 \\ 6 & 5 & 8 & 7 \\ 7 & 8 & 5 & 6 \\ 3 & 4 & 1 & 2 \end{vmatrix}$;

   (b) $\begin{vmatrix} 1 & 3 & -4 \\ 2 & 8 & 3 \\ 0 & -2 & -5 \end{vmatrix} = \begin{vmatrix} 1 & 2 & 0 \\ 3 & 8 & -2 \\ -4 & 3 & -5 \end{vmatrix}$;

   (c) $\begin{vmatrix} 2x & 3x & 4x \\ 5x & 6x & 7x \\ 8x & 9x & 9x \end{vmatrix} = x \begin{vmatrix} 2 & 3 & 4 \\ 5 & 6 & 7 \\ 8 & 9 & 9 \end{vmatrix}$, if $x \ne 0$ and $x \ne \pm 1$.

2. Without expanding, by making use of two theorems, evaluate

   $$\begin{vmatrix} 2 & 4 & 6 & 4 \\ 0 & 4 & 6 & 9 \\ 2 & 1 & 4 & 0 \\ 1 & 2 & 3 & 2 \end{vmatrix}.$$

3. Given: $\det A = 8$, and B is a matrix that is the same as A except that the first and fourth rows have been interchanged. What is the value of $\det B$? Justify your answer.

4. If the order of A is greater than 1, and if $|A| \ne 0$, does $2 \det A = \det(2A)$? Why?

5. If $\det[a_{ij}]_{(3,3)} = 4$, find $\det(3[a_{ij}]_{(3,3)})$. Give reasons for your answer.

6. Show that $\begin{vmatrix} x+y & -z(x+y) \\ z+x & y(z+x) \end{vmatrix} = (x+y)(z+x)(y+z)$, by using Theorem 8.

7. Without expanding, show that

$$\begin{vmatrix} x^2 - y^2 & x+y & x \\ x - y & 1 & 1 \\ x - y & 1 & y \end{vmatrix} = 0.$$

Give reasons for your steps.

8. Change the form but not the value of $|A| = \begin{vmatrix} -1 & 1 & 2 & 0 \\ -2 & 1 & 3 & 1 \\ 1 & 0 & 2 & -1 \\ 2 & 1 & -1 & 2 \end{vmatrix}$ so that zeros occur everywhere in the first column except in the third row.

9. Evaluate $|A|$ in Exercise 8.

10. Using the determinant $\begin{vmatrix} 2 & 3 & 4 \\ 5 & 6 & 7 \\ 8 & 9 & 1 \end{vmatrix}$, illustrate Theorem 7.

11. Evaluate each determinant.

(a) $\begin{vmatrix} 3 & -1 & 3 \\ 2 & 5 & -3 \\ 5 & 4 & -1 \end{vmatrix}$; (b) $\begin{vmatrix} 1 & 1 & 1 & 1 \\ 1 & 0 & -1 & 0 \\ 0 & 1 & 1 & -1 \\ 2 & 0 & -1 & -3 \end{vmatrix}$; (c) $\begin{vmatrix} 2 & -2 & 1 & 3 \\ 0 & 2 & -1 & -1 \\ 2 & -3 & 2 & 4 \\ 0 & -1 & 1 & 1 \end{vmatrix}$;

(d) $\det \begin{bmatrix} 2 & 1 & 5 & 2 \\ 2 & 2 & 3 & 0 \\ 2 & 0 & 2 & 1 \\ 3 & 2 & 1 & 0 \end{bmatrix}$; (e) $\det \begin{bmatrix} 3 & -3 & 0 & 0 \\ 3 & 2 & 1 & 2 \\ 0 & 2 & 0 & 3 \\ -3 & 0 & -2 & 0 \end{bmatrix}$.

12. Evaluate each of the determinants of Exercise 11 by the method presented in Example 11.

13. If E_1, E_2, and E_3 are n by n elementary C-matrices, show that

$$|E_1 E_2 E_3| = |E_1| \cdot |E_2| \cdot |E_3|.$$

14. Use $A = \begin{bmatrix} 2 & 3 \\ 1 & 4 \end{bmatrix}$ to illustrate the corollaries of Theorems 2, 8, and 9.

15. State corollaries similar to the three in this section for the case of postmultiplication of A by E.

16. Evaluate the determinant of the triangular matrix $A = [a_{ij}]_{(n,n)}$.

17. Prove that if $A = \begin{bmatrix} A_{11} & A_{12} \\ \hline 0 & A_{22} \end{bmatrix}$, where A_{11} is invertible and A_{22} is square, then

$$\det A = (\det A_{11})(\det A_{22}).$$

$\left(\text{Hint: } A = \begin{bmatrix} A_{11} & 0 \\ \hline 0 & I \end{bmatrix} \begin{bmatrix} I & 0 \\ \hline 0 & A_{22} \end{bmatrix} \begin{bmatrix} I & A_{11}^{-1} A_{12} \\ \hline 0 & I \end{bmatrix}. \right)$

5.4 Proofs of Determinant Theorems (optional)

Theorem 2 will be proved first because it is needed in the proof of Theorem 1.

Theorem 2. *If a matrix B is formed from an n by n C-matrix A by the interchange of two parallel lines (rows or columns), then $|A| = -|B|$.*

Proof. Assume that two rows are interchanged, say the kth and pth rows, and let $k < p$.

STATEMENT	REASON
(1) $|A| = \sum_{(j)} (-1)^{t_1} a_{1j_1} \cdots a_{kj_k} \cdots a_{pj_p} \cdots a_{nj_n}$.	(1) By definition of a determinant.
(2) $|B| = \sum_{(j)} (-1)^{t_2} a_{1j_1} \cdots a_{pj_p} \cdots a_{kj_k} \cdots a_{nj_n}$.	(2) Given that the kth and pth rows are interchanged.
(3) Except for the exponents t_1 and t_2, the right sides of statements 1 and 2 are the same.	(3) Multiplication of complex numbers is associative and commutative.
(4) $(-1)^{t_1} = -(-1)^{t_2}$.	(4) If t_1 interchanges will restore $j_1, \ldots, j_k, \ldots, j_p, \ldots, j_n$ to natural order, then $t_1 + 1$ interchanges will restore $j_1, \ldots, j_p, \ldots, j_k, \ldots, j_n$ to natural order. Hence if t_1 is odd, then t_2 is even, and vice-versa.
(5) $|B| = -|A|$.	(5) By Statements 3 and 4. ∎

A similar proof can be made if two columns are interchanged.

Theorem 1. *If $A = [a_{ij}]_{(n,n)}$ is a C-matrix ($n \geq 2$), then*

$$\det A = a_{i1} A_{i1} + a_{i2} A_{i2} + \cdots + a_{in} A_{in},$$

and

$$\det A = a_{1j} A_{1j} + a_{2j} A_{2j} + \cdots + a_{nj} A_{nj}.$$

Proof of the first part. By definition we know that each term of the expansion of $|A|$ contains precisely one factor from row i. If all of the terms of the expansion of $|A|$ that contain the entry a_{ij} are collected, and if this is done for each j (as we did in Section 5.2 for the justification of this theorem when $n = 3$), then

$$|A| = a_{i1}p_{i1} + a_{i2}p_{i2} + \cdots + a_{in}p_{in}$$

results, where p_{ij} represents what is left after a_{ij} has been factored out. It should be evident that we must now show that

$$p_{ij} = A_{ij} = (-1)^{i+j}M_{ij}$$

where M_{ij} is the minor of a_{ij}. This can be accomplished in three steps.

FIRST: We show that $a_{11}p_{11} = a_{11}M_{11}$. From the definition of a determinant, it is apparent that the corresponding terms of these two quantities are numerically the same. Moreover the signs of the terms of $a_{11}p_{11}$ are the same as the signs of the corresponding terms of $a_{11}M_{11}$ because the number of interchanges required to reduce the column subscripts to natural order in the terms of the expansion of M_{11} is not changed by premultiplying each term by a_{11}.

SECOND: Consider any arbitrary entry $a_{ij} = m$. This entry can be moved to the original position of a_{11} by performing $i - 1$ interchanges of adjacent rows and $j - 1$ interchanges of adjacent columns in such a way that the minor of the entry m remains unchanged. By the proof of Theorem 2 the determinant of this new matrix B is

$$|B| = (-1)^{i-1+j-1}|A| = (-1)^{i+j}|A|,$$

and, moreover, each term of $|B|$ is equal to $(-1)^{i+j}$ times a corresponding term of $|A|$.

THIRD: By the first part of our proof (because m is in the upper left-hand corner), the sum of all the terms involving m in the expansion of $|B|$ is equal to m times its minor in B. But in the second part of our proof, the minor of m in B is the same as the minor of m in A, that is, M_{ij}. Hence, the sum of the terms involving m in the expansion of $|A|$ is equal to $(-1)^{i+j}mM_{ij}$, that is,

$$a_{ij}p_{ij} = (-1)^{i+j}a_{ij}M_{ij},$$

or

$$p_{ij} = A_{ij} = (-1)^{i+j}M_{ij},$$

as we set out to prove. □

Theorem 3. *The determinants of an n by n C-matrix and its transpose are equal; that is* $|A| = |A^T|$.

5.4 PROOFS OF DETERMINANT THEOREMS (OPTIONAL)

Proof. First, we will verify that $|A| = \sum_{(i)} (-1)^t a_{i_1 1} \cdots a_{i_n n}$. Recall that $|A|$ was defined as $\sum_{(j)} (-1)^t a_{1 j_1} \cdots a_{n j_n}$. Because each of these expansions is the sum of terms consisting of exactly one factor from each row and each column, there is a one-to-one correspondence between the terms of the two expansions. These corresponding terms can differ only by a sign. However, when we interchange the factors $a_{1 j_1} \cdots a_{n j_n}$ to put the column subscripts in natural order, we are simultaneously interchanging the row subscripts away from natural order. Thus the sign of the term $a_{1 j_1} \cdots a_{n j_n}$ is the same as the sign of the corresponding term $a_{i_1 1} \cdots a_{i_n n}$ because the interchanges necessary to reduce $j_1, \ldots, j_n$ to natural order are the same interchanges, in reverse order, necessary to bring $i_1, \ldots, i_n$ back to natural order.

Theorem 3 now follows immediately because, by definition, $|A| = \sum_{(j)} (-1)^t a_{1 j_1} \cdots a_{n j_n}$ and $|A^T| = \sum_{(j)} (-1)^t a_{i_1 1} \cdots a_{i_n n}$; and we have just shown that these are equal. ◻

Theorem 4. *If all of the entries of any line of an n by n C-matrix A are zero, then $|A| = 0$.*

Proof. Expansion about the particular line which consists of all zeros yields 0. (Theorem 1). ◻

The proof of Theorem 5 is left (as Exercise 1, page 130) for the reader, with the hint to make use of Theorem 1.

Theorem 6. *The determinant of an n by n C-matrix with two identical parallel lines is zero.*

Proof. Let the kth and pth lines be identical in an n by n C-matrix A.

STATEMENT	REASON
(1) $|A| = |B|$, where B is the matrix obtained by interchanging the kth and pth lines of A.	(1) $B = A$.
(2) $|B| = -|A|$.	(2) By Theorem 2.
(3) $|A| = -|A|$.	(3) Substitute (1) into (2).
(4) $2|A| = 0$, $|A| = 0$.	(4) By the properties of the algebra of complex numbers. ◻

A proof of Theorem 7 is left (as Exercise 2, page 130) for the reader, with the hint to use Theorem 5 and then Theorem 6.

Theorem 8 is to be proved by the reader in Exercise 3 on page 130.

Theorem 9. *If a matrix B is obtained from an n by n C-matrix A by adding to each entry of a line of A a constant complex multiple of the corresponding entry of a parallel line, then $|B| = |A|$.*

Proof. Let
$$A = \begin{bmatrix} a_{11} & \cdots & a_{1k} & \cdots & a_{1p} & \cdots & a_{1n} \\ \vdots & & \vdots & & \vdots & & \vdots \\ a_{n1} & \cdots & a_{nk} & \cdots & a_{np} & \cdots & a_{nn} \end{bmatrix}.$$

Construct
$$B = \begin{bmatrix} a_{11} & \cdots & a_{1k} & \cdots & (a_{1p} + ca_{1k}) & \cdots & a_{1n} \\ \vdots & & \vdots & & \vdots & & \vdots \\ a_{n1} & \cdots & a_{nk} & \cdots & (a_{np} + ca_{nk}) & \cdots & a_{nn} \end{bmatrix}.$$

If we expand $|B|$ about the pth column, we obtain

$$\begin{aligned} |B| &= (a_{1p} + ca_{1k})A_{1p} + \cdots + (a_{np} + ca_{nk})A_{np} \\ &= (a_{1p}A_{1p} + \cdots + a_{np}A_{np}) + c(a_{1k}A_{1p} + \cdots + a_{nk}A_{np}) \\ &= \det A + 0 \end{aligned}$$

by Theorem 1 and Theorem 7. A similar proof could be made if a constant multiple of each entry of some *row* were added to the corresponding entry of another *row*. □

Theorem 10. *The determinant of the product of an elementary matrix E and a square C-matrix A of the same order is equal to the product of the determinants of the two matrices; that is, $|EA| = |E||A|$ and $|AE| = |A||E|$.*

Proof. The proof follows from the corollaries of Theorems 2, 8, and 9, and Exercise 15 of Section 5.3. □

EXERCISES

1. Prove Theorem 5.   2. Prove Theorem 7.   3. Prove Theorem 8.
4. Prove the following theorem: If any two parallel lines of an n by n C-matrix A are proportional, then $\det A = 0$.
5. Prove that the converse of the theorem in Exercise 4 is not true.
6. Prove that $\begin{vmatrix} a & b & c \\ d & e & f \\ g+h & i+j & k+l \end{vmatrix} = \begin{vmatrix} a & b & c \\ d & e & f \\ g & i & k \end{vmatrix} + \begin{vmatrix} a & b & c \\ d & e & f \\ h & j & l \end{vmatrix}$.
7. Rewrite the proof of Theorem 3 in Statement-Reason form.
8. Verify in detail the proof of Theorem 10.
9. Prove that if A is invertible (A^{-1} exists), then $|A| \neq 0$. (*Hint:* Use Theorem 23 of Chapter 3.) (Answer is provided.)

5.5 Rank of a Matrix

Determinants can be used to define the *rank* of a matrix; the matrix can be either rectangular or square. The usefulness of rank will become apparent in later sections.

Definition 4. *The rank of a C-matrix A is the greatest integer r for which A has an rth order submatrix whose determinant is not zero. When all of the entries are zero, the rank is zero.*

Example 1. Given

$$A = \begin{bmatrix} 2 & -1 & 1 \\ 1 & 4 & 5 \\ 3 & 2 & 5 \end{bmatrix}.$$

The rank is not 3, because $\det A = 0$. The rank is 2, because the determinant of at least one of the 2 by 2 submatrices is not zero.

Example 2. Given

$$A = \begin{bmatrix} 2 & 1 & 3 & 0 \\ 0 & 0 & 0 & 2 \\ 1 & 0 & 0 & 0 \end{bmatrix}.$$

Since there are no 4 by 4 submatrices, the rank is at most 3. To find whether it is 3, we must start calculating the determinants of the four 3 by 3 submatrices. As soon as we find one (if we do) whose determinant is not zero, we can stop with assurance that the rank is 3.

$$\det \begin{bmatrix} 2 & 1 & 3 & 0 \\ 0 & 0 & 0 & 2 \\ 1 & 0 & 0 & 0 \end{bmatrix} = 2 \neq 0,$$

therefore the rank is 3. One of the other third-order submatrices has a nonzero determinant, and hence it could also have been used to reach the same conclusion.

If a matrix has many rows and columns, it is difficult to find its rank directly from the definition. A better method will be developed using the next theorem.

Theorem 11. *Equivalent C-matrices have the same rank.*

Outline of Proof. The proof is split into two parts.

FIRST: Show that the first two elementary operations (that is, (1) the interchange of any two rows or any two columns, and (2) the multiplication of any row or column by a nonzero complex number) do not change a nonzero determinant of a submatrix to zero, nor change a zero determinant of a submatrix to one that is nonzero, and therefore do not alter the rank of a matrix.

SECOND: Show that the third elementary operation ((3) the addition to any line of a complex multiple of any parallel line) does not increase the rank of a matrix A. If the rank r of A is as large as the order of A permits, then r cannot be increased. If the rank r of A is not as large as the order permits, then consider all submatrices of order $r + 1$ after elementary operation (3) has been applied. By a generalization of Exercise 6 of Section 5.4 and by Theorem 8 of Section 5.3, the determinants of these submatrices can be expressed as $|S| + k|R|$, where S is a submatrix of A of order $r + 1$. Also, R is a matrix of order $r + 1$ that is the same as a submatrix of A except that either one line is out of place or two lines are identical. In any event $|R| = 0$; also $|S| = 0$ because the rank of A was r. Therefore all submatrices of order $r + 1$ of the transformed matrix have a rank no greater than r, and thus the rank of A is no greater than r. Neither is the rank less than r because the inverse transformation (of the same type) would cause an increase in rank, which we have just shown is impossible. ∎

Example 3. Find the rank of $A = \begin{bmatrix} 2 & 1 & 3 & 1 \\ 1 & 0 & 1 & 1 \\ 2 & 1 & 3 & 1 \\ -1 & 0 & -1 & -1 \end{bmatrix}$. We will use elementary row and column operations to reduce A to normal form (see Definition 7 of Section 3.3 for a definition of normal form). But A is equivalent to its normal form and therefore has the same rank as that form.

$$A \underset{C_1 \leftrightarrow C_2}{\sim} \begin{bmatrix} 1 & 2 & 3 & 1 \\ 0 & 1 & 1 & 1 \\ 1 & 2 & 3 & 1 \\ 0 & -1 & -1 & -1 \end{bmatrix} \underset{-R_1 + R_3}{\sim} \begin{bmatrix} 1 & 2 & 3 & 1 \\ 0 & 1 & 1 & 1 \\ 0 & 0 & 0 & 0 \\ 0 & -1 & -1 & -1 \end{bmatrix}$$

$$\underset{\substack{-2C_1 + C_2 \\ -3C_1 + C_3 \\ -C_1 + C_4}}{\sim} \begin{bmatrix} 1 & 0 & 0 & 0 \\ 0 & 1 & 1 & 1 \\ 0 & 0 & 0 & 0 \\ 0 & -1 & -1 & -1 \end{bmatrix} \underset{R_2 + R_4}{\sim} \begin{bmatrix} 1 & 0 & 0 & 0 \\ 0 & 1 & 1 & 1 \\ 0 & 0 & 0 & 0 \\ 0 & 0 & 0 & 0 \end{bmatrix}$$

$$\underset{\substack{-C_2 + C_3 \\ -C_2 + C_4}}{\sim} \begin{bmatrix} 1 & 0 & 0 & 0 \\ 0 & 1 & 0 & 0 \\ 0 & 0 & 0 & 0 \\ 0 & 0 & 0 & 0 \end{bmatrix} = \begin{bmatrix} I_2 & 0 \\ \hline 0 & 0 \end{bmatrix}.$$

Thus the rank of the normal form is obviously 2. Therefore, by Theorem 11, the rank of A is also 2.

APPLICATIONS

Example 4. On pages 239–245 of [23] (Noble) several illustrations are given of the relevancy of rank of a matrix in an applied problem. One of the illustrations deals with a mixture of CO, H_2, and CH_4 placed in a furnace and burned with oxygen to form CO, CO_2, and H_2O. On page 242 of [23] the use of rank in a discussion of systems of linear equations is illustrated. "Rank and linear systems" is the topic of our next section.

EXERCISES

1. Are the following statements true or false? Give your reasons.
   (a) If $[a_{ij}]_{(4,5)}$ has rank 3, then the determinant of every fourth-order submatrix is 0, and, conversely, if every such fourth-order determinant is zero, the rank of the matrix is 3.
   (b) The rank of A is equal to the rank of A^T.
   (c) If $[a_{ij}]_{(3,3)}$ is skew-symmetric, its rank is less than 3.

2. What is the largest possible rank of an m by n matrix under the assumption that $m > n$?

3. Using the definition, find the rank of:

   (a) $\begin{bmatrix} 1 & 2 & -1 \\ 4 & 1 & 5 \\ 3 & -1 & 6 \end{bmatrix}$; (b) $\begin{bmatrix} 3 & 1 & 4 & 4 \\ 2 & 1 & 3 & 1 \\ 0 & 2 & 2 & 0 \end{bmatrix}$; (c) $\begin{bmatrix} 2 & 4 \\ -4 & -8 \\ 1 & 2 \end{bmatrix}$;

   (d) $\begin{bmatrix} 0 & 0 & 0 \\ 0 & 0 & 0 \end{bmatrix}$; (e) $\begin{bmatrix} 0 & 0 & 2 \\ 0 & 0 & 0 \end{bmatrix}$.

4. Using Theorem 11, find the rank of each matrix in Exercise 3.

In each of the Exercises 5–10, find the rank of the given matrix by reducing it to normal form.

5. $\begin{bmatrix} 3 & 2 & -1 \\ 7 & 8 & 0 \\ 4 & 6 & 1 \end{bmatrix}$.

6. $\begin{bmatrix} 0 & 1 & 3 & 0 \\ 0 & 4 & 0 & 2 \\ 1 & 0 & 3 & 0 \\ -1 & 1 & 0 & 0 \end{bmatrix}$.

7. $\begin{bmatrix} 3 & 1 & 4 & 6 & 2 \\ 4 & 1 & 3 & 9 & 6 \end{bmatrix}$.

8. $\begin{bmatrix} 3 & 2 & 7 \\ 4 & -3 & -2 \\ 0 & 1 & 2 \\ 6 & 1 & 8 \end{bmatrix}$.

9. $\begin{bmatrix} 4 & 9 & -3 & 1 \\ 6 & 9 & -4 & 0 \\ 2 & 9 & -2 & 2 \\ -2 & 0 & 1 & 1 \end{bmatrix}$.

10. $\begin{bmatrix} -1 & 0 & -1 & 3 & 2 \\ -1 & 4 & 0 & 0 & 1 \\ 0 & 0 & 0 & 4 & 2 \\ 1 & 0 & 1 & 1 & 0 \\ 0 & 4 & 1 & 1 & 1 \end{bmatrix}$.

11. What is the normal form of a 4 by 6 C-matrix of rank 2? Express the answer using submatrices.

12. Is $\begin{bmatrix} 2 & 1 & 4 \\ 0 & 1 & 3 \\ 2 & 0 & 1 \end{bmatrix}$ equivalent to I_3? Why?

13. For what value of x is the rank of the following matrix 3? Is it possible for the rank to be 1? Why?
$$\begin{bmatrix} 2 & 4 & 2 \\ 2 & 1 & 2 \\ 1 & 0 & x \end{bmatrix}.$$

14. Find the ranks of the augmented matrix and the coefficient matrix of the system
$$\begin{cases} x + y = 6, \\ x + 3y = 4, \\ 2x - y = 2. \end{cases}$$

15. Make up an example of a 4 by 4 matrix that has rank 2.

16. If the second column of $A = [a_{ij}]_{(4,3)}$ consists entirely of zeros, what can be said about the rank of A?

17. Let $A = [a_{ij}]_{(4,4)}$.
    (a) If det $A = 0$, what can be said about the rank of A?
    (b) If det $A \neq 0$, what can be said about the rank of A?

18. In three-dimensional space, show that the cubic space curve with parametric equations $x = t$, $y = t^2$, and $z = t^3$ is the locus of points (x, y, z) where the matrix $\begin{bmatrix} x & y & z \\ 1 & x & y \end{bmatrix}$ has rank 1.

19. (a) Let (x_1, y_1), $(4, 5)$, and $(7, 6)$ be three fixed points in a plane. Show that these three points are collinear if and only if the rank of the matrix
$$\begin{bmatrix} x_1 & y_1 & 1 \\ 4 & 5 & 1 \\ 7 & 6 & 1 \end{bmatrix}$$ is less than 3.

   (b) Repeat part (a) for the points (x_1, y_1), (x_2, y_2), and (x_3, y_3), and the matrix $\begin{bmatrix} x_1 & y_1 & 1 \\ x_2 & y_2 & 1 \\ x_3 & y_3 & 1 \end{bmatrix}$.

20. In Section 3.2 the statement was made that the number of nonzero rows of the reduced echelon form of a square C-matrix is the same no matter what sequence of elementary row operations is used to obtain the reduced echelon form. Prove this statement.

21. In Section 3.3 the statement was made that for any C-matrix, the equivalent normal form is unique. Prove this statement.

5.6 Rank and Linear Systems

As we learned in Chapter 2, systems of linear equations over the complex numbers may have (1) exactly one solution, or (2) an infinite number of solutions, or (3) no solution.

The following Theorems 12 and 13 enable one to discover which of the previous situations exists for a given system by finding the ranks of the coefficient and augmented matrices.

Theorem 12. *A system of linear equations $AX = B$ over the complex numbers is consistent if and only if the rank of the augmented matrix is equal to the rank of the coefficient matrix. (This common value, if it exists, will be denoted by r and is often called the* **rank of the system.**)

The proofs of this theorem and of the next will be outlined at the end of this section.

Theorem 13. *A consistent system of linear equations $AX = B$ over the complex numbers has a unique solution if and only if $r = n$, where n equals the number of unknowns of the system.*

Example 1. The system $AX = B$, or

$$\begin{cases} x_1 + 3x_2 + x_3 = 2, \\ x_1 - 2x_2 - 2x_3 = 3, \\ 2x_1 + x_2 - x_3 = 6, \end{cases}$$

in which $n = m = 3$, is inconsistent, because the rank of $[A \vdots B]$ is 3 and the rank of A is 2. The reader is reminded that m represents the number of equations.

Example 2. The system $AX = B$, or

$$\begin{cases} -4x_1 - 10x_2 - 4x_3 = 2, \\ x_1 - 2x_2 + x_3 = 1, \\ x_1 + x_2 + x_3 = 0, \\ 2x_1 - x_2 + 2x_3 = 1, \end{cases}$$

in which $m = 4$, $n = 3$, is consistent because the rank of $[A \vdots B]$ is 2, and the rank of A is 2. The solution is not unique (hence there are an infinite number of solutions), because $r = 2$ and $n = 3$.

Complete formal proofs of the two theorems of this section are left as exercises. Outlines are given as an aid to the student.

Outline of proof of Theorem 12. By performing the same suitable elementary row transformations on both $AX = B$ and $[A \vdots B]$, and, if necessary, by properly rearranging the subscripts of the unknowns, we eventually obtain the equivalent system

$$\begin{cases} x_1 \quad\; + c_{1,k+1}x_{k+1} + c_{1,k+2}x_{k+2} + \cdots + c_{1n}x_n = d_1, \\ \quad x_2 \quad + c_{2,k+1}x_{k+1} + c_{2,k+2}x_{k+2} + \cdots + c_{2n}x_n = d_2, \\ \quad\cdots\cdots\cdots\cdots\cdots\cdots\cdots\cdots\cdots\cdots\cdots\cdots \\ \quad\quad x_k + c_{k,k+1}x_{k+1} + c_{k,k+2}x_{k+2} + \cdots + c_{kn}x_n = d_k, \\ \quad\quad\quad\quad 0x_{k+1} + \quad\; 0x_{k+2} + \cdots + \;\; 0x_n = d_{k+1}, \\ \quad\quad\quad\quad 0x_{k+1} + \quad\; 0x_{k+2} + \cdots + \;\; 0x_n = d_{k+2}, \\ \quad\cdots\cdots\cdots\cdots\cdots\cdots\cdots\cdots\cdots\cdots\cdots\cdots \\ \quad\quad\quad\quad 0x_{k+1} + \quad\; 0x_{k+2} + \cdots + \;\; 0x_n = d_m, \end{cases}$$

and the corresponding augmented matrix. If any one of the numbers $d_{k+1}, \ldots, d_m$ is different from zero, we obtain a contradiction of one of the last $m - k$ equations, and the above system is inconsistent. Hence, the equivalent original system is inconsistent. But if $d_{k+1} = d_{k+2} = \cdots = d_m = 0$, the last $m - k$ equations are satisfied and so are the first k equations, since we can solve them for $x_1, \ldots, x_k$ in terms of $x_{k+1}, \ldots, x_n$. Therefore the system is consistent if and only if

$$d_{k+1} = d_{k+2} = \cdots = d_m = 0.$$

Since $|I_k| \neq 0$, the rank of the coefficient matrix is equal to k, and so is the rank of the augmented matrix when $d_{k+1} = d_{k+2} = \cdots = d_m = 0$. But if any one of these $m - k$ numbers is different from zero, the rank of the augmented matrix is greater than k. Hence the system is consistent if and only if the ranks of the coefficient and augmented matrices are the same. ☐

Outline of proof of Theorem 13. In the outline of the above proof of Theorem 12, when $AX = B$ is consistent and the rank equals k, we can solve for certain x_j in terms of d_j and $n - k$ remaining unknowns, that is,

$$x_j = d_j - c_{j,k+1}x_{k+1} - c_{j,k+2}x_{k+2} - \cdots - c_{jn}x_n, \quad (j = 1, 2, \ldots, k).$$

If the system has a unique solution, then $n - k = 0$ and hence $k = n$, which means that the rank of the system equals the number of unknowns. Conversely, if the rank of the system equals the number of unknowns, then $k = n$ and hence $x_j = d_j$, which is a unique solution for $j = 1, 2, \ldots, n$. ☐

APPLICATIONS

Example 3. Suppose that a certain manufacturer produces two brands of fertilizer—P and Q. Each pound of brand P contains 20 units of chemical A and 10 units of chemical B; each pound of brand Q contains 30 units of chemical

5.6 RANK AND LINEAR SYSTEMS

A and 50 units of chemical B. Suppose there is a limited supply of 1,200 units of chemical A and 900 units of chemical B, but that all other ingredients are in ample supply. How many pounds of each brand of fertilizer can be produced within the restriction imposed by the limited supply of chemicals?

Solution: If we let x_1 represent the number of pounds of P produced and x_2 represent the number of pounds of Q produced, then the problem posed above can be restated as a system of linear inequalities:

$$\begin{cases} 20x_1 + 30x_2 \leq 1200, \\ 10x_1 + 50x_2 \leq 900, \\ x_1 \geq 0, \\ x_2 \geq 0. \end{cases}$$

In this particular problem, the solution set is a region bounded by the system of lines

$$\begin{cases} 20x_1 + 30x_2 = 1200, \\ 10x_1 + 50x_2 = 900, \\ x_1 = 0, \\ x_2 = 0. \end{cases}$$

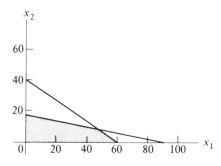

Figure 5.6.1

This system of equations is inconsistent (notice that the rank of the augmented matrix is 3 and that the rank of the coefficient matrix is 2); the graph of the system of equations is shown in Figure 5.6.1. The solution set, called the *feasible set*, for the original system of inequalities is indicated by the shaded area in Figure 5.6.1; this is the region common to the four half-planes that satisfy the four given inequalities. Feasible amounts of the two brands of fertilizers that can be produced are represented by the coordinates of any point in the feasible set, for example, $x_1 = 40$ (pounds of P) and $x_2 = 10$ (pounds of Q), or $x_1 = 20$ (pounds of P) and $x_2 = 14$ (pounds of Q).

Example 4. Two examples from the engineering context of fluid flow and heat transfer are given on pages 234–237 of [23] (Noble). These examples illustrate the use of the material of this section.

EXERCISES

1. By use of Theorems 12 and 13, determine whether the following systems are inconsistent, consistent with a unique solution, or consistent with an infinite number of solutions. Find the rank of the system, r, if possible, and state the relation between r and n.

(a) $\begin{cases} x_1 + 3x_2 + x_3 = 4, \\ x_1 + x_2 - x_3 = 1, \\ 2x_1 + 4x_2 = 0. \end{cases}$
(b) $\begin{cases} x_1 - x_2 + 6 = 0, \\ x_1 + 2x_2 - 5 = 0, \\ 3x_1 + 3x_2 - 4 = 0. \end{cases}$

(c) $\begin{cases} x_1 + 2x_2 + x_3 = 6, \\ x_1 - x_2 = 2, \\ x_1 - x_2 + x_3 = 0, \\ 3x_1 + 2x_3 = 8. \end{cases}$
(d) $\begin{cases} x_1 + x_2 + x_3 - 5 = 0, \\ x_1 - x_2 - x_3 - 4 = 0. \end{cases}$

(e) $\begin{bmatrix} 1 & 1 & 1 \\ 1 & -1 & 0 \\ 2 & 0 & 1 \end{bmatrix} \begin{bmatrix} x_1 \\ x_2 \\ x_3 \end{bmatrix} = \begin{bmatrix} 0 \\ 0 \\ 0 \end{bmatrix}$.

(f) $AX = \mathbf{0}$, where $A = \begin{bmatrix} 2 & 1 & 0 \\ 1 & -1 & 1 \\ 0 & 1 & -1 \end{bmatrix}$ and $X = \begin{bmatrix} x_1 \\ x_2 \\ x_3 \end{bmatrix}$.

(g) $\begin{cases} x_1 + x_2 + 2x_3 - x_4 = 2, \\ 2x_1 + 2x_3 = 0, \\ -x_1 - x_2 + x_3 = 0, \\ 2x_1 - x_3 + x_4 = -2. \end{cases}$

(h) $AX = B$, where $B = \begin{bmatrix} 2 \\ 0 \\ 4 \end{bmatrix}$, $A = \begin{bmatrix} 1 & 1 & -1 \\ 2 & 2 & -1 \\ 1 & 1 & 0 \end{bmatrix}$, $X = \begin{bmatrix} x_1 \\ x_2 \\ x_3 \end{bmatrix}$.

2. Which of the systems in Exercise 1 are homogeneous?

3. A certain system of linear equations with six unknowns is known to be consistent and to have a unique solution. What can be said about the ranks of the augmented and coefficient matrices? What can be said about the number of equations? Give reasons.

4. The rank of the coefficient matrix of a certain inconsistent system of linear equations with five unknowns is found to be 4. What can be said about the rank of the augmented matrix? What can be said about the number of equations?

5. Suppose we are given 5 linear homogeneous equations in 5 unknowns. If A is the coefficient matrix, and if $\det A \neq 0$, what can we say about a solution?

5.7 A FORMULA FOR A^{-1}

6. Determine whether or not $x_1 = \frac{1}{2}$, $x_2 = -\frac{3}{2}$, $x_3 = 1$ is a solution of the following system:

$$\begin{cases} x_1 + x_2 + x_3 = 0, \\ x_1 - 3x_2 - 2x_3 = 3, \\ 2x_1 + 2x_2 - x_3 = 2. \end{cases}$$

Can a solution be found?

7. The system $AX = B$ consists of m linear equations (which may or may not be homogeneous) in n unknowns. Assume that the rank of A equals the rank of $[A \vdots B]$, and that this common rank is r for this system. Why is it always true that $r \leq m$ and $r \leq n$?

8. Give a complete proof in Statement-Reason form for: (a) Theorem 12; (b) Theorem 13.

5.7 A Formula for A^{-1}

In Chapter 3, a method of calculating A^{-1} was developed. The purpose of this section is to find an alternate method which links A^{-1} with the determinant of A. We will also prove two very important results in the theory of matrices; namely, that A is invertible if and only if $|A| \neq 0$, and that $|AB| = |A||B|$. The following useful terminology is needed first, however.

Definition 5. *Let $A = [a_{ij}]_{(n,n)}$ be a C-matrix where $n \geq 2$. The **cofactor matrix** of A, designated by $\operatorname{cof} A$, is the matrix of order n whose entry in row i and column j is A_{ij}, the cofactor of a_{ij} in A.*

Example 1. Let

$$A = \begin{bmatrix} 2 & 4 & 0 \\ 0 & 2 & 1 \\ 3 & 0 & 2 \end{bmatrix}.$$

$$\operatorname{cof} A = \begin{bmatrix} +\begin{vmatrix} 2 & 1 \\ 0 & 2 \end{vmatrix} & -\begin{vmatrix} 0 & 1 \\ 3 & 2 \end{vmatrix} & +\begin{vmatrix} 0 & 2 \\ 3 & 0 \end{vmatrix} \\ -\begin{vmatrix} 4 & 0 \\ 0 & 2 \end{vmatrix} & +\begin{vmatrix} 2 & 0 \\ 3 & 2 \end{vmatrix} & -\begin{vmatrix} 2 & 4 \\ 3 & 0 \end{vmatrix} \\ +\begin{vmatrix} 4 & 0 \\ 2 & 1 \end{vmatrix} & -\begin{vmatrix} 2 & 0 \\ 0 & 1 \end{vmatrix} & +\begin{vmatrix} 2 & 4 \\ 0 & 2 \end{vmatrix} \end{bmatrix} = \begin{bmatrix} 4 & 3 & -6 \\ -8 & 4 & 12 \\ 4 & -2 & 4 \end{bmatrix}.$$

Definition 6. *The **adjoint matrix**, designated by $\operatorname{adj} A$, of a square C-matrix A (of order $n \geq 2$) is the transpose of $\operatorname{cof} A$.*

Example 2. Using the matrix A of the previous example, we have

$$\operatorname{adj} A = (\operatorname{cof} A)^T = \begin{bmatrix} 4 & 3 & -6 \\ -8 & 4 & 12 \\ 4 & -2 & 4 \end{bmatrix}^T = \begin{bmatrix} 4 & -8 & 4 \\ 3 & 4 & -2 \\ -6 & 12 & 4 \end{bmatrix}.$$

Theorem 14. *For an n by n C-matrix A, the inverse, A^{-1}, exists if and only if $|A| \neq 0$. Moreover, if $n \geq 2$ and if A^{-1} exists, then*

$$A^{-1} = \frac{1}{|A|} \operatorname{adj} A.$$

Proof. Since the first assertion contains the double conditional "if and only if," we must prove the two assertions (1) $|A| \neq 0$ implies the existence of A^{-1}, and, conversely, (2) the existence of A^{-1} implies that $|A| \neq 0$.

Assertion (1). Assume that $|A| \neq 0$, and prove that A^{-1} exists. If $n = 1$, the proof is trivial; if $n \geq 2$, we have the following argument:

STATEMENT	REASON
(1) $A(\operatorname{adj} A) =$ $\begin{bmatrix} a_{11} & a_{12} & \cdots & a_{1n} \\ a_{21} & a_{22} & \cdots & a_{2n} \\ \vdots & \vdots & & \vdots \\ a_{n1} & a_{n2} & \cdots & a_{nn} \end{bmatrix} \begin{bmatrix} A_{11} & A_{21} & \cdots & A_{n1} \\ A_{12} & A_{22} & \cdots & A_{n2} \\ \vdots & \vdots & & \vdots \\ A_{1n} & A_{2n} & \cdots & A_{nn} \end{bmatrix}$ $= \begin{bmatrix} |A| & 0 & 0 & 0 & \cdots & 0 \\ 0 & |A| & 0 & 0 & \cdots & 0 \\ 0 & 0 & |A| & 0 & \cdots & 0 \\ \vdots & \vdots & \vdots & \vdots & & \vdots \\ 0 & 0 & 0 & 0 & \cdots & |A| \end{bmatrix}.$	(1) Definition 6 (adjoint matrix) and Theorems 1 and 7 after matrix multiplication.
(2) $A(\operatorname{adj} A) = |A|I_n$.	(2) Definition 6, page 11, of Chapter 1 (multiplication of a matrix by a scalar).
(3) $|A| \neq 0$.	(3) Assumption.
(4) $A \dfrac{\operatorname{adj} A}{|A|} = I_n$.	(4) Statements (2) and (3).
(5) If we define $A_1 = \dfrac{\operatorname{adj} A}{|A|}$, then A_1 exists and $AA_1 = I_n$.	(5) Statement (4).
(6) Also $A_1 A = I_n$.	(6) An argument similar to Statements (1) to (5).
(7) $A^{-1} = A_1 = \dfrac{\operatorname{adj} A}{|A|}$.	(7) Statements (5) and (6) and Definition 4, page 53, of Chapter 3 (inverse matrix).

5.7   A FORMULA FOR A^{-1}

Thus the first part of the proof is complete. Moreover, we have developed the formula of Theorem 14 for calculating A^{-1} if $n \geq 2$.

Assertion (2). Assume that A^{-1} exists, and prove that $|A| \neq 0$. See answer to Exercise 9 of Section 5.4. ☐

The following example illustrates the use of the formula given in Theorem 14 for finding the inverse of a given invertible matrix.

Example 3. In Examples 1 and 2 we found that if

$$A = \begin{bmatrix} 2 & 4 & 0 \\ 0 & 2 & 1 \\ 3 & 0 & 2 \end{bmatrix}, \quad \text{then adj } A = \begin{bmatrix} 4 & -8 & 4 \\ 3 & 4 & -2 \\ -6 & 12 & 4 \end{bmatrix}.$$

We find also that $|A| = 20$; therefore,

$$A^{-1} = \frac{\text{adj } A}{|A|} = \begin{bmatrix} \frac{4}{20} & -\frac{8}{20} & \frac{4}{20} \\ \frac{3}{20} & \frac{4}{20} & -\frac{2}{20} \\ -\frac{6}{20} & \frac{12}{20} & \frac{4}{20} \end{bmatrix}.$$

To check our result, we find that

$$AA^{-1} = A^{-1}A = I_3.$$

The next theorem will be used to prove Theorem 16, a very important theorem.

Theorem 15. *If A and B are n by n C-matrices, and if $|A| = 0$, then $|AB| = 0$.*

Proof.   Case I: (A is the null matrix.)

$$A = 0 \Rightarrow AB = 0 \Rightarrow |AB| = 0.$$

Case II: (A is not the null matrix.)

STATEMENT	REASON
(1) A can be reduced by elementary operations to normal form, N.	(1) Theorem 15, page 62, of Chapter 3.
(2) There exist invertible matrices P and Q such that $A = PNQ$.	(2) Theorem 25, page 71, of Chapter 3 and Exercise 17(b) of Section 3.3, page 67.
(3) $AB = PNQB$.	(3) Postmultiplication of both sides of (2) by B.
(4) $P^{-1}(AB) = NQB$.	(4) Premultiplication of both sides of (3) by P^{-1}.
(5) $P^{-1} = E_1 E_2 \cdots E_i$ where the E's are elementary row transformation matrices.	(5) Theorems 23 and 9, pages 70 and 55, of Chapter 3.

(6) $|P^{-1}(AB)| = |(E_1 E_2 \cdots E_i)(AB)|$
    $= |E_1 E_2 \cdots E_i| |AB|$
    $= |P^{-1}| |AB|$.

(6) By repeated use of Theorem 10 of this Chapter.

(7) $|P^{-1}| |AB| = |NQB|$.

(7) By (4) and (6).

(8) $N \neq I_n$, and hence the last row of N consists entirely of zeros.

(8) From (1), $A \sim N$, by hypothesis $|A| = 0$, and by Theorem 11, page 131, the rank of A must equal the rank of N.

(9) The last row of NQ consists entirely of zeros and consequently so does the last row of NQB.

(9) By multiplication.

(10) $|NQB| = 0$.

(10) By Theorem 4, page 121.

(11) $|P^{-1}| |AB| = 0$.

(11) By (7) and (10).

(12) $|AB| = 0$.

(12) $|P^{-1}| \neq 0$ by Theorem 14. □

Example 4. Let $A = \begin{bmatrix} 2 & 4 \\ 1 & 2 \end{bmatrix}$ and $B = \begin{bmatrix} 1 & 3 \\ 2 & 7 \end{bmatrix}$. Notice that $|A| = 0$. As we expect from Theorem 15, $\det(AB) = \det\left(\begin{bmatrix} 2 & 4 \\ 1 & 2 \end{bmatrix} \begin{bmatrix} 1 & 3 \\ 2 & 7 \end{bmatrix}\right) = \det \begin{bmatrix} 10 & 34 \\ 5 & 17 \end{bmatrix} = 0$.

Using the last two theorems, we can now develop a very important property in matrix theory.

Theorem 16. *If A and B are n by n C-matrices, then $|AB| = |A| |B|$.*

Proof. Case I: ($|A| = 0$.) By Theorem 15, $|AB| = 0$, and obviously $|A| |B| = 0$; hence $|AB| = |A| |B|$.

Case II: ($|A| \neq 0$.) By Theorem 14, A is invertible; hence, by Theorem 23 of Chapter 3, page 70, matrix A can be expressed as a product of elementary matrices. Therefore

$$|AB| = |E_1 \cdots E_p B|.$$

Then by repeated use of Theorem 10, page 130,

$$|AB| = |E_1| |E_2 \cdots E_p B|$$
$$\vdots$$
$$= |E_1| \cdots |E_p| |B|$$
$$\vdots$$
$$= |E_1 \cdots E_p| |B|$$
$$= |A| |B|. \quad \square$$

APPLICATIONS

Example 5. This example presents another system of linear equations that can be solved by use of an inverse matrix; the example is very important in applied statistics. In the applied sciences it is frequently desirable to write an equation of a straight line $y = ax + b$ as an estimate of a possible relationship between two variables when pairs of values of these variables are known. One method of obtaining such an equation is the so-called method of least squares. We shall present a matrix approach to the method of least squares. As an illustration of the problem that confronts us, suppose that we have the following data

(5.7.1)

x	1	3	4	6
y	3	7	8	8

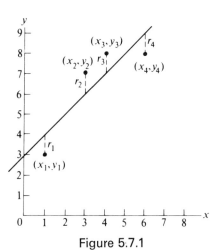

Figure 5.7.1

We wish to find the numbers a and b in the equations

(5.7.2) $$y_i = ax_i + b + r_i \quad (i = 1, 2, 3, 4)$$

in such a way that $S = \sum_{i=1}^{4} (r_i)^2$ is a minimum (whence the name "least squares"). We shall call the quantity r_i the *residual*. The four residuals of this example are pictured in Figure 5.7.1. In general, m equations like (5.7.2) can be written in matrix notation as

(5.7.3) $$Y = C\alpha + \rho$$

where

$$Y = \begin{bmatrix} y_1 \\ y_2 \\ \vdots \\ y_m \end{bmatrix}, \quad C = \begin{bmatrix} x_1 & 1 \\ x_2 & 1 \\ \vdots & \vdots \\ x_m & 1 \end{bmatrix}, \quad \alpha = \begin{bmatrix} a \\ b \end{bmatrix}, \quad \rho = \begin{bmatrix} r_1 \\ r_2 \\ \vdots \\ r_m \end{bmatrix}.$$

Remember that we wish to minimize

$$S = r_1^2 + r_2^2 + \cdots + r_m^2,$$

which we recognize as the entry of the 1 by 1 matrix $\rho^T \rho$.

We will prove later that S assumes its minimum value only if

(5.7.4) $\qquad\qquad (C^T C)\alpha = C^T Y,$

which we can solve for the unknown vector α. We illustrate the use of (5.7.4) by finding the equation of the line shown in Figure 5.7.1.

Using the data of (5.7.1), $C^T C \alpha = C^T Y$ becomes

$$\begin{bmatrix} 1 & 3 & 4 & 6 \\ 1 & 1 & 1 & 1 \end{bmatrix} \begin{bmatrix} 1 & 1 \\ 3 & 1 \\ 4 & 1 \\ 6 & 1 \end{bmatrix} \begin{bmatrix} a \\ b \end{bmatrix} = \begin{bmatrix} 1 & 3 & 4 & 6 \\ 1 & 1 & 1 & 1 \end{bmatrix} \begin{bmatrix} 3 \\ 7 \\ 8 \\ 8 \end{bmatrix},$$

or

$$\begin{bmatrix} 62 & 14 \\ 14 & 4 \end{bmatrix} \begin{bmatrix} a \\ b \end{bmatrix} = \begin{bmatrix} 104 \\ 26 \end{bmatrix},$$

whose solution can be found by use of the inverse matrix,

$$\begin{bmatrix} a \\ b \end{bmatrix} = \begin{bmatrix} 62 & 14 \\ 14 & 4 \end{bmatrix}^{-1} \begin{bmatrix} 104 \\ 26 \end{bmatrix}$$

$$= \begin{bmatrix} \frac{2}{26} & \frac{-7}{26} \\ \frac{-7}{26} & \frac{31}{26} \end{bmatrix} \begin{bmatrix} 104 \\ 26 \end{bmatrix} = \begin{bmatrix} 1 \\ 3 \end{bmatrix}.$$

Hence if there is a minimum S, then the desired equation of the line in Figure 5.7.1 is

$$y = x + 3.$$

If for some reason it is desired to weight one or more of the squared residuals then this can be done by minimizing

$$S_w = w_1 r_1^2 + \cdots + w_m r_m^2.$$

It can be shown that such a minimum can be found by solving

$$(C^T W C)\alpha = C^T W Y,$$

where W is the diagonal matrix

$$W = \begin{bmatrix} w_1 & \cdots & 0 \\ \vdots & \ddots & \vdots \\ 0 & \cdots & w_m \end{bmatrix}.$$

We can further generalize the problem by using n independent variables $x_{(1)}, \ldots, x_{(n)}$ rather than just the one independent variable x. The problem

5.7 A FORMULA FOR A^{-1}

becomes: Find the coefficients and y-intercept of

$$y = a_1 x_{(1)} + a_2 x_{(2)} + \cdots + a_n x_{(n)} + b$$

in such a way that we minimize

$$S = r_1^2 + \ldots + r_m^2,$$

where

(5.7.5)
$$\begin{cases} y_1 = a_1 x_{(1)_1} + a_2 x_{(2)_1} + \cdots + a_n x_{(n)_1} + b + r_1, \\ y_2 = a_1 x_{(1)_2} + a_2 x_{(2)_2} + \cdots + a_n x_{(n)_2} + b + r_2, \\ \vdots \\ y_m = a_1 x_{(1)_m} + a_2 x_{(2)_m} + \cdots + a_n x_{(n)_m} + b + r_m. \end{cases}$$

Equations (5.7.5) can be expressed in matrix notation as

$$Y = C\alpha + \rho,$$

where

$$Y = \begin{bmatrix} y_1 \\ \vdots \\ y_m \end{bmatrix}, \quad C = \begin{bmatrix} x_{(1)_1} & \cdots & x_{(n)_1} & 1 \\ \vdots & & \vdots & \vdots \\ x_{(1)_m} & \cdots & x_{(n)_m} & 1 \end{bmatrix}, \quad \alpha = \begin{bmatrix} a_1 \\ \vdots \\ a_n \\ b \end{bmatrix}, \quad \rho = \begin{bmatrix} r_1 \\ \vdots \\ r_m \end{bmatrix}.$$

We can find the unknown α for given C and Y by simply solving, as before, the system

$$C^T C \alpha = C^T Y.$$

In summary, data such as (5.7.1) generally produces an inconsistent system $C\alpha = Y$; by premultiplying both sides of this matrix equation by C^T (or $C^T W$) and by solving for α, we obtain the least squares solution of the hyperplane of best fit if one exists. We now prove why this is true.

★When there is one independent variable we can prove that S attains its minimum only if $C^T C \alpha = C^T Y$ by using a theorem from calculus to assert that minimum S implies $\dfrac{\partial S}{\partial a} = \dfrac{\partial S}{\partial b} = 0$. Hence if

$$S = r_1^2 + \cdots + r_m^2 = (y_1 - ax_1 - b)^2 + \cdots + (y_m - ax_m - b)^2,$$

then, remembering that the x_i and y_i are known, whereas a and b are unknown, we get

$$\begin{cases} 0 = \dfrac{\partial S}{\partial a} = 2(y_1 - ax_1 - b)(-x_1) + \cdots + 2(y_m - ax_m - b)(-x_m), \\ 0 = \dfrac{\partial S}{\partial b} = 2(y_1 - ax_1 - b)(-1) + \cdots + 2(y_m - ax_m - b)(-1). \end{cases}$$

Dividing both equations by 2 and then rearranging the terms, we get

$$\begin{cases} (x_1^2 + \cdots + x_m^2)a + (x_1 + \cdots + x_m)b = (x_1 y_1 + \cdots + x_m y_m), \\ (x_1 + \cdots + x_m)a + \qquad\qquad mb = (y_1 + \cdots + y_m), \end{cases}$$

which can be expressed as the matrix equation

$$\begin{bmatrix} (x_1^2 + \cdots + x_m^2) & (x_1 + \cdots + x_m) \\ (x_1 + \cdots + x_m) & m \end{bmatrix} \begin{bmatrix} a \\ b \end{bmatrix} = \begin{bmatrix} (x_1 y_1 + \cdots + x_m y_m) \\ (y_1 + \cdots + y_m) \end{bmatrix};$$

or,

$$\left(\begin{bmatrix} x_1 & \cdots & x_m \\ 1 & \cdots & 1 \end{bmatrix} \begin{bmatrix} x_1 & 1 \\ \vdots & \vdots \\ x_m & 1 \end{bmatrix} \right) \begin{bmatrix} a \\ b \end{bmatrix} = \begin{bmatrix} x_1 & \cdots & x_m \\ 1 & \cdots & 1 \end{bmatrix} \begin{bmatrix} y_1 \\ \vdots \\ y_m \end{bmatrix}.$$

The last equation is simply

$$(C^T C)\alpha = C^T Y.$$

Example 6. A very interesting illustration of the use of the material presented in the last example may be found in [23] (Noble) on pages 225–232. The application deals with a new method of determining the positions of stars from overlapping photographs. This reference should be particularly interesting to readers who are interested in astronomy or its relevance to space navigation.

EXERCISES

1. Calculate the adjoint matrices for the following matrices:

$$A = \begin{bmatrix} 2 & 3 \\ -4 & 1 \end{bmatrix}; \quad B = \begin{bmatrix} 1 & 3 & 0 \\ 2 & 1 & 0 \\ 0 & 1 & -1 \end{bmatrix}; \quad C = \begin{bmatrix} 1 & 0 & 0 & 1 \\ 0 & 0 & 1 & 0 \\ 0 & 2 & 0 & 0 \\ 0 & 0 & 0 & 2 \end{bmatrix}.$$

2. Making use of the answers to the preceding exercise, calculate the inverses of the given matrices.

3. Calculate the inverses of the following matrices, if possible. Check by $AA^{-1} = A^{-1}A = I_n$, or by use of the results of Exercise 17 of this section.

$$A = \begin{bmatrix} 2 & -1 \\ 4 & 3 \end{bmatrix}; \quad B = \begin{bmatrix} 4 & 2 \\ 2 & 1 \end{bmatrix}; \quad C = \begin{bmatrix} 2 & 0 & 3 \\ -1 & 0 & 2 \\ 0 & 1 & 1 \end{bmatrix}; \quad D = \begin{bmatrix} 3 & 2 & 1 \\ 2 & -1 & -1 \\ 1 & 4 & 0 \end{bmatrix}.$$

4. Using theorems of this section prove that if A and B are invertible of order n, then AB is invertible. (A similar statement could be made for the product of three or more invertible matrices of the same order.) Notice that this theorem was stated and proved in Chapter 3 (Theorem 7).

5. (a) What can be said about the rank of an invertible matrix of order 3?
   (b) What can be said about the rank of a noninvertible matrix of order 3?

6. Suppose A is a noninvertible matrix. Can we tell whether A^T will also be noninvertible? Explain.

7. If A is symmetric, how do cof A and adj A compare?

5.7 A FORMULA FOR A^{-1}

8. Prove that $\det(\text{adj } A) = (\det A)^{n-1}$, and then verify for $A = \begin{bmatrix} 2 & 1 & 0 \\ 0 & 2 & 1 \\ 3 & 0 & 2 \end{bmatrix}$.

   (See illustrative Examples 1 and 2 for the calculation of adj A.)

9. If $G = \begin{bmatrix} a & b \\ c & d \end{bmatrix}$ is an invertible C-matrix show that $\det G = \det(\text{adj } G)$.

10. For a square C-matrix A of order $n \geq 2$, does $(\text{cof } A)^T = \text{cof}(A^T)$?

11. Verify Theorem 16 for each of the following:

    (a) $A = \begin{bmatrix} 3 & 5 \\ 2 & -8 \end{bmatrix}$, $B = \begin{bmatrix} -1 & -3 \\ -4 & 6 \end{bmatrix}$;

    (b) $A = \begin{bmatrix} 0 & 1 & 1 \\ -2 & 2 & 3 \\ 1 & 0 & -1 \end{bmatrix}$, $B = \begin{bmatrix} 1 & 3 & 1 \\ 0 & 0 & 2 \\ -4 & -5 & -6 \end{bmatrix}$.

12. For C-matrices $A = [a_{ij}]_{(n,\,n)}$, $B = [b_{ij}]_{(n,\,n)}$, and $C = [c_{ij}]_{(n,\,n)}$, use Theorem 16 to prove that $|ABC| = |A|\,|B|\,|C|$.

13. Prove or disprove the following conjecture:

    $$\det(A + B) = \det A + \det B.$$

14. If $C = \begin{bmatrix} 2 & x \\ 4 & 2 \end{bmatrix}$, under what condition is C^{-2} defined? Find C^{-2} under that condition.

15. Prove that if $\det A \neq 0$, then $\det(A^{-1}) = (\det A)^{-1}$. (*Hint:* Make use of Theorem 16.) Note the different meanings of the exponents in this exercise.

16. If A is a fifth-order matrix and A^{-1} exists, do we know what the rank of A is? Why?

17. Theorem 13 of Chapter 3 stated that if $AB = I_n$ and if A or B is invertible, then $A = B^{-1}$ and $B = A^{-1}$. Prove that if $AB = I_n$ and if A is square, then both A and B must be invertible, thus simplifying the hypothesis of Theorem 13. (*Hint:* Use Theorem 16 of this Chapter.)

18. The material presented in this section can be used to prove Cramer's Rule (page 119): For the system $AX = B$ of n linear equations in n unknowns over the complex numbers, let $(^J A)$ denote the matrix obtained from A by replacing the jth column of A by the vector B. If $\det A \neq 0$, then the system $AX = B$ has exactly one solution; this solution is

    $$x_j = \frac{\det(^J A)}{\det A}, \quad j = 1, 2, \ldots, n.$$

    A proof of this theorem begins as follows:

    Proof. For the system $AX = B$, where A is invertible, we can show that $X = A^{-1}B$ (see page 57). By Theorem 14 and by definition of the adjoint matrix

    $$X = A^{-1}B = \frac{\text{adj } A}{|A|} B = \frac{1}{|A|} (\text{cof } A)^T B.$$

    Complete the proof.

19. Let $AX = B$ be a homogeneous system of n linear equations in n unknowns, where A is a C-matrix.
(a) If A is invertible, how many solutions does the system have?
(b) If the rank of A is less than n, how many solutions does the system have?

20. (a) How would Theorem 15 follow as a corollary of Theorem 16? (Note, however, that Theorem 15 was used in the proof of Theorem 16.)
(b) How would assertion (2) of the proof of Theorem 14 follow as a corollary of Theorem 16? (Note, however, that Theorem 14 was used in the proof of Theorem 16.)

21. Prove: The linear system

$$\begin{cases} a_{11}x_1 + \cdots + a_{1n}x_n = b_1, \\ \quad\vdots \qquad\qquad\qquad \vdots \\ a_{n1}x_1 + \cdots + a_{nn}x_n = b_n, \end{cases}$$

over C has a unique solution if and only if the coefficient matrix is invertible.

NEW VOCABULARY

determinant function   5.1
determinant of A 5.1
expansion of a determinant   5.1
interchange   5.1
order of a determinant   5.1
minor of an entry   5.2
cofactor of an entry   5.2

cofactor expansion   5.2
Cramer's Rule   5.2
line of a matrix   5.3
rank of a matrix   5.5
rank of a system   5.6
cofactor matrix   5.7
adjoint matrix   5.7

SPECIAL PROJECT

Start with the one-to-one correspondence between the set C of all complex numbers $a + bi$ and the set M of all R-matrices $\begin{bmatrix} a & b \\ -b & a \end{bmatrix}$. Investigate whether the one-to-one correspondence is preserved (1) for sums of corresponding elements in the respective sets, (2) for products of corresponding elements in the respective sets, (3) for real multiples of the elements of the respective sets. Investigate $\det \begin{bmatrix} a & b \\ -b & a \end{bmatrix}$; also investigate a matrix representation for the polar form of $a + bi$, and DeMoivre's Theorem in matrix form. Other conjectures may develop as the proposed items are investigated.

6

ALGEBRAIC SYSTEMS

6.1 Groups

Consider the set of all m by n C-matrices and the operation of matrix addition. Consider also the set of all integers and the operation of integer addition. Strange as it may seem at first, these two mathematical structures have some things in common. First, in both structures the addition of any two elements of the set produces a unique result that also belongs to the set; formally we say that the set is *closed* under addition, or that addition is a *binary operation over the set*—that is, for a set S, addition is a particular kind of mapping of the Cartesian product[1] $S \times S$ into S. Secondly, the associative law for addition is valid for both structures. Thirdly, both sets contain an identity element for addition. Fourthly, in each set, every element has an inverse element (with respect to addition) that also belongs to the set. We formalize these observations with the following definitions.

Definition 1. *An **algebraic system** consists of a nonempty set of elements S and one or more operations over S, which satisfy certain postulates.*

Definition 2. *A **group** is an algebraic system consisting of:*
  (A) *a nonempty set of elements S, and*

[1] For those not familiar with the term Cartesian product, see Appendix A on page A1.

(B) a binary operation ∘ over S, subject to the postulates:
  (1) ∘ is associative;
  (2) S contains an identity element for ∘;
  (3) each element in S has an inverse element in S with respect to ∘.

We have already observed that the algebraic systems formed by *the set of all m by n C-matrices with the binary operation of matrix addition and the set of all integers with the binary operation of integer addition* satisfy the requirements of Definition 2; hence *each* of them is an example of a group. The set of all real numbers with the binary operation of multiplication is an example of algebraic system that is *not* a group; the third postulate is *not* valid since 0 does not have an inverse. The study of *abstract* algebraic systems such as groups is very important in mathematics.[2] One reason is that by establishing general properties of a group then these properties must apply to any specific system that can be shown to be a group.

Definition 3. *A **commutative** (or **Abelian**) group is a group that satisfies the additional postulate that ∘ is commutative.*

Example 1. Both the group of all m by n C-matrices with respect to addition and the group of integers with respect to addition are commutative groups.

Further examples of groups are given next, and other examples may be found in the exercises.

Example 2. The system of nonzero real numbers, where ∘ is ordinary multiplication, is a commutative group. Notice that the product of two nonzero real numbers is a nonzero real number, the number 1 is an identity element, and the inverse of a nonzero real number x is $1/x$.

Example 3. The system of coordinate vectors with n complex components under addition is a commutative group.

Definition 4. *If n is a nonnegative integer, and if $a_0, a_1, \ldots, a_n$ are complex numbers, then*

$$p(x) = a_0 + a_1 x + \cdots + a_n x^n$$

*is said to be a **polynomial over the complex numbers**. If $a_n \neq 0$, then $p(x)$ is of*

---

[2] A French mathematician, Evariste Galois (1811–1832), had a great influence on the early development of group theory. His contributions to mathematics undoubtedly would have been much greater if he had lived beyond the age of twenty. (He was killed in a duel at that age.)

degree n. If every coefficient a_i is zero, then $p(x)$ is called the **zero polynomial**, which has no degree. If $a_0, a_1, \ldots, a_n$ are real numbers then $p(x)$ is a **polynomial over the real numbers.**

We assume that if the term $a_j x^j$ is omitted, then $a_j = 0$ ($j = 1, 2, \cdots$). Likewise, if a_0 is omitted then $a_0 = 0$.

If
$$p(x) = a_0 + a_1 x + \cdots + a_n x^n$$
and
$$q(x) = b_0 + b_1 x + \cdots + b_m x^m$$
are polynomials of degrees n and m, respectively, over the complex numbers, then $p(x) = q(x)$ if $a_i = b_i$ for all $i = 0, 1, 2, \cdots$. Assuming that $n \geq m$, addition of two polynomials is defined according to the formula

$$p(x) + q(x) = (a_0 + a_1 x + \cdots + a_m x^m + \cdots + a_n x^n)$$
$$+ (b_0 + b_1 x + \cdots + b_m x^m)$$
$$= (a_0 + b_0) + (a_1 + b_1)x + \cdots + (a_m + b_m)x^m + \cdots + a_n x^n.$$

The same formula is used to define addition when one of the polynomials is the zero polynomial.

Example 4. The following expressions are examples of polynomials over the complex numbers.

$p_1(x) = 2 + 3x + 0x^2 + 7x^3$, degree 3,
$p_2(x) = 2 + 3x + 7x^3$, degree 3,
$p_3(x) = 6 + ix + 6x^2$, degree 2,
$p_4(x) = 3$, degree 0,
$p_5(x) = 3 + 0x + 0x^2$, degree 0,
$p_6(x) = 0$ (the zero polynomial), no degree.

According to our assumption that an omitted term implies that the coefficient of that term is zero, and by the definition of equality, then

$$p_1(x) = p_2(x) \quad \text{and} \quad p_5(x) = p_4(x).$$

Also, by the definition of polynomial addition,

$$p_1(x) + p_3(x) = 8 + (3 + i)x + 6x^2 + 7x^3.$$

Obviously the zero polynomial is the identity element for polynomial addition. It is easy to verify that the set of *all* polynomials over the complex numbers forms a commutative group if the operation ∘ is polynomial addition.

Example 5. The system of all invertible n by n C-matrices with respect to matrix multiplication is an example of a group that is not commutative.

The set of elements forming a group need not be an infinite set as the examples so far might suggest. Some examples of *finite groups* follow.

Example 6. Let a set S consist of the set of numbers 1, 2, ..., 12 on a clock face, and let the operation $\circ$ represent the addition of hours obtained by the movement of the hour hand. In this group $5 \circ 9 = 2$ and $11 \circ 10 = 9$. The set S together with the operation $\circ$ forms a group.

Example 7. The set of numbers $\{1, -1, i, -i\}$, where $i = \sqrt{-1}$, together with ordinary multiplication, forms a finite group. The matrix representations of these complex numbers $\begin{bmatrix} 1 & 0 \\ 0 & 1 \end{bmatrix}, \begin{bmatrix} -1 & 0 \\ 0 & -1 \end{bmatrix}, \begin{bmatrix} 0 & 1 \\ -1 & 0 \end{bmatrix}, \begin{bmatrix} 0 & -1 \\ 1 & 0 \end{bmatrix}$ (see page 148) under multiplication forms a group.

Example 8. The set S of cube roots of 1 under multiplication forms a group. Here S consists of

$$1, \quad \frac{-1 + i\sqrt{3}}{2}, \quad \frac{-1 - i\sqrt{3}}{2}.$$

Likewise, the matrix representations of these complex numbers under multiplication form a group.

Example 9. Consider an equilateral triangle. Let the elements of S consist of a rotation of 120°, a rotation of 240°, and a rotation of 360°. Let $\circ$ be the operation "followed by." As an illustration, (a rotation of 240°) $\circ$ (a rotation of 120°) = (a rotation of 360°). S forms a group under $\circ$.

As stated in Exercise 12, a group can have only one identity with respect to the group operation $\circ$.

APPLICATIONS

Example 10. Subgroups, permutation groups, and cyclic groups are defined on pages 300–305 of [18] (Kemeny et al.), and the use of matrices in the study of cyclic groups is illustrated. These concepts, along with powers and inverses of matrices, are used in a discussion of marriage rules in primitive societies on pages 424–433 of [18]; basic references for this sociological study are given on page 458 of [18].

Example 11. Although a much deeper study of group theory is ordinarily required to develop applications, the titles of the following books hopefully will indicate to the reader some areas of applications and furnish references as well.

Cotton, F. A., *Chemical Applications of Group Theory*, Interscience Publishers, Inc., New York, 1963.

Hammermesh, M., *Group Theory and its Application to Physical Problems*, Addison-Wesley Publishing Co. Inc., Reading, Mass., 1962.

Hollingsworth, Charles A., *Vectors, Matrices, and Group Theory for Scientists and Engineers*, McGraw-Hill, New York, 1967.

Wigner, E. P., *Group Theory and its Application to Quantum Mechanics and Atomic Spectra*, Academic Press Inc., New York, 1959.

EXERCISES

1. Determine which of the following sets together with the specified operations are examples of groups. If one of them does not form a group, then state which postulates do not hold.
   (a) All odd integers under addition.
   (b) All odd integers under multiplication.
   (c) All n-dimensional real coordinate vectors under addition.
   (d) All even integers under addition.
   (e) All even integers under multiplication.
   (f) All integers under subtraction.
   (g) 1, w, w^2, under multiplication if $w^3 = 1$.
   (h) All 2 by 2 R-matrices under matrix multiplication.
   (i) All 2 by 3 R-matrices under matrix addition.
   (j) All rational numbers under multiplication.
   (k) All nonzero complex numbers under multiplication.
   (l) The elements a and b under operation ∘, where ∘ is defined by the following table:

∘	a	b
a	a	b
b	b	a

2. Which of the groups in Exercise 1 are commutative groups?

3. Which of the groups in Exercise 1 are finite groups?

4. Does the set of all 2 by 2 symmetric C-matrices form a group if matrix addition is the operation? Explain your answer.

5. Does the set of all 2 by 2 skew-symmetric R-matrices form a group with matrix addition as the operation? Explain your answer.

6. Does the set of all 2 by 2 Hermitian C-matrices form a group with matrix addition as the operation? Explain your answer.

7. Does the set of all third-order triangular C-matrices form a group with:
   (a) matrix addition as the operation? Explain.
   (b) multiplication as the operation? Explain.

8. Does the set of all third-order diagonal C-matrices form a group with:
   (a) matrix addition as the operation? Explain.
   (b) multiplication as the operation? Explain.

9. Does the set of all 2 by 2 orthogonal R-matrices form a group with:
   (a) matrix addition as the operation? Explain.
   (b) multiplication as the operation? Explain.
10. Does the set of all 2 by 2 unitary C-matrices form a group with:
    (a) matrix addition as the operation? Explain.
    (b) multiplication as the operation? Explain.
11. Is it possible for a group to contain only one element?
12. Prove that a group can have only one identity with respect to the group operation ∘. (Answer provided in booklet of Answers to Selected Even Numbered Exercises.)
13. Prove that if a, b, and c are elements of a group and $a \circ b = a \circ c$, then $b = c$ (cancellation property).

6.2 Rings

In the last section, a particular type of algebraic system with one operation was discussed; it seems natural to inquire next about systems involving a set of elements and *two* binary operations. Consider the following two structures: (*i*) the set S of all n by n C-matrices together with the operation of matrix addition, and (*ii*) the set S of all integers together with the operation of addition. Suppose that second operations, matrix multiplication and multiplication of integers, are added to the respective systems; here again, observe that these two systems have some common properties. First of all notice that in both systems if any two elements are combined by either operation, the result or image of the operation belongs to the original set; we say that each multiplication is a binary operation over the set, that is, a mapping of $S \times S$ into S. Secondly, both multiplications are associative. Thirdly, in both systems multiplication is right and left distributive with respect to addition. Fourthly, both systems have a multiplicative identity. Again we formalize these observations with the following definitions.

Definition 5. *A ring is an algebraic system consisting of:*
 (A) *a nonempty set of elements, S,*
 (B) *two binary operations $\oplus$ and $\odot$, over S, subject to the postulates:*
  (1) *S forms a commutative group with respect to $\oplus$;*
  (2) *$\odot$ is associative;*
  (3) *$\odot$ is right and left distributive with respect to $\oplus$.*

[DISTRIBUTIVITY]

Definition 6. *A **ring with unity** is a ring with the additional property: S contains an identity element for $\odot$.*

Both the ring of integers and the ring of n by n C-matrices are examples of rings with unity. Next notice that multiplication is commutative over the set of integers, but Cayley multiplication is *not* commutative over the set of n by n matrices. Thus only one of these systems is an example of a *commutative ring*, according to the following definition.

Definition 7. *A **commutative ring** is a ring with the additional property:* $\odot$ *is commutative.*

Definition 8. *A **division ring** is a ring with unity, with at least two elements, and with the additional property: every element except the identity for* $\oplus$ *has an inverse with respect to* $\odot$.

Further examples of rings are given below.

Example 1. The system of all even integers where $\oplus$ and $\odot$ are addition and multiplication, respectively.

Example 2. The system of all rational numbers with respect to addition and multiplication.

Example 3. The system of all real numbers with respect to addition and multiplication.

Example 4. The system of all complex numbers with respect to addition and multiplication.

Example 5. The system of all 2 by 2 C-matrices of the form $\begin{bmatrix} a & b \\ -b & a \end{bmatrix}$ with respect to matrix addition and multiplication.

Example 6. The set of all subsets of a given nonempty set U together with the operations of $\oplus$ and $\odot$, which are defined as follows: If a and b are arbitrary subsets of U then $a \oplus b$ is the set of all elements in the union but not in the intersection of a and b; $a \odot b$ is the intersection of the two subsets a and b.

EXERCISES

1. Determine which of the following are examples of a ring. If the given structure is not a ring, state why it is not.
   (a) The system of odd integers, where $\oplus$ is addition and $\odot$ is multiplication.
   (b) The system of nth order real column vectors, where $\oplus$ is addition and $\odot$ is the standard inner product.

(c) The system of pure imaginary numbers, where $\oplus$ is addition and $\odot$ is multiplication.

(d) The system of integers that are integral multiples of 3, where $\oplus$ is addition and $\odot$ is multiplication.

(e) The system of all 2 by 2 matrices over the integers, where $\oplus$ is matrix addition and $\odot$ is matrix multiplication.

(f) The system of all 1 by 1 matrices over the integers, where $\oplus$ is matrix addition and $\odot$ is matrix multiplication.

2. Which of the rings in Exercise 1 are commutative rings?

3. (a) Which of the rings in Exercise 1 are rings with unity?
   (b) Which of the rings in Exercise 1 are division rings?

4. Suppose that a set S of elements together with operations $\oplus$ and $\odot$ forms a ring. It may happen that a certain subset of S together with the same operations $\oplus$ and $\odot$ also meets all the requirements for a ring. When this is true, we have what is called a *subring* of the original ring. Determine which of the following subsets of the set of real numbers form a subring of the ring of real numbers. ($\oplus$ is "+" and $\odot$ is "·".)

   (a) The set of all integers.
   (b) The set of all positive integers.
   (c) The set of all even integers.
   (d) The set of numbers of the form $a \oplus (b \odot \sqrt{2})$, where a and b are integers.
   (e) The set of numbers of the form $a \oplus (b \odot \sqrt[3]{2})$, where a and b are integers.

5. Given that the elements w, u, v, and x with operations $\oplus$ and $\odot$ form a ring when the elements combine as shown in the following tables:

$\oplus$	w	u	v	x
w	u	w	x	v
u	w	u	v	x
v	x	v	u	w
x	v	x	w	u

$\odot$	w	u	v	x
w	w	u	w	u
u	u	u	u	u
v	w	u	v	x
x	u	u	x	x

The idea of these tables is analogous to the idea of addition and multiplication tables in arithmetic. Here $x \oplus v = w$, $x \odot v = x$, etc.

(a) Why is it clear from the tables that the set of elements is closed under the binary operations $\oplus$ and $\odot$?

(b) Verify that $u \oplus v = v \oplus u$. What significance is there in the fact that the table for $\oplus$ is symmetric about its main diagonal?

(c) Verify each of the following by using the tables:

$$u \oplus (w \oplus v) = (u \oplus w) \oplus v,$$
$$x \odot (v \odot u) = (x \odot v) \odot u,$$
$$x \odot (w \oplus x) = (x \odot w) \oplus (x \odot x).$$

Why would it be very tedious to verify completely the associative and distributive postulates?

(d) Why is u the identity element for $\oplus$?
(e) What is the inverse of x with respect to $\oplus$?
(f) Does this ring have an identity element for $\odot$?

6. Which of the following sets form a ring under matrix addition and Cayley multiplication? Give reasons for your answer.
    (a) The set of all m by m symmetric R-matrices.
    (b) The set of all m by m Hermitian C-matrices.
    (c) The set of all m by m diagonal C-matrices.
    (d) The set of all m by m upper triangular C-matrices.
    (e) The set of all m by m orthogonal R-matrices.
    (f) The set of all m by m unitary C-matrices.
7. Does the set of n by n C-matrices under matrix addition and Cayley multiplication form a ring? Give reasons for your answer.
8. Does the set of n by n C-matrices under matrix addition and Jordan multiplication (defined on page 52) form a ring? Give reasons for your answer.
9. Verify that the system of Example 5 really is a ring.
10. Verify that the system of Example 6 really is a ring.
11. Let a set S with binary operations $\oplus$ and $\odot$ be a ring, and suppose that y is the identity element for $\oplus$. Prove that if x is any element of S, then $x \odot y = y$ and $y \odot x = y$.

6.3 Fields

The reader undoubtedly is very familiar with the system of rational numbers, the system of real numbers, and the system of complex numbers. Each of these systems has certain properties in common: each one is a commutative division ring, or, as it is called, a *field*.[3] The definition of a field may also be made as follows:

Definition 9. *A field is an algebraic system consisting of:*
   (A) *a set S containing at least two elements,*
   (B) *two binary operations $\oplus$ and $\odot$ over S, subject to the postulates:*
       (1) *S forms a commutative group with respect to $\oplus$;*
       (2) *S, without the identity for $\oplus$, forms a commutative group with respect to $\odot$;*
       (3) *$\odot$ is right and left distributive with respect to $\oplus$.*

The set of n by n C-matrices does not form a field, because, in general, matrix multiplication is not commutative; also an n by n matrix does not

[3] The set of quaternions which are discussed in Special Project II at the end of this chapter is an example of a set that forms a division ring but not a field.

always have a multiplicative inverse. There are, however, a few sets of special matrices that form fields. For example, the set of R-matrices of the form $\begin{bmatrix} a & b \\ -b & a \end{bmatrix}$ forms a field under matrix addition and Cayley multiplication; the reader who has investigated the Special Project on page 148 will recognize the elements of this set as the matrix representations of the complex numbers. The set of n by n scalar C-matrices forms a field under matrix addition and Cayley multiplication.

Other less familiar examples of fields are:

Example 1. The system formed by all numbers of the form $a + b\sqrt{2}$, where a and b are rational numbers, and the operations are addition and multiplication.

Example 2. The system formed by the set of numbers $\{0, 1\}$ and the operations $\oplus$, $\odot$, which are defined according to the following operation tables:

$\oplus$	0	1
0	0	1
1	1	0

$\odot$	0	1
0	0	0
1	0	1

Example 3. The system formed by the finite set of consecutive integers $S = \{0, 1, \ldots, p-1\}$, where p is prime, and where the operations $\oplus$ and $\odot$ are defined as follows:

$$a \oplus b = c,$$

where c belongs to S and $c = a + b - np$ for some nonnegative integer n (illustration: if $p = 5$, then $4 \oplus 3 = 4 + 3 - 1 \cdot 5 = 2$);

$$a \odot b = d,$$

where d belongs to S and $d = a \cdot b - mp$ for some nonnegative integer m (illustration: if $p = 5$, then $4 \odot 4 = 4 \cdot 4 - 3 \cdot 5 = 1$).

Now we are in a position to examine more closely the role of an entry of a matrix in matrix theory. For reference we restate the definition of a matrix given at the beginning of this book.

Definition 2 of Chapter 1. *Let m and n be positive integers; a rectangular array of entries arranged in m rows and n columns*

$$A = \begin{bmatrix} a_{11} & a_{12} & \cdots & a_{1n} \\ a_{21} & a_{22} & \cdots & a_{2n} \\ \vdots & \vdots & & \vdots \\ a_{m1} & a_{m2} & \cdots & a_{mn} \end{bmatrix},$$

6.3 **FIELDS** 159

where the entries are elements of some algebraic system S, is called an **m by n matrix over S**.

Recall that so far we have restricted S to the field of complex numbers or the field of real numbers; notice in the definition, however, that a matrix may be defined over any algebraic system. Of course the properties developed for a set of m by n matrices over one system are likely to differ from the properties developed for a set of m by n matrices over a different system. For instance, the following example illustrates a fundamental property of matrices over a field that is not necessarily valid for matrices over a ring.

Example 4. Consider the following conjecture: If A is an m by n matrix over S and c is an element of S, then $cA = Ac$. By definition of cA and Ac we have $cA = [ca_{ij}]_{(m, n)}$ and $Ac = [a_{ij}c]_{(m, n)}$. But ca_{ij} does not necessarily equal $a_{ij}c$ if S is a ring, whereas $ca_{ij} = a_{ij}c$ if S is a field. Thus the conjecture is valid in one case and not in the other.

In Section 6.6 we shall develop a few properties of matrices with entries from still another algebraic system. The point we wish to emphasize is that *any theory of matrices that is developed depends upon the algebraic system to which the entries belong.*

The theory developed so far in this book has been limited to C-matrices and R-matrices. We have observed, however, that the real and complex number systems are fields, and it is natural to consider developing a theory of matrices over any field rather than just over the complex number field (see Special Project III at the end of this chapter). Such a development would differ little from what has been done so far in this book; for this reason, and because we wish to consider the theory of matrices over a field (where feasible), in the remainder of this book we will adopt the following convention.

When theorems and definitions concerning C-matrices that have been stated in the first five chapters are used as a reference in proving later theorems concerning matrices over an arbitrary field, we shall mean that the quoted theorems or definitions are valid for matrices over an arbitrary field.

For example, in Chapter 9 we will develop certain concepts that will require the use of the determinant of a matrix with entries from an *arbitrary* field F; we will refer the reader to Definition 1 of Chapter 5 (definition of determinant of a C-matrix) with the understanding that this definition can be extended to include any square matrix over an arbitrary field F. Likewise we will refer the reader to certain theorems of Chapter 5 with the understanding that when we make such a reference, the theorems concerning determinants of C-matrices can also be proved for matrices over an arbitrary field F.

EXERCISES

1. Determine which of the following are examples of fields; if one is not a field, state which of the postulates do not hold (let operations $\oplus$ be "+" and $\odot$ be "·" for the first three sets):
   (a) the set of all integers;
   (b) the set of all rational numbers;
   (c) the set of all pure imaginary numbers;
   (d) the set of all four-dimensional real vectors, where $\oplus$ is vector addition and $\odot$ is the dot product;
   (e) the set of all 1 by 1 matrices with rational entries where $\oplus$ is matrix addition and $\odot$ is Cayley multiplication;
   (f) the set of all 2 by 2 invertible C-matrices, where $\oplus$ is matrix addition and $\odot$ is matrix multiplication.

2. Suppose we did not require a field to have at least two elements (see part (A) of the definition of a field). Verify that in this case the single rational number zero, where $\oplus$ is "+" and $\odot$ is "·", would form a field. Notice that zero would be both the identity element for addition and the identity element for multiplication. It can be shown that if a field is required to have more than one element then the identity elements for $\oplus$ and $\odot$ must be different. It seems desirable to have different identity elements for $\oplus$ and $\odot$, which explains why we require a field to have at least two elements.

3. Consider the system formed by the set $S = \{0, 1, 2\}$ with $\oplus$ and $\odot$ defined as follows:

$\oplus$	0	1	2
0	0	1	2
1	1	2	0
2	2	0	1

$\odot$	0	1	2
0	0	0	0
1	0	1	2
2	0	2	1

   Give reasons for your answers to the following questions.
   (a) Are both $\oplus$ and $\odot$ binary operations over S?
   (b) What are the respective identity elements for $\oplus$ and $\odot$?
   (c) What is the inverse of each element with respect to $\oplus$ and $\odot$?
   (d) Are both $\oplus$ and $\odot$ commutative?
   (e) $\odot$ is distributive with respect to $\oplus$; does S under $\oplus$ and $\odot$ form a field?

4. Consider the system formed by the set $S = \{0, 1, 2, 3\}$ with $\oplus$ and $\odot$ defined as follows:

$\oplus$	0	1	2	3
0	0	1	2	3
1	1	2	3	0
2	2	3	0	1
3	3	0	1	2

$\odot$	0	1	2	3
0	0	0	0	0
1	0	1	2	3
2	0	2	0	2
3	0	3	2	1

   Answer the questions of Exercise 3.

5. Prove that in a field the cancellation law holds for both $\oplus$ and $\odot$.
6. Convince yourself, without doing any writing, that the set of rational numbers with operations of ordinary addition and multiplication is an example of a field.
7. (a) Is every field also a ring? Why?
   (b) Is every ring also a field? Why?
8. Let S be the set of all 2 by 2 matrices whose entries are integers with the binary operations of matrix addition and Cayley multiplication. What type of algebraic system do we have? How would the situation be changed if S were further restricted to symmetric matrices?
9. (a) In Example 1, show that multiplication is a binary operation over the given set; that is, show that the set is closed under multiplication.
   (b) In Example 1, what is the multiplicative inverse of $5 + \sqrt{2}$?
10. Prove that within a field the commutative and associative laws for $\odot$ do apply to the entire set S. (*Hint:* Use Exercise 11 of Section 6.2.) Notice that this is not assumed for postulate (2) of Definition 9.

6.4 Vector Spaces

The purpose of this section is to define another algebraic system involving a set of elements, called *vectors*, and an *associated field*.

Example 1. Consider the commutative group of all m by n C-matrices under matrix addition. We can associate with this group the field of complex numbers and then define a scalar multiplication as we did in Section 1.3: Let k be a complex number; then
$$k[a_{ij}]_{(m,\,n)} = [ka_{ij}]_{(m,\,n)}.$$
Obviously the result of this multiplication is an m by n C-matrix.

Example 2. Consider the commutative group under addition of all polynomials $a_0 + a_1 x + \cdots + a_n x^n$ with complex coefficients. Associate with this group the field of complex numbers, and then define a scalar multiplication as follows: Let k belong to the field of complex numbers; then
$$k(a_0 + a_1 x + \cdots + a_n x^n) = (ka_0) + (ka_1)x + \cdots + (ka_n)x^n.$$
The result is also a polynomial of the group.

Notice that, in both examples, *two* sets of elements were involved, along with a scalar multiplication which combined elements from both sets.

More precisely, scalar multiplication is a mapping $F \times V \to V$, where V is the set of elements of the commutative group, and where F is the set of elements of the associated field.

The structures begun in Examples 1 and 2 have certain common properties, because both are examples of another, and comparatively more sophisticated, algebraic system known as a **vector space**. In the following definition notice that the structures of Examples 1 and 2 do satisfy the requirements of a vector space.

Definition 10. *A set of elements V forms a **vector space** over a field F if:*

(A) *V forms a commutative group with respect to a binary operation,*

(B) *any element of V can be premultiplied by an arbitrary element of the associated field F and the result of this scalar multiplication is a unique element of V (that is, there exists a mapping $F \times V \to V$),*

(C) *the following postulates are valid: If α and β are arbitrary elements of V, if k_1 and k_2 are elements of the associated field F, if $\oplus$ is the group operation, if $+$ and $\cdot$ are the field operations, and if scalar multiplication is denoted by placing the elements involved in juxtaposition, then*

(1) $(k_1 + k_2)\alpha = k_1\alpha \oplus k_2\alpha$;
(2) $(k_1 \cdot k_2)\alpha = k_1(k_2\alpha)$;
(3) $k_1(\alpha \oplus \beta) = k_1\alpha \oplus k_1\beta$;
(4) $1\alpha = \alpha$ where 1 *is the multiplicative identity of F.*

The elements of V are called **vectors** and the elements of F are called **scalars**. The vector space is designated $V(F)$.

Unfortunately the terminology used here often confuses beginning students because they often think of the concepts of "vector" and "scalar" in a much less abstract sense. Hopefully the next few pages will help remove the difficulty.

Example 3. For a geometric example of a vector space consider the set of all two-dimensional *geometric vectors* as directed coplanar line segments, each of which originates at the origin of the Cartesian coordinate system and terminates at some point in the plane; the vector that originates and terminates at the origin is called the *zero geometric vector*. Let the sum of two geometric vectors α and β be defined as follows: If α and β are not collinear, construct a parallelogram with α and β as two adjacent sides; the sum $\alpha + \beta$ is the geometric vector that originates at the origin and terminates at the opposite vertex of the parallelogram (Figure 6.4.1). If α and β are collinear and in the same direction, with lengths $\|\alpha\|$ and $\|\beta\|$ respectively, then $\alpha + \beta$ has length $\|\alpha\| + \|\beta\|$ in the same direction as both α and β (Figure 6.4.2). If α and β are collinear and in opposite directions, then $\alpha + \beta$ has a length equal to the absolute value of $\|\alpha\| - \|\beta\|$ in the direction of the longer geometric vector, α or β (Figure 6.4.3); of course, $\alpha + \beta$ is the zero geometric vector if $\|\alpha\| - \|\beta\| = 0$. If either α or β is the zero geometric vector, then $\alpha + \beta$ is equal to the other geometric vector.

We can easily show that this set of vectors with the operation of vector addition forms a commutative group. Furthermore we can define the multiplication of a geometric vector by a real number as follows: If α is a geometric vector with

6.4 VECTOR SPACES

length $\|\alpha\|$, and if k is a real number, then $k\alpha$ is a geometric vector with length $|k|\,\|\alpha\|$ in the direction of α if k is positive and in the opposite direction if k is negative (the zero geometric vector results if $\alpha = \mathbf{0}$ or if $k = 0$). See Figure 6.4.4. It is then easy to verify that the postulates of a vector space are satisfied, and hence this group of geometric vectors is a vector space over the field of real numbers. Any element of this vector space can be pictured as a unique arrow in two-space, and addition and scalar multiplication can be visualized as shown in Figures 6.4.1 through 6.4.4.

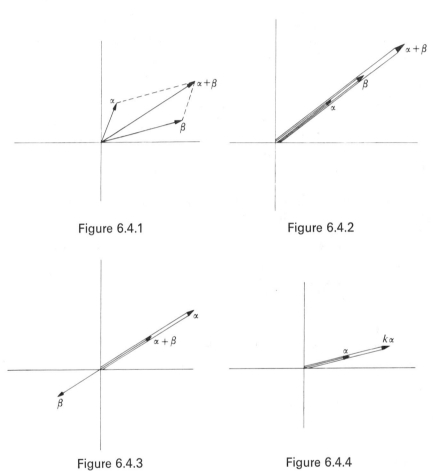

Figure 6.4.1   Figure 6.4.2

Figure 6.4.3   Figure 6.4.4

Example 4. It is easy to show that the set V of all coordinate vectors with n real components forms a commutative group with respect to addition. From Section 4.3 we know that multiplication of a coordinate vector $(a_1, a_2, \ldots, a_n)$ by a real number k is

$$k(a_1, a_2, \ldots, a_n) = (ka_1, ka_2, \ldots, ka_n),$$

which also belongs to the set V. From this definition it is not difficult to verify that the postulates of a vector space are satisfied, and hence this commutative group of real coordinate vectors with n components forms a vector space over the field of real numbers. Similarly, it can be shown that the commutative group with respect to addition of coordinate vectors with n components from a field F forms a vector space over that same field F; this vector space will be very important in the remainder of this book, and the special notation $V_n(F)$ will be reserved for it.

Example 5. The set of coordinate vectors with two integral components does not form a vector space over the field of rational numbers, because part (B) of Definition 10 is not satisfied. For example, $\frac{2}{3}(6, 4) = (4, \frac{8}{3})$ does not belong to the set, because $\frac{8}{3}$ is not an integer.

It is left as an exercise to verify that the structures begun in Examples 1 and 2 are examples of a vector space. (Theorems stated in previous chapters will suffice in Example 1.) Notice that in Example 1 the elements of V are m by n matrices and in Example 2 the elements of V are polynomials. According to Definition 10 these elements of a vector space are vectors; it is emphasized that the word vector is used here in a general sense. For example, a geometric vector is only a special example of the general concept of a vector. Likewise an ordered n-tuple, or coordinate vector, is only one example of the general concept of a vector. *The elements of any commutative group that satisfy the postulates for a vector space are vectors.* This includes the group of polynomials, the group of m by n matrices, and many others as well as the group of geometric vectors and the group of coordinate vectors.

The field F associated with the commutative group V is often referred to as the **underlying field**; the vector space notation $V(F)$ is customarily used to reveal this underlying field F. If the underlying field is the system of real numbers R, then a vector space over R is designated $V(R)$, or, if the underlying field is the system of complex numbers C, then the notation $V(C)$ is used.

Theorem 1. *If α is an arbitrary element of a vector space $V(F)$ and k is an arbitrary element of the field F, then*
  (i) $k\mathbf{0} = \mathbf{0}$;
  (ii) $0\alpha = \mathbf{0}$;
  (iii) $(-1)\alpha = -\alpha$.

Proof of part (i)   (Reasons are left as Exercise 14).
$$k\mathbf{0} = k\mathbf{0} + \mathbf{0} = k\mathbf{0} + (k\mathbf{0} + \{-(k\mathbf{0})\})$$
$$= (k\mathbf{0} + k\mathbf{0}) + \{-(k\mathbf{0})\} = k(\mathbf{0} + \mathbf{0}) + \{-(k\mathbf{0})\}$$
$$= k\mathbf{0} + \{-(k\mathbf{0})\} = \mathbf{0}.$$

The proof of parts (ii) and (iii) are left as Exercises 15 and 16, respectively. □

APPLICATIONS

★**Example 6.** Differential equations and their solutions are very important in many applied disciplines. One important use of the concept of a vector space arises from the fact that the solutions of the matrix linear differential equation

$$\begin{bmatrix} D_t y_1(t) \\ \vdots \\ D_t y_n(t) \end{bmatrix} = \begin{bmatrix} a_{11}(t) & \cdots & a_{1n}(t) \\ \vdots & & \vdots \\ a_{n1}(t) & \cdots & a_{nn}(t) \end{bmatrix} \begin{bmatrix} y_1(t) \\ \vdots \\ y_n(t) \end{bmatrix}$$

form a vector space called the *solution space*; a discussion of this statement may be found in Section 3.2 of [6] (Cole).

EXERCISES

In Exercises 1–4, determine which of the following sets form a vector space over the specified field F.

1. The set of all R-matrices of the form $\begin{bmatrix} a & b \\ -b & a \end{bmatrix}$ under matrix addition, where F is the field of real numbers.
2. The set of all real coordinate vectors (a, b) under addition, where F is the field of real numbers.
3. The set of all polynomials $ax^2 + bx + c$ of degree 2 or less (and the zero polynomial) with real coefficients under addition, where F is the field of real numbers.
4. The set of all $2n$ by $2n$ diagonal block C-matrices of the form

$$\begin{bmatrix} A & 0 \\ 0 & B \end{bmatrix},$$

where each block is n by n, under matrix addition. Let F be the field of complex numbers.
5. Verify that the commutative group V of Example 1 is a vector space, $V(C)$. (*Hint:* Use previous theorems.)
6. Verify that the commutative group V of Example 2 is a vector space, $V(C)$.
7. (*a*) Given the commutative group of coordinate vectors with n complex components under addition. Does this group form a vector space over the field of real numbers?
   (*b*) Given the commutative group of coordinate vectors with n real components under addition. Does this group form a vector space over the field of complex numbers? Why?
8. Will the set of three vectors $(2, 3)$, $(1, 1)$, and $(0, 0)$ form a vector space (usual operations)? Give sufficient reasons for your answer.

9. Verify that the vector (0, 0) forms a vector space over R under the usual operations of addition and scalar multiplication of a vector. Will any one real vector with two components other than (0, 0) form a vector space? Why?
10. Show that the set of real numbers forms a vector space over the real numbers where $\oplus$ is ordinary addition of real numbers.
11. Show that the set of n by n matrices with entries belonging to a field F forms a vector space over F, where $\oplus$ is matrix addition.
12. Using Example 3 as a model, construct the vector space of all geometric vectors in three-space over the field of real numbers.
13. Show that the set of vectors of Example 3 forms a commutative group under the operation of addition.
14. Supply the reasons for part (i) of the proof of Theorem 1.
15. Prove part (ii) of Theorem 1.
16. Prove part (iii) of Theorem 1.

6.5 Linear Algebras (optional)[4]

In the last section we learned that a vector space required a binary operation $\oplus$ over the set of vectors V. In this section we consider those vector spaces that have a second binary operation defined over V, that is, another mapping $V \times V \to V$.

Example 1. Consider the vector space $V(C)$ of n by n matrices with entries belonging to C. (See Exercise 11 of Section 6.4.) Early in this text we defined Cayley multiplication of any pair of n by n matrices with entries in C and the result was also an n by n matrix with entries in C; hence Cayley multiplication is a second binary operation over V.

Example 2. Consider the vector space $V(C)$ of all polynomials $a_0 + a_1 x + \cdots + a_n x^n$ with complex coefficients. If two arbitrary elements of this vector space are, say, of degree n and degree m, respectively, their multiplication can be defined as follows:

$$(a_0 + a_1 x + \cdots + a_n x^n)(b_0 + b_1 x + \cdots + b_m x^m) = (a_0 b_0) + (a_0 b_1 + a_1 b_0)x$$
$$+ (a_0 b_2 + a_1 b_1 + a_2 b_0)x^2 + \cdots + \left(\sum_{i=0}^{k} a_i b_{k-i}\right)x^k + \cdots + (a_n b_m)x^{n+m}.$$

Notice that the result is also a polynomial with complex coefficients; hence polynomial multiplication is a second binary operation over V.

---
[4] This section is listed as optional since it is a prerequisite only for the optional Section 8.8; the unifying concepts presented here, however, should be considered, even though time may not permit a study in depth.

6.5 LINEAR ALGEBRAS (OPTIONAL)

Using Examples 1 and 2 as illustrations we now define a new and richer abstract system called a linear algebra; we thereby demonstrate the common properties of the two structures in Examples 1 and 2 as well as any other structures that can be classified as *linear algebras*.

Definition 11.[5] *A **linear algebra** (or **algebra**) is a vector space $V(F)$ with:*
  (A) *a second binary operation $\odot$ over the set of vectors,*
  (B) *the following postulates where u, v, and w are arbitrary elements of $V(F)$ and k is an element of F:*
   (1) $u \odot (v \oplus w) = (u \odot v) \oplus (u \odot w)$ *(left distributive law);*
   (2) $(u \oplus v) \odot w = (u \odot w) \oplus (v \odot w)$ *(right distributive law);*
   (3) $k(u \odot v) = (ku) \odot v = u \odot (kv)$.
Because a linear algebra is a vector space we also refer to the elements of an algebra as vectors.

Definition 12. *An **associative algebra** is a linear algebra with the additional property:*
$\odot$ *is associative.*

Definition 13. *A **commutative algebra** is a linear algebra with the additional property:*
$\odot$ *is commutative.*

Definition 14. *An **algebra with unity** is a linear algebra which contains a vector that is an identity element for $\odot$.*

Definition 15. *A **division algebra** is a linear algebra with unity such that every vector except the identity for $\oplus$ has an inverse element with respect to $\odot$.*

Example 3. We may call upon theorems from previous chapters to prove that the vector space $V(C)$ of all n by n C-matrices is a linear algebra. Moreover, two more theorems from earlier chapters will establish that this vector space is an associative algebra with unity. Why is it not a division algebra?

Example 4. With some lengthy but straightforward work, it is possible to justify that the vector space $V(C)$ of all polynomials over C is a linear algebra by showing that each postulate of Definition 11 is satisfied, using the definition of polynomial multiplication as given in Example 2. Moreover it can be shown that $V(C)$ is a commutative and associative algebra with unity.

---

[5] The definition of a linear algebra is not uniform in the literature on the subject. Some authors prefer to require associativity of $\odot$ for a linear algebra and then to refer to a system without associativity as a nonassociative algebra. (See Definition 12.)

Example 5. The commutative group of complex numbers under addition, together with the associated field of real numbers, form a vector space $V(R)$ that is a linear algebra where $\oplus$ is ordinary addition of complex numbers and $\odot$ is ordinary multiplication of complex numbers. The reader undoubtedly is familiar with the algebra of complex numbers, and the justification of the properties required by Definitions 10 and 11 should be fairly easy. Moreover, it can be shown that $V(R)$ is a commutative and associative division algebra.

In retrospect let us examine the abstract algebraic systems developed in this chapter. We can see that we have developed algebraic systems that were increasingly more sophisticated. We began with a rather simple structure, known as a group, involving only *one* set of elements with *one* binary operation over the set. Then we defined a system, known as a ring, involving *two* operations rather than one. Next we found that a field is essentially a ring with a few more properties. The next system, known as a vector space, involved *two* sets, V and F, a binary operation $\oplus$ over V, and a scalar multiplication which combines elements of F with elements of V. Finally we found that a linear algebra is basically a vector space involving a *second* binary operation $\odot$ over V.

EXERCISES

In Exercises 1–4 of the last section, all of the sets formed vector spaces. Which of the four following vector spaces together with the specified multiplications are linear algebras?

1. The vector space $V(R)$ of all R-matrices of the form $\begin{bmatrix} a & b \\ -b & a \end{bmatrix}$ where $\odot$ is Cayley multiplication.

2. The vector space $V(R)$ of the real coordinate vectors (a, b) where $\odot$ is defined as follows

    $$(a, b) \odot (c, d) = (ac - bd, ad + bc).$$

3. The vector space $V(R)$ of all polynomials of degree 2 or less with real coefficients and the zero polynomial, where $\odot$ is defined as the multiplication of Example 2.

4. The vector space $V(C)$ of all $2n$ by $2n$ diagonal block C-matrices of the form

    $$\begin{bmatrix} A & 0 \\ \hline 0 & B \end{bmatrix},$$

    where each block is n by n and where $\odot$ is Cayley multiplication.

5. In Exercises 1–4, identify all associative algebras.

6. In Exercises 1–4, identify all commutative algebras.
7. In Exercises 1–4, identify all linear algebras with unity.
8. In Exercises 1–4, identify all division algebras.
9. Prove that the vector space of Example 3 is an associative algebra with unity. Why is it not a division algebra?
10. Prove that the vector space of Example 5 is a linear algebra.
11. For the vector space $V_n(R)$ of coordinate vectors with n real components, will the "dot product of" serve as the second operation $\odot$ in order for $V_n(R)$ to be a linear algebra? Give reasons for your answer.
12. For the vector space $V(C)$ of m by n C-matrices, will Cayley multiplication serve as $\odot$ in order for $V(C)$ to be a linear algebra? Give reasons for your answer.
13. Is the vector space $V(R)$ of n by n R-matrices a linear algebra if $\odot$ is Jordan multiplication? (See Exercise 24 of Section 3.1 for definition of Jordan multiplication.)
14. Is the vector space $V(R)$ of n by n R-matrices a linear algebra if $\odot$ is Lie multiplication? (See Exercise 24 of Section 3.1 for definition of Lie multiplication.)
15. Prove that the vector space of Example 1 is a linear algebra.
16. Prove that the vector space of Example 2 is a linear algebra.

6.6 Boolean Algebras (optional)

Another algebraic system of considerable interest is known as a Boolean Algebra (named after George Boole, English, 1815–1864).

Definition 16. *A **Boolean algebra** is an algebraic system consisting of:*
   (A) *a nonempty set S of elements,*
   (B) *two binary operations $\cup$ and $\cap$ (read "cup" and "cap") over S, subject to the postulates:*
   (1) *$\cup$ and $\cap$ are commutative;*
   (2) *$\cup$ is distributive with respect to $\cap$ on S and $\cap$ is distributive with respect to $\cup$ on S;*
   (3) *S contains identity elements 0 for $\cup$ and 1 for $\cap$;*
   (4) *for every element x in S there is an element designated x' also in S such that*
$$x \cup x' = 1 \quad \text{and} \quad x \cap x' = 0.$$

Thus if x, y, and z are elements of S:
   (1) $x \cup y = y \cup x;$ $x \cap y = y \cap x.$
   (2) $x \cup (y \cap z) = (x \cup y) \cap (x \cup z);$ $x \cap (y \cup z) = (x \cap y) \cup (x \cap z).$
   (3) $x \cup 0 = x;$ $x \cap 1 = x.$

Example 1. The set of all subsets of a given set U, with $\cup$ representing the union of two sets and $\cap$ representing the intersection of two sets, forms a Boolean algebra. The empty set (or null set) $\varnothing$ is the identity element for $\cup$, and the set U is the identity element for $\cap$.

Example 2. Let the set S consist of only two elements 0 and 1, define $\cup$ and $\cap$ according to

$$0 \cup 0 = 0, \qquad 0 \cap 0 = 0,$$
$$0 \cup 1 = 1 \cup 0 = 1, \qquad 0 \cap 1 = 1 \cap 0 = 0,$$
$$1 \cup 1 = 1, \qquad 1 \cap 1 = 1,$$

and define $0' = 1$, $1' = 0$; then we have another example of a Boolean algebra. Because there are only two elements, this structure is known as a *binary Boolean algebra*.

Definition 17. *Matrices whose entries are elements of a Boolean algebra are called* **Boolean matrices**.

Because a Boolean algebra is not a field, the structure that we have constructed for matrices over a field will not necessarily apply in its entirety to Boolean matrices. We must start all over again; we begin with two binary operations over a set S of m by n Boolean matrices.

Definition 18. *For two m by n Boolean matrices,*

$$A = [a_{ij}]_{(m,n)} \text{ and } B = [b_{ij}]_{(m,n)},$$

we define $A \cup B = [(a_{ij} \cup b_{ij})]_{(m,n)}$ *and* $A \cap B = [(a_{ij} \cap b_{ij})]_{(m,n)}$. *The results of these binary operations are called the* **union** *and* **intersection**, *respectively, of A and B.*

Example 3. Consider Boolean matrices over a binary Boolean algebra; from Example 2 we have

$$0 \cup 0 = 0, \quad 0 \cup 1 = 1 \cup 0 = 1, \quad 1 \cup 1 = 1,$$

and

$$0 \cap 0 = 0, \quad 0 \cap 1 = 1 \cap 0 = 0, \quad 1 \cap 1 = 1.$$

Hence

$$\begin{bmatrix} 1 & 0 & 1 & 0 \\ 0 & 1 & 0 & 1 \end{bmatrix} \cup \begin{bmatrix} 1 & 0 & 0 & 1 \\ 1 & 1 & 0 & 0 \end{bmatrix} = \begin{bmatrix} 1 & 0 & 1 & 1 \\ 1 & 1 & 0 & 1 \end{bmatrix}.$$

Also,

$$\begin{bmatrix} 1 & 0 & 1 & 0 \\ 0 & 1 & 0 & 1 \end{bmatrix} \cap \begin{bmatrix} 1 & 0 & 0 & 1 \\ 1 & 1 & 0 & 0 \end{bmatrix} = \begin{bmatrix} 1 & 0 & 0 & 0 \\ 0 & 1 & 0 & 0 \end{bmatrix}.$$

6.6 BOOLEAN ALGEBRAS (OPTIONAL)

It would be useful for the reader to show that the commutative and associative laws are valid for both the operations $\cup$ and $\cap$ over the set of matrices whose entries are elements of a binary Boolean algebra. Both distributive laws

$$A \cap (B \cup C) = (A \cap B) \cup (A \cap C)$$

and

$$A \cup (B \cap C) = (A \cup B) \cap (A \cup C)$$

are also valid. The identity element for $\cup$ is the null matrix; that is,

$$A \cup \mathbf{0} = \mathbf{0} \cup A = A.$$

The identity element for $\cap$ is the matrix U for which every entry is 1, that is,

$$A \cap U = U \cap A = A,$$

where $U = [1]_{(m, n)}$.

Figure 6.6.1

APPLICATIONS

Example 4. If x and y represent elements of the set $\{0, 1\}$, the function

$$f(x, y) = x \cup (x \cap y)$$

is an example of a class of functions known as binary Boolean functions. This particular function $f(x, y)$ defined over the set of switches represents the electrical circuit shown in Figure 6.6.1; this circuit is closed if x is closed (whether or not y is closed). Over the set of propositions, this function represents the statement

[proposition x] or [(proposition x) and (proposition y)];

this statement is true if proposition x is valid (whether or not proposition y is valid).

The question arises as to whether this function $f(x, y) = x \cup (x \cap y)$ can be represented by a matrix. If we let the variable x be represented by the matrix $\begin{bmatrix} 1 & 0 \\ 1 & 0 \end{bmatrix}$ and y be represented by the matrix $\begin{bmatrix} 1 & 1 \\ 0 & 0 \end{bmatrix}$, then we obtain results under the matrix operations $\cup$ and $\cap$ that are consistent with those obtained without the use of matrices. Using the basic definitions of Example 2, we can show that $x \cup (x \cap y) = x$. From the matrices defined above, we obtain

$$f(x, y) = x \cup (x \cap y)$$
$$= \begin{bmatrix} 1 & 0 \\ 1 & 0 \end{bmatrix} \cup \left(\begin{bmatrix} 1 & 0 \\ 1 & 0 \end{bmatrix} \cap \begin{bmatrix} 1 & 1 \\ 0 & 0 \end{bmatrix} \right)$$
$$= \begin{bmatrix} 1 & 0 \\ 1 & 0 \end{bmatrix} \cup \begin{bmatrix} 1 & 0 \\ 0 & 0 \end{bmatrix} = \begin{bmatrix} 1 & 0 \\ 1 & 0 \end{bmatrix},$$

and we notice that the result is x; thus, we have a matrix representation for $f(x, y)$, and moreover a simplification has been accomplished by means of matrix operations.

Consider another binary Boolean function

$$f(x, y) = (x \cap y') \cup \{(x' \cup y) \cap x\},$$

which represents the arrangement of switches shown in Figure 6.6.2. Using the

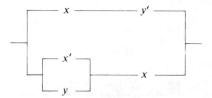

Figure 6.6.2

matrix representations for x and y and defining

$$x' = \begin{bmatrix} 0 & 1 \\ 0 & 1 \end{bmatrix} \quad \text{and} \quad y' = \begin{bmatrix} 0 & 0 \\ 1 & 1 \end{bmatrix},$$

(which are the opposites of x and y and satisfy the postulates of Definition 16), we have

$$f(x, y) = (x \cap y') \cup \{(x' \cup y) \cap x\}$$
$$= \left(\begin{bmatrix} 1 & 0 \\ 1 & 0 \end{bmatrix} \cap \begin{bmatrix} 0 & 0 \\ 1 & 1 \end{bmatrix} \right) \cup \left\{ \left(\begin{bmatrix} 0 & 1 \\ 0 & 1 \end{bmatrix} \cup \begin{bmatrix} 1 & 1 \\ 0 & 0 \end{bmatrix} \right) \cap \begin{bmatrix} 1 & 0 \\ 1 & 0 \end{bmatrix} \right\}$$
$$= \begin{bmatrix} 0 & 0 \\ 1 & 0 \end{bmatrix} \cup \left\{ \begin{bmatrix} 1 & 1 \\ 0 & 1 \end{bmatrix} \cap \begin{bmatrix} 1 & 0 \\ 1 & 0 \end{bmatrix} \right\}$$
$$= \begin{bmatrix} 0 & 0 \\ 1 & 0 \end{bmatrix} \cup \begin{bmatrix} 1 & 0 \\ 0 & 0 \end{bmatrix} = \begin{bmatrix} 1 & 0 \\ 1 & 0 \end{bmatrix},$$

which is x. Hence we have demonstrated that the single switch x is equivalent to the whole arrangement of switches shown in Figure 6.6.2. A manufacturer should install only switch x with the obvious advantage in savings in the cost, space requirements, and maintenance of the circuit. We have also demonstrated that a single proposition x is equivalent to a whole statement of propositions.

6.6 BOOLEAN ALGEBRAS (OPTIONAL)

Important simplifications or analyses of laws, contracts, and judicial decisions become possible.[6]

Notice that x and y are defined so that the following intersections,

$$x \cap y = \begin{bmatrix} 1 & 0 \\ 0 & 0 \end{bmatrix}, \quad x \cap y' = \begin{bmatrix} 0 & 0 \\ 1 & 0 \end{bmatrix},$$

$$x' \cap y = \begin{bmatrix} 0 & 1 \\ 0 & 0 \end{bmatrix}, \quad x' \cap y' = \begin{bmatrix} 0 & 0 \\ 0 & 1 \end{bmatrix},$$

are represented by unique matrices each with exactly one nonzero entry. To define matrices representing x, y, and z for functions of three variables, matrices with eight entries are needed. The reader should verify that using the following definitions

$$x = \begin{bmatrix} 1 & 1 & 0 & 0 \\ 1 & 1 & 0 & 0 \end{bmatrix}, \quad y = \begin{bmatrix} 1 & 1 & 1 & 1 \\ 0 & 0 & 0 & 0 \end{bmatrix}, \quad z = \begin{bmatrix} 0 & 1 & 1 & 0 \\ 0 & 1 & 1 & 0 \end{bmatrix},$$

the following statements are valid:

$$x \cap (y \cap z) = \begin{bmatrix} 0 & 1 & 0 & 0 \\ 0 & 0 & 0 & 0 \end{bmatrix}, \quad \text{and} \quad x' \cap (y \cap z') = \begin{bmatrix} 0 & 0 & 0 & 1 \\ 0 & 0 & 0 & 0 \end{bmatrix}.$$

A more detailed discussion of Boolean matrices and their applications may be found in [10] (Flegg).

Figure 6.6.3

Example 5. An interesting application of Boolean matrices in logic may be found in an article by Parker [24]; there is a discussion of this same article on page 51 of [9] (Eves). For the sake of brevity we shall introduce this application in terms of an oversimplified and small logical system. Let there be a set of undefined and defined terms, and a set of statements $\{A_1, A_2, A_3, A_4\}$ involving those terms. Then make a set of assumptions $A_3 \Rightarrow A_4$, $A_4 \Rightarrow A_2$, $A_2 \Rightarrow A_1$, and $A_2 \Rightarrow A_3$ as illustrated in Figure 6.6.3. The Cayley product of binary Boolean matrices can be used to determine which propositions will and which will not follow from the four basic assumptions. Construct the Boolean matrix

$$M = \begin{bmatrix} 1 & 0 & 0 & 0 \\ 1 & 1 & 1 & 0 \\ 0 & 0 & 1 & 1 \\ 0 & 1 & 0 & 1 \end{bmatrix}, \text{ where } m_{ij} = 1 \text{ if } A_i \Rightarrow A_j \text{ or if } i = j; \text{ otherwise } m_{ij} = 0.$$

---

[6] Kort, Fred, "Simultaneous Equations and Boolean Algebra in the Analysis of Judicial Decisions," *Jurimetrics*, Hans Baade, ed., Basic Books, New York, 1963.

Then calculate the Cayley product M^2 under the rules of combination for the binary Boolean entries (that is, $0+0=0$, $1+0=0+1=1$, $1+1=1$, $0\cdot 0=0$, $0\cdot 1=1\cdot 0=0$, and $1\cdot 1=1$). We find $M^2 = \begin{bmatrix} 1 & 0 & 0 & 0 \\ 1 & 1 & 1 & 1 \\ 0 & 1 & 1 & 1 \\ 1 & 1 & 1 & 1 \end{bmatrix}$.

If the (ij)th entry of M^2 is 1, then either $i=j$ or $A_i \Rightarrow A_j$ was a postulate or $A_i \Rightarrow A_k \Rightarrow A_j$ for at least one value of k. Finally, calculate $M^3 = \begin{bmatrix} 1 & 0 & 0 & 0 \\ 1 & 1 & 1 & 1 \\ 1 & 1 & 1 & 1 \\ 1 & 1 & 1 & 1 \end{bmatrix}$.

Because we began with only 4 statements, the entries of M^3 determine whether or not Statement A_j follows from Statement A_i. For a deeper discussion the reader is encouraged to read Parker's article [24].

EXERCISES

1. Consider the set of 2 by 2 matrices with entries that are elements of a binary Boolean algebra.
   (a) What are the union and intersection of
   $$\begin{bmatrix} 0 & 1 \\ 1 & 0 \end{bmatrix} \text{ and } \begin{bmatrix} 0 & 0 \\ 1 & 1 \end{bmatrix}?$$
   (b) What are the identity elements for $\cup$ and $\cap$?
   (c) Prove that the commutative law for $\cup$ is valid.
2. Justify the binary Boolean algebra theorems
   $$(x \cup y)' = x' \cap y' \text{ and } (x \cap y)' = x' \cup y',$$
   using Boolean matrices. (De Morgan's laws, named after Augustus De Morgan, English, 1806–1871.)

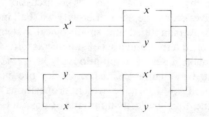

Figure 6.6.4

3. (a) Express the arrangement of switches shown in Figure 6.6.4 as a binary Boolean function.
   (b) Simplify the function using Boolean matrices.
4. For a binary Boolean function of four variables $f(x, y, z, t)$, how many entries must a Boolean matrix representing x, y, z, or t contain?

NEW VOCABULARY

binary operation over a set   6.1
closed set   6.1
algebraic system   6.1
group   6.1
commutative (or Abelian) group   6.1
polynomial over the complex numbers   6.1
degree of a polynomial   6.1
zero polynomial   6.1
polynomial over the real numbers   6.1
finite group   6.1
ring   6.2
ring with unity   6.2
commutative ring   6.2
division ring   6.2
subring   6.2
field   6.3

vector space over a field   6.4
vector   6.4
associated scalar field of a vector space   6.4
geometric vector   6.4
zero geometric vector   6.4
underlying field of a vector space   6.4
linear algebra   6.5
associative algebra   6.5
commutative algebra   6.5
algebra with unity   6.5
division algebra   6.5
Boolean algebra   6.6
binary Boolean algebra   6.6
Boolean matrix   6.6
union of Boolean matrices   6.6
intersection of Boolean matrices   6.6

SPECIAL PROJECTS

I. Start with the definition: *If A is an n by n matrix over a field F, where $n \geq 2$ and $A = \text{cof } A$, then A is a **cofactoral matrix over F**.* Investigate such things as det A, A^{-1}, does the set of all cofactoral matrices form a group under multiplication, does the set of all 2 by 2 cofactoral matrices form a field, is real A orthogonal? Other conjectures should develop as the proposed items are investigated; write a paper consisting of theorems and proofs that follow from the definition given above; then refer to the article Cofactoral Matrices by Bucher, R., and Godbole, S., in *Mathematics Magazine*, Vol, 42, No. 3, May, 1969, pp. 142–145.

II. Study the history and development of **quaternions** by consulting general encyclopedias, and books on algebra such as Birkoff, G., and MacLane, S., *A Survey of Modern Algebra, Third Edition*, Macmillan, New York, 1965, pp. 222–223. Then study how matrices can be used to represent quaternions by 2 by 2 *C*-matrices as shown in the SMSG Book *Introduction to Matrix Algebra*, Yale University Press, New Haven, 1961, or by 4 by 4

R-matrices as shown in Davis, P. J., *The Mathematics of Matrices*, Blaisdell Publishing Company, New York, 1965, pp. 319–321. Write a short paper reflecting your study.

III. State the generalized form of all of the theorems and definitions in the first five chapters of this book that can be generalized to use matrices over a *field* rather than the C-matrices or R-matrices that were used.

IV. State the generalized form of all of the theorems and definitions in the first five chapters of this book that can be generalized to use matrices over a *ring* rather than the C-matrices or R-matrices that were used.

7

VECTOR SPACES

7.1 Linear Dependence and Independence

In Section 6.4, a vector space is defined to be an algebraic system satisfying certain specified properties, and the elements of a vector space are called vectors. As illustrations, the reader is reminded here of a few of the many sets of vectors; in each case listed below, addition is the vector operation, and in each case the associated field must be chosen carefully to maintain closure when an element of the set is multiplied by an element of the associated field:

(1) The set of all three-dimensional (or two-dimensional) geometric vectors.
(2) The set of all coordinate vectors with n real (or complex) components.
(3) The set of all polynomials in x with real (or complex) coefficients.
(4) The set of all polynomials of degree n or less with real (or complex) coefficients, and the zero polynomial.
(5) The set of all m by n matrices with real (or complex) entries.

Keep in mind that other sets of elements exist which can form vector spaces and that the word vector is used here in an abstract sense to mean an element of a vector space. We will denote vectors with Greek letters (α, β, γ, etc.), unless otherwise stated. Remember that for a set of vectors V the symbol $V(F)$ designates a vector space over the field F; if we wish to restrict ourselves to the field of real numbers R, then we will designate the vector space as $V(R)$.

For later illustrative purposes we now point out that the vector space of three-dimensional geometric vectors can be used as a geometric representation of the vector space of coordinate vectors with three real components. This is possible because to every three-dimensional geometric vector there corresponds a unique three-dimensional real coordinate vector and conversely, and this one-to-one correspondence is preserved under addition and scalar multiplication. Likewise two-dimensional geometric vectors can be used to represent two-dimensional real coordinate vectors. See Figures 7.1.1 and 7.1.2 for examples of the preservation of the one-to-one correspondence under addition and scalar multiplication in two dimensions.

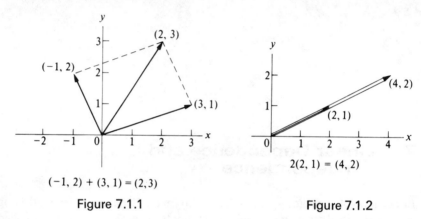

$(-1, 2) + (3, 1) = (2, 3)$

Figure 7.1.1

$2(2, 1) = (4, 2)$

Figure 7.1.2

Certain vectors in a vector space can be combined to produce other vectors in the same vector space.

Example 1. Any two-dimensional coordinate vector $\gamma = (c_1, c_2)$ can be produced by a proper choice of scalar coefficients in the following equation,

$$\gamma = k_1(1, 2) + k_2(3, 4);$$

the proper k_1 and k_2 can be found by solving

$$\begin{cases} k_1 + 3k_2 = c_1, \\ 2k_1 + 4k_2 = c_2, \end{cases}$$

for k_1 and k_2. We say that γ is a *linear combination* of $(1, 2)$ and $(3, 4)$.

Definition 1. *If $\{\alpha_1, \alpha_2, \ldots, \alpha_n\}$ is a set of vectors of a vector space $V(F)$, if the symbol "$+$" represents the vector operation, and if $k_1, k_2, \ldots, k_n$ are scalars belonging to F, then the vector*

$$\gamma = k_1\alpha_1 + k_2\alpha_2 + \cdots + k_n\alpha_n$$

*is said to be a **linear combination** of the set of vectors*

$$\{\alpha_1, \alpha_2, \ldots, \alpha_n\}.$$

7.1 LINEAR DEPENDENCE AND INDEPENDENCE

Definition 2. *A set of vectors* $\{\alpha_1, \alpha_2, \ldots, \alpha_n\}$ *is said to* **span** *or* **generate** *a vector space* $V(F)$ *if and only if each element of the set belongs to* $V(F)$ *and every vector in* $V(F)$ *can be expressed as a linear combination of the set.*

Example 2. The coordinate vectors $\alpha_1 = (1, 0, 0)$, $\alpha_2 = (0, 1, 0)$, and $\alpha_3 = (0, 0, 1)$ *span* the vector space $V(R)$ consisting of all three-dimensional real coordinate vectors, because, by appropriate choice of coefficients, any three-dimensional real coordinate vector γ can be written as a linear combination of α_1, α_2, and α_3. See Figure 7.1.3.

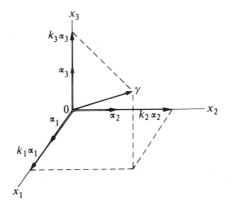

Figure 7.1.3

Example 3. The vectors $\beta_1 = (1, 0, 0)$, $\beta_2 = (0, 1, 0)$, and $\beta_3 = (1, 1, 0)$ do not span the vector space $V(R)$ consisting of all three-dimensional real coordinate vectors, because there is no choice of the k's that will enable us to express any vector with a nonzero third component in terms of β_1, β_2, and β_3. That is, any linear combination γ of $(1, 0, 0)$, $(0, 1, 0)$, and $(1, 1, 0)$ will be in the x_1x_2-plane (Fig. 7.1.4). Therefore, these three vectors do not span three-space $V(R)$. (It can

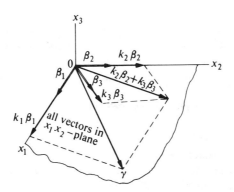

Figure 7.1.4

be verified, however, that the set of all three-dimensional real coordinate vectors which terminate in the x_1x_2-plane forms a vector space. Notice that β_1, β_2, β_3 span this vector space; that is, each vector in the x_1x_2-plane can be expressed as a linear combination of β_1, β_2, and β_3.)

Example 3 provides us with an illustration of what is known as a *subspace*. Notice that the set of coordinate vectors that can be written as a linear combination of β_1, β_2, and β_3 is a subset W of the set of all three-dimensional real coordinate vectors V, and moreover the subset W forms a vector space over the real numbers. We say that $W(R)$ is a *subspace* of $V(R)$.

Definition 3. *For a vector space $V(F)$, let W be a subset of V. If W forms a vector space $W(F)$ under the same operations, then $W(F)$ is a* **subspace** *of $V(F)$.*

Theorem 1. *If $\alpha_1, \ldots, \alpha_k$ are elements of a vector space $V(F)$, then the set of all linear combinations of $\alpha_1, \ldots, \alpha_k$ is a subspace of $V(F)$.*

The proof is left as Exercise 13.

In Examples 2 and 3 we have illustrations of two fundamental concepts in the study of vector spaces. Notice that in Example 2 each vector of the set is *independent* of the other two vectors, that is, no vector of the set can be written as a linear combination of the others. On the other hand, in Example 3, there is a vector of the set that is *dependent* on the other two vectors in the set, that is, there is a vector that can be written as a linear combination of the others. These two concepts are formalized in Definitions 4 and 5. (In what follows, a bold face "**0**" will designate the identity element of the set of vectors. For example, in the set of coordinate vectors, **0** represents the zero coordinate vector, and, in the set of polynomials, **0** represents the zero polynomial.)

Definition 4. *For a vector space $V(F)$ with the vector operation designated by the symbol "$+$" the n vectors $\alpha_1, \alpha_2, \ldots, \alpha_n$ of $V(F)$ are said to be* **linearly dependent** *if there exists a set of scalars $k_1, k_2, \ldots, k_n$ of F, not all zero (where zero is the additive identity of F), such that*

$$k_1\alpha_1 + k_2\alpha_2 + \cdots + k_n\alpha_n = \mathbf{0}.$$

It follows from the last equation that if the vectors are linearly dependent, any vector α_i whose coefficient is not zero can be expressed as a linear combination of the other vectors by transposing them to the other side of the equation and then multiplying both sides by the multiplicative inverse of k_i.

7.1 LINEAR DEPENDENCE AND INDEPENDENCE

Definition 5. *For a vector space $V(F)$ the vectors $\alpha_1, \alpha_2, \ldots, \alpha_n$ of $V(F)$ that are not linearly dependent are said to be **linearly independent**. That is, $\alpha_1, \alpha_2, \ldots, \alpha_n$ are linearly independent if and only if*

$$k_1\alpha_1 + k_2\alpha_2 + \cdots + k_n\alpha_n = 0,$$

implies that $k_1 = k_2 = \cdots = k_n = 0$.

Example 4. The coordinate vectors $\alpha_1 = (1, 1, 0)$, $\alpha_2 = (3, 2, 1)$, and $\alpha_3 = (2, 1, 1)$ are linearly dependent because there exists a set of scalars k_1, k_2, k_3, not all zero, such that $k_1(1, 1, 0) + k_2(3, 2, 1) + k_3(2, 1, 1) = (0, 0, 0)$; such a set of k_i's can be found by considering the system of homogeneous equations resulting from equating corresponding components.

$$\begin{cases} k_1 + 3k_2 + 2k_3 = 0, \\ k_1 + 2k_2 + k_3 = 0, \\ k_2 + k_3 = 0. \end{cases}$$

First of all we know that some nonzero k_i's exist because the rank of the system is 2, and hence there is a solution other than $(0, 0, 0)$ for the homogeneous system. We can find such a solution by first finding a complete solution, $\begin{cases} k_1 = k_3, \\ k_2 = -k_3. \end{cases}$ Let $k_3 = 1$; then a particular solution is $k_1 = 1$, $k_2 = -1$, $k_3 = 1$. Thus $(1)\alpha_1 + (-1)\alpha_2 + (1)\alpha_3 = 0$, and any of these vectors can be written as a linear combination of the other two. For example, $\alpha_1 = \alpha_2 - \alpha_3$.

Example 5. The coordinate vectors $(3, 2, 1)$, $(0, 1, 2)$, and $(1, 0, 2)$ are linearly independent because if $k_1(3, 2, 1) + k_2(0, 1, 2) + k_3(1, 0, 2) = (0, 0, 0)$, then necessarily $k_1 = k_2 = k_3 = 0$. We know this because the linear homogeneous system

$$\begin{cases} 3k_1 \phantom{{}+ k_2} + k_3 = 0, \\ 2k_1 + k_2 \phantom{{}+ 2k_3} = 0, \\ k_1 + 2k_2 + 2k_3 = 0, \end{cases}$$

has rank 3, and hence the only solution is $k_1 = 0$, $k_2 = 0$, $k_3 = 0$.

A *set of vectors* is **linearly dependent** or **linearly independent** according to whether all of its elements are linearly dependent or linearly independent.

The reader should observe at this point that if the number of coordinate vectors exceeds the number of components of the coordinate vectors involved, then the set of vectors must be linearly dependent. For example, the vectors $(1, 1)$, $(3, 2)$, $(1, 2)$ must be linearly dependent because

$$k_1(1, 1) + k_2(3, 2) + k_3(1, 2) = (0, 0),$$

yields the homogeneous linear system

$$\begin{cases} k_1 + 3k_2 + k_3 = 0, \\ k_1 + 2k_2 + 2k_3 = 0, \end{cases}$$

for which the rank must be less than 3, and therefore a solution other than $k_1 = k_2 = k_3 = 0$ must exist.

The following theorem will be useful in the proof of a later theorem.

Theorem 2. *Let $n \geq 2$. A set of nonzero vectors $\{\alpha_1, \ldots, \alpha_n\}$ of $V(F)$ is linearly dependent if and only if one vector of the set, say α_i, where $i \geq 2$, can be written as a linear combination of the preceding vectors, $\alpha_1, \alpha_2, \ldots, \alpha_{i-1}$.*

Proof. Reasons for the statements in this proof are left as an exercise for the reader (Exercise 14). Let the elements k_i be elements of F.

PART I. $\alpha_i = k_1 \alpha_1 + \cdots + k_{i-1} \alpha_{i-1}$
$\Rightarrow k_1 \alpha_1 + \cdots + k_{i-1} \alpha_{i-1} + (-1)\alpha_i + 0\alpha_{i+1} + \cdots + 0\alpha_n = 0$
$\Rightarrow \{\alpha_1, \ldots, \alpha_n\}$ is linearly dependent.

PART II. $\{\alpha_1, \ldots, \alpha_n\}$ is linearly dependent
$\Rightarrow$ there exist scalars $k'_1, \ldots, k'_n$, of F not all zero, such that $k'_1 \alpha_1 + \cdots + k'_n \alpha_n = 0$
$\Rightarrow$ either $k'_n \neq 0$, or there exists some $k'_i \neq 0$ such that

$$k'_{i+1} = k'_{i+2} = \cdots = k'_n = 0,$$

where $2 \leq i \leq n - 1$

$\Rightarrow \alpha_i = -k'_1(k'_i)^{-1}\alpha_1 - \cdots - k'_{i-1}(k'_i)^{-1}\alpha_{i-1}$, where $2 \leq i < n$

$\Rightarrow \alpha_i = k_1 \alpha_1 + \cdots + k_{i-1} \alpha_{i-1}$. □

APPLICATIONS

Example 6. In an article on an application of linear algebra to petrologic problems, Perry [25] makes use of the concept of linear independence in mineral classification.

The term molecular member is defined and is represented by an m by 1 column vector A_i. The maximum number of molecular members necessary to represent a mineral is asserted to be the number of linearly independent vectors in the set of all A_i ($i = 1, 2, \ldots, p$) of a given mineral group; that is, the rank of $[A_1 | \cdots | A_p]$. It is further asserted that there are at most $m - 1$ linearly independent vectors; the n linearly independent vectors are then used as the columns of an m by n matrix A, where $n < m$. A system of linear equations $AX = B$ is formed for which a mineral composition X is sought. But since $m > n$ the system can be solved if and only if the rank of $[A | B]$ equals the rank of A which equals the number of linearly independent columns of A.

EXERCISES

1. Determine whether the following sets of coordinate vectors are linearly dependent or independent. If they are linearly dependent, find a set of k_i's, and then express one of the vectors as a linear combination of the others.
   (a) $\{(2, -1), (-4, 2)\}$;
   (b) $\{(1, 1), (2, 1)\}$;
   (c) $\{(2, 1, 0), (1, 3, 2), (0, 9, 1)\}$;
   (d) $\{(1, 0, -3), (3, 1, 1), (2, 1, 4)\}$;
   (e) $\{(2, 1), (1, 4), (6, 9)\}$;
   (f) $\{(2, 1, 6), (4, 5, 0)\}$;
   (g) $\{(6, 9, 1), (0, 1, 0), (1, 0, 0)\}$;
   (h) $\{(2, 0, 3), (0, 1, 1), (2, 1, 4)\}$.

2. Express $\begin{bmatrix} 3 \\ 4 \end{bmatrix}$ as a linear combination of $\begin{bmatrix} 2 \\ 1 \end{bmatrix}$ and $\begin{bmatrix} 1 \\ 1 \end{bmatrix}$. Illustrate your answer graphically.

3. Show how any real coordinate vector (a, b, c) can be expressed as a linear combination of $(1, 0, 0)$, $(0, 1, 0)$, and $(0, 0, 1)$.

4. Do the coordinate vectors $(2, 3)$ and $(2, -1)$ span the vector space consisting of all real coordinate vectors in two-space? Would $(2, 3)$ and $(-4, -6)$ span the same space?

5. Do the coordinate vectors $(3, 4, 6)$ and $(4, 9, 1)$ span the vector space $V(R)$ consisting of all real coordinate vectors in three-space? Why? Do these vectors span a subspace of $V(R)$?

6. Do the coordinate vectors $(4, 9, 1, 0)$, $(0, 4, 9, 1)$, and $(1, 0, 0, 1)$ span the vector space, $V(R)$, consisting of all real coordinate vectors with four components? Do these vectors span a subspace of $V(R)$?

7. (a) Is it possible for three coordinate vectors in two-space to be linearly independent?
   (b) Is it possible for three coordinate vectors in two-space to be linearly dependent?
   (c) Is it possible for two coordinate vectors in three-space to be linearly independent?
   (d) Is it possible for two coordinate vectors in three-space to be linearly dependent?

8. Show that $(4, 2)$ cannot be expressed as a linear combination of $(-3, 2)$ and $(9, -6)$.

9. In each part, describe geometrically the vector space spanned by the given set of vectors:

   (a) $\left\{ \begin{bmatrix} 0 \\ 2 \\ 1 \end{bmatrix}, \begin{bmatrix} 1 \\ 1 \\ 1 \end{bmatrix}, \begin{bmatrix} 2 \\ 1 \\ 4 \end{bmatrix} \right\}$;
   (b) $\left\{ \begin{bmatrix} 0 \\ 1 \\ 4 \end{bmatrix}, \begin{bmatrix} 0 \\ 3 \\ 2 \end{bmatrix}, \begin{bmatrix} 0 \\ 2 \\ 4 \end{bmatrix} \right\}$;
   (c) $\left\{ \begin{bmatrix} 2 \\ 1 \\ 3 \end{bmatrix}, \begin{bmatrix} 4 \\ 1 \\ 6 \end{bmatrix} \right\}$.

10. If a set of vectors spans a vector space, must it be a linearly independent set? One part of Exercise 9 illustrates the correct answer to this question. Which part is it?

11. Would the vector $(0, 0)$ by itself be linearly independent or linearly dependent? Answer the same question for the vector $(1, 2)$.

12. Write out a proof for the fact that a set of nonzero vectors of a given vector space is linearly dependent if and only if one of them can be expressed as a linear combination of the others.
13. Prove Theorem 1.
14. Give reasons for the statements in the proof of Theorem 2.
15. Consider the vector space of Exercise 3 of Section 6.4.
    (a) Determine whether
    $$\{2x^2 + x + 1,\ x^2 + 2x + 3,\ 3x^2 + 2x + 2\}$$
    is a linearly dependent set of vectors. Does this set span the space? If not, describe the subspace that is spanned or generated.
    (b) Repeat (a) with the third vector changed to $3x^2 + 3x + 4$.
    (c) Repeat (a) for $\{3x + 2, 2x + 5\}$ and also for $\{4x^2 + 2, 2x^2 + 1\}$.
16. Describe all subspaces of the vector space of all two-dimensional geometric vectors.
17. If possible, write one of the coordinate vectors in each case as a linear combination of the other vectors:
    (a) $\{(2, 3, 5), (6, 9, 15), (3, -1, 1)\}$;
    (b) $\{(3, 5), (7, 1), (8, 2)\}$;
    (c) $\{(1, 0, 0), (3, 2, 4), (10, 4, 8)\}$;
    (d) $\{(1, 0, 0), (0, 2, 0), (0, 0, 3)\}$.
18. Why is every set of real four-dimensional coordinate vectors which includes $(0, 0, 0, 0)$ a linearly dependent set? Would every such set which does not include $(0, 0, 0, 0)$ be a linearly independent set? Explain.
19. Let S_1 and S_2 be sets of vectors from a vector space $V(F)$, and suppose that S_1 is a subset of S_2.
    (a) Prove that if S_2 is linearly independent, then S_1 is linearly independent, or, equivalently, prove that if S_1 is linearly dependent, then S_2 is linearly dependent.
    (b) Why might part (a) be incorrect if the words independent and dependent were exchanged?
20. Prove that a square matrix over F is invertible if and only if its columns (or rows) are linearly independent.
21. Prove that n coordinate vectors span the vector space of n-dimensional coordinate vectors over F if and only if the n coordinate vectors are linearly independent.

7.2 Basis of a Vector Space

In the last section we learned that a given set of vectors spans or generates a certain vector space when every vector in the space can be expressed as a linear combination of the given set. Obviously there can be more than one set that can span a certain vector space; moreover the number of vectors in each generating set can vary.

7.2 BASIS OF A VECTOR SPACE

Example 1. The set $\{(1, 0), (0, 1)\}$ spans the vector space $V(R)$ of all two-dimensional real coordinate vectors; also, the set $\{(1, 2), (2, 1), (3, 3)\}$ spans the same vector space. Every vector in the space can be expressed as a linear combination of each of the given sets, and therefore each set is a spanning set even though the sets are different in both content and number of elements.

There is, however, a distinction between two kinds of spanning sets; that is, for a given vector space, some spanning sets are linearly independent and some are linearly dependent. Those spanning sets that are linearly independent are very important in the study of linear algebra; such a set arranged in a specified order is called a ***basis*** of the vector space.

Definition 6. *A **basis** of a vector space is any ordered set of vectors of the space that:*
  (1) *are linearly independent; and*
  (2) *span the vector space.*

Example 2. The set $\{(1, 0, 0), (0, 1, 0), (0, 0, 1)\}$ is a basis for the vector space of all coordinate vectors with three real components over the field of real numbers because the set is linearly independent and the set spans the space.

Example 3. The set $\{(0, 1), (1, 0), (1, 1)\}$ is *not* a basis for the vector space of all two-dimensional real coordinate vectors over the field of real numbers, because the set is not linearly independent. The set does, however, span the space.

Example 4. The set $\left\{ \begin{bmatrix} 1 & 0 \\ 0 & 1 \end{bmatrix}, \begin{bmatrix} 0 & 1 \\ 1 & 0 \end{bmatrix} \right\}$ is *not* a basis for the vector space $V(R)$ of all 2 by 2 R-matrices because the set does not span the space. The set is, however, linearly independent. (A justification of these assertions is left for Exercise 13.)

Example 5. The set of polynomials $\{x^2, x, 1\}$ is a basis for the vector space $V(R)$ consisting of the zero polynomial and all polynomials of second degree or less over the real numbers. It is easy to see that each polynomial in the space can be expressed as a linear combination of the elements of the basis, that is

$$a(x^2) + b(x) + c(1).$$

Also, the given set is linearly independent because

$$k_1(x^2) + k_2(x) + k_3(1) = 0x^2 + 0x + 0$$

if and only if $k_1 = k_2 = k_3 = 0$. Of course there are other sets which also form a basis of the same vector space (see Exercise 14).

Example 6. The set $\left\{ \begin{bmatrix} 1 & 0 \\ 0 & 0 \end{bmatrix}, \begin{bmatrix} 0 & 1 \\ 0 & 0 \end{bmatrix}, \begin{bmatrix} 0 & 0 \\ 1 & 0 \end{bmatrix}, \begin{bmatrix} 0 & 0 \\ 0 & 1 \end{bmatrix} \right\}$ is a basis of the vector space $V(R)$ of all 2 by 2 R-matrices. Certainly it is obvious that the set generates the vector space. Also, the set is linearly independent, because if

$$k_1 \begin{bmatrix} 1 & 0 \\ 0 & 0 \end{bmatrix} + k_2 \begin{bmatrix} 0 & 1 \\ 0 & 0 \end{bmatrix} + k_3 \begin{bmatrix} 0 & 0 \\ 1 & 0 \end{bmatrix} + k_4 \begin{bmatrix} 0 & 0 \\ 0 & 1 \end{bmatrix} = \begin{bmatrix} 0 & 0 \\ 0 & 0 \end{bmatrix},$$

then

$$\begin{bmatrix} k_1 & k_2 \\ k_3 & k_4 \end{bmatrix} = \begin{bmatrix} 0 & 0 \\ 0 & 0 \end{bmatrix},$$

and, by definition of matrix equality, the last statement can be true only if $k_1 = k_2 = k_3 = k_4 = 0$. There are, of course, other sets which serve as a basis of the given vector space.

Theorem 3. *If $\{\beta_1, \ldots, \beta_n\}$ is a basis of a vector space $V(F)$, then each vector in $V(F)$ can be expressed uniquely as a linear combination of $\{\beta_1, \ldots, \beta_n\}$.*

Proof. Because $\{\beta_1, \ldots, \beta_n\}$ spans the space, an arbitrary vector α can be expressed as a linear combination of $\{\beta_1, \ldots, \beta_n\}$; assume that

$$\alpha = a_1\beta_1 + \cdots + a_n\beta_n \text{ and } \alpha = b_1\beta_1 + \cdots + b_n\beta_n$$

are two such expressions of α, and show that $a_i = b_i$ for $i = 1, \ldots, n$. Reasons are left for Exercise 15.

STATEMENTS

(1) $a_1\beta_1 + \cdots + a_n\beta_n = b_1\beta_1 + \cdots + b_n\beta_n$.
(2) $(a_1 - b_1)\beta_1 + \cdots + (a_n - b_n)\beta_n = \mathbf{0}$.
(3) $(a_i - b_i) = 0$ for $i = 1, \ldots, n$.
(4) $a_i = b_i$ for $i = 1, \ldots, n$. □

Since it is apparent that there can be more than one basis for a given vector space, it is natural to raise the question: If $\{\beta_1, \ldots, \beta_n\}$ is a basis of a vector space, will each other basis contain the same number of vectors? The following theorem provides an affirmative answer to that question; *the unique number n of vectors in a basis is called the* **dimension** *of the vector space*.

Theorem 4. *If a set of n vectors is a basis of a given vector space $V(F)$, then every basis of $V(F)$ has exactly n vectors.*

7.2 BASIS OF A VECTOR SPACE

Proof. Let $\{\alpha_1, \ldots, \alpha_s\}$ and $\{\beta_1, \ldots, \beta_n\}$ be two bases and prove that $s = n$.

STATEMENT	REASON
(1) $\{\alpha_1, \beta_1, \ldots, \beta_n\}$ is a linearly dependent set.	(1) Theorem 3, and Exercise 12 of Section 7.1.
(2) There is some vector in the set $\{\alpha_1, \beta_1, \ldots, \beta_n\}$ other than α_1 that can be expressed as a linear combination of the preceding vectors of the set. Relabel the vectors, if necessary, so that this is the one called β_n.	(2) By Statement (1) and Theorem 2.
(3) After deleting β_n the set $\{\alpha_1, \beta_1, \ldots, \beta_{n-1}\}$ spans $V(F)$.	(3) Because $\{\beta_1, \ldots, \beta_n\}$ is a basis, and by Statement (2).
(4) If $1 < n < s$, then α_2 exists, and $\{\alpha_2, \alpha_1, \beta_1, \ldots, \beta_{n-1}\}$ is linearly dependent. If $1 = n < s$, then proceed to statement (8).	(4) Because $\{\alpha_1, \beta_1, \ldots, \beta_{n-1}\}$ spans $V(F)$ and α_2 belongs to $V(F)$.
(5) There is some vector in $\{\alpha_2, \alpha_1, \beta_1, \ldots, \beta_{n-1}\}$ other than α_1 and α_2 which can be expressed as a linear combination of the preceding vectors. Relabel the vectors if necessary so that this is the one called β_{n-1}.	(5) By Statement (4), Theorem 2, and because $\{\alpha_2, \alpha_1\}$ is linearly independent.
(6) After deleting β_{n-1} the set $\{\alpha_2, \alpha_1, \beta_1, \ldots, \beta_{n-2}\}$ spans $V(F)$.	(6) By (5).
(7) $\{\alpha_n, \ldots, \alpha_1\}$ spans $V(F)$.	(7) if $n < s$, repetition (if necessary) of the procedure used in previous statements produces the set $\{\alpha_n, \ldots, \alpha_1\}$ which spans $V(F)$.
(8) But then α_{n+1} can be expressed as a linear combination of $\{\alpha_n, \ldots, \alpha_1\}$ which implies that $\{\alpha_s, \ldots, \alpha_n, \ldots, \alpha_1\}$ is linearly dependent, contrary to the hypothesis of the Theorem.	(8) By (7) (or (4) if $n = 1$), Definition 2, and then by Exercise 19(a) of Section 7.1.

(9) Therefore $n \geq s$. (9) By contradiction of the assumption that $n < s$.

(10) But also $n \leq s$. (10) By reversing the roles of the two assumed bases and repeating the argument used in Statements (1)–(9).

(11) Therefore $n = s$. (11) By Statements (9) and (10). □

Definition 7. *Let n be a positive integer. If a basis of a vector space $V(F)$ has n vectors, then $V(F)$ has **dimension n**. The vector space consisting of the zero vector only is said to have **dimension zero**.*

So far, in the theorems and assertions that we have made concerning bases of a vector space, we have assumed the existence of a basis with n vectors. Certainly not all vector spaces have a basis consisting of a finite number of vectors; for example, consider the vector space $V(R)$ of all polynomials over the real numbers. If one attempts to construct a spanning set, one finds that no matter what finite number of vectors is placed in the set, there always exists a polynomial of degree greater than any vector in the set. If, however, there does exist a finite spanning set S of a vector space $V(F)$, then it is easy to prove (Exercise 19) that there is a linearly independent subset of S that also spans $V(F)$ and thereby guarantees the existence of a basis of $V(F)$. A distinction, therefore, is made between the vector spaces that are spanned by a finite set and hence are *finite-dimensional* and the vector spaces that are not spanned by a finite set. In this text we are concerned with finite-dimensional vector spaces, hence in what follows *unless stated otherwise an arbitrary vector space will be assumed to be of finite dimension*. A discussion of the existence of a basis for arbitrary $V(F)$ can be found in more advanced texts.[1]

Definition 8. *The vector space $V(F)$ consisting of the set of all linear combinations of columns of a matrix A over F is called the **column space** of A.*

Example 7. Consider each column of the R-matrix A as a vector.

$$A = \begin{bmatrix} 2 & 6 & 1 & 0 & 5 \\ 0 & 2 & 0 & 1 & 6 \end{bmatrix}.$$

The set of vectors that can be written as a linear combination of these vectors

[1] Dean, R. A., *Elements of Abstract Algebra*, John Wiley & Sons Inc., New York, 1966, pages 183–184.

7.2 BASIS OF A VECTOR SPACE

forms a two-dimensional vector space $V(R)$, called the *column space* of A. The third and fourth vectors form a basis of this column space; these vectors generate or span the column space. Of the five columns of A, is this the only pair that forms a basis of the column space of A?

Let $V(R)$ represent the vector space of all real coordinate vectors with n components, and then observe that any set $\{\alpha_1, \ldots, \alpha_n\}$ of vectors from $V(R)$ that includes the zero vector must be linearly dependent; this follows from the fact that the coefficient of combination of $\alpha_i = 0$ in $k_1\alpha_1 + \cdots + k_n\alpha_n = 0$ never need be zero. Hence, a basis of $V(R)$ never includes the zero vector. These ideas can be generalized to any vector space.

APPLICATIONS

Example 8. The concept of a basis of a vector space is fundamental in a study of linear programming. This example will furnish some indication and background for such an application. Suppose we are given a consistent system of equations

$$\begin{cases} x_1 + x_2 + 3x_3 + x_4 &= 6, \\ x_1 - 3x_2 - x_3 + x_5 &= 1, \\ x_1 + x_3 + x_6 &= 2. \end{cases}$$

This system may be written as

$$x_1 \begin{bmatrix} 1 \\ 1 \\ 1 \end{bmatrix} + x_2 \begin{bmatrix} 1 \\ -3 \\ 0 \end{bmatrix} + x_3 \begin{bmatrix} 3 \\ -1 \\ 1 \end{bmatrix} + x_4 \begin{bmatrix} 1 \\ 0 \\ 0 \end{bmatrix} + x_5 \begin{bmatrix} 0 \\ 1 \\ 0 \end{bmatrix} + x_6 \begin{bmatrix} 0 \\ 0 \\ 1 \end{bmatrix} = \begin{bmatrix} 6 \\ 1 \\ 2 \end{bmatrix}.$$

The vectors $\begin{bmatrix} 1 \\ 0 \\ 0 \end{bmatrix}$, $\begin{bmatrix} 0 \\ 1 \\ 0 \end{bmatrix}$, and $\begin{bmatrix} 0 \\ 0 \\ 1 \end{bmatrix}$ form a basis for the column space of the augmented matrix of the system. The vector $\begin{bmatrix} 6 \\ 1 \\ 2 \end{bmatrix}$ on the right-hand side can be expressed as a linear combination of this basis by assigning values to the scalar multipliers, that is $x_1 = 0$, $x_2 = 0$, $x_3 = 0$, $x_4 = 6$, $x_5 = 1$, $x_6 = 2$.

If we add three times the members of the first equation to the members of the second equation, we get

$$\begin{cases} x_1 + x_2 + 3x_3 + x_4 &= 6, \\ 4x_1 + 8x_3 + 3x_4 + x_5 &= 19, \\ x_1 + x_3 + x_6 &= 2, \end{cases}$$

or

$$x_1 \begin{bmatrix} 1 \\ 4 \\ 1 \end{bmatrix} + x_2 \begin{bmatrix} 1 \\ 0 \\ 0 \end{bmatrix} + x_3 \begin{bmatrix} 3 \\ 8 \\ 1 \end{bmatrix} + x_4 \begin{bmatrix} 1 \\ 3 \\ 0 \end{bmatrix} + x_5 \begin{bmatrix} 0 \\ 1 \\ 0 \end{bmatrix} + x_6 \begin{bmatrix} 0 \\ 0 \\ 1 \end{bmatrix} = \begin{bmatrix} 6 \\ 19 \\ 2 \end{bmatrix}.$$

Now another solution may be obtained by treating the second, fifth, and sixth vectors as a basis and expressing the vector $\begin{bmatrix} 6 \\ 19 \\ 2 \end{bmatrix}$ as a linear combination of them by assigning the values $x_1 = 0$, $x_2 = 6$, $x_3 = 0$, $x_4 = 0$, $x_5 = 19$, $x_6 = 2$. Note that the new member of the basis was introduced by an elementary row transformation.

Problems similar to Example 8 will be pursued further in Chapter 11.

★**Example 9.** Differential equations and their solutions are very important in many applied disciplines. Using elementary techniques, a solution of the differential equation $y'' + y' - 12y = 0$ is found to be $y_1 = e^{3t}$; the reader can verify that y_1 is a solution, by substituting y_1'', y_1' and y_1 into the differential equation. It turns out that $y_2 = e^{-4t}$ is also a solution of the given differential equation. We form a linear combination

$$y = k_1 y_1 + k_2 y_2$$

of the two solutions y_1 and y_2; it can be verified that any solution of the original differential equation can be expressed as such a linear combination of y_1 and y_2. It also can be verified that the set of these solutions forms a vector space, and, as we have just seen, y_1 and y_2 span this vector space. Therefore, because y_1 and y_2 are also linearly independent, the set $\{y_1, y_2\}$ is a basis of the solution space. Use of the concepts pointed out in this example is becoming more prevalent in elementary studies of differential equations.

EXERCISES

1. Verify that the vectors $(2, 1)$ and $(4, 0)$ form a basis for the vector space $V(R)$ of all coordinate vectors with two real components.
2. Express $(6, 1)$ in terms of the basis of the preceding exercise.
3. Which of the following sets of vectors are not a basis of the vector space $V(R)$ consisting of all three-dimensional real coordinate vectors? State why they are not a basis.
   (a) $\{(1, 0, 1), (2, 0, 1), (1, 1, 1)\}$;
   (b) $\{(2, 1, 4), (1, 1, 1)\}$;
   (c) $\{(1, 2, 1), (3, 1, 0), (1, 0, 0)\}$;
   (d) $\{(1, 1, 1), (0, 0, 1), (1, 0, 0), (0, 1, 0)\}$;
   (e) $\{(1, 2, 1), (1, 3, 0), (0, 1, -1)\}$.
4. Write two sets of columns that will serve as a basis of the column space of A:

$$A = \begin{bmatrix} 2 & -4 & 1 & 2 \\ 1 & -2 & 0 & 0 \end{bmatrix}.$$

Find two sets of columns that will not serve, and state why they will not serve.

5. Would any three vectors of a three-dimensional vector space be a basis of that vector space? Why?
6. A certain vector space is spanned by four linearly independent vectors. What can be said about the dimension of this vector space?
7. A certain vector space is spanned by four linearly dependent vectors. What can be said about the dimension of this vector space?
8. What is the dimension of the vector space spanned by the vectors (2, 1, 4) and (2, 1, 1)?
9. What is the dimension of the column space of each of the following matrices?

(a) $\begin{bmatrix} 1 \\ 2 \end{bmatrix}$; (b) $\begin{bmatrix} 2 & -1 \\ 1 & -1 \end{bmatrix}$; (c) $\begin{bmatrix} 0 & 2 & -1 \\ 0 & 4 & -2 \end{bmatrix}$;

(d) $\begin{bmatrix} 1 & 1 \\ 1 & 2 \\ 1 & 3 \end{bmatrix}$; (e) $\begin{bmatrix} 1 & 0 & 0 \\ 0 & 0 & 1 \\ 0 & 3 & 0 \end{bmatrix}$; (f) I_4.

10. For the system
$$\begin{cases} x_1 + 2x_2 + x_3 = 4, \\ x_1 - x_2 + x_4 = 5, \end{cases}$$
write a set of columns which forms a basis for the column space of the coefficient matrix A. Then assign values to x_1, x_2, x_3, and x_4 which will enable you to write $\begin{bmatrix} 4 \\ 5 \end{bmatrix}$ as a linear combination of the columns of A.

11. Define the *row space* $V(F)$ of a matrix over F.
12. In the last section we stated that if a subset of the elements of a vector space forms a vector space under the original operations, then this subset forms a subspace.
    (a) For the vector space $V(R)$ consisting of all coordinate vectors with two real components, give two examples of subspaces.
    (b) Give two examples of subsets of the same set of vectors as in (a) which do not form subspaces.
    (c) How would the dimension of a subspace compare with the dimension of the whole vector space?
    (d) Would a basis of a vector space also be a basis for an arbitrary subspace of that vector space?
13. Justify the statements of Example 4.
14. Show that the set $\{x^2 + x, x, 1\}$ is a basis of the vector space in Example 5.
15. Give the reason for each statement in the proof of Theorem 3.
16. Is the set of R-matrices of the form $\begin{bmatrix} x & x \\ 0 & 0 \end{bmatrix}$ a subspace of the vector space $V(R)$ of all 2 by 2 R-matrices? If so, give the dimension of this subspace, and also state one basis of this subspace. Answer the same questions for the set of R-matrices of the form $\begin{bmatrix} x & x+1 \\ 0 & 0 \end{bmatrix}$.

17. Describe geometrically the subspaces of the vector space of all three-dimensional geometric vectors; specify the dimension.

18. $\{(1, 4, 2), (3, 0, -1), (2, -1, -1)\}$ is a basis of the vector space $V(R)$ of all real coordinate vectors with three components. Express $(17, 3, 19)$ as a unique linear combination of these basis vectors.

19. Prove that if there is a finite set S of vectors that span a vector space $V(F)$, then there is a linearly independent subset of S that spans $V(F)$.

7.3 Coordinate Vector Representation of Abstract Vectors

The fact that every vector in a given vector space can be expressed as a unique linear combination of the vectors of a specific basis suggests that we write the elements of an arbitrary vector space as a coordinate vector whose components are the coefficients in the linear combination.

Example 1. It was pointed out in Example 5 on page 185 that the set $\{x^2, x, 1\}$ is a basis of the vector space $V(R)$ consisting of the zero polynomial and all polynomials of degree 2 or less over the real numbers. It was also pointed out that each element of $V(R)$ could be expressed as a linear combination of the basis, that is, as

$$ax^2 + bx + c.$$

Now, since the basis vectors are ordered in a specified way, then to each element of $V(R)$, $ax^2 + bx + c$, there corresponds a unique coordinate vector (a, b, c) with respect to the basis $\{x^2, x, 1\}$, and vice versa. Moreover this one-to-one correspondence is preserved under addition and scalar multiplication. For example, the polynomial $x^2 + 3x + 6$ corresponds to the coordinate vector $(1, 3, 6)$, which can be expressed in column matrix form as $\begin{bmatrix} 1 \\ 3 \\ 6 \end{bmatrix}$, and the polynomial $5x^2 - 6x + 2$ corresponds to $(5, -6, 2)$ or $\begin{bmatrix} 5 \\ -6 \\ 2 \end{bmatrix}$, and vice versa.

The sum (result of addition) of the two polynomials is $6x^2 - 3x + 8$, which corresponds to the sum of the two column vectors

$$\begin{bmatrix} 1 \\ 3 \\ 6 \end{bmatrix} + \begin{bmatrix} 5 \\ -6 \\ 2 \end{bmatrix} = \begin{bmatrix} 6 \\ -3 \\ 8 \end{bmatrix},$$

and vice versa. Also if k is a real number, then $k(x^2 + 3x + 6) = kx^2 + 3kx + 6k$, which corresponds to

$$k \begin{bmatrix} 1 \\ 3 \\ 6 \end{bmatrix} = \begin{bmatrix} k \\ 3k \\ 6k \end{bmatrix},$$

and vice versa.

A one-to-one correspondence between two sets of vectors which preserves the results of addition and scalar multiplication, as shown in the last example, is an illustration of an *isomorphism* (iso meaning "the same," and "morphos" meaning "form"). These ideas are formalized by the following definition and theorem.

Definition 9. *A vector space $V(F)$ is **isomorphic** to a vector space $V'(F)$ over the same field if there exists a one-to-one mapping from $V(F)$ onto $V'(F)$ such that the following are true. (Let α and β be arbitrary elements of $V(F)$).*

(1) If α' is the image of α, and if β' is the image of β, then $(\alpha + \beta)' = \alpha' + \beta'$; that is, the image of the sum is equal to sum of the images;

(2) If k belongs to F then $(k\alpha)' = k\alpha'$; that is, the image of $k\alpha$ is equal to k times the image of α.

*Such a one-to-one mapping is called an **isomorphic mapping** or an **isomorphism from $V(F)$ onto $V'(F)$**.*

Notice that Example 1 is an illustration of Definition 9. It is easy to see that *if $V(F)$ is isomorphic to $V'(F)$, then $V'(F)$ is isomorphic to $V(F)$*. The concept of isomorphic vector spaces is very important, because such vector spaces have the same algebraic properties. The next theorem establishes the importance of *the vector space of all n-dimensional coordinate vectors with components in F over the same field F; this vector space will be designated $V_n(F)$ hereafter*.

Theorem 5. *Let $\{\alpha_1, \alpha_2, \ldots, \alpha_n\}$ be a basis for an arbitrary vector space $V(F)$, and let $V_n(F)$ designate the vector space of all n-dimensional coordinate vectors with components in F; then $V(F)$ is isomorphic to $V_n(F)$.*

Proof.

(1) Because $\{\alpha_1, \ldots, \alpha_n\}$ is a basis of $V(F)$ then an arbitrary vector α of $V(F)$ can be expressed uniquely as $\alpha = k_1\alpha_1 + \cdots + k_n\alpha_n$ according to Theorem 3.

(2) Thus there exists a one-to-one correspondence

$$(k_1\alpha_1 + \cdots + k_n\alpha_n) \leftrightarrow (k_1, \ldots, k_n)$$

or

which associates each vector of $V(F)$ with a unique element of $V_n(F)$, and vice versa.

(3) Also, if $\beta = b_1\alpha_1 + \cdots + b_n\alpha_n$ is in $V(F)$ with the image $\beta' = (b_1, \ldots, b_n)$ in $V_n(F)$, then

$$\alpha + \beta = (k_1 + b_1)\alpha_1 + \cdots + (k_n + b_n)\alpha_n \leftrightarrow (k_1 + b_1, \ldots, k_n + b_n) = \alpha' + \beta'.$$

(4) Also, if c belongs to F then

$$c\alpha = ck_1\alpha_1 + \cdots + ck_n\alpha_n \leftrightarrow (ck_1, \ldots, ck_n) = c\alpha'. \quad \square$$

The attention of the reader is called to the fact that in the proof of Theorem 5, as well as in Definition 9, the same symbol " $+$ " is used to represent perhaps different vector operations of $V(F)$ and $V_n(F)$; for example if $V(R)$ is an n-dimensional vector space of polynomials, then $+$ represents the addition of polynomials over $V(R)$, whereas $+$ represents coordinate vector addition over $V_n(R)$. Also, it should be emphasized that the elements in the basis must be ordered.

Theorem 5 is very useful because the isomorphism implies that any arbitrary vector space of dimension n will have the same algebraic properties as the corresponding n-dimensional vector space of coordinate vectors. Practically, it means that, if we choose, we may work with the coordinate vector images rather than with the given vectors. As the proof of Theorem 5 bears out, the components of the image of a vector α are simply the coefficients of the unique expression of α as a linear combination of a basis.

Example 2. Consider the vector space $V(R)$ of 2 by 2 R-matrices. If we select as a basis the ordered set

$$\left\{ \begin{bmatrix} 1 & 0 \\ 0 & 0 \end{bmatrix}, \begin{bmatrix} 0 & 1 \\ 0 & 0 \end{bmatrix}, \begin{bmatrix} 0 & 0 \\ 1 & 0 \end{bmatrix}, \begin{bmatrix} 0 & 0 \\ 0 & 1 \end{bmatrix} \right\},$$

then the matrix $\begin{bmatrix} 6 & 4 \\ 3 & 2 \end{bmatrix}$ can be uniquely expressed as

$$\begin{bmatrix} 6 & 4 \\ 3 & 2 \end{bmatrix} = 6\begin{bmatrix} 1 & 0 \\ 0 & 0 \end{bmatrix} + 4\begin{bmatrix} 0 & 1 \\ 0 & 0 \end{bmatrix} + 3\begin{bmatrix} 0 & 0 \\ 1 & 0 \end{bmatrix} + 2\begin{bmatrix} 0 & 0 \\ 0 & 1 \end{bmatrix}.$$

Hence the coordinate vector corresponding to $\begin{bmatrix} 6 & 4 \\ 3 & 2 \end{bmatrix}$ is (6, 4, 3, 2), which can be written if desired as a row or column matrix. In general, with respect to the basis above, the coordinate vector representation of $\begin{bmatrix} a_{11} & a_{12} \\ a_{21} & a_{22} \end{bmatrix}$ is $(a_{11}, a_{12}, a_{21}, a_{22})$. Of course, if the basis changes, so does the coordinate vector representation. For instance, suppose another basis is chosen, such as

$$\left\{ \begin{bmatrix} 1 & 1 \\ 0 & 0 \end{bmatrix}, \begin{bmatrix} 1 & 0 \\ 1 & 0 \end{bmatrix}, \begin{bmatrix} 1 & 1 \\ 1 & 0 \end{bmatrix}, \begin{bmatrix} 0 & 0 \\ 0 & 1 \end{bmatrix} \right\}.$$

7.3 REPRESENTATION OF ABSTRACT VECTORS

Then the same 2 by 2 matrix is expressed as a linear combination of the new basis as

$$\begin{bmatrix} 6 & 4 \\ 3 & 2 \end{bmatrix} = k_1 \begin{bmatrix} 1 & 1 \\ 0 & 0 \end{bmatrix} + k_2 \begin{bmatrix} 1 & 0 \\ 1 & 0 \end{bmatrix} + k_3 \begin{bmatrix} 1 & 1 \\ 1 & 0 \end{bmatrix} + k_4 \begin{bmatrix} 0 & 0 \\ 0 & 1 \end{bmatrix}.$$

The last equation yields the system

$$\begin{cases} k_1 + k_2 + k_3 & = 6, \\ k_1 + k_3 & = 4, \\ k_2 + k_3 & = 3, \\ \phantom{k_1 + k_2 + k_3 = {}} k_4 & = 2. \end{cases}$$

The unique solution of the system is (3, 2, 1, 2), which is the coordinate vector corresponding to $\begin{bmatrix} 6 & 4 \\ 3 & 2 \end{bmatrix}$ with respect to the new basis.

Now that we have established that every element α of an arbitrary n-dimensional vector space $V(F)$ can be represented by a unique coordinate vector, then notation for this representation also should be established. The task of doing so is complicated somewhat, however, because a coordinate vector may be expressed as a column matrix, a row matrix, an n-tuple, or in any way that maintains the order of the n components. A further complication arises from the fact that a coordinate vector representation of α depends upon the choice of basis of $V(F)$. Keeping these matters in mind, *we arbitrarily adopt the following notation for this book.*

Definition 10. *Let α be an element of an arbitrary n-dimensional vector space $V(F)$ with basis $b = \{\beta_1, \ldots, \beta_n\}$. For the unique expression*

$$\alpha = a_1 \beta_1 + \cdots + a_n \beta_n,$$

*the elements $a_1, \ldots, a_n$ of F are called **coordinates** of α with respect to the basis b, and the column matrix*

$$[\alpha]_b = \begin{bmatrix} a_1 \\ \vdots \\ a_n \end{bmatrix}$$

*is called the **coordinate matrix** of α with respect to b.*

Example 3. In the last example, for the basis

$$b = \left\{ \begin{bmatrix} 1 & 0 \\ 0 & 0 \end{bmatrix}, \begin{bmatrix} 0 & 1 \\ 0 & 0 \end{bmatrix}, \begin{bmatrix} 0 & 0 \\ 1 & 0 \end{bmatrix}, \begin{bmatrix} 0 & 0 \\ 0 & 1 \end{bmatrix} \right\},$$

the element $\alpha = \begin{bmatrix} 6 & 4 \\ 3 & 2 \end{bmatrix}$ can be expressed as the coordinate matrix $[\alpha]_b = \begin{bmatrix} 6 \\ 4 \\ 3 \\ 2 \end{bmatrix}$.

For the second basis

$$e = \left\{ \begin{bmatrix} 1 & 1 \\ 0 & 0 \end{bmatrix}, \begin{bmatrix} 1 & 0 \\ 1 & 0 \end{bmatrix}, \begin{bmatrix} 1 & 1 \\ 1 & 0 \end{bmatrix}, \begin{bmatrix} 0 & 0 \\ 0 & 1 \end{bmatrix} \right\},$$

the coordinate matrix with respect to e is

$$[\alpha]_e = \begin{bmatrix} 3 \\ 2 \\ 1 \\ 2 \end{bmatrix}.$$

Example 4. Consider the vector space $V_2(R)$ (coordinate vectors with two real components). Let $\alpha = (2, 4)$ and $b = \{(1, 0), (0, 1)\}$. Clearly

$$\alpha = (2, 4) = 2(1, 0) + 4(0, 1) = 2\beta_1 + 4\beta_2.$$

Hence the coordinates of α with respect to b are 2 and 4, and the coordinate matrix of α is

$$[\alpha]_b = \begin{bmatrix} 2 \\ 4 \end{bmatrix}.$$

If, however, we choose another basis $e = \{(1, 1), (0, 2)\}$, then

$$\alpha = (2, 4) = c_1(1, 1) + c_2(0, 2) = c_1\varepsilon_1 + c_2\varepsilon_2,$$

where the coordinates c_1 and c_2 must be determined. This can be done by solving the system formed by equating corresponding components:

$$\begin{cases} 2 = 1c_1 + 0c_2, \\ 4 = 1c_1 + 2c_2. \end{cases}$$

The solution is $c_1 = 2$ and $c_2 = 1$. Thus

$$[\alpha]_e = \begin{bmatrix} 2 \\ 1 \end{bmatrix}.$$

The expression of α with respect to the two different bases is illustrated in Figure 7.3.1.

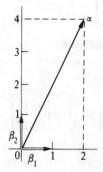
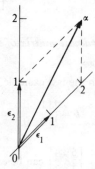

Figure 7.3.1

Hereafter, the notation $P_n(F)$ will be reserved to represent the vector space consisting of the zero polynomial and all polynomials of degree n or less over a field F. Notice that $P_n(F)$ has dimension $n + 1$ and hence is isomorphic to $V_{n+1}(F)$.

Example 5. Consider the vector space $P_2(R)$ with basis $b = \{1, x, x^2\}$. Let $\alpha = 2 + 3x + 5x^2$; then

$$[\alpha]_b = \begin{bmatrix} 2 \\ 3 \\ 5 \end{bmatrix}.$$

If the basis is $e = \{1 + x, x, x^2\}$, then

$$\alpha = 2 + 3x + 5x^2 \\ = 2(1 + x) + 1(x) + 5(x^2),$$

and, consequently,

$$[\alpha]_e = \begin{bmatrix} 2 \\ 1 \\ 5 \end{bmatrix}.$$

EXERCISES

1. For the vector space $V_2(R)$ (all coordinate vectors with two real components), write $[\alpha]_b$ if:
   (a) $\alpha = (7, 2)$ and $b = \{(1, 0), (0, 1)\}$;
   (b) $\alpha = (7, 2)$ and $b = \{(1, 0), (1, 1)\}$;
   (c) $\alpha = (-1, 3)$ and $b = \{(1, 0), (1, 1)\}$.

2. For the vector space $V_3(R)$ (all coordinate vectors with three real components), write $[\alpha]_b$ if:
   (a) $\alpha = (6, 3, 4)$ and $b = \{(1, 0, 0), (0, 1, 0), (0, 0, 1)\}$;
   (b) $\alpha = (6, 3, 4)$ and $b = \{(1, 1, 0), (0, 1, 0), (0, 1, 1)\}$;
   (c) $\alpha = (1, 2, 7)$ and $b = \{(1, 1, 0), (0, 1, 0), (0, 1, 1)\}$.

3. For the vector space $P_1(R)$ (the zero polynomial and all polynomials of degree 1 or less over R), write $[\alpha]_b$ if:
   (a) $\alpha = 2x + 3$ and $b = \{x, 1\}$;
   (b) $\alpha = 2x + 3$ and $b = \{x + 1, 2x - 3\}$;
   (c) $\alpha = 5x + 4$ and $b = \{x + 1, 2x - 3\}$.

4. For the vector space $P_2(R)$ (the zero polynomial and all polynomials of degree 2 or less over R), write $[\alpha]_b$ if:
   (a) $\alpha = x^2 + 3x - 6$ and $b = \{x^2, x, 1\}$;
   (b) $\alpha = x^2 + 3x - 6$ and $b = \{x^2 + x, x, x + 1\}$;
   (c) $\alpha = x^2 + 6$ and $b = \{x^2 + x, x, x + 1\}$.

5. For the vector space $V(R)$ of all 2 by 2 R-matrices, write $[\alpha]_b$ if:

(a) $\alpha = \begin{bmatrix} 1 & 2 \\ 3 & 4 \end{bmatrix}$ and $b = \left\{ \begin{bmatrix} 1 & 0 \\ 0 & 0 \end{bmatrix}, \begin{bmatrix} 0 & 1 \\ 0 & 0 \end{bmatrix}, \begin{bmatrix} 0 & 0 \\ 1 & 0 \end{bmatrix}, \begin{bmatrix} 0 & 0 \\ 0 & 1 \end{bmatrix} \right\}$;

(b) $\alpha = \begin{bmatrix} 1 & 2 \\ 3 & 4 \end{bmatrix}$ and $b = \left\{ \begin{bmatrix} 1 & 0 \\ 0 & 0 \end{bmatrix}, \begin{bmatrix} 1 & 2 \\ 0 & 0 \end{bmatrix}, \begin{bmatrix} 1 & 0 \\ 2 & 0 \end{bmatrix}, \begin{bmatrix} 0 & 0 \\ 0 & 1 \end{bmatrix} \right\}$;

(c) $\alpha = \begin{bmatrix} 1 & 1 \\ 3 & 3 \end{bmatrix}$ and $b = \left\{ \begin{bmatrix} 1 & 0 \\ 0 & 0 \end{bmatrix}, \begin{bmatrix} 1 & 2 \\ 0 & 0 \end{bmatrix}, \begin{bmatrix} 1 & 0 \\ 2 & 0 \end{bmatrix}, \begin{bmatrix} 0 & 0 \\ 0 & 1 \end{bmatrix} \right\}$.

6. For the vector space $V(R)$ of all 3 by 3 R-matrices, write $[\alpha]_b$ if $\alpha = \begin{bmatrix} 1 & 2 & 0 \\ 6 & 0 & 4 \\ 3 & 5 & 7 \end{bmatrix}$ and b is determined by the reader.

7. For the vector space $P_1(C)$ (the zero polynomial and all polynomials of degree 1 or less over the complex numbers), write $[\alpha]_b$ if:
(a) $\alpha = x + i$ and $b = \{x, 1\}$;
(b) $\alpha = x + i$ and $b = \{x + 1, ix - i\}$;
(c) $\alpha = ix + 3$ and $b = \{x + 1, ix - i\}$.

8. For the vector space $P_2(C)$ (the zero polynomial and all polynomials of degree 2 or less over the complex numbers), write $[\alpha]_b$ if:
(a) $\alpha = x^2 + 2x + i$ and $b = \{x^2, x, 1\}$;
(b) $\alpha = x^2 + 2x + i$ and $b = \{x^2, x, i\}$;
(c) $\alpha = ix^2 + 3ix + 2$ and $b = \{x^2, x, i\}$.

9. Consider the vector spaces $V_2(R)$ and $V_3(R)$. Find a subspace of $V_3(R)$ which is isomorphic to $V_2(R)$, and explain why the subspace is isomorphic to $V_2(R)$.

10. Consider the vector spaces $P_2(R)$ and $V_2(R)$. Find a subspace of $P_2(R)$ which is isomorphic to $V_2(R)$, and explain why the subspace is isomorphic to $V_2(R)$.

7.4 Inner Product Spaces (optional)[2]

A vector space involves a set of vectors and an associated field. If we restrict this underlying field to be either the field of complex numbers or the field of real numbers, and if we define a certain mapping, known as the *inner product function* with respect to the given vector space, then we create still another important algebraic system known as an *inner product space*. Such a special vector space is very useful because within this structure we can define abstractly such common concepts as "length," "distance," and "angle."

---
[2] This section, however, is a prerequisite to the optional Sections 7.5, 7.6, 10.4, 10.5, and 10.6.

7.4 INNER PRODUCT SPACES (OPTIONAL)

Actually we have already considered one specific example of an inner product function; in Section 4.3 we defined the dot product (or standard inner product) of two coordinate vectors: *If α and β are coordinate vectors with complex components $a_1, a_2, \ldots, a_n$ and $b_1, b_2, \ldots, b_n$, respectively, then the dot product (or standard inner product) $\alpha \cdot \beta$ is the complex number*

$$\alpha \cdot \beta = a_1 \bar{b}_1 + a_2 \bar{b}_2 + \cdots + a_n \bar{b}_n = \sum_{i=1}^{n} a_i \bar{b}_i,$$

where $\bar{b}_i$ is the conjugate of the complex number b_i. (*If both α and β are real vectors, then the dot product $\alpha \cdot \beta$ is the real number*

$$\alpha \cdot \beta = a_1 b_1 + a_2 b_2 + \cdots + a_n b_n = \sum_{i=1}^{n} a_i b_i,$$

since $\bar{b}_i = b_i$ for real b_i.) Notice that an ordered pair of coordinate vectors (α, β) is mapped into a complex number $\alpha \cdot \beta$; in Examples 1 and 2, we will demonstrate that this mapping, known as the *dot product function*, is simply a special case of the *inner product function* that is to be defined next in Definition 11. The complex number image of the inner product function is called the *inner product*, and it is designated $\langle \alpha, \beta \rangle$ where α and β are elements of a vector space $V(C)$. Since $\langle \alpha, \beta \rangle$ is a complex number, we will use the notation $\overline{\langle \alpha, \beta \rangle}$ to represent the complex conjugate of $\langle \alpha, \beta \rangle$.

Definition 11. *Let α, β, and γ be arbitrary elements of a vector space $V(C)$, and let k_1 and k_2 be complex numbers; an **inner product function** is a mapping $V \times V \to C$ which has the properties*

(1) $\qquad\qquad \langle \alpha, \beta \rangle = \overline{\langle \beta, \alpha \rangle},$
(2) $\qquad \langle (k_1 \alpha + k_2 \beta), \gamma \rangle = k_1 \langle \alpha, \gamma \rangle + k_2 \langle \beta, \gamma \rangle,$
(3) $\qquad\qquad \alpha \neq 0 \Rightarrow \langle \alpha, \alpha \rangle > 0.$

*If α, β and γ are elements of a vector space $V(R)$, and if k_1 and k_2 are real numbers, then an **inner product function** is a mapping $V \times V \to R$ subject to properties* (1), (2), (3). *Of course, in the real field, $\overline{\langle \beta, \alpha \rangle} = \langle \beta, \alpha \rangle$.*

Example 1. Let the vector space be $V_n(R)$. The dot product function of any ordered pair (α, β) of these vectors is an example of an inner product function because $(\alpha, \beta) \mapsto \alpha \cdot \beta$ defines a mapping $V \times V \to R$ with properties

$$\alpha \cdot \beta = \beta \cdot \alpha,$$
$$(k_1 \alpha + k_2 \beta) \cdot \gamma = k_1 (\alpha \cdot \gamma) + k_2 (\beta \cdot \gamma),$$

where k_1 and k_2 are real numbers, and

$$\alpha \neq 0 \Rightarrow \alpha \cdot \alpha > 0.$$

In Exercise 9 the reader will be asked to verify these assertions. Thus we see that $\alpha \cdot \beta$ is one example of an inner product $\langle \alpha, \beta \rangle$. As an illustration of the second of these properties, let $\alpha = (1, 5)$, $\beta = (3, 4)$, $\gamma = (2, 2)$, $k_1 = 7$, $k_2 = 8$; then

$$(k_1\alpha + k_2\beta) \cdot \gamma = ((7, 35) + (24, 32)) \cdot (2, 2)$$
$$= (31, 67) \cdot (2, 2)$$
$$= 196,$$

and

$$k_1(\alpha \cdot \gamma) + k_2(\beta \cdot \gamma) = 7((1, 5) \cdot (2, 2)) + 8((3, 4) \cdot (2, 2))$$
$$= 84 + 112$$
$$= 196.$$

Example 2. Let the vector space be $V_n(C)$. The dot product function of any pair (α, β) of the vectors is an example of an inner product function because $(\alpha, \beta) \mapsto \alpha \cdot \beta$ defines a mapping $V \times V \to C$ with properties

$$\alpha \cdot \beta = \overline{\beta \cdot \alpha},$$
$$(k_1\alpha + k_2\beta) \cdot \gamma = k_1(\alpha \cdot \gamma) + k_2(\beta \cdot \gamma),$$

where k_1 and k_2 are complex numbers, and

$$\alpha \neq \mathbf{0} \Rightarrow \alpha \cdot \alpha > 0.$$

In Exercise 10 the reader will be asked to verify these assertions. As an illustration of the first of these properties let $\alpha = (i, 2)$ and $\beta = (1, -3i)$.

$$\alpha \cdot \beta = (i)(\overline{1}) + (2)(\overline{-3i})$$
$$= 7i,$$

and

$$\beta \cdot \alpha = (1)(\overline{i}) + (-3i)(\overline{2})$$
$$= -7i,$$

and therefore

$$\overline{\beta \cdot \alpha} = \overline{-7i} = 7i = \alpha \cdot \beta.$$

Example 3. Let the vector space be $V_2(R)$. For arbitrary $\alpha = (a_1, a_2)$ and $\beta = (b_1, b_2)$ define $\alpha * \beta$ (read α star β) by setting

$$\alpha * \beta = a_1 b_1 - a_2 b_1 - a_1 b_2 + 2 a_2 b_2.$$

Thus $(\alpha, \beta) \mapsto \alpha * \beta$ defines a mapping $V \times V \to R$, and it can be shown (Exercise 11) that the following properties are valid:

$$\alpha * \beta = \beta * \alpha,$$
$$(k_1\alpha + k_2\beta) * \gamma = k_1(\alpha * \gamma) + k_2(\beta * \gamma),$$

7.4 INNER PRODUCT SPACES (OPTIONAL)

where k_1 and k_2 are real numbers, and

$$\alpha \neq \mathbf{0} \Rightarrow \alpha * \alpha > 0.$$

Thus $\alpha * \beta$ as defined here satisfies the conditions of Definition 11 and is our second example of an inner product $\langle \alpha, \beta \rangle$ with respect to the vector space $V_2(R)$. As a numerical illustration of this inner product, let $\alpha = (6, 1)$, $\beta = (3, 4)$, and $\gamma = (5, 7)$; then

$$\alpha * \beta = 6 \cdot 3 - 1 \cdot 3 - 6 \cdot 4 + 2(1 \cdot 4) = -1$$

and

$$\beta * \alpha = 3 \cdot 6 - 4 \cdot 6 - 3 \cdot 1 + 2(4 \cdot 1) = -1.$$

If $k_1 = -1$ and $k_2 = 3$, then

$$(k_1\alpha + k_2\beta) * \gamma = (3, 11) * (5, 7) = 93$$

and

$$k_1(\alpha * \gamma) + k_2(\beta * \gamma) = -1(-3) + 3(30) = 93.$$

Also

$$\alpha * \alpha = 6^2 - 6 \cdot 1 - 1 \cdot 6 + 2 \cdot 1^2 = 26 > 0.$$

Example 4. The *trace* of a C-matrix $A = [a_{ij}]_{(n,n)}$ is defined to be the sum of the entries on the main diagonal, that is, $\sum_{i=1}^{n} a_{ii}$; the trace of a matrix A is designated tr A. Let $V(C)$ be the vector space of all m by n C-matrices; then $(A, B) \mapsto \text{tr}(B^*A)$ defines a mapping $V \times V \to C$. For example, if $m = 1$ and $n = 2$,

$$\text{tr}([i \quad 2]^* [3 \quad 2i]) = \text{tr}\left(\begin{bmatrix} -i \\ 2 \end{bmatrix} [3 \quad 2i]\right)$$

$$= \text{tr}\begin{bmatrix} -3i & 2 \\ 6 & 4i \end{bmatrix} = -3i + 4i = i.$$

The following properties of the mapping $V \times V \to C$ defined by $(A, B) \mapsto \text{tr}(B^*A)$ are valid:

$$\text{tr}(B^*A) = \overline{\text{tr}(A^*B)},$$

$$\text{tr}\{D^*(k_1A + k_2B)\} = k_1 \text{tr}(D^*A) + k_2 \text{tr}(D^*B),$$

where k_1 and k_2 are complex numbers, and

$$A \neq \mathbf{0} \Rightarrow \text{tr}(A^*A) > 0.$$

In Exercise 12 the reader will be asked to verify these properties. Thus we see that $\text{tr}(B^*A)$ is another example of an inner product $\langle A, B \rangle$.

Example 5. Because $B^* = B^T$ when B is an R-matrix, it follows from Example 4 that $\text{tr}(B^T A)$ is an example of an inner product $\langle A, B \rangle$ for the vector space $V(R)$ of all m by n R-matrices. As an illustration of the inner product $\langle A, B \rangle = \text{tr}(B^T A)$ for $V(R)$, let

$$A = \begin{bmatrix} 2 & 3 \\ -1 & 0 \end{bmatrix} \text{ and } B = \begin{bmatrix} 6 & 0 \\ 3 & 4 \end{bmatrix}.$$

$$\langle A, B \rangle = \text{tr}(B^T A) = \text{tr}\left(\begin{bmatrix} 6 & 3 \\ 0 & 4 \end{bmatrix} \begin{bmatrix} 2 & 3 \\ -1 & 0 \end{bmatrix}\right) = \text{tr}\begin{bmatrix} 9 & 18 \\ -4 & 0 \end{bmatrix} = 9 + 0 = 9.$$

$$\langle B, A \rangle = \text{tr}(A^T B) = \text{tr}\left(\begin{bmatrix} 2 & -1 \\ 3 & 0 \end{bmatrix} \begin{bmatrix} 6 & 0 \\ 3 & 4 \end{bmatrix}\right) = \text{tr}\begin{bmatrix} 9 & -4 \\ 18 & 0 \end{bmatrix} = 9 + 0 = 9.$$

★**Example 6.** If $V(R)$ is the vector space of all real continuous functions on the interval $a \leq x \leq b$, then

$$(f, g) \mapsto \int_a^b f(x) g(x) \, dx$$

defines a mapping $V \times V \to R$, and it can be shown that the following properties are valid:

$$\int_a^b f(x) g(x) \, dx = \int_a^b g(x) f(x) \, dx,$$

$$\int_a^b (k_1 f(x) + k_2 g(x)) h(x) \, dx = k_1 \int_a^b f(x) h(x) \, dx + k_2 \int_a^b g(x) h(x) \, dx,$$

$$(f(x) \neq 0 \text{ on } a \leq x \leq b) \Rightarrow \int_a^b (f(x))^2 \, dx > 0.$$

For those who have studied calculus these properties are easy to verify, and hence

$$\langle f, g \rangle = \int_a^b f(x) g(x) \, dx$$

is still another example of an inner product.

Definition 12. *A **real inner product space**, or a **Euclidean space**, is a vector space $V(R)$ for which an inner product function is defined.*

The vector spaces of Examples 1, 3, 5, and 6 are Euclidean spaces. Notice that the associated field of a Euclidean space is the field of real numbers, whereas in the following Definition 13 the associated field is the field of complex numbers.

Definition 13. *A **complex inner product space** or a **unitary space** is a vector space $V(C)$ for which an inner product function is defined.*

The vector space $V_n(C)$ with a dot product function is a unitary space (Example 2). Also, the vector space of Example 4 is a unitary space.

EXERCISES

1. For the vector space $V_4(R)$ of all four-dimensional real coordinate vectors, find $\langle \alpha, \beta \rangle = \alpha \cdot \beta$ if $\alpha = (6, 3, 4, 6)$, $\beta = (6, 9, 4, 3)$.
2. In Example 2, find $\langle \alpha, \beta \rangle$ if $n = 3$ and $\alpha = (i, 6, 2 + i)$ and $\beta = (i, 7, 1)$.
3. In Example 3, find $\langle \alpha, \beta \rangle$ if $\alpha = (2, 4)$, $\beta = (6, 2)$.
4. Find the trace of $A = \begin{bmatrix} 6 & 3 & 4 \\ 2 & 1 & 0 \\ 9 & 6 & 5 \end{bmatrix}$.
5. Find the trace of AB if $A = \begin{bmatrix} 6 & 3 & 2 \\ 1 & 4 & 0 \end{bmatrix}$ and $B = \begin{bmatrix} 6 & 9 \\ 0 & 2 \\ 3 & 1 \end{bmatrix}$.
6. In Example 4 find $\langle A, B \rangle$ if $m = n = 2$ and $A = \begin{bmatrix} 2 & i \\ 4 & 0 \end{bmatrix}$, and $B = \begin{bmatrix} i & 6 \\ 0 & 1 \end{bmatrix}$.
7. In Example 5 find $\langle A, B \rangle$ if $m = n = 2$ and $A = \begin{bmatrix} 2 & 4 \\ 3 & 1 \end{bmatrix}$, and $B = \begin{bmatrix} 7 & 0 \\ 1 & 1 \end{bmatrix}$.
★8. In Example 6 find $\langle f, g \rangle$ if $a = 0$, $b = 1$, $f(x) = x^2$, and $g(x) = x^2 + x$.
9. Verify the assertions made in Example 1 that the dot product function possesses the necessary properties to be an inner product function for the vector space of that example.
10. Verify the assertions made in Example 2 that the dot product function possesses the necessary properties to be an inner product function for the vector space of that example.
11. Verify the assertions made in Example 3 that the function defined there possesses the necessary properties to be an inner product function for the vector space of that example.
12. Verify the assertions made in Example 4 that the function defined there possesses the necessary properties to be an inner product function for the vector space of that example.
13. Verify that $\operatorname{tr}(B^T A)$ is an inner product for the vector space of Example 5.
★14. Verify the assertions made in Example 6 that $\int_a^b f(x)g(x)\,dx$ is an inner product for the vector space of that example.

7.5 Length, Distance, and Angle (optional)[3]

Within an inner product space (real or complex) **length** *(or* **magnitude** *or* **norm***) is defined as* $\|\alpha\| = \sqrt{\langle \alpha, \alpha \rangle}$. Length within the inner product space of coordinate vectors has been presented in Section 4.3 and at that time probably

---
[3] This section, however, is a prerequisite to the optional Sections 7.6, 10.4, 10.5, and 10.6.

seemed compatible with the reader's geometric notions of length. It is emphasized, however, that length as presented in Section 4.3 is just one example of the more abstract version just defined.

Example 1. In Example 4 of the last section the trace of a square C-matrix A was defined to be the sum of the entries of the main diagonal of A. Furthermore it was established that $\operatorname{tr}(B^*A)$ is an inner product $\langle A, B \rangle$ for the vector space of all m by n C-matrices. Therefore

$$\|A\| = \sqrt{\langle A, A \rangle} = \sqrt{\operatorname{tr}(A^*A)}.$$

Consider the 2 by 3 matrix

$$A = \begin{bmatrix} i & 0 & 2 \\ 1 & 3 & 4 \end{bmatrix}.$$

$$\operatorname{tr}(A^*A) = \operatorname{tr}\left(\begin{bmatrix} -i & 1 \\ 0 & 3 \\ 2 & 4 \end{bmatrix}\begin{bmatrix} i & 0 & 2 \\ 1 & 3 & 4 \end{bmatrix}\right) = \operatorname{tr}\begin{bmatrix} 2 & 3 & 4-2i \\ 3 & 9 & 12 \\ 4+2i & 12 & 20 \end{bmatrix} = 31;$$

therefore, the length of A is $\|A\| = \sqrt{\operatorname{tr}(A^*A)} = \sqrt{31}$. It should be observed that $\operatorname{tr}(A^*A)$ must be real and nonnegative.

★**Example 2.** From Example 6 of the last section we find

$$\|f\| = \sqrt{\langle f, f \rangle} = \sqrt{\int_a^b [f(x)]^2\, dx}.$$

If $f(x) = x + 1$, if $a = 0$, and if $b = 1$, then

$$\langle f, f \rangle = \int_0^1 (x+1)^2\, dx = \int_0^1 (x^2 + 2x + 1)\, dx = \left.(\tfrac{1}{3}x^3 + x^2 + x)\right|_0^1 = \tfrac{7}{3}.$$

The length of f is

$$\|f\| = \sqrt{\int_0^1 (x+1)^2\, dx} = \sqrt{\tfrac{7}{3}} = \tfrac{1}{3}\sqrt{21}.$$

If $\|\alpha\| = 1$, that is, if $\langle \alpha, \alpha \rangle = \sqrt{1}$, then α is a **unit vector** or a **normalized vector**.

Also, within an inner product space (real or complex) we define **distance between two vectors** α **and** β as $\|\alpha - \beta\|$; that is, $\|\alpha + (-\beta)\|$. For the dot product defined for $V_3(R)$

$$\|\alpha - \beta\| = \sqrt{(\alpha - \beta) \cdot (\alpha - \beta)}$$
$$= \sqrt{(a_1 - b_1)^2 + (a_2 - b_2)^2 + (a_3 - b_3)^2},$$

which the reader recognizes as the familiar distance formula. Other examples of the abstract concept of distance follow.

Example 3. For the unitary space (complex inner product space) $V(C)$ of m by n C-matrices for which $\langle A, B \rangle$ is $\operatorname{tr}(B^*A)$, the distance between A and B is

$$\|A - B\| = \sqrt{\langle (A - B), (A - B) \rangle} = \sqrt{\operatorname{tr}\{(A - B)^*(A - B)\}}.$$

7.5 LENGTH, DISTANCE, AND ANGLE (OPTIONAL)

★**Example 4.** For the inner product space $V(R)$ of real continuous functions on the interval $a \leq x \leq b$ for which the inner product $\langle f, g \rangle$ is $\int_a^b f(x)g(x)\, dx$, the distance between f and g is

$$\|f - g\| = \sqrt{\int_a^b [f(x) - g(x)]^2 \, dx}.$$

An *angle* θ *between two nonzero elements* of a real inner product space is defined by the following formula:

$$\cos\theta = \frac{\langle \alpha, \beta \rangle}{\|\alpha\|\, \|\beta\|}.$$

The *Schwarz inequality*[4] assures us that $-1 \leq (\langle \alpha, \beta \rangle / \|\alpha\|\, \|\beta\|) \leq 1$, which is the range of the cosine function; moreover, this fraction is the cosine of a unique angle between 0 and 180° inclusive. Notice that the inequality

$$-1 \leq \frac{\langle \alpha, \beta \rangle}{\|\alpha\|\, \|\beta\|} \leq 1$$

is not valid if $\langle \alpha, \beta \rangle$ is imaginary; for this reason we have considered only the definition of angle for a *real* inner product space.

Example 5. For the real inner product space $V_2(R)$ for which the dot product is defined, it is easy to verify that

$$\cos\theta = \frac{\alpha \cdot \beta}{\|\alpha\|\, \|\beta\|}.$$

Refer to Figure 7.5.1.

$$\theta = A - B,$$
$$\cos\theta = \cos(A - B)$$
$$= \cos A \cos B + \sin A \sin B$$
$$= \frac{a_1}{\|\alpha\|}\frac{b_1}{\|\beta\|} + \frac{a_2}{\|\alpha\|}\frac{b_2}{\|\beta\|}$$
$$= \frac{\alpha \cdot \beta}{\|\alpha\|\, \|\beta\|}.$$

---

[4] The Schwarz inequality states that within an inner product space

$$|\langle \alpha, \beta \rangle| \leq \|\alpha\|\, \|\beta\|;$$

a proof may be found in Hoffman and Kunze, *Linear Algebra*, Prentice-Hall, New York, 1961, page 228.

Example 6. For the Euclidean space (real inner product space) of m by n R-matrices for which

$$\langle A, B \rangle = \text{tr}(B^T A),$$

if A and B are nonzero,

$$\cos \theta = \frac{\text{tr}(B^T A)}{\sqrt{\text{tr}(A^T A)\,\text{tr}(B^T B)}}.$$

Definition 14. *Two nonzero vectors α and β of an inner product space (real or complex) are said to be **orthogonal** if $\langle \alpha, \beta \rangle = 0$. For a given set $\{\alpha_1, \alpha_2, \ldots, \alpha_n\}$ of nonzero vectors, if $\langle \alpha_i, \alpha_j \rangle = 0$ for all i and j where $i \neq j$, then the set is an **orthogonal set**. If all of the vectors in an orthogonal set are unit vectors, then the set is an **orthonormal set**.*

For a real inner product space, since $\cos \theta = \langle \alpha, \beta \rangle / \|\alpha\| \|\beta\|$, then $\langle \alpha, \beta \rangle = 0$ implies that $\cos \theta = 0$ or $\theta = \tfrac{1}{2}\pi$. Notice that the discussion of orthogonality given in Section 4.3 is a special case of the more general version presented here.

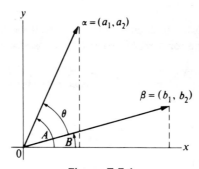

Figure 7.5.1

APPLICATIONS

Example 7. In the study of elementary physics we learn that, as a particle moves through a certain distance, the work done by a constant force acting on the particle in the direction of movement is given by the equation

(work) = (force)(distance),

where the unit of work (such as foot-pound, dyne-centimeter, etc.) is expressed in terms of the units of force and distance. Suppose we want to find the work done (in a plane) in moving a particle along the path designated by $\alpha = (3, 1)$ by means of a force $\beta = (1, 2)$. (See Figure 7.5.2.) Since the total magnitude of the force represented by the vector β does not act directly in the direction of the particle's movement, we are interested only in the component of the force β that

7.5 LENGTH, DISTANCE, AND ANGLE (OPTIONAL)

acts in the direction of α. From Figure 7.5.2 we see that this component is $\|\overrightarrow{OM}\| = \|\beta\| \cos \theta$. The distance moved is $\|\alpha\|$; therefore the total work done is $\|\alpha\| \|\beta\| \cos \theta$. We have proved that, in two dimensions, $\|\alpha\| \|\beta\| \cos \theta$ is simply $\alpha \cdot \beta$. Thus, we have found that

$$(\text{work}) = \alpha \cdot \beta = (3, 1) \cdot (1, 2) = 3 + 2 = 5.$$

Since the magnitudes of α and β are measured in different kinds of units, say, $\|\alpha\|$ in feet (centimeters) and $\|\beta\|$ in pounds (dynes), there is actually no connection between the lengths of the two lines OP and OM. There is no reason why α and β need be drawn with coordinate axes of the same scale, except that in order to

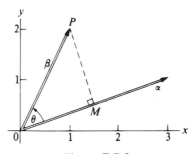

Figure 7.5.2

find θ we need to have their directions relative to each other. Let us, then, repeat this Example with the substitution $\beta' = (7, 14)$ with β' having the same direction as $\beta = (1, 2)$. The angle θ is therefore the same as before, and the new solution is

$$(\text{work}) = \alpha \cdot \beta' = (3, 1) \cdot (7, 14) = 21 + 14 = 35.$$

As we expected, the work is 7 times as great as it was before, since all conditions remain unchanged except that the magnitude of the force is multiplied by 7.

Example 8. Consider the data of Example 8 of Section 4.1 in which four judges vote on a motion to favor an appeal for each of six different appeals. The responses were *yes* or *no* or *abstain* on each vote. Suppose the voting data are reflected in four matrices representing each of the four judges, rather than in three matrices representing each of the three responses, as was the case in Example 8 of Section 4.1. For each "Judge matrix," $q_{ij} = 1$ if that judge gave the ith response to the jth appeal, and $q_{ij} = 0$ otherwise. Rearranging the data of Example 8 of Section 4.1 we have

APPEALS

$$Q_1 = \begin{bmatrix} \#1 & \#2 & \#3 & \#4 & \#5 & \#6 \\ 1 & 0 & 0 & 1 & 1 & 0 \\ 0 & 1 & 1 & 0 & 0 & 1 \\ 0 & 0 & 0 & 0 & 0 & 0 \end{bmatrix} \begin{matrix} yes \\ no \\ abs. \end{matrix} \quad \text{RESPONSES}$$

APPEALS

$$Q_2 = \begin{bmatrix} \#1 & \#2 & \#3 & \#4 & \#5 & \#6 \\ 1 & 0 & 0 & 1 & 0 & 0 \\ 0 & 1 & 1 & 0 & 0 & 1 \\ 0 & 0 & 0 & 0 & 1 & 0 \end{bmatrix} \begin{matrix} yes \\ no \\ abs. \end{matrix} \quad \text{RESPONSES}$$

APPEALS

$$Q_3 = \begin{bmatrix} \#1 & \#2 & \#3 & \#4 & \#5 & \#6 \\ 0 & 0 & 1 & 0 & 0 & 1 \\ 1 & 0 & 0 & 1 & 0 & 0 \\ 0 & 1 & 0 & 0 & 1 & 0 \end{bmatrix} \begin{matrix} yes \\ no \\ abs. \end{matrix} \quad \text{RESPONSES}$$

APPEALS

$$Q_4 = \begin{bmatrix} \#1 & \#2 & \#3 & \#4 & \#5 & \#6 \\ 1 & 1 & 1 & 0 & 1 & 0 \\ 0 & 0 & 0 & 1 & 0 & 1 \\ 0 & 0 & 0 & 0 & 0 & 0 \end{bmatrix} \begin{matrix} yes \\ no \\ abs. \end{matrix} \quad \text{RESPONSES}$$

Suppose we define the *alignment index* of Judge h and Judge k to be

$$\alpha_{hk} = \frac{\text{tr}(Q_h^T Q_k)}{\|Q_h\| \|Q_k\|}$$

which we recognize from Example 6 as the cosine of a generalized angle between Q_h and Q_k. The alignment indices can be calculated easily from the matrix

$$E = \begin{bmatrix} Q_1^T \\ \hline Q_2^T \\ \hline Q_3^T \\ \hline Q_4^T \end{bmatrix} [Q_1 \mid Q_2 \mid Q_3 \mid Q_4]$$

$$= \begin{bmatrix} Q_1^T Q_1 & Q_1^T Q_2 & Q_1^T Q_3 & Q_1^T Q_4 \\ \hline Q_2^T Q_1 & Q_2^T Q_2 & Q_2^T Q_3 & Q_2^T Q_4 \\ \hline Q_3^T Q_1 & Q_3^T Q_2 & Q_3^T Q_3 & Q_3^T Q_4 \\ \hline Q_4^T Q_1 & Q_4^T Q_2 & Q_4^T Q_3 & Q_4^T Q_4 \end{bmatrix}$$

7.5 LENGTH, DISTANCE, AND ANGLE (OPTIONAL)

if we recall that in general $\|Q_n\| = \sqrt{\mathrm{tr}(Q_n^T Q_n)}$. Using our data we find that

$$\alpha_{12} = \tfrac{5}{6}, \quad \alpha_{13} = 0, \quad \alpha_{14} = \tfrac{1}{2}, \quad \alpha_{23} = \tfrac{1}{6}, \quad \alpha_{24} = \tfrac{1}{3}, \quad \alpha_{34} = \tfrac{1}{3}.$$

The alignment index α_{hh} of a judge with himself naturally is 1. Notice that in this problem each α_{hk} is simply the ratio of "the number of times a specific pair voted alike" to "the number of times the pair voted." Obviously such information can be obtained by other methods but the fascination here is: (1) the relative efficiency (although this may not be apparent at first) with which all of the α's can be found in a larger problem with vast amounts of data; and (2) the fact that we have a concrete interpretation of the cosine of a generalized angle. A similar observation may be found in considering the generalized distance between Q_h and Q_k, which we will call the *index of separation* σ_{hk} of Judges h and k. We define

$$\sigma_{hk} = \sqrt{\mathrm{tr}(Q_h - Q_k)^T(Q_h - Q_k)}.$$

For the data of our problem we find

$$\sigma_{12} = \sqrt{2}, \quad \sigma_{13} = \sqrt{12}, \quad \sigma_{14} = \sqrt{6}, \quad \sigma_{23} = \sqrt{10}, \quad \sigma_{24} = \sqrt{8}, \quad \sigma_{34} = \sqrt{8}.$$

These indices of separation appear to indicate that Judges #1 and #2 have relatively little separation in voting patterns, while Judges #1 and #3 have the greatest separation. Analysis of the basic data will verify these results for this problem.

EXERCISES

1. For the real inner product space $V_2(R)$, where $\langle \alpha, \beta \rangle = \alpha \cdot \beta$, $\alpha = (1, 2)$, and $\beta = (6, 4)$, find $\|\alpha\|$, $\|\alpha - \beta\|$, and $\cos \theta$ for θ between α and β. Are α and β orthogonal?

2. For the complex inner product space $V_3(C)$ where $\langle \alpha, \beta \rangle = \alpha \cdot \bar{\beta}$, and $\alpha = (i, 2, 0)$, and $\beta = (0, -i, 6)$, find $\|\alpha\|$ and $\|\alpha - \beta\|$. Are α and β orthogonal?

3. For the Euclidean space (real inner product space) of all 2 by 2 R-matrices where $\langle A, B \rangle = \mathrm{tr}(B^T A)$,

$$A = \begin{bmatrix} 6 & 3 \\ 0 & 1 \end{bmatrix}, \text{ and } B = \begin{bmatrix} -1 & 0 \\ 2 & 1 \end{bmatrix},$$

find $\|A\|$, $\|A - B\|$, and $\cos \theta$ for θ between A and B. Is $\{A, B\}$ an orthogonal set? Is $\{A, B\}$ an orthonormal set?

4. For the unitary space (complex inner product space) of all 2 by 2 C-matrices where

$$\langle A, B \rangle = \mathrm{tr}(B^*A),$$

if

$$A = \begin{bmatrix} i & i \\ 2 & 0 \end{bmatrix}, \text{ and } B = \begin{bmatrix} 2 & i \\ -i & 2 \end{bmatrix},$$

find $\|A\|$ and $\|A - B\|$. Is $\{A, B\}$ an orthogonal set? Is $\{A, B\}$ an orthonormal set?

★5. For the real inner product space of all real continuous functions over the interval $0 \leq x \leq 2$ where

$$\langle f, g \rangle = \int_0^2 f(x)g(x)\,dx, \quad \text{with} \quad f(x) = x^2, \quad \text{and} \quad g(x) = x + 2,$$

find $\|f\|$, $\|f - g\|$, and $\cos\theta$ for θ between f and g. Are f and g orthogonal?

7.6 Gram-Schmidt Orthogonalization (optional)[5]

The basis $\{(1, 0), (0, 1)\}$ of the vector space $V_2(C)$ of all two-dimensional coordinate vectors, and the basis $\{(1, 0, 0), (0, 1, 0), (0, 0, 1)\}$ of the vector space $V_3(C)$ of all three-dimensional coordinate vectors, both have the property that the standard inner product of every pair of vectors within each basis is zero; because of this property these two bases are illustrations of orthogonal bases. In both of these examples a geometric interpretation is that for each orthogonal pair of vectors the corresponding geometric vectors are perpendicular to each other.

Definition 15. *A basis of a finite-dimensional inner product space is an **orthogonal basis** if every pair of elements of the basis is orthogonal (if the basis has only one vector, then the basis is orthogonal). Moreover, if the basis is orthogonal, and if every basis vector has unit length, then the basis is an **orthonormal basis**.*

Example 1. Consider the real inner product space (or Euclidean space) of all 2 by 2 R-matrices with the inner product defined according to

$$\langle A, B \rangle = \operatorname{tr}(B^T A).$$

The basis $b = \left\{ \begin{bmatrix} 1 & 0 \\ 0 & 0 \end{bmatrix}, \begin{bmatrix} 0 & 1 \\ 0 & 0 \end{bmatrix}, \begin{bmatrix} 0 & 0 \\ 1 & 0 \end{bmatrix}, \begin{bmatrix} 0 & 0 \\ 0 & 1 \end{bmatrix} \right\} = \{\beta_1, \beta_2, \beta_3, \beta_4\}$ is an orthogonal basis because $\langle \beta_i, \beta_j \rangle = 0$ if $i \neq j$; we will illustrate one of the six inner products that must be considered to verify this conclusion, leaving the others as an exercise for the reader (Exercise 9).

$$\langle \beta_4, \beta_3 \rangle = \operatorname{tr}(\beta_3^T \beta_4) = \operatorname{tr}\left(\begin{bmatrix} 0 & 0 \\ 1 & 0 \end{bmatrix}^T \begin{bmatrix} 0 & 0 \\ 0 & 1 \end{bmatrix} \right)$$

$$= \operatorname{tr}\left(\begin{bmatrix} 0 & 1 \\ 0 & 0 \end{bmatrix} \begin{bmatrix} 0 & 0 \\ 0 & 1 \end{bmatrix} \right) = \operatorname{tr}\begin{bmatrix} 0 & 1 \\ 0 & 0 \end{bmatrix} = 0 + 0 = 0.$$

---
[5] This section, however, is a prerequisite to the optional Sections 10.4, 10.5, and 10.6.

7.6 GRAM-SCHMIDT ORTHOGONALIZATION (OPTIONAL)

Moreover, the basis b is orthonormal because each element of the basis has length 1; we illustrate one of the four inner products that must be considered to verify this conclusion.

$$\|\beta_2\| = \langle \beta_2, \beta_2 \rangle^{1/2} = \{\text{tr}(\beta_2^T \beta_2)\}^{1/2}$$

$$= \left\{\text{tr}\left(\begin{bmatrix} 0 & 0 \\ 1 & 0 \end{bmatrix}\begin{bmatrix} 0 & 1 \\ 0 & 0 \end{bmatrix}\right)\right\}^{1/2}$$

$$= \left\{\text{tr}\begin{bmatrix} 0 & 0 \\ 0 & 1 \end{bmatrix}\right\}^{1/2} = \{1\}^{1/2} = 1.$$

Since it is often desirable to work with an orthogonal basis we now illustrate how to construct one from an arbitrary basis of an n-dimensional real or complex inner product space; the process that we will use to make this construction is known as the **Gram-Schmidt orthogonalization process**.

Let $b = \{\beta_1, \ldots, \beta_n\}$ be an arbitrary basis of an inner product space (real or complex); a new orthogonal basis $e = \{\varepsilon_1, \ldots, \varepsilon_n\}$ can be constructed according to the following **Gram-Schmidt formulas**:

(7.6.1)
$$\begin{cases} \varepsilon_1 = \beta_1, \\ \varepsilon_2 = \beta_2 - \left(\dfrac{\langle \beta_2, \varepsilon_1 \rangle}{\|\varepsilon_1\|^2}\right)\varepsilon_1, \\ \varepsilon_3 = \beta_3 - \left(\dfrac{\langle \beta_3, \varepsilon_2 \rangle}{\|\varepsilon_2\|^2}\right)\varepsilon_2 - \left(\dfrac{\langle \beta_3, \varepsilon_1 \rangle}{\|\varepsilon_1\|^2}\right)\varepsilon_1, \\ \vdots \\ \varepsilon_n = \beta_n - \left(\dfrac{\langle \beta_n, \varepsilon_{n-1} \rangle}{\|\varepsilon_{n-1}\|^2}\right)\varepsilon_{n-1} - \cdots - \left(\dfrac{\langle \beta_n, \varepsilon_1 \rangle}{\|\varepsilon_1\|^2}\right)\varepsilon_1. \end{cases}$$

These formulas can be stated in the abbreviated form

$$\varepsilon_1 = \beta_1 \quad \text{and} \quad \varepsilon_i = \beta_i - \sum_{k=1}^{i-1}\left(\frac{\langle \beta_i, \varepsilon_k \rangle}{\|\varepsilon_k\|^2}\right)\varepsilon_k \quad (i = 2, \ldots, n).$$

Example 2. Let $b = \{(1, 2), (4, 3)\}$ be a basis of the real inner product space $V_2(R)$, and let the inner product be defined as the standard inner product

$$\langle \alpha, \gamma \rangle = \alpha \cdot \gamma$$
$$= a_1 c_1 + a_2 c_2.$$

According to the formulas (7.6.1) of the Gram-Schmidt process

$$\begin{cases} \varepsilon_1 = \beta_1, \\ \varepsilon_2 = \beta_2 - \left(\dfrac{\beta_2 \cdot \varepsilon_1}{\|\varepsilon_1\|^2}\right)\varepsilon_1. \end{cases}$$

Therefore
$$\varepsilon_1 = (1, 2),$$
and
$$\varepsilon_2 = (4, 3) - \left(\frac{(4, 3) \cdot (1, 2)}{1^2 + 2^2}\right)(1, 2)$$
$$= (4, 3) - 2(1, 2)$$
$$= (2, -1).$$
Hence
$$e = \{(1, 2), (2, -1)\}.$$

Example 3. Let $b = \{(1, 2, 0), (10, 0, 4), (0, 1, 1)\}$ be a basis of the real inner product space $V_3(R)$, and let the inner product be defined as the standard inner product
$$\langle \alpha, \gamma \rangle = \alpha \cdot \gamma$$
$$= a_1 c_1 + a_2 c_2 + a_3 c_3.$$

According to the formulas (7.6.1) of the Gram-Schmidt process
$$\begin{cases} \varepsilon_1 = \beta_1, \\ \varepsilon_2 = \beta_2 - \left(\dfrac{\beta_2 \cdot \varepsilon_1}{\|\varepsilon_1\|^2}\right) \varepsilon_1, \\ \varepsilon_3 = \beta_3 - \left(\dfrac{\beta_3 \cdot \varepsilon_2}{\|\varepsilon_2\|^2}\right) \varepsilon_2 - \left(\dfrac{\beta_3 \cdot \varepsilon_1}{\|\varepsilon_1\|^2}\right) \varepsilon_1. \end{cases}$$

Therefore
$$\varepsilon_1 = (1, 2, 0),$$
and
$$\varepsilon_2 = (10, 0, 4) - \left(\frac{(10, 0, 4) \cdot (1, 2, 0)}{1^2 + 2^2 + 0^2}\right)(1, 2, 0)$$
$$= (10, 0, 4) - \tfrac{10}{5}(1, 2, 0)$$
$$= (8, -4, 4).$$

Then
$$\varepsilon_3 = (0, 1, 1) - \left(\frac{(0, 1, 1) \cdot (8, -4, 4)}{8^2 + (-4)^2 + 4^2}\right)(8, -4, 4)$$
$$\quad - \left(\frac{(0, 1, 1) \cdot (1, 2, 0)}{1^2 + 2^2 + 0^2}\right)(1, 2, 0)$$
$$= (0, 1, 1) - 0(8, -4, 4) - \tfrac{2}{5}(1, 2, 0)$$
$$= (-\tfrac{2}{5}, \tfrac{1}{5}, 1).$$

7.6   GRAM-SCHMIDT ORTHOGONALIZATION (OPTIONAL)

It is natural to wonder how the Gram-Schmidt formulas were developed and why they produce an orthogonal basis. It is an iterative process in which we begin by accepting the first vector β_1 of the old basis as the first vector ε_1 in the new basis. Secondly, as the second element ε_2 of our new basis, we choose a linear combination of ε_1 and β_2 in such a way that the result ε_2 is orthogonal to ε_1. In other words, we must choose the coefficients k_1 and k_2 of a linear combination of ε_1 and β_2 so that $\langle (k_1\varepsilon_1 + k_2\beta_2), \varepsilon_1 \rangle = 0$. But by the definition of an inner product

$$0 = \langle (k_1\varepsilon_1 + k_2\beta_2), \varepsilon_1 \rangle = k_1\langle \varepsilon_1, \varepsilon_1 \rangle + k_2\langle \beta_2, \varepsilon_1 \rangle.$$

Since $\langle \varepsilon_1, \varepsilon_1 \rangle = \|\varepsilon_1\|^2$, and by solving this equation for k_1/k_2, we have

$$\frac{k_1}{k_2} = \frac{-\langle \beta_2, \varepsilon_1 \rangle}{\|\varepsilon_1\|^2}.$$

No generality is lost by assigning the value $k_2 = 1$; hence

$$\varepsilon_2 = k_1\varepsilon_1 + k_2\beta_2 = \beta_2 - \left(\frac{\langle \beta_2, \varepsilon_1 \rangle}{\|\varepsilon_1\|^2}\right)\varepsilon_1.$$

Thirdly, in order to find the third element ε_3 of the basis (if needed) we choose a linear combination of β_3 and the two vectors ε_1 and ε_2 already selected for the new basis in such a way that the result is orthogonal to both ε_1 and ε_2. In other words we must choose coefficients k_1, k_2, and k_3 in such a way that

(7.6.2) $$\begin{cases} \langle (k_1\varepsilon_1 + k_2\varepsilon_2 + k_3\beta_3), \varepsilon_1 \rangle = 0, \\ \langle (k_1\varepsilon_1 + k_2\varepsilon_2 + k_3\beta_3), \varepsilon_2 \rangle = 0. \end{cases}$$

But, by definition of an inner product, (7.6.2) gives us

(7.6.3) $$\begin{cases} k_1\langle \varepsilon_1, \varepsilon_1 \rangle + k_2\langle \varepsilon_2, \varepsilon_1 \rangle + k_3\langle \beta_3, \varepsilon_1 \rangle = 0, \\ k_1\langle \varepsilon_1, \varepsilon_2 \rangle + k_2\langle \varepsilon_2, \varepsilon_2 \rangle + k_3\langle \beta_3, \varepsilon_2 \rangle = 0. \end{cases}$$

But ε_1 is orthogonal to ε_2, hence $\langle \varepsilon_2, \varepsilon_1 \rangle = \langle \varepsilon_1, \varepsilon_2 \rangle = 0$ and (7.6.3) becomes

(7.6.4) $$\begin{cases} \dfrac{k_1}{k_3} = \dfrac{-\langle \beta_3, \varepsilon_1 \rangle}{\|\varepsilon_1\|^2}, \\ \dfrac{k_2}{k_3} = \dfrac{-\langle \beta_3, \varepsilon_2 \rangle}{\|\varepsilon_2\|^2}. \end{cases}$$

No generality is lost by letting $k_3 = 1$ in (7.6.4), hence

$$\varepsilon_3 = \beta_3 - \left(\frac{\langle \beta_3, \varepsilon_2 \rangle}{\|\varepsilon_2\|^2}\right)\varepsilon_2 - \left(\frac{\langle \beta_3, \varepsilon_1 \rangle}{\|\varepsilon_1\|^2}\right)\varepsilon_1.$$

If n is the dimension of the inner product space, then we continue to construct new vectors of the basis in the manner just described until ε_n is found. Each vector ε_i in the new basis is a linear combination of β_i and the previously selected vectors in the new basis; the coefficients of combination are selected in such a way that ε_i will be orthogonal to each element in $\{\varepsilon_1, \ldots, \varepsilon_{i-1}\}$.

Using the ideas just presented, mathematical induction can be used to verify the general formula of the Gram-Schmidt process and to establish[6] the following theorem.

Theorem 6. *Every subspace of dimension 1 or more of a finite-dimensional inner product space possesses an orthogonal basis.*

★**Example 4.** Let $\{1, x, x^2\}$ be a basis of the inner product space $P_2(R)$ (all polynomials of degree 2 or less over the real numbers and the zero polynomial). Let the inner product be defined as

$$\langle p(x), q(x)\rangle = \int_0^1 p(x)q(x)\,dx.$$

We know that an orthogonal basis of $P_2(R)$ exists, hence we proceed to find one. First we choose $\varepsilon_1 = 1$; then

$$\varepsilon_2 = \beta_2 - \left(\frac{\int_0^1 \beta_2 \varepsilon_1\,dx}{\int_0^1 \varepsilon_1^2\,dx}\right)\varepsilon_1$$

$$= x - \left(\frac{\int_0^1 x\,dx}{\int_0^1 dx}\right)(1) = x - \tfrac{1}{2}.$$

Then

$$\varepsilon_3 = \beta_3 - \left(\frac{\int_0^1 \beta_3 \varepsilon_2\,dx}{\int_0^1 \varepsilon_2^2\,dx}\right)\varepsilon_2 - \left(\frac{\int_0^1 \beta_3 \varepsilon_1\,dx}{\int_0^1 \varepsilon_1^2\,dx}\right)\varepsilon_1$$

$$= x^2 - \left(\frac{\int_0^1 (x^3 - \tfrac{1}{2}x^2)\,dx}{\int_0^1 (x^2 - x + \tfrac{1}{4})\,dx}\right)(x - \tfrac{1}{2}) - \left(\frac{\int_0^1 x^2\,dx}{\int_0^1 dx}\right)(1)$$

$$= x^2 - \left(\frac{\tfrac{1}{4}x^4 - \tfrac{1}{6}x^3 \big|_0^1}{\tfrac{1}{3}x^3 - \tfrac{1}{2}x^2 + \tfrac{1}{4}x \big|_0^1}\right)(x - \tfrac{1}{2}) - \left(\frac{\tfrac{1}{3}x^3 \big|_0^1}{x \big|_0^1}\right)(1)$$

$$= x^2 - (1)(x - \tfrac{1}{2}) - \tfrac{1}{3}$$

$$= x^2 - x + \tfrac{1}{6}.$$

Therefore, an orthogonal basis of the given vector space is

$$\{1, x - \tfrac{1}{2}, x^2 - x + \tfrac{1}{6}\}.$$

---

[6] For a real inner product space see Moore, *Elements of Linear Algebra and Matrix Theory*, McGraw-Hill, New York, 1968, pages 194–196. For a complex inner product space see MacLane and Birkhoff, *Algebra*, MacMillan, New York, 1967, page 409.

Each vector can be normalized by multiplication of the reciprocal of its length. For example the length of ε_2 is given by

$$\|\varepsilon_2\| = \langle \varepsilon_2, \varepsilon_2 \rangle^{1/2}$$
$$= \sqrt{\int_0^1 (x - \tfrac{1}{2})^2 \, dx}$$
$$= \sqrt{\int_0^1 (x^2 - x + \tfrac{1}{4}) \, dx}$$
$$= \sqrt{(\tfrac{1}{3}x^3 - \tfrac{1}{2}x^2 + \tfrac{1}{4}x)\big|_0^1}$$
$$= \sqrt{\tfrac{1}{12}}.$$

EXERCISES

1. Consider the inner product space $V_3(R)$ with standard inner product
$$\langle \alpha, \beta \rangle = \alpha \cdot \beta = a_1 b_1 + a_2 b_2 + a_3 b_3.$$
Determine which of the following bases are orthogonal, and then determine which are orthonormal.
 (a) $b = \{(1, 2, 3), (1, 0, 2), (-2, -1, 3)\}$;
 (b) $b = \{(\tfrac{1}{2}\sqrt{2}, 0, -\tfrac{1}{2}\sqrt{2}), (0, 1, 0), (\tfrac{1}{2}\sqrt{2}, 0, \tfrac{1}{2}\sqrt{2})\}$;
 (c) $b = \{(1, -1, 2), (-2, 0, 1), (1, 5, 2)\}$.

2. Consider the inner product space $V_2(C)$ with standard inner product
$$\langle \alpha, \beta \rangle = \alpha \cdot \beta = a_1 \bar{b}_1 + a_2 \bar{b}_2.$$
Determine which of the following bases are orthogonal, and then determine which are orthonormal.
 (a) $b = \{(1, 0), (0, i)\}$;
 (b) $b = \{(i, 1), (2, i)\}$;
 (c) $b = \{(2, i), (-1, 2i)\}$.

3. Consider the inner product space $V_2(R)$ with inner product defined as in Example 3 of Section 7.4, namely,
$$\langle \alpha, \beta \rangle = a_1 b_1 - a_2 b_1 - a_1 b_2 + 2 a_2 b_2.$$
Determine which of the following bases are orthogonal, and then determine which are orthonormal.
 (a) $b = \{(1, 0), (0, 1)\}$;
 (b) $b = \{(1, 0), (1, 1)\}$;
 (c) $b = \{(2, 1), (0, -2)\}$.

★4. Consider the inner product space $P_1(R)$, and let the inner product be defined by
$$\langle \alpha, \beta \rangle = \int_0^2 \alpha \beta \, dx.$$
Determine which of the following bases are orthogonal, and then determine which are orthonormal.
 (a) $b = \{1, x\}$;
 (b) $b = \{1, x - 2\}$;
 (c) $b = \{x - \tfrac{2}{3}, \tfrac{1}{2}x\}$.

5. Using the basis of part (a) of Exercise 1, construct an orthogonal basis by the Gram-Schmidt process.
6. Using the basis of part (b) of Exercise 2, construct an orthogonal basis by the Gram-Schmidt process.
7. Using the basis of part (a) of Exercise 3, construct an orthogonal basis by the Gram-Schmidt process.
8. Using the basis of part (a) of Exercise 4, construct an orthogonal basis by the Gram-Schmidt process.
9. In Example 1 on page 210 verify that b is an orthogonal basis.
10. In formulas (7.6.4) on page 213, explain why k_3 cannot equal zero.

NEW VOCABULARY

linear combination of a set of vectors   7.1
span (or generate)   7.1
subspace of a vector space   7.1
linearly dependent vectors   7.1
linearly independent vectors   7.1
basis of a vector space   7.2
dimension of a vector space   7.2
finite-dimensional vector space   7.2
column space of a matrix   7.2
row space of a matrix   7.2
isomorphic vector spaces   7.3
isomorphic mapping (or isomorphism)   7.3
coordinates of a vector   7.3
coordinate matrix   7.3
inner product of vectors   7.4
trace   7.4
real inner product space (or Euclidean space)   7.4
complex inner product space (or unitary space)   7.4
length (or magnitude) of a vector   7.5
unit vector (normalized vector)   7.5
distance between two vectors   7.5
angle between two vectors   7.5
Schwarz inequality   7.5
orthogonal vectors   7.5
orthogonal set of vectors   7.5
orthonormal set of vectors   7.5
orthogonal basis   7.6
orthonormal basis   7.6
Gram-Schmidt formulas   7.6

SPECIAL PROJECT

Let X and Y be elements of the vector space $V_2(R)$ and define a product $X * Y$ according to $X * Y = XAY$ where A is a 2 by 2 R-matrix and $|A| > 0$ and $a_{11} > 0$. Determine whether $*$ is or is not an inner product function for $V_2(R)$, and prove the truth of your answer. Then similarly define and investigate an analogous product over $V_3(R)$. If possible, investigate length, distance, and angles with respect to $*$. Report your findings in a short paper.

8

LINEAR TRANSFORMATIONS

8.1 Definition and Examples

In the first section of this book we pointed out that a transformation is a mapping or a function that associates with every element of the domain of the transformation a unique element of a codomain of the transformation.

Example 1. Associated with every real number x there is another unique real number $\sin x$; thus we say that "the sine of" is a transformation from the domain of real numbers into the codomain of real numbers. Of course the range of the transformation is a subset of the codomain consisting of the set of images; in this example, the range is the set of real numbers between -1 and $+1$, inclusive.

Recall that if A is the domain, and if B is a codomain, then a transformation T from A into B is designated $T: A \to B$, or $A \xrightarrow{T} B$; the effect of T on the elements of A is designated by $x \xmapsto{T} T(x)$. In the last example the transformation was $T: R \to R$ or $R \xrightarrow{T} R$ where the effect of T on the elements of R was $x \xmapsto{T} \sin x$, or $T(x) = \sin x$.

Example 2. Let the domain be the set of real coordinate vectors V with n components, and let the codomain be the set of real numbers, R. To every element α of the domain there corresponds a unique element $\|\alpha\|$ of the codomain. In symbols, $V \xrightarrow{T} R$ where $\alpha \xmapsto{T} \|\alpha\|$ or $T(\alpha) = \|\alpha\|$. The range of the transformation is the subset of all nonnegative real numbers.

Example 3. Let the domain and codomain be the set of all coordinate vectors with two real components. We have shown that this set forms a vector space $V_2(R)$ with respect to vector addition. We now define several transformations of the given vector space into itself, that is, $V_2(R) \xrightarrow{T} V_2(R)$. Because of the isomorphism between the given vector space $V_2(R)$ and the vector space of two-dimensional geometric vectors, these transformations are frequently called *transformations of the plane* and may be illustrated by Figures 8.1.1–8.1.6.

(1) Let T_1 be the "stretching" or "contraction" transformation such that $\alpha \xmapsto{T_1} k\alpha$ where $k > 0$. In Figure 8.1.1 notice that T_1 maps each nonzero vector α into a vector that has the same direction but k times the magnitude of α; the zero vector is mapped into itself.

(2) Let T_2 be the transformation called "reflection through the origin" such that $\alpha \xmapsto{T_2} -\alpha$. In Figure 8.1.2 notice that T_2 maps a nonzero vector α into a vector having the same magnitude as α but the opposite direction; the zero vector is mapped into itself.

(3) Let T_3 be the transformation called "reflection through the x-axis" such that $(a_1, a_2) \xmapsto{T_3} (a_1, -a_2)$. In Figure 8.1.3 notice that if $a_2 \neq 0$, then T_3 reflects each vector (a_1, a_2) through the x-axis; if $a_2 = 0$, then the vector is mapped into itself.

(4) Let T_4 be the transformation of rotation through $90°$, for which $(a_1, a_2) \xmapsto{T_4} (-a_2, a_1)$; see Figure 8.1.4. Notice that the zero vector is mapped into itself.

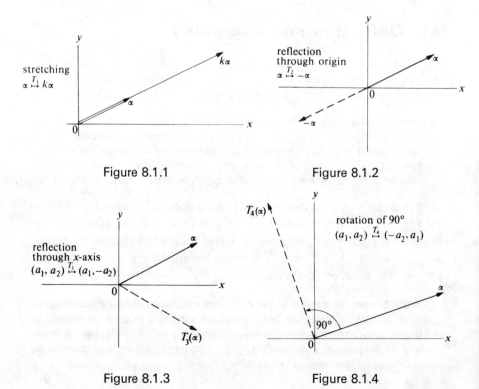

Figure 8.1.1

Figure 8.1.2

Figure 8.1.3

Figure 8.1.4

8.1   DEFINITIONS AND EXAMPLES

(5) Let T_5 be the "shearing" transformation such that $(a_1, a_2) \stackrel{T_5}{\mapsto} (a_1 + ka_2, a_2)$ where $k \neq 0$. In Figure 8.1.5, notice that if $a_2 \neq 0$, then T_5 slides the terminal point of the geometric vector parallel to the x-axis by a distance of ka_2 units while the initial point remains fixed; if $a_2 = 0$, then the vector is mapped into itself.

(6) Let T_6 be the transformation called "projection on the x-axis" such that $(a_1, a_2) \stackrel{T_6}{\mapsto} (a_1, 0)$. See Figure 8.1.6; notice that if $a_2 = 0$ then the vector is mapped into itself.

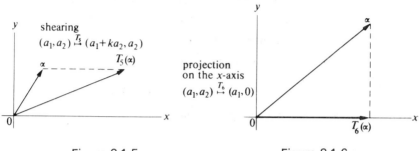

Figure 8.1.5   Figure 8.1.6

In this chapter we limit our considerations to a special type of transformation. First of all we shall require that both the domain and codomain be the elements of a vector space (not necessarily the same vector space). Furthermore we shall impose the so-called linearity requirements according to the following definition.

Definition 1. *A transformation T from a vector space $V(F)$ into a vector space $U(F)$ is said to be a **linear transformation** if*

$$T(\alpha + \beta) = T(\alpha) + T(\beta) \quad \text{and} \quad T(k\alpha) = kT(\alpha),$$

where α and β are arbitrary elements of $V(F)$ and k is an element of F.

Notice that in the left side of

$$T(\alpha + \beta) = T(\alpha) + T(\beta)$$

the symbol " $+$ " designates addition over $V(F)$ whereas in the right side it designates addition over $U(F)$.

The transformation "sine of" in Example 1 is not linear, because $\sin(x + y) \neq \sin x + \sin y$ for arbitrary real numbers x and y; moreover $\sin kx \neq k \sin x$. The transformation of Example 2 is not linear, because $\|\alpha + \beta\| \neq \|\alpha\| + \|\beta\|$ for arbitrary real coordinate vectors α and β. The six transformations of Example 3 are all linear; the proof of this is left as an exercise.

Example 4. For the vector space $V_2(R)$, let the transformation $T: V_2(R) \to V_2(R)$ be defined in such a way that $(a, b) \overset{T}{\mapsto} (a + 2, b)$. This transformation is not linear because

$$T\{(a_1, b_1) + (a_2, b_2)\} = T\{(a_1 + a_2, b_1 + b_2)\}$$
$$= (a_1 + a_2 + 2, b_1 + b_2),$$

whereas

$$T\{(a_1, b_1)\} + T\{(a_2, b_2)\} = (a_1 + 2, b_1) + (a_2 + 2, b_2)$$
$$= (a_1 + a_2 + 4, b_1 + b_2),$$

and these two results are not equal. Furthermore,

$$T\{k(a, b)\} = T\{(ka, kb)\} = (ka + 2, kb),$$

whereas

$$kT(a, b) = k(a + 2, b) = (ka + k2, kb),$$

and in general these two results are not equal.

★**Example 5.** Consider the vector space $P_n(R)$ (polynomials of degree n or less over the real numbers and the zero polynomial). If we take the derivative of an element of $P_n(R)$ we will obtain a unique image which also is an element of $P_n(R)$, thus the derivative D is a transformation $T: P_n(R) \to P_n(R)$. The fact that *D is a linear transformation* follows from the well-known theorems of calculus that for arbitrary differentiable functions, f and g, defined over a common domain, and for a real number k,

$$D(f(x) + g(x)) = Df(x) + Dg(x) \quad \text{and} \quad Dkf(x) = kDf(x).$$

★**Example 6.** Let $P_n(R)$ be the vector space of polynomials defined in Example 5, and let $P_{n+1}(R)$ be a second vector space consisting of polynomials of degree $n + 1$ or less and the zero polynomial. If we take the indefinite integral, $\int_a^x (\ ) \, dx$ (a and x real), of any element of $P_n(R)$ we will obtain a unique image which belongs to $P_{n+1}(R)$; thus the indefinite integral $\int_a^x (\ ) \, dx$ is a transformation $T: P_n(R) \to P_{n+1}(R)$. The fact that *the transformation is linear* follows from the well-known theorems of calculus that for arbitrary functions f and g that are continuous over R and for a real number k

$$\int_a^x (f(x) + g(x)) \, dx = \int_a^x f(x) \, dx + \int_a^x g(x) \, dx$$

and

$$\int_a^x kf(x) \, dx = k \int_a^x f(x) \, dx.$$

★**Example 7.** Let $P_n(R)$ be the vector space of polynomials defined in Example 5 and used again in Example 6. Let $R(R)$ be the vector space of real numbers over the field of real numbers. If we take the definite integral, $\int_a^b (\ ) \, dx$ (a and b real), of any element of $P_n(R)$, we will obtain a unique image in $R(R)$; thus "the definite integral of" is a transformation $T: P_n(R) \to R(R)$. The fact that *the transformation is linear* follows from the well-known theorems of calculus that for functions f and g that are continuous over R and for a real number k

$$\int_a^b (f(x) + g(x)) \, dx = \int_a^b f(x) \, dx + \int_a^b g(x) \, dx,$$

and

$$\int_a^b kf(x) \, dx = k \int_a^b f(x) \, dx.$$

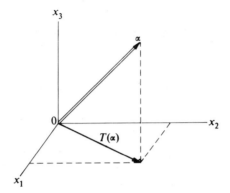

Figure 8.1.7

Example 8. Let $V(R)$ be the vector space of all n by 1 column R-matrices, and let $U(R)$ be the vector space of all m by 1 column R-matrices. If any element of $V(R)$ is premultiplied by an m by n R-matrix, A, then we will obtain a unique image which is an element of $U(R)$ (a consequence of the definition of matrix multiplication); thus the specified matrix multiplication, T, is a transformation $T: V(R) \to U(R)$. The fact that this very important transformation is linear will be guaranteed by Theorem 1 below. If $n = 3$, $m = 2$, $A = \begin{bmatrix} 1 & 0 & 0 \\ 0 & 1 & 0 \end{bmatrix}$, and if

$X = \begin{bmatrix} x_1 \\ x_2 \\ x_3 \end{bmatrix}$ is an arbitrary element of $V(R)$, then the matrix multiplication

$$\begin{bmatrix} 1 & 0 & 0 \\ 0 & 1 & 0 \end{bmatrix} \begin{bmatrix} x_1 \\ x_2 \\ x_3 \end{bmatrix} = \begin{bmatrix} x_1 \\ x_2 \end{bmatrix}$$

maps any element of $V(R)$ into a unique element of $U(R)$; geometrically, a premultiplication by the matrix $\begin{bmatrix} 1 & 0 & 0 \\ 0 & 1 & 0 \end{bmatrix}$ can be thought of as effecting a projection of any three-dimensional geometric vector onto the $x_1 x_2$-plane as shown in Figure 8.1.7. This example is an illustration of the next definition.

Definition 2. *Let α be an arbitrary element of the vector space $V(F)$ of all n by 1 matrices over a field F, let $U(F)$ be the vector space of all m by 1 matrices over F, and let A be an m by n matrix over F. A transformation $T: V(F) \to U(F)$, where $\alpha \xrightarrow{T} A\alpha$, is called a **matrix transformation** from $V(F)$ into $U(F)$.*

Theorem 1. *If $U(F)$ and $V(F)$ are the vector spaces of Definition 2, then every matrix transformation from $V(F)$ into $U(F)$ is a linear transformation from $V(F)$ into $U(F)$.*

Proof. Let α and β be arbitrary elements of $V(F)$, let k be an arbitrary element of F, and let A be an m by n matrix over F.

STATEMENT	REASON
(1) $A(\alpha + \beta) = A\alpha + A\beta$.	(1) Distributive law (generalization of Theorem 2 of Section 3.1 for matrices over a field.)
(2) $A(k\alpha) = k(A\alpha)$.	(2) Generalization of Theorem 4 of Section 3.1 for matrices over a field. □

Example 8 and Theorem 1 touch on one of the most important uses of matrices, that is, that matrix multiplication effects linear transformations of column vectors. This becomes particularly significant if one recalls that every n-dimensional vector space is isomorphic to a vector space of n by 1 column vectors.

It is then natural to wonder whether *every* linear transformation from one arbitrary n-dimensional vector space to another (or to the same) vector space, like those illustrated in Examples 3, 5, 6, and 7, can be accomplished by some unique *matrix* transformation from the vector space of n by 1 matrices into the vector space of m by 1 matrices. For example, can a derivative of a polynomial be found by means of matrix multiplication? Affirmative answers to these questions will be established in the next sections.

APPLICATIONS

Example 9. Cryptography is a subject that is concerned with invertible transformations. For example, suppose we set up the mapping

$$\begin{array}{cccccccc} A & B & C & \cdots & X & Y & Z \\ \updownarrow & \updownarrow & \updownarrow & \cdots & \updownarrow & \updownarrow & \updownarrow \\ 26 & 25 & 24 & \cdots & 3 & 2 & 1 \end{array}.$$

The message GO TO PARIS according to this transformation would be

$$20, \ 12, \ 7, \ 12, \ 11, \ 26, \ 9, \ 18, \ 8.$$

Once the message is transmitted, it is decoded by the inverse transformation. Such a code (transformation) would be easy to crack, however, because of known frequencies of letters in the various languages. One method that attempts to avoid this difficulty is to arrange the message and the corresponding numbers by columns in an m by n matrix A and then perform an invertible matrix transformation T on the columns of A. The matrix TA is then transmitted and decoded by $T^{-1}(TA) = A$. As an illustration, arrange the message GO TO PARIS as a sequence of triples GOT OPA RIS and use these triples as columns of the matrix

$$\begin{bmatrix} G & O & R \\ O & P & I \\ T & A & S \end{bmatrix}.$$

The numerical counterpart according to the preliminary code is

$$A = \begin{bmatrix} 20 & 12 & 9 \\ 12 & 11 & 18 \\ 7 & 26 & 8 \end{bmatrix}.$$

Now choose an arbitrary[1] invertible matrix T. Suppose that

$$T = \begin{bmatrix} 1 & 2 & 1 \\ -1 & -1 & 1 \\ 0 & 1 & 3 \end{bmatrix}.$$

Then we have the transformation

$$\begin{bmatrix} G & O & R \\ O & P & I \\ T & A & S \end{bmatrix} \rightarrow \begin{bmatrix} 1 & 2 & 1 \\ -1 & -1 & 1 \\ 0 & 1 & 3 \end{bmatrix} \begin{bmatrix} 20 & 12 & 9 \\ 12 & 11 & 18 \\ 7 & 26 & 8 \end{bmatrix} = \begin{bmatrix} 51 & 60 & 53 \\ -25 & 3 & -19 \\ 33 & 89 & 42 \end{bmatrix},$$

which is transmitted and decoded by premultiplying by T^{-1}.

$$\begin{bmatrix} -4 & -5 & 3 \\ 3 & 3 & -2 \\ -1 & -1 & 1 \end{bmatrix} \begin{bmatrix} 51 & 60 & 53 \\ -25 & 3 & -19 \\ 33 & 89 & 42 \end{bmatrix} = \begin{bmatrix} 20 & 12 & 9 \\ 12 & 11 & 18 \\ 7 & 26 & 8 \end{bmatrix} \rightarrow \begin{bmatrix} G & O & R \\ O & P & I \\ T & A & S \end{bmatrix}.$$

For those interested in cryptography, four references and a more sophisticated discussion of certain problems of cryptography are given in [20] (Levine). An elementary discussion similar to that given here may be found on pages 211–214 in [22] (Nahikian), and on pages 331–333 in [7] (Davis). (The latter gives additional references.)

Example 10. Let forces of f_1, f_2, and f_3 be exerted on a beam at three points as shown in Figure 8.1.8. Making use of Hooke's Law, the deflections D_i of the beam at the respective points can be found according to the matrix transformation

$$\begin{bmatrix} D_1 \\ D_2 \\ D_3 \end{bmatrix} = \begin{bmatrix} d_{11} & d_{12} & d_{13} \\ d_{21} & d_{22} & d_{23} \\ d_{31} & d_{32} & d_{33} \end{bmatrix} \begin{bmatrix} f_1 \\ f_2 \\ f_3 \end{bmatrix},$$

---

[1] Frequently T is chosen so that $T = T^{-1}$. In this case the coding matrix and the decoding matrix would be the same.

where d_{ij} represents the deflection at point i on the beam when a force of 1 lb is exerted at point j; by the Reciprocal Theorem $d_{ij} = d_{ji}$, and hence the transformation matrix is symmetric. The associated matrix transformation

$$\begin{bmatrix} f_1 \\ f_2 \\ f_3 \end{bmatrix} = \begin{bmatrix} d_{11} & d_{12} & d_{13} \\ d_{21} & d_{22} & d_{23} \\ d_{31} & d_{32} & d_{33} \end{bmatrix}^{-1} \begin{bmatrix} D_1 \\ D_2 \\ D_3 \end{bmatrix}$$

allows us to consider the matrix of forces as the image of the matrix of deflections under the matrix transformation $[d_{ij}]^{-1}$. Matrix $[d_{ij}]$ is called the *flexibility matrix*, whereas $[d_{ij}]^{-1}$ is called the *stiffness matrix*. A more general discussion of this example can be found on pages 68–70 and page 121 of [7] (Davis), and on pages 206–210 of [23] (Noble).

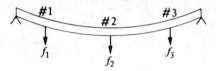

Figure 8.1.8

EXERCISES

1. For the vector space $V_2(R)$, find the image of the given element β under the transformation $T: V_2(R) \to V_2(R)$ if $\beta \stackrel{T}{\mapsto} -2\beta$, or $T(\beta) = -2\beta$. Graph the geometric vectors corresponding to β and its image.
   (a) $\beta = (1, -2)$;   (b) $\beta = (0, 3)$.

In each of Exercises 2-9, choose a domain A and a codomain B such that the given T is a transformation from A into B. Also state the range of the transformation corresponding to the chosen domain. Let x and y represent real numbers.

2. $T(x) = x^2 + 1$.   3. $x \stackrel{T}{\mapsto} (x, 2x)$.

4. $T(\beta) = \beta^2$, where $\beta = (x, y)$.   5. $\begin{bmatrix} x \\ y \end{bmatrix} \stackrel{T}{\mapsto} \begin{bmatrix} 0 \\ y \end{bmatrix}$.

6. $T(A) = A^2$, where $A = \begin{bmatrix} x & x \\ x & x \end{bmatrix}$.   7. $T(x) = \sqrt{1-x^2}$.

8. $T(x) = -\sqrt{1-x^2}$.   9. $T(x) = \sqrt{x^2-1}+1$.

Let the vector space be $V_2(R)$. In each of Exercises 10–15, graph the geometric vectors corresponding to one element of the domain and its image under the given transformation $T: V_2(R) \to V_2(R)$.

10. $(x_1, x_2) \stackrel{T}{\mapsto} (2x_1, x_2)$.   11. $T(\alpha) = -\tfrac{1}{2}\alpha$, where $\alpha = (a_1, a_2)$.
12. $(x_1, x_2) \stackrel{T}{\mapsto} (2x_1, 0)$.   13. $(x_1, x_2) \stackrel{T}{\mapsto} (x_1 + x_2, x_2)$.
14. $(x_1, x_2) \stackrel{T}{\mapsto} (x_1 + 2, x_2 + 3)$.   15. $(x_1, x_2) \stackrel{T}{\mapsto} (-2x_2, 2x_1)$.

8.1 DEFINITION AND EXAMPLES

16. Prove that T_1, T_2, and T_3 of Example 3 are linear transformations.

17. Prove that T_4, T_5, and T_6 of Example 3 are linear transformations.

In each of Exercises 18–23, let $V(R)$ be the vector space of all 2 by 1 R-matrices, and determine the image of the element α of $V(R)$ under the matrix transformation $T: V(R) \to V(R)$ where $\alpha \overset{T}{\mapsto} A\alpha$. Graph the geometric vectors corresponding to $\alpha = \begin{bmatrix} a_1 \\ a_2 \end{bmatrix}$ and its image under the given matrix transformation.

18. $A = \begin{bmatrix} 1 & 0 \\ 0 & 1 \end{bmatrix}$.   **19.** $A = \begin{bmatrix} 2 & 0 \\ 0 & 1 \end{bmatrix}$.   **20.** $A = \begin{bmatrix} 0 & 0 \\ 0 & 1 \end{bmatrix}$.

21. $A = \begin{bmatrix} 0 & -3 \\ 3 & 0 \end{bmatrix}$.   **22.** $A = \begin{bmatrix} 1 & 1 \\ 0 & 1 \end{bmatrix}$.   **23.** $A = \begin{bmatrix} 1 & 1 \\ 1 & 1 \end{bmatrix}$.

In each of Exercises 24–29, let the vector space be $V_2(R)$. Which of the following transformations $T: V_2(R) \to V_2(R)$ are linear? State reasons for your answer.

24. $(a, b) \overset{T}{\mapsto} (a + 1, b)$.   **25.** $(a, b) \overset{T}{\mapsto} (-2a, 2b)$.

26. $(a, b) \overset{T}{\mapsto} (a, 2)$.   **27.** $(a, b) \overset{T}{\mapsto} (0, 0)$.

28. $(a, b) \overset{T}{\mapsto} (3a, 2b)$.   **29.** $\alpha \overset{T}{\mapsto} \|\alpha\|\alpha$.

In each of Exercises 30–31, let $V(R)$ be the vector space of all 2 by 1 R-matrices. Which of the following transformations $T: V(R) \to V(R)$ are linear? State reasons for your answer.

30. $\begin{bmatrix} a \\ b \end{bmatrix} \overset{T}{\mapsto} \begin{bmatrix} a+b \\ a-b \end{bmatrix}$.   **31.** $\begin{bmatrix} a \\ b \end{bmatrix} \overset{T}{\mapsto} \frac{1}{2}\left(\begin{bmatrix} a \\ b \end{bmatrix} + \begin{bmatrix} 2 \\ 1 \end{bmatrix}\right)$.

★32. Consider the vector space $P_3(R)$ and the vector space $R(R)$ of real numbers. Is the transformation $T: P_3(R) \to R(R)$ where $p(x) \overset{T}{\mapsto} \int_0^1 p(x)\,dx$ a linear transformation? Give reasons for your answer.

★33. Let the vector space be $P_n(R)$. Is the transformation $T: P_n(R) \to P_n(R)$ where $p(x) \overset{T}{\mapsto} D_x^2(p(x))$ a linear transformation? Give reasons for your answer.

34. In Section 3.4 we learned that an m by n C-matrix B can be obtained from an equivalent m by n C-matrix A by means of elementary operations. We also learned that the elementary operations can be accomplished by matrix multiplication. That is,

$$B = PAQ,$$

where P is a certain m by m invertible C-matrix and Q is a certain n by n invertible C-matrix. Let $V(C)$ be the vector space of all m by n C-matrices and define a transformation $T: V(C) \to V(C)$ such that $A \overset{T}{\mapsto} PAQ$. Prove that T is linear.

35. Let $V(R)$ be the vector space of all n by n R-matrices, and let P be a fixed n by n invertible R-matrix. We define a transformation $T: V(R) \to V(R)$ such that $A \overset{T}{\mapsto} P^{-1}AP$. Prove that T is linear.

36. Let $V(R)$ be the vector space of all n by n R-matrices, and let P be a fixed n by n invertible R-matrix. Define a transformation $T: V(R) \to V(R)$ such that $A \xmapsto{T} P^T A P$. Prove that T is linear.
37. Let T be a linear transformation from a vector space $V(F)$ into a vector space $U(F)$. Prove that the image of the zero vector of $V(F)$ is the zero vector of $U(F)$.
38. Using the transformations of Example 9, write a matrix representing the message "BUY CAR". How can the method be varied to send a seven-letter message?

8.2 Matrix Representation of a Linear Transformation

At the outset of this section we wish to emphasize that here we consider only a single finite-dimensional vector space $V(F)$ with a fixed basis $b = \{\beta_1, \ldots, \beta_n\}$. We will show that corresponding to every linear transformation T from $V(F)$ into $V(F)$ there is a unique n by n matrix over F, constructed in a certain way and designated $[T]_b$, and that, conversely, corresponding to every n by n matrix $[T]_b$ over F there is a unique linear transformation T from $V(F)$ into $V(F)$. The matrix $[T]_b$, that we shall learn to construct, is called the *matrix representation of T* or the *transformation matrix*. Then in the next section we shall establish that any linear transformation T of vectors of $V(F)$ may be obtained by a matrix transformation using $[T]_b$ as the operator. Later, in Section 8.6, we will study linear transformations from one vector space into a *different* vector space. Before developing these significant ideas, however, the reader is reminded of some very important concepts that were established in Section 7.3.

Each element α of a vector space $V(F)$ can be expressed as a unique linear combination of the elements of a basis $\{\beta_1, \beta_2, \ldots, \beta_n\}$,

$$\alpha = a_1 \beta_1 + a_2 \beta_2 + \cdots + a_n \beta_n.$$

The elements $a_1, a_2, \ldots, a_n$ of F are called the *coordinates* of α with respect to the given basis. The column matrix of coefficients,

$$[\alpha]_b = \begin{bmatrix} a_1 \\ a_2 \\ \vdots \\ a_n \end{bmatrix},$$

is a coordinate vector corresponding to α and will be called the *coordinate matrix* of α with respect to the basis $b = \{\beta_1, \ldots, \beta_n\}$. We remind the reader that the basis is an ordered set. Obviously, if a different basis is chosen, the coordinates and the coordinate matrix of α will change; hence, in what follows it is imperative that we keep in mind what basis we are using.

8.2 REPRESENTATION OF A LINEAR TRANSFORMATION

Example 1. Consider the vector space $P_2(R)$. Suppose we choose to let our basis elements, $\beta_1, \beta_2, \beta_3$, be 1, x, and x^2, respectively; then the element $3x^2 + 6x + 2$ has coordinates 2, 6, and 3 with respect to the chosen basis b. The column matrix

$$[\alpha]_b = \begin{bmatrix} 2 \\ 6 \\ 3 \end{bmatrix}$$

is the coordinate matrix of $\alpha = 3x^2 + 6x + 2$ with respect to the chosen basis. The reader may verify that if we choose another basis such as

$$e = \{1, 1+x, 1+x+x^2\}$$

then a different coordinate matrix results. For example if the same polynomial is expressed as a linear combination of the new basis, we obtain

$$3x^2 + 6x + 2 = v_1(1) + v_2(1+x) + v_3(1+x+x^2)$$
$$= (v_3)x^2 + (v_2 + v_3)x + (v_1 + v_2 + v_3)1;$$

it follows that the coordinates v_1, v_2, v_3 with respect to the new basis are $-4, 3, 3$, respectively. The new coordinate matrix is

$$[\alpha]_e = \begin{bmatrix} -4 \\ 3 \\ 3 \end{bmatrix}.$$

We now proceed to explain the one-to-one correspondence between the set of linear transformations from a vector space $V(F)$ into itself and the set of n by n matrices over F. Let T be a linear transformation such that the images of the basis vectors $\{\beta_1, \ldots, \beta_n\}$ are given by

(8.2.1)
$$\begin{cases} T(\beta_1) = t_{11}\beta_1 + t_{12}\beta_2 + \cdots + t_{1n}\beta_n, \\ T(\beta_2) = t_{21}\beta_1 + t_{22}\beta_2 + \cdots + t_{2n}\beta_n, \\ \vdots \quad\quad \vdots \quad\quad \vdots \quad\quad\quad\quad \vdots \\ T(\beta_n) = t_{n1}\beta_1 + t_{n2}\beta_2 + \cdots + t_{nn}\beta_n. \end{cases}$$

The uniqueness of the coefficients of (8.2.1) is guaranteed by Theorem 3 of Chapter 7. Thus, with respect to the same basis, the coordinates of the image of β_1 are $(t_{11}, t_{12}, \ldots, t_{1n})$, the coordinates of the image of β_2 are $(t_{21}, t_{22}, \ldots, t_{2n})$, etc.; hence the coordinate matrices are:

$$[T(\beta_1)]_b = \begin{bmatrix} t_{11} \\ t_{12} \\ \vdots \\ t_{1n} \end{bmatrix}, \quad [T(\beta_2)]_b = \begin{bmatrix} t_{21} \\ t_{22} \\ \vdots \\ t_{2n} \end{bmatrix}, \quad \ldots, \quad [T(\beta_n)]_b = \begin{bmatrix} t_{n1} \\ t_{n2} \\ \vdots \\ t_{nn} \end{bmatrix}.$$

For reasons that will become apparent later, we form a matrix consisting of $[T(\beta_1)]_b, \ldots, [T(\beta_n)]_b$ as columns, and we designate this matrix $[T]_b$. Thus

$$[T]_b = \begin{bmatrix} t_{11} & t_{21} & \cdots & t_{n1} \\ t_{12} & t_{22} & \cdots & t_{n2} \\ \vdots & \vdots & & \vdots \\ t_{1n} & t_{2n} & \cdots & t_{nn} \end{bmatrix},$$

which we notice is simply the transpose of the coefficient matrix of (8.2.1). We have therefore determined that corresponding to any linear transformation T of a given vector space there is a unique n by n matrix $[T]_b$.

Example 2. For the vector space $V_2(R)$ of vectors $\alpha = (x, y)$, let $T: V_2(R) \to V_2(R)$ be the shearing transformation such that

$$(x, y) \xmapsto{T} (x + ky, y).$$

In Exercise 17 of the last section we showed that the shearing transformation is linear. Let the basis be $b = \{(1, 0), (0, 1)\}$. The effect of T on the basis vectors is given by

$$\begin{cases} T(\beta_1) = T(1, 0) = (1, 0) = 1(1, 0) + 0(0, 1), \\ T(\beta_2) = T(0, 1) = (k, 1) = k(1, 0) + 1(0, 1); \end{cases}$$

therefore

(8.2.2)
$$\begin{cases} T(\beta_1) = 1\beta_1 + 0\beta_2, \\ T(\beta_2) = k\beta_1 + 1\beta_2. \end{cases}$$

With respect to the same basis, the coordinates of the images of β_1 and β_2 are $(1, 0)$ and $(k, 1)$ respectively. The coordinate matrices of the images are

$$[T(\beta_1)]_b = \begin{bmatrix} 1 \\ 0 \end{bmatrix} \quad \text{and} \quad [T(\beta_2)]_b = \begin{bmatrix} k \\ 1 \end{bmatrix};$$

hence

$$[T]_b = \begin{bmatrix} 1 & k \\ 0 & 1 \end{bmatrix}.$$

Notice that $[T]_b$ is the transpose of the coefficient matrix of (8.2.2).

★**Example 3.** Consider the vector space $P_2(R)$. Let $T: P_2(R) \to P_2(R)$ be the derivative

$$p(x) \xmapsto{T} D_x p(x),$$

and let the basis be $b = \{1, x, x^2\}$. Then the images of the basis vectors are

$$\begin{cases} T(\beta_1) = D_x(1) = 0, \\ T(\beta_2) = D_x(x) = 1, \\ T(\beta_3) = D_x(x^2) = 2x. \end{cases}$$

8.2 REPRESENTATION OF A LINEAR TRANSFORMATION

Therefore,

(8.2.3)
$$\begin{cases} T(\beta_1) = 0\beta_1 + 0\beta_2 + 0\beta_3, \\ T(\beta_2) = 1\beta_1 + 0\beta_2 + 0\beta_3, \\ T(\beta_3) = 0\beta_1 + 2\beta_2 + 0\beta_3. \end{cases}$$

The coordinate matrices of the images are

$$[T(\beta_1)]_b = \begin{bmatrix} 0 \\ 0 \\ 0 \end{bmatrix}, \quad [T(\beta_2)]_b = \begin{bmatrix} 1 \\ 0 \\ 0 \end{bmatrix}, \quad [T(\beta_3)]_b = \begin{bmatrix} 0 \\ 2 \\ 0 \end{bmatrix};$$

hence

$$[T]_b = \begin{bmatrix} 0 & 1 & 0 \\ 0 & 0 & 2 \\ 0 & 0 & 0 \end{bmatrix}.$$

Notice that $[T]_b$ is the transpose of the coefficient matrix of (8.2.3).

Another way of saying that we have a linear transformation from a vector space $V(F)$ into itself is to say that we have a *linear transformation over $V(F)$*. Now we proceed with the more difficult task of proving that to every n by n matrix $[T]_b$ over F there corresponds a unique linear transformation T over $V(F)$. This is important because otherwise the matrix $[T]_b$ which we have learned to calculate could be used as a representation of more than one linear transformation. We must use the following theorem.

Theorem 2. *If $b = \{\beta_1, \beta_2, \ldots, \beta_n\}$ is a basis of a vector space $V(F)$, and if $\gamma_1, \gamma_2, \ldots, \gamma_n$ are arbitrary vectors of $V(F)$, then there exists exactly one linear transformation $T: V(F) \to V(F)$ such that*

$$T(\beta_i) = \gamma_i \quad (i = 1, 2, \ldots, n).$$

For the proof of this, see Theorem A.8.2, page A3, Appendix B.

We are now in a position to prove the primary result of this section, which is one of the most important theorems in linear algebra.

Theorem 3. *For a finite-dimensional vector space $V(F)$ with a fixed basis $b = \{\beta_1, \beta_2, \ldots, \beta_n\}$, there is a one-to-one correspondence between the linear transformations T from $V(F)$ into $V(F)$ and the n by n matrices*

$$[T]_b = [[T(\beta_1)]_b \; \vdots \; [T(\beta_2)]_b \; \vdots \; \cdots \; \vdots \; [T(\beta_n)]_b]$$

over F.

Proof. Since for an arbitrary element of $V(F)$, say, $\alpha = k_1\beta_1 + \cdots + k_n\beta_n$, we have

$$T(\alpha) = T(k_1\beta_1 + \cdots + k_n\beta_n) = k_1 T(\beta_1) + \cdots + k_n T(\beta_n),$$

we see that a linear transformation T of arbitrary α of $V(F)$ is determined by the effect of T on the basis.

PART I. To every linear transformation T over $V(F)$ there corresponds a unique n by n matrix $[T]_b$ over F. This part is left as Exercise 18.

PART II. To every n by n matrix $[T]_b$ over F there corresponds a unique linear transformation T over $V(F)$.

(1) Associated with any matrix $A = [a_{ij}]_{(n,n)}$, there is a set of n vectors $\{\gamma_1, \gamma_2, \ldots, \gamma_n\}$ formed by using the entries of the ith column of A as the coordinates of γ_i ($i = 1, \ldots, n$).

$$\begin{cases} \gamma_1 = a_{11}\beta_1 + a_{21}\beta_2 + \cdots + a_{n1}\beta_n, \\ \gamma_2 = a_{12}\beta_1 + a_{22}\beta_2 + \cdots + a_{n2}\beta_n, \\ \vdots \\ \gamma_n = a_{1n}\beta_1 + a_{2n}\beta_2 + \cdots + a_{nn}\beta_n. \end{cases}$$

(2) To the set $\{\gamma_1, \gamma_2, \ldots, \gamma_n\}$ there corresponds a *unique* linear transformation $T: V(F) \to V(F)$ such that $\gamma_i = T(\beta_i)$, according to Theorem 2. Therefore, to every n by n matrix over F there corresponds a set $\{\gamma_1, \ldots, \gamma_n\}$ which in turn corresponds to a unique linear transformation from $V(F)$ into $V(F)$. □

Example 4. This example illustrates Part II of the proof of Theorem 3. Let the vector space be $V_2(R)$ with basis $b = \{(1, 0), (0, 1)\}$. Consider a 2 by 2 matrix $A = \begin{bmatrix} 1 & 4 \\ 0 & 1 \end{bmatrix}$. Matrix A can be used to form a set of vectors $\{\gamma_1, \gamma_2\}$. Let the first column be the coordinates of γ_1 and let the second column be the coordinates of γ_2. Then

$$\begin{cases} \gamma_1 = 1\beta_1 + 0\beta_2, \\ \gamma_2 = 4\beta_1 + 1\beta_2, \end{cases}$$

or

$$\begin{cases} \gamma_1 = 1(1, 0) + 0(0, 1) = (1, 0), \\ \gamma_2 = 4(1, 0) + 1(0, 1) = (4, 1). \end{cases}$$

According to Theorem 2, there exists a unique linear transformation

$$T: V_2(R) \to V_2(R)$$

such that

$$\gamma_1 = T(\beta_1) \quad \text{and} \quad \gamma_2 = T(\beta_2),$$

or

$$\beta_1 \overset{T}{\mapsto} \gamma_1 \quad \text{and} \quad \beta_2 \overset{T}{\mapsto} \gamma_2,$$

or

$$(1, 0) \overset{T}{\mapsto} (1, 0) \quad \text{and} \quad (0, 1) \overset{T}{\mapsto} (4, 1).$$

This example is simple enough to see by trial and error that T is the shear defined by $(x, y) \overset{T}{\mapsto} (x + 4y, y)$. Significantly, Theorem 2 also assures us that T is unique.

EXERCISES

1. For the vector space $V_3(R)$, write the coordinate matrix of $(7, 8, 1)$ with respect to each of the following bases.
   (a) $b = \{(1, 0, 0), (0, 1, 0), (0, 0, 1)\}$;
   (b) $b = \{(0, 1, 0), (1, 0, 0), (0, 0, 1)\}$. Notice that the order of the vectors in the basis is important;
   (c) $b = \{(1, 1, 0), (0, 1, 0), (0, 1, 1)\}$.

2. For the vector space $P_2(R)$, write the coordinate matrix of $4x^2 + 3x + 9$ with respect to each of the following bases.
   (a) $b = \{1, x, x^2\}$;
   (b) $b = \{x^2, x, 1\}$;
   (c) $b = \{x + 1, x, x^2\}$.

3. Rework Example 2 of this section if the basis is $b = \{(1, 0), (1, 2)\}$ rather than the basis that was given. This exercise emphasizes that $[T]_b$ depends upon the basis as well as upon the linear transformation.

4. Consider the vector space $V_2(R)$ with basis $b = \{(1, 0), (0, 1)\}$ and the transformation $T: V_2(R) \to V_2(R)$ where $(x, y) \xmapsto{T} (x + 1, y)$. Find $[T]_b$, if possible; if it is not possible, give reasons.

In each of Exercises 5–9, consider the vector space $V_2(R)$ with a basis $b = \{(1, 0), (0, 1)\}$. Write the matrix representation of the given transformation $T: V_2(R) \to V_2(R)$.

5. Let T be a stretching transformation,
$$\alpha \xmapsto{T} 4\alpha \quad \text{or} \quad T(\alpha) = 4\alpha.$$

6. Let T be a reflection through the origin,
$$\alpha \xmapsto{T} -\alpha \quad \text{or} \quad T(\alpha) = -\alpha.$$

7. Let T be a reflection through the x-axis,
$$(a_1, a_2) \xmapsto{T} (a_1, -a_2) \quad \text{or} \quad T(\alpha) = (a_1, -a_2).$$

8. Let T be a rotation of $90°$,
$$(a_1, a_2) \xmapsto{T} (-a_2, a_1) \quad \text{or} \quad T(\alpha) = (-a_2, a_1).$$

9. Let T be a projection on the x-axis,
$$(a_1, a_2) \xmapsto{T} (a_1, 0) \quad \text{or} \quad T(\alpha) = (a_1, 0).$$

In each of Exercises 10–13, let the vector space be $V_3(R)$ with a basis $b = \{(1, 0, 0), (0, 1, 0), (0, 0, 1)\}$. Write the matrix representation of the given transformation $T: V_3(R) \to V_3(R)$.

10. Let T be a contraction,
$$\alpha \xmapsto{T} \tfrac{1}{3}\alpha \quad \text{or} \quad T(\alpha) = \tfrac{1}{3}\alpha.$$

11. Let T be a reflection through the $x_1 x_2$-plane,
$$(a_1, a_2, a_3) \xmapsto{T} (a_1, a_2, -a_3) \quad \text{or} \quad T(\alpha) = (a_1, a_2, -a_3).$$

12. Let T be a 90° rotation about the x_1-axis,
$$(a_1, a_2, a_3) \stackrel{T}{\mapsto} (a_1, -a_3, a_2) \quad \text{or} \quad T(\alpha) = (a_1, -a_3, a_2).$$

13. Let T be a projection on the $x_1 x_2$-plane,
$$(a_1, a_2, a_3) \stackrel{T}{\mapsto} (a_1, a_2, 0) \quad \text{or} \quad T(\alpha) = (a_1, a_2, 0).$$

In each of Exercises 14–16, let the vector space be $P_3(R)$ consisting of all polynomials $p(x) = a_3 x^3 + a_2 x^2 + a_1 x + a_0$ with basis $b = \{1, x, x^2, x^3\}$. Write the matrix representation $[T]_b$ of the given transformation $T: P_3(R) \to P_3(R)$.

★14. $p(x) \stackrel{T}{\mapsto} 4 D_x(p(x))$, or $T(p(x)) = 4 D_x(p(x))$.

★15. $p(x) \stackrel{T}{\mapsto} D_x^2(p(x))$, or $T(p(x)) = D_x^2(p(x))$.

★16. $p(x) \stackrel{T}{\mapsto} D_x^3(p(x))$, or $T(p(x)) = D_x^3(p(x))$.

17. Over the vector space $V_2(R)$, the general rotation transformation $T: V_2(R) \to V_2(R)$ through an angle θ is defined by
$$(x, y) \stackrel{T}{\mapsto} (x \cos \theta - y \sin \theta, x \sin \theta + y \cos \theta).$$
Find the matrix representation $[T]_b$ if $b = \{(1, 0), (0, 1)\}$.

18. Give Part I of the proof of Theorem 3 in formal Statement-Reason form.

In each of Exercises 19–22, let the vector space be $P_2(R)$ with basis $b = \{1, x + x^2, x^2\}$. Write the matrix representation $[T]_b$ of the given linear transformation $T: P_2(R) \stackrel{T}{\to} P_2(R)$. Let $p(x) = a_2 x^2 + a_1 x + a_0$.

★19. $p(x) \stackrel{T}{\mapsto} 3 D_x(p(x))$.

★20. $p(x) \stackrel{T}{\mapsto} (D_x + 1)(p(x))$.

★21. $p(x) \stackrel{T}{\mapsto} (D_x^2 + 1)(p(x))$.

★22. $p(x) \stackrel{T}{\mapsto} (2 D_x^2 + 3 D_x + 1)(p(x))$.

23. Let the vector space be $V_2(R)$ with basis $b = \{(1, 0), (0, 1)\}$. Give a geometric description of T if $[T]_b = \begin{bmatrix} 0 & -1 \\ 1 & 0 \end{bmatrix}$. Use the method of Example 4.

In each of Exercises 24–27, let $V(R)$ be the vector space of functions f generated by the basis $b = \{1, e^x, x, xe^x\}$. Write the matrix representation $[T]_b$ of the given linear transformation $T: V(R) \to V(R)$.

★24. $f(x) \stackrel{T}{\mapsto} D_x f(x)$.

★25. $f(x) \stackrel{T}{\mapsto} D_x^2 f(x)$.

★26. $f(x) \stackrel{T}{\mapsto} (a_1 D_x + a_0) f(x)$.

★27. $f(x) \stackrel{T}{\mapsto} (a_2 D_x^2 + a_1 D_x + a_0) f(x)$.

8.3 Linear Operators

Linear transformations which map a vector space into itself are called **linear operators** on the space. We have established that any linear operator T on a finite-dimensional vector space $V(F)$ with basis b can be uniquely represented

8.3 LINEAR OPERATORS

by a matrix $[T]_b$; in this section we prove that for an arbitrary vector α of $V(F)$, the image of the matrix transformation, $[T]_b[\alpha]_b$, is the coordinate matrix of $T(\alpha)$. In other words, the image of an arbitrary vector α of a vector space under a linear operator T can be found by means of a unique matrix transformation

$$[T]_b[\alpha]_b = [T(\alpha)]_b.$$

Example 1. Consider the vector space $V_2(R)$ of vectors (x, y); let $T: V_2(R) \to V_2(R)$ be the shearing transformation defined by $(x, y) \stackrel{T}{\mapsto} (x + ky, y)$, and let the basis be $b = \{(1, 0), (0, 1)\}$. In Example 2 of the last section we found that the transformation matrix is

$$[T]_b = \begin{bmatrix} 1 & k \\ 0 & 1 \end{bmatrix}.$$

If the coordinate matrix of an arbitrary vector $\alpha = (a_1, a_2)$ of the vector space is premultiplied by the transformation matrix $[T]_b$, we obtain

$$[T]_b[\alpha]_b = \begin{bmatrix} 1 & k \\ 0 & 1 \end{bmatrix} \begin{bmatrix} a_1 \\ a_2 \end{bmatrix} = \begin{bmatrix} a_1 + ka_2 \\ a_2 \end{bmatrix},$$

which is the coordinate matrix of the image $T(\alpha) = (a_1 + ka_2, a_2)$. In other words, in this example we have obtained the image of an arbitrary α under T by means of matrix multiplication; the following important theorem assures us that this result was no accident. In the proof, the reason for defining $[T]_b$ as the matrix representation of T will become apparent.

Theorem 4. *If T is a linear transformation of an n-dimensional vector space $V(F)$ into itself, and if α is an element of $V(F)$, then, with respect to the basis $b = \{\beta_1, \ldots, \beta_n\}$, the coordinate matrix of α premultiplied by the unique matrix representation of T is equal to the coordinate matrix of the image $T(\alpha)$; that is,*

$$[T]_b[\alpha]_b = [T(\alpha)]_b.$$

Proof.

STATEMENT	REASON
(1) Let $$\alpha = a_1\beta_1 + a_2\beta_2 + \cdots + a_n\beta_n;$$ then $$T(\alpha) = a_1 T(\beta_1) + a_2 T(\beta_2) + \cdots + a_n T(\beta_n),$$ where $a_1, \ldots, a_n$ are unique elements of F.	(1) Because α can be expressed as a linear combination of a basis in only one way, and because T is a linear transformation.

(2) $T(\alpha) = a_1(t_{11}\beta_1 + t_{12}\beta_2 + \cdots + t_{1n}\beta_n)$
$+ a_2(t_{21}\beta_1 + t_{22}\beta_2 + \cdots + t_{2n}\beta_n)$
$+ \cdots$
$+ a_n(t_{n1}\beta_1 + t_{n2}\beta_2 + \cdots + t_{nn}\beta_n).$

(2) By substituting the expressions for each $T(\beta_j)$. (See page 227, Equations (8.2.1).) The t_{ij} are unique elements of F by Theorem 3 of Chapter 7 (page 186).

(3) $T(\alpha) = (t_{11}a_1 + t_{21}a_2 + \cdots + t_{n1}a_n)\beta_1$
$+ (t_{12}a_1 + t_{22}a_2 + \cdots + t_{n2}a_n)\beta_2$
$+ \cdots$
$+ (t_{1n}a_1 + t_{2n}a_2 + \cdots + t_{nn}a_n)\beta_n.$

(3) By rearranging the terms of (2).

(4) Therefore the coordinate matrix of $T(\alpha)$ is

$$[T(\alpha)]_b = \begin{bmatrix} (t_{11}a_1 + t_{21}a_2 + \cdots + t_{n1}a_n) \\ (t_{12}a_1 + t_{22}a_2 + \cdots + t_{n2}a_n) \\ \vdots \\ (t_{1n}a_1 + t_{2n}a_2 + \cdots + t_{nn}a_n) \end{bmatrix}.$$

(4) Any vector of the given space can be written as a unique linear combination of the basis vectors and the coefficients of the basis vectors are the coordinates with respect to that basis.

(5) $$[T(\alpha)]_b = \begin{bmatrix} t_{11} & t_{21} & \cdots & t_{n1} \\ t_{12} & t_{22} & \cdots & t_{n2} \\ \vdots & \vdots & & \vdots \\ t_{1n} & t_{2n} & \cdots & t_{nn} \end{bmatrix} \begin{bmatrix} a_1 \\ a_2 \\ \vdots \\ a_n \end{bmatrix}.$$

(5) By matrix multiplication.

(6) $[T(\alpha)]_b = [T]_b [\alpha]_b.$

(6) By definition of $[T]_b$ and by recognizing the coordinate matrix of α from Statement (1). □

Example 2. Let the vector space be $V_3(R)$ with basis $b = \{\beta_1, \beta_2, \beta_3\}$, and let T be the linear operator that rotates the basis vectors $\beta_1 = (1, 0, 0)$ into $(\frac{1}{2}\sqrt{2}, \frac{1}{2}\sqrt{2}, 0)$, and $\beta_2 = (0, 1, 0)$ into $(-\frac{1}{2}\sqrt{2}, \frac{1}{2}\sqrt{2}, 0)$ while $\beta_3 = (0, 0, 1)$ remains the same. Theorem 2 of Section 8.2 assures us that T is linear and unique. Geometrically this amounts to a 45° rotation about the x_3-axis. The coordinate matrices of the images of the basis vectors are

$$[T(\beta_1)]_b = \begin{bmatrix} \frac{1}{2}\sqrt{2} \\ \frac{1}{2}\sqrt{2} \\ 0 \end{bmatrix}, \quad [T(\beta_2)]_b = \begin{bmatrix} -\frac{1}{2}\sqrt{2} \\ \frac{1}{2}\sqrt{2} \\ 0 \end{bmatrix}, \quad [T(\beta_3)]_b = \begin{bmatrix} 0 \\ 0 \\ 1 \end{bmatrix},$$

from which we obtain the transformation matrix

$$[T]_b = \begin{bmatrix} \frac{1}{2}\sqrt{2} & -\frac{1}{2}\sqrt{2} & 0 \\ \frac{1}{2}\sqrt{2} & \frac{1}{2}\sqrt{2} & 0 \\ 0 & 0 & 1 \end{bmatrix}.$$

8.3   LINEAR OPERATORS

By Theorem 4 the coordinate matrix of the image of an arbitrary vector $\alpha = (a_1, a_2, a_3)$ of the space can be found by

$$[T(\alpha)]_b = [T]_b[\alpha]_b = \begin{bmatrix} \tfrac{1}{2}\sqrt{2} & -\tfrac{1}{2}\sqrt{2} & 0 \\ \tfrac{1}{2}\sqrt{2} & \tfrac{1}{2}\sqrt{2} & 0 \\ 0 & 0 & 1 \end{bmatrix} \begin{bmatrix} a_1 \\ a_2 \\ a_3 \end{bmatrix} = \begin{bmatrix} \tfrac{1}{2}\sqrt{2}(a_1 - a_2) \\ \tfrac{1}{2}\sqrt{2}(a_1 + a_2) \\ a_3 \end{bmatrix}.$$

As a check, we calculate

$$T(\alpha) = T(a_1, a_2, a_3) = T(a_1\beta_1 + a_2\beta_2 + a_3\beta_3)$$
$$= a_1 T(\beta_1) + a_2 T(\beta_2) + a_3 T(\beta_3)$$
$$= a_1(\tfrac{1}{2}\sqrt{2}, \tfrac{1}{2}\sqrt{2}, 0) + a_2(-\tfrac{1}{2}\sqrt{2}, \tfrac{1}{2}\sqrt{2}, 0) + a_3(0, 0, 1)$$
$$= (\tfrac{1}{2}\sqrt{2}(a_1 - a_2), \tfrac{1}{2}\sqrt{2}(a_1 + a_2), a_3).$$

★**Example 3.** Consider the vector space $P_2(R)$ of polynomials. Let T be the derivative operator such that $p(x) \overset{T}{\mapsto} D_x p(x)$, and let $b = \{1, x, x^2\}$. In Example 3 of the last section we found that

$$[T]_b = \begin{bmatrix} 0 & 1 & 0 \\ 0 & 0 & 2 \\ 0 & 0 & 0 \end{bmatrix},$$

and, by Theorem 4, the derivatives of an arbitrary element $ax^2 + bx + c$ of the vector space can be obtained by

$$[T]_b \begin{bmatrix} c \\ b \\ a \end{bmatrix} = \begin{bmatrix} 0 & 1 & 0 \\ 0 & 0 & 2 \\ 0 & 0 & 0 \end{bmatrix} \begin{bmatrix} c \\ b \\ a \end{bmatrix} = \begin{bmatrix} b \\ 2a \\ 0 \end{bmatrix},$$

which is the coordinate matrix of $b + 2ax$. We see that the linear operator $[T]_b$ over the vector space of all 3 by 1 column R-matrices serves as a matrix representation of the linear operator D_x over the given vector space of polynomials.

APPLICATIONS

Example 4. Consider a hypothetical legislature with two parties, composed of representatives elected for 2-year terms, and, moreover, a legislator can serve a maximum of 3 terms. The procedures illustrated in this example, with slight modification, may be used to examine realistic legislatures; the restrictions made here simply reduce the size of the problem. Also, the breakdown need not be made according to party lines, nor need it be into only two groups; the procedure developed will handle n ideological groups just as well.

Let the parties be called D and R, and let

x_i be the number of members of D who are serving in their ith term at time t;
y_i be the number of members of R who are serving in their ith term at time t.

Construct a column matrix $[L_t]_b = \begin{bmatrix} x_1 \\ x_2 \\ x_3 \\ y_1 \\ y_2 \\ y_3 \end{bmatrix}$ which belongs to $V_6(R)$ with basis

$b = \left\{ \begin{bmatrix} 1 \\ 0 \\ 0 \\ 0 \\ 0 \\ 0 \end{bmatrix}, \begin{bmatrix} 0 \\ 1 \\ 0 \\ 0 \\ 0 \\ 0 \end{bmatrix}, \begin{bmatrix} 0 \\ 0 \\ 1 \\ 0 \\ 0 \\ 0 \end{bmatrix}, \begin{bmatrix} 0 \\ 0 \\ 0 \\ 1 \\ 0 \\ 0 \end{bmatrix}, \begin{bmatrix} 0 \\ 0 \\ 0 \\ 0 \\ 1 \\ 0 \end{bmatrix}, \begin{bmatrix} 0 \\ 0 \\ 0 \\ 0 \\ 0 \\ 1 \end{bmatrix} \right\}.$ We will construct a transformation matrix $[T]_b$ for the preceding five elections in which the president of the state belonged to party D and was not up for reelection; suppose that there have been 100 first term seats belonging to party D and that they have been transformed by the five elections according to

$$\begin{bmatrix} 100 \\ 0 \\ 0 \\ 0 \\ 0 \\ 0 \end{bmatrix} \stackrel{T}{\mapsto} \begin{bmatrix} 10 \\ 50 \\ 0 \\ 40 \\ 0 \\ 0 \end{bmatrix}, \text{ or } \begin{bmatrix} 1 \\ 0 \\ 0 \\ 0 \\ 0 \\ 0 \end{bmatrix} \stackrel{T}{\mapsto} \begin{bmatrix} \frac{1}{10} \\ \frac{1}{2} \\ 0 \\ \frac{2}{5} \\ 0 \\ 0 \end{bmatrix},$$

or $\beta_1 \stackrel{T}{\mapsto} \frac{1}{10}\beta_1 + \frac{1}{2}\beta_2 + 0\beta_3 + \frac{2}{5}\beta_4 + 0\beta_5 + 0\beta_6$. Using a similar procedure for 2nd and 3rd term D's and for 1st, 2nd, and 3rd term R's, suppose we find the transformations of the other basis vectors and write all of them:

(8.3.1) $\begin{cases} \beta_1 \stackrel{T}{\mapsto} \frac{1}{10}\beta_1 + \frac{1}{2}\beta_2 + 0\beta_3 + \frac{2}{5}\beta_4 + 0\beta_5 + 0\beta_6, \\ \beta_2 \stackrel{T}{\mapsto} \frac{1}{10}\beta_1 + 0\beta_2 + \frac{4}{5}\beta_3 + \frac{1}{10}\beta_4 + 0\beta_5 + 0\beta_6, \\ \beta_3 \stackrel{T}{\mapsto} \frac{2}{5}\beta_1 + 0\beta_2 + 0\beta_3 + \frac{3}{5}\beta_4 + 0\beta_5 + 0\beta_6, \\ \beta_4 \stackrel{T}{\mapsto} \frac{1}{10}\beta_1 + 0\beta_2 + 0\beta_3 + \frac{3}{10}\beta_4 + \frac{3}{5}\beta_5 + 0\beta_6, \\ \beta_5 \stackrel{T}{\mapsto} \frac{1}{10}\beta_1 + 0\beta_2 + 0\beta_3 + \frac{1}{10}\beta_4 + 0\beta_5 + \frac{4}{5}\beta_6, \\ \beta_6 \stackrel{T}{\mapsto} \frac{1}{2}\beta_1 + 0\beta_2 + 0\beta_3 + \frac{1}{2}\beta_4 + 0\beta_5 + 0\beta_6. \end{cases}$

From Theorem 3 (page 229) we have

$$[T]_b = \begin{bmatrix} \frac{1}{10} & \frac{1}{10} & \frac{2}{5} & \frac{1}{10} & \frac{1}{10} & \frac{1}{2} \\ \frac{1}{2} & 0 & 0 & 0 & 0 & 0 \\ 0 & \frac{4}{5} & 0 & 0 & 0 & 0 \\ \frac{2}{5} & \frac{1}{10} & \frac{3}{5} & \frac{3}{10} & \frac{1}{10} & \frac{1}{2} \\ 0 & 0 & 0 & \frac{3}{5} & 0 & 0 \\ 0 & 0 & 0 & 0 & \frac{4}{5} & 0 \end{bmatrix}.$$

Suppose that the legislature at a certain time t is constituted according to

$$[L_t]_b = \begin{bmatrix} 5 \\ 15 \\ 10 \\ 10 \\ 5 \\ 5 \end{bmatrix}.$$

If $[L_t]_b$ is transformed by $[T]_b$, the image $T(L_t)$ is

$$[T]_b[L_t]_b = \begin{bmatrix} 10 \\ 2.5 \\ 12 \\ 15.5 \\ 6 \\ 4 \end{bmatrix}.$$

The image obtained by this method may well differ from the actual results of the election (in the case of fractions in $T(L_t)$, the actual results, of course, must differ). Variations of the procedure outlined in this example might be used: (1) to predict outcomes of elections based on the present composition and past trends; (2) to study so-called Critical Elections[2]; (3) to study the stability of legislatures (not necessarily according to tenure or along party lines but perhaps according to various ideological differences); (4) to study non-political situations such as the survival and reproduction of animals or plants and their population stability. See Example 9 on page 275 of Section 9.1.

EXERCISES

1. In the vector space $V_2(R)$ with basis $b = \{(1, 0), (0, 1)\}$, let $T: V_2(R) \to V_2(R)$ be the shearing transformation for which

$$(x, y) \stackrel{T}{\mapsto} (x + ky, y).$$

   Find the image of $(6, 4)$ under the transformation with and without the use of $[T]_b$. The calculation of $[T]_b$ for this exercise may be found in Example 2 of Section 8.2, if needed.

★2. In the vector space $P_2(R)$ of polynomials $p(x)$, let the basis be $b = \{1, x, x^2\}$, and let T be the derivative operator for which

$$p(x) \stackrel{T}{\mapsto} D_x p(x).$$

   Find the image of $6 + 7x + 5x^2$ under the transformation with and without the use of $[T]_b$. The calculation of $[T]_b$ for this exercise may be found in Example 3 of Section 8.2, if needed.

[2] Key, V. O., Jr., "A Theory of Critical Elections" *The Journal of Politics*, Vol. 17 (1955).

3. For the vector space $V_2(R)$ with basis $b = \{(1, 0), (0, 1)\}$, a linear operator T exists such that

$$(1, 0) \overset{T}{\mapsto} (0, 1) \quad \text{and} \quad (0, 1) \overset{T}{\mapsto} (1, 0).$$

Using a matrix transformation find $T(\alpha)$ if $\alpha = (2, 5)$. As this problem illustrates, the basis is an ordered set of vectors, and any rearrangement of the order of the basis vectors constitutes a linear transformation.

4. For the vector space $V_3(R)$ with basis $b = \{(1, 0, 0), (0, 1, 0), (0, 0, 1)\}$ a linear operator T exists such that

$$(1, 0, 0) \overset{T}{\mapsto} (0, 1, 0), \quad (0, 1, 0) \overset{T}{\mapsto} (0, 0, 1), \quad (0, 0, 1) \overset{T}{\mapsto} (1, 0, 0).$$

Find $T(\alpha)$ if $\alpha = (1, 3, 2)$.

5-9. In each of Exercises 5-9 of Section 8.2, find the image, $T(\alpha)$, of the vector $\alpha = (a_1, a_2)$ by using the matrix transformation

$$[T]_b[\alpha]_b = [T(\alpha)]_b.$$

10-13. In each of Exercises 10-13 of Section 8.2, find the image, $T(\alpha)$, of the vector $\alpha = (a_1, a_2, a_3)$ by using the matrix transformation

$$[T]_b[\alpha]_b = [T(\alpha)]_b.$$

★14-16. In each of Exercises 14-16 of Section 8.2, find the image of the element

$$p(x) = 5x^3 + 6x^2 + 9x + 4$$

by using the matrix transformation

$$[T]_b[p(x)]_b = [T(p(x))]_b.$$

17. In Exercise 17 of Section 8.2, find the image of $\alpha = (5, 2)$ if $\theta = 30°$ by using the matrix transformation

$$[T]_b[\alpha]_b = [T(\alpha)]_b.$$

Graph α and its image.

18. Repeat Exercise 17 if the basis is $b = \{(1, 1), (0, -1)\}$ rather than the one given.

★19-22. In each of Exercises 19-22 of Section 8.2, find the image of the element

$$p(x) = x^2 - 6x + 5$$

by using the matrix transformation

$$[T]_b[p(x)]_b = [T(p(x))]_b.$$

★23. In Exercise 27 of Section 8.2, find the image of the element $f(x) = 3x + 2xe^x$ by using the matrix transformation

$$[T]_b[f(x)]_b = [T(f(x))]_b.$$

24. Prove that the matrix $[T]_b$ in Theorem 4 is the only matrix for which $[T]_b[\alpha]_b = [T(\alpha)]_b$.

8.4 Change of Basis

We have seen that a linear transformation of a vector space into itself can be represented by a matrix. It was emphasized that this matrix representation depends on the basis that we choose for the vector space. *Many very important applications of linear algebra depend upon our ability to change the basis of the vector space in such a way that the matrix representation of the linear transformation assumes a certain desired form.*

Suppose we wish to change the basis of a given vector space from the set $b = \{\beta_1, \beta_2, \ldots, \beta_n\}$ to the set $e = \{\varepsilon_1, \varepsilon_2, \ldots, \varepsilon_n\}$. Because the vectors of a basis e belong to the vector space, they can be expressed uniquely in terms of the original basis according to

(8.4.1)
$$\begin{cases} \varepsilon_1 = p_{11}\beta_1 + p_{12}\beta_2 + \cdots + p_{1n}\beta_n, \\ \varepsilon_2 = p_{21}\beta_1 + p_{22}\beta_2 + \cdots + p_{2n}\beta_n, \\ \quad \cdot \quad \cdot \quad \cdot \quad \cdot \quad \cdot \quad \cdot \quad \cdot \quad \cdot \quad \cdot \\ \varepsilon_n = p_{n1}\beta_1 + p_{n2}\beta_2 + \cdots + p_{nn}\beta_n. \end{cases}$$

The coordinate matrices of $\varepsilon_1, \varepsilon_2, \ldots, \varepsilon_n$ with respect to the original basis $\{\beta_1, \beta_2, \ldots, \beta_n\}$ are therefore

$$[\varepsilon_1]_b = \begin{bmatrix} p_{11} \\ p_{12} \\ \vdots \\ p_{1n} \end{bmatrix}, \quad [\varepsilon_2]_b = \begin{bmatrix} p_{21} \\ p_{22} \\ \vdots \\ p_{2n} \end{bmatrix}, \quad \ldots, \quad [\varepsilon_n]_b = \begin{bmatrix} p_{n1} \\ p_{n2} \\ \vdots \\ p_{nn} \end{bmatrix}.$$

We will prove that the matrix formed by using the coordinate matrices of $\varepsilon_1, \varepsilon_2, \ldots, \varepsilon_n$ as columns can be used to relate the coordinate matrix of a vector with respect to a new basis to the coordinate matrix of the same vector with respect to the old basis. For this reason we make the following formal definition.

Definition 3. *The matrix*

$$P = [[\varepsilon_1]_b \mid [\varepsilon_2]_b \mid \cdots \mid [\varepsilon_n]_b] = \begin{bmatrix} p_{11} & p_{21} & \cdots & p_{n1} \\ p_{12} & p_{22} & \cdots & p_{n2} \\ \vdots & \vdots & & \vdots \\ p_{1n} & p_{2n} & \cdots & p_{nn} \end{bmatrix}$$

is called the **transition matrix** *from the original basis b to the new basis e of a vector space. The transition matrix is the transpose of the matrix of coefficients of* (8.4.1).

Example 1. Consider the vector space $V(R)$ of all 2 by 1 column R-matrices with basis $b = \left\{ \begin{bmatrix} 1 \\ 0 \end{bmatrix}, \begin{bmatrix} 0 \\ 1 \end{bmatrix} \right\}$. Suppose we wish to change the basis to $e = \left\{ \begin{bmatrix} 1 \\ 1 \end{bmatrix}, \begin{bmatrix} 0 \\ 2 \end{bmatrix} \right\}$. Since

$$\varepsilon_1 = \begin{bmatrix} 1 \\ 1 \end{bmatrix} = (1)\begin{bmatrix} 1 \\ 0 \end{bmatrix} + (1)\begin{bmatrix} 0 \\ 1 \end{bmatrix} = \beta_1 + \beta_2,$$

and

$$\varepsilon_2 = \begin{bmatrix} 0 \\ 2 \end{bmatrix} = (0)\begin{bmatrix} 1 \\ 0 \end{bmatrix} + (2)\begin{bmatrix} 0 \\ 1 \end{bmatrix} = 0\beta_1 + 2\beta_2,$$

then the coordinate matrices of ε_1 and ε_2 with respect to the original basis are

$$[\varepsilon_1]_b = \begin{bmatrix} 1 \\ 1 \end{bmatrix} \quad \text{and} \quad [\varepsilon_2]_b = \begin{bmatrix} 0 \\ 2 \end{bmatrix}.$$

Consequently $P = \begin{bmatrix} 1 & 0 \\ 1 & 2 \end{bmatrix}$. The transition matrix P can also be recognized as the transpose of the matrix of coefficients of the system

$$\begin{cases} \varepsilon_1 = \beta_1 + \beta_2, \\ \varepsilon_2 = 0\beta_1 + 2\beta_2. \end{cases}$$

We now state how the transition matrix relates the coordinate matrix of an arbitrary vector with respect to one basis to the coordinate matrix of the same vector with respect to another basis.

Theorem 5. *If α is an arbitrary element of a vector space, and if P is the transition matrix from the basis $b = \{\beta_1, \ldots, \beta_n\}$ to the basis $e = \{\varepsilon_1, \ldots, \varepsilon_n\}$, then*

$$P[\alpha]_e = [\alpha]_b.$$

The proof is left as Exercise 13.

Corollary. *A transition matrix, P, which effects a change of basis of a vector space is invertible.*

Proof.

STATEMENT	REASON
(1) $P[\alpha]_e = [\alpha]_b$.	(1) By Theorem 5.
(2) Let $[\alpha]_b = 0$ and let $[\alpha]_e = X$; then we have the homogenous system $PX = 0.$	(2) By substitution into Statement (1).

8.4 CHANGE OF BASIS

(3) $PX = 0$ has a unique solution $X = 0$.

(3) Because b and e are bases, they are linearly independent. Any linear combination of either set equals the zero vector if and only if the coefficients of the linear combination are zero. Hence

$$([\alpha]_b = 0) \Leftrightarrow (\alpha = 0) \Leftrightarrow ([\alpha]_e = 0).$$

(4) Therefore P is invertible.

(4) Exercise 21 of Section 5.7. □

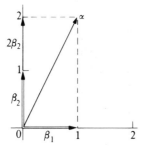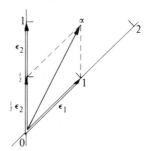

Figure 8.4.1

Example 2. Consider again the same vector space of Example 1; we found that the transition matrix from the basis $b = \left\{\begin{bmatrix}1\\0\end{bmatrix}, \begin{bmatrix}0\\1\end{bmatrix}\right\}$ to the basis $e = \left\{\begin{bmatrix}1\\1\end{bmatrix}, \begin{bmatrix}0\\2\end{bmatrix}\right\}$ was $P = \begin{bmatrix}1 & 0\\1 & 2\end{bmatrix}$. Let $\alpha = \begin{bmatrix}1\\2\end{bmatrix}$, then $[\alpha]_b = \begin{bmatrix}1\\2\end{bmatrix}$ also. According to Theorem 5 and its Corollary,

$$P[\alpha]_e = [\alpha]_b, \quad \text{or} \quad [\alpha]_e = P^{-1}[\alpha]_b;$$

hence

$$[\alpha]_e = \begin{bmatrix}1 & 0\\1 & 2\end{bmatrix}^{-1}\begin{bmatrix}1\\2\end{bmatrix} = \begin{bmatrix}1 & 0\\-\tfrac{1}{2} & \tfrac{1}{2}\end{bmatrix}\begin{bmatrix}1\\2\end{bmatrix} = \begin{bmatrix}1\\\tfrac{1}{2}\end{bmatrix}.$$

This result can be verified by observing that

$$\alpha = \begin{bmatrix}1\\2\end{bmatrix} = 1\begin{bmatrix}1\\1\end{bmatrix} + \tfrac{1}{2}\begin{bmatrix}0\\2\end{bmatrix} = \varepsilon_1 + \tfrac{1}{2}\varepsilon_2;$$

hence the numbers 1 and $\tfrac{1}{2}$ are the coordinates of α with respect to e, and the coordinate matrix is

$$[\alpha]_e = \begin{bmatrix}1\\\tfrac{1}{2}\end{bmatrix}.$$

A geometric interpretation of this example is given in Figure 8.4.1.

Example 3. Consider the vector space $V(R)$ of all 3 by 1 column R-matrices with the basis $b = \left\{ \begin{bmatrix} 1 \\ 0 \\ 1 \end{bmatrix}, \begin{bmatrix} 0 \\ 1 \\ 0 \end{bmatrix}, \begin{bmatrix} 0 \\ 0 \\ 1 \end{bmatrix} \right\}$, and suppose that we wish to change the basis to $e = \left\{ \begin{bmatrix} 1 \\ 0 \\ 0 \end{bmatrix}, \begin{bmatrix} 3 \\ 4 \\ 5 \end{bmatrix}, \begin{bmatrix} 0 \\ 0 \\ 1 \end{bmatrix} \right\}$. Then $\varepsilon_1, \varepsilon_2, \varepsilon_3$ can be written as

$$\varepsilon_1 = \begin{bmatrix} 1 \\ 0 \\ 0 \end{bmatrix} = p_{11}\begin{bmatrix} 1 \\ 0 \\ 1 \end{bmatrix} + p_{12}\begin{bmatrix} 0 \\ 1 \\ 0 \end{bmatrix} + p_{13}\begin{bmatrix} 0 \\ 0 \\ 1 \end{bmatrix},$$

$$\varepsilon_2 = \begin{bmatrix} 3 \\ 4 \\ 5 \end{bmatrix} = p_{21}\begin{bmatrix} 1 \\ 0 \\ 1 \end{bmatrix} + p_{22}\begin{bmatrix} 0 \\ 1 \\ 0 \end{bmatrix} + p_{23}\begin{bmatrix} 0 \\ 0 \\ 1 \end{bmatrix},$$

$$\varepsilon_3 = \begin{bmatrix} 0 \\ 0 \\ 1 \end{bmatrix} = p_{31}\begin{bmatrix} 1 \\ 0 \\ 1 \end{bmatrix} + p_{32}\begin{bmatrix} 0 \\ 1 \\ 0 \end{bmatrix} + p_{33}\begin{bmatrix} 0 \\ 0 \\ 1 \end{bmatrix},$$

from which we obtain

(8.4.2) $\qquad \begin{cases} \varepsilon_1 = 1\beta_1 + 0\beta_2 - 1\beta_3, \\ \varepsilon_2 = 3\beta_1 + 4\beta_2 + 2\beta_3, \\ \varepsilon_3 = 0\beta_1 + 0\beta_2 + 1\beta_3. \end{cases}$

The coordinate matrices of $\varepsilon_1, \varepsilon_2,$ and ε_3 with respect to b are

$$[\varepsilon_1]_b = \begin{bmatrix} 1 \\ 0 \\ -1 \end{bmatrix}, \quad [\varepsilon_2]_b = \begin{bmatrix} 3 \\ 4 \\ 2 \end{bmatrix}, \quad [\varepsilon_3]_b = \begin{bmatrix} 0 \\ 0 \\ 1 \end{bmatrix};$$

hence, by Definition 3,

$$P = \begin{bmatrix} 1 & 3 & 0 \\ 0 & 4 & 0 \\ -1 & 2 & 1 \end{bmatrix}.$$

Of course P can also be recognized as the transpose of the coefficient matrix of (8.4.2).

Now consider the vector $[\alpha]_b = \begin{bmatrix} 6 \\ 7 \\ 8 \end{bmatrix}$ of the same vector space. Let $[\alpha]_e = \begin{bmatrix} a_1 \\ a_2 \\ a_3 \end{bmatrix}$; knowing $[\alpha]_b$, we want to find $[\alpha]_e$. By the last theorem we have

$$P[\alpha]_e = [\alpha]_b,$$

or

$$\begin{bmatrix} 1 & 3 & 0 \\ 0 & 4 & 0 \\ -1 & 2 & 1 \end{bmatrix} \begin{bmatrix} a_1 \\ a_2 \\ a_3 \end{bmatrix} = \begin{bmatrix} 6 \\ 7 \\ 8 \end{bmatrix}.$$

8.4 CHANGE OF BASIS

Then

$$\begin{bmatrix} a_1 \\ a_2 \\ a_3 \end{bmatrix} = \begin{bmatrix} 1 & 3 & 0 \\ 0 & 4 & 0 \\ -1 & 2 & 1 \end{bmatrix}^{-1} \begin{bmatrix} 6 \\ 7 \\ 8 \end{bmatrix} = \begin{bmatrix} 1 & -\tfrac{3}{4} & 0 \\ 0 & \tfrac{1}{4} & 0 \\ 1 & -\tfrac{5}{4} & 1 \end{bmatrix} \begin{bmatrix} 6 \\ 7 \\ 8 \end{bmatrix} = \begin{bmatrix} \tfrac{3}{4} \\ \tfrac{7}{4} \\ \tfrac{21}{4} \end{bmatrix}.$$

Thus by premultiplying by P^{-1} we can find the coordinate matrix of any element of the vector space when the basis of that vector space has been changed.

It is very useful and important to realize that *if A is invertible* then the matrix equation $A[\alpha]_b = B$ can be interpreted in two different ways. First of all, it can represent a matrix transformation where $A = [T]_b$ and $B = [T(\alpha)]_b$ according to Theorem 4. Secondly, the matrix equation $A[\alpha]_b = B$ may represent a change in basis from b to e, where $A = P^{-1}$ and $B = [\alpha]_e$ according to Theorem 5 and its Corollary. In other words, a matrix equation $A[\alpha]_b = B$, where A is invertible, may be used to express the image of α under a linear transformation with respect to the same basis, or it may be used to express α itself with respect to a different basis. Geometrically, in two dimensions, the two interpretations may be expressed as a movement of vectors in one case and a renaming of fixed vectors in the other case.

Example 4. For the vector space $V_2(R)$ with basis $b = \{(1, 0), (0, 1)\}$, the linear transformation T, "rotation through an angle θ," is defined by $\alpha = (x, y) \xmapsto{T} ((x \cos \theta - y \sin \theta), (x \sin \theta + y \cos \theta))$. The corresponding matrix transformation is

$$[T]_b [\alpha]_b = [T(\alpha)]_b,$$

where, according to Exercise 17 of Section 8.2,

$$[T]_b = \begin{bmatrix} \cos \theta & -\sin \theta \\ \sin \theta & \cos \theta \end{bmatrix}.$$

Specifically, let $\theta = 30°$, and let $\alpha = (2, 0)$; then $[\alpha]_b = \begin{bmatrix} 2 \\ 0 \end{bmatrix}$ and we obtain

(8.4.3)
$$\begin{bmatrix} \tfrac{1}{2}\sqrt{3} & -\tfrac{1}{2} \\ \tfrac{1}{2} & \tfrac{1}{2}\sqrt{3} \end{bmatrix} \begin{bmatrix} 2 \\ 0 \end{bmatrix} = \begin{bmatrix} \sqrt{3} \\ 1 \end{bmatrix};$$

the coordinate matrix $\begin{bmatrix} \sqrt{3} \\ 1 \end{bmatrix}$ of (8.4.3) may be interpreted as the matrix representation of the image of α under the rotation transformation. Geometrically, this can be thought of as a counterclockwise rotation of 30° of the geometric vector α shown in Figure 8.4.2.

By Theorem 5, however, the matrix transformation (8.4.3) may be interpreted as

$$P^{-1}[\alpha]_b = [\alpha]_e,$$

since the matrix operator is invertible. Hence

$$P^{-1} = \begin{bmatrix} \frac{1}{2}\sqrt{3} & -\frac{1}{2} \\ \frac{1}{2} & \frac{1}{2}\sqrt{3} \end{bmatrix}, \text{ and } P = \begin{bmatrix} \frac{1}{2}\sqrt{3} & \frac{1}{2} \\ -\frac{1}{2} & \frac{1}{2}\sqrt{3} \end{bmatrix}.$$

By the definition of P this means that

$$[\varepsilon_1]_b = \begin{bmatrix} \frac{1}{2}\sqrt{3} \\ -\frac{1}{2} \end{bmatrix} \text{ and } [\varepsilon_2]_b = \begin{bmatrix} \frac{1}{2} \\ \frac{1}{2}\sqrt{3} \end{bmatrix}.$$

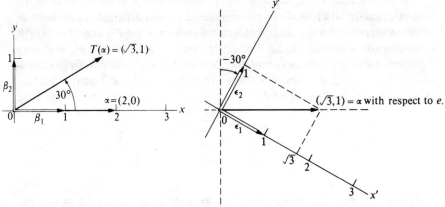

Figure 8.4.2   Figure 8.4.3

Under this interpretation of the matrix transformation (8.4.3) we observe that

$$[\alpha]_e = \begin{bmatrix} \sqrt{3} \\ 1 \end{bmatrix},$$

which represents the original vector α with respect to the new basis e. Geometrically, $[\alpha]_e = \begin{bmatrix} \sqrt{3} \\ 1 \end{bmatrix}$ can be thought of as the coordinate matrix representation of α expressed as a linear combination of ε_1 and ε_2 as shown in Figure 8.4.3. The new basis vectors can be thought of as orthogonal unit vectors in a new $x'y'$-coordinate system, and $(\sqrt{3}, 1)$ represents the coordinates of α in the new coordinate system.

Example 5. The purpose of this example is to illustrate how to use Theorem 5 to find the coordinates of the new basis vectors with respect to the old basis. We shall use the problem of the last example for this illustration. Obviously

$$\varepsilon_1 = 1\varepsilon_1 + 0\varepsilon_2 \text{ and } \varepsilon_2 = 0\varepsilon_1 + 1\varepsilon_2;$$

therefore

$$[\varepsilon_1]_e = \begin{bmatrix} 1 \\ 0 \end{bmatrix} \text{ and } [\varepsilon_2]_e = \begin{bmatrix} 0 \\ 1 \end{bmatrix}.$$

By Theorem 5, $P[\varepsilon_i]_e = [\varepsilon_i]_b$, hence

$$[\varepsilon_1]_b = \begin{bmatrix} \frac{1}{2}\sqrt{3} & \frac{1}{2} \\ -\frac{1}{2} & \frac{1}{2}\sqrt{3} \end{bmatrix} \begin{bmatrix} 1 \\ 0 \end{bmatrix} = \begin{bmatrix} \frac{1}{2}\sqrt{3} \\ -\frac{1}{2} \end{bmatrix}$$

and

$$[\varepsilon_2]_b = \begin{bmatrix} \frac{1}{2}\sqrt{3} & \frac{1}{2} \\ -\frac{1}{2} & \frac{1}{2}\sqrt{3} \end{bmatrix} \begin{bmatrix} 0 \\ 1 \end{bmatrix} = \begin{bmatrix} \frac{1}{2} \\ \frac{1}{2}\sqrt{3} \end{bmatrix}.$$

Notice that these results are consistent with our observation in the last example that $[\varepsilon_1]_b$ and $[\varepsilon_2]_b$ are the columns of P according to the definition of P.

In this book we will frequently refer to a matrix transformation $[\alpha]_b = P[\alpha]_e$ (or $X = PY$) effecting a basis change as a ***transition***.

APPLICATIONS

Example 6. In crystallography the periodic motif in a crystal can be represented by a primitive "space lattice" defined by the basis vectors **a**, **b**, and **c**. However, sometimes it is convenient to define a new set of basis vectors **A**, **B**, and **C** in terms of the original set in order to gain insight into restrictions imposed on the X-ray diffraction record of the crystal. For example, we can define a non-primitive C-centered space lattice in terms of the original primitive space lattice by the following transition:

$$\begin{cases} \mathbf{A} = \mathbf{a} - \mathbf{b}, \\ \mathbf{B} = \mathbf{a} + \mathbf{b}, \\ \mathbf{C} = \qquad \mathbf{c}, \end{cases}$$

as illustrated in Figure 8.4.4.

Therefore the transition matrix from basis {**a**, **b**, **c**} to basis {**A**, **B**, **C**} is

$$P = \begin{bmatrix} 1 & 1 & 0 \\ -1 & 1 & 0 \\ 0 & 0 & 1 \end{bmatrix}.$$

Moreover, any atom located by coordinates (x, y, z) with respect to the basis {**a**, **b**, **c**} has coordinates

$$P^{-1} \begin{bmatrix} x \\ y \\ z \end{bmatrix}$$

with respect to the new basis {**A**, **B**, **C**} according to Theorem 5 and its Corollary.

In passing, we point out that the volume of the parallelepiped (unit cell) outlined by **A**, **B**, and **C** is equal to the determinant of the transition matrix times the volume of the parallelepiped outlined by **a**, **b**, and **c**. That is,

$$V_{\mathbf{ABC}} = \left(\det \begin{bmatrix} 1 & 1 & 0 \\ -1 & 1 & 0 \\ 0 & 0 & 1 \end{bmatrix} \right) V_{\mathbf{abc}} = 2 V_{\mathbf{abc}}.$$

(See pages 25–28 of Buerger [4] for a verification.)

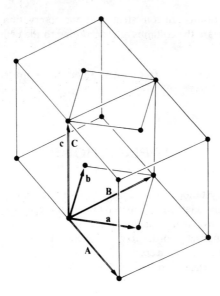

Figure 8.4.4

Example 7. This example is based on Example 6, which deals with a change of basis for a space lattice representation of a crystal. With respect to a basis {**a**, **b**, **c**}, it is customary to represent a certain system of parallel planes of atoms in a crystal by the equation

(8.4.4) $$\frac{x}{\|\mathbf{a}\|/h} + \frac{y}{\|\mathbf{b}\|/k} + \frac{z}{\|\mathbf{c}\|/l} = 1,$$

where h, k, and l are integers. From equation (8.4.4) we can see that intercepts of the first plane of the system are $\|\mathbf{a}\|/h$, $\|\mathbf{b}\|/k$, and $\|\mathbf{c}\|/l$. Thus the planes of the system intersect each basis vector in such a way as to divide each of them into h, k, and l parts respectively. (See Figure 8.4.5.) The numbers h, k, and l completely specify a system of planes and are called the Miller indices of the system of planes. When we define a new set of basis vectors, the orientations of planes of atoms in the crystal remain unchanged, whereas the characterizing intercepts and indices of these planes depend on the bases chosen.

8.4 CHANGE OF BASIS

It can be shown (see Chapter 2 of Buerger [4]) that if the integers (h, k, l) define a system of planes with respect to the basis $\{\mathbf{a}, \mathbf{b}, \mathbf{c}\}$ then, with respect to the new basis $\{\mathbf{A}, \mathbf{B}, \mathbf{C}\}$, the new indices (H, K, L) of the same plane are given by

$$\begin{bmatrix} H \\ K \\ L \end{bmatrix} = P^T \begin{bmatrix} h \\ k \\ l \end{bmatrix},$$

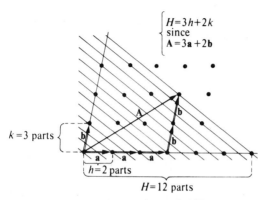

Figure 8.4.5

where P is the transition matrix from basis $\{\mathbf{a}, \mathbf{b}, \mathbf{c}\}$ to basis $\{\mathbf{A}, \mathbf{B}, \mathbf{C}\}$. Specifically, for the change in basis shown in Example 6 we have

$$\begin{bmatrix} H \\ K \\ L \end{bmatrix} = \begin{bmatrix} 1 & -1 & 0 \\ 1 & 1 & 0 \\ 0 & 0 & 1 \end{bmatrix} \begin{bmatrix} h \\ k \\ l \end{bmatrix},$$

or

$$\begin{cases} H = h - k, \\ K = h + k, \\ L = l. \end{cases}$$

By forming the sum $H + K = (h - k) + (h + k) = 2h$ and remembering that $h, k,$ and l are integers, we find a restriction for (H, K, L) is that $H + K$ must be an even number. When the X-ray diffraction record of this crystal is examined, the restriction that $H + K$ is even tells the mineralogist-geologist that he is dealing with a crystal whose crystal structure is based on a C-center space lattice.

In a study of the diffraction record, the crystal is sometimes rotated in an X-ray beam about a selected zone axis $[UVW]$. We can use the inverse of the transition matrix from basis $\{\mathbf{a}, \mathbf{b}, \mathbf{c}\}$ to basis $\{\mathbf{A}, \mathbf{B}, \mathbf{C}\}$ to define this axis with respect to the basis $\{\mathbf{A}, \mathbf{B}, \mathbf{C}\}$. For example, let the crystal of Figure 8.4.4 be rotated about the zone axis [1 1 1] defined by the vector $\mathbf{r} = \mathbf{a} + \mathbf{b} + \mathbf{c}$; this zone axis with respect to basis {A, B, C} is obtained by the transformation

$$[\mathbf{r}]_{\{A,B,C\}} = P^{-1}[\mathbf{r}]_{\{a,b,c\}},$$

or

$$\begin{bmatrix} 0 \\ 1 \\ 1 \end{bmatrix} = \begin{bmatrix} \tfrac{1}{2} & -\tfrac{1}{2} & 0 \\ \tfrac{1}{2} & \tfrac{1}{2} & 0 \\ 0 & 0 & 1 \end{bmatrix} \begin{bmatrix} 1 \\ 1 \\ 1 \end{bmatrix},$$

or $\mathbf{r} = \mathbf{B} + \mathbf{C}$, which can be verified in Figure 8.4.4, and which defines the zone axis [1 1 1] in the C-centered space lattice.

EXERCISES

1. In Example 1, find the transition matrix P if the basis is changed from $b = \left\{ \begin{bmatrix} 1 \\ 0 \end{bmatrix}, \begin{bmatrix} 0 \\ 1 \end{bmatrix} \right\}$ to $e = \left\{ \begin{bmatrix} 0 \\ 1 \end{bmatrix}, \begin{bmatrix} 1 \\ 0 \end{bmatrix} \right\}$. Also, if $[\alpha]_b = \begin{bmatrix} 3 \\ 4 \end{bmatrix}$, then find $[\alpha]_e$ by using Theorem 5.

2. In Example 1, find the transition matrix P if the basis is changed from b to $e = \left\{ \begin{bmatrix} 0 \\ 2 \end{bmatrix}, \begin{bmatrix} 1 \\ 1 \end{bmatrix} \right\}$. Also, if $[\alpha]_b = \begin{bmatrix} 2 \\ 5 \end{bmatrix}$, then find $[\alpha]_e$ by using Theorem 5.

In each of Exercises 3–6, let the vector space be $V_2(R)$. Find the transition matrix P and $[\alpha]_e$ if $[\alpha]_b = \begin{bmatrix} 8 \\ 6 \end{bmatrix}$. Graph the geometric vectors corresponding to α and to the old and new basis vectors.

3. $b = \{(1, 0), (0, 1)\}, \quad e = \{(1, 1), (0, 1)\}$.
4. $b = \{(1, 0), (0, 1)\}, \quad e = \{(1, 1), (1, -1)\}$.
5. $b = \{(1, 1), (1, -1)\}, \quad e = \{(1, 0), (0, 1)\}$.
6. $b = \{(2, 1), (1, 2)\}, \quad e = \{(3, 1), (1, 3)\}$.

In each of Exercises 7–10, let the vector space be $P_1(R)$. Find the transition matrix P and $[\alpha]_e$ if $[\alpha]_b = \begin{bmatrix} 6 \\ 2 \end{bmatrix}$.

7. $b = \{1, x\}, \quad e = \{x, 1\}$.
8. $b = \{1, x\}, \quad e = \{1 + x, x\}$.
9. $b = \{-1, 2x\}, \quad e = \{1, 1 + x\}$.
10. $b = \{x + 1, x\}, \quad e = \{x, 1\}$.
11. In Example 3, find the transition matrix P if the basis is changed from
$b = \left\{ \begin{bmatrix} 1 \\ 0 \\ 1 \end{bmatrix}, \begin{bmatrix} 0 \\ 1 \\ 0 \end{bmatrix}, \begin{bmatrix} 0 \\ 0 \\ 1 \end{bmatrix} \right\}$ to $e = \left\{ \begin{bmatrix} 1 \\ 0 \\ 0 \end{bmatrix}, \begin{bmatrix} 0 \\ 1 \\ 0 \end{bmatrix}, \begin{bmatrix} 0 \\ 0 \\ 1 \end{bmatrix} \right\}$. Also, if $[\alpha]_b = \begin{bmatrix} 6 \\ 7 \\ 8 \end{bmatrix}$, find $[\alpha]_e$.

12. In Example 4, let $\theta = 60°$, and find $T(\alpha)$. Give two different interpretations of your answer.
13. Prove Theorem 5. *Hint:* The proof is somewhat similar to the proof of Theorem 4. Begin with the statement $\alpha = a_1 \varepsilon_1 + a_2 \varepsilon_2 + \cdots + a_n \varepsilon_n$. (Statements of proof are given in Answers.)
14. For the vector space $V(R)$ of all 2 by 1 R-matrices with basis $b = \left\{ \begin{bmatrix} 1 \\ 0 \end{bmatrix}, \begin{bmatrix} 0 \\ 1 \end{bmatrix} \right\}$ and the matrix equation

$$\begin{bmatrix} \cos 45° & -\sin 45° \\ \sin 45° & \cos 45° \end{bmatrix} [\alpha]_b = B,$$

give two different interpretations of the equation if $[\alpha]_b = \begin{bmatrix} 1 \\ 1 \end{bmatrix}$.

15. For the vector space of all 2 by 1 R-matrices with basis $b = \left\{ \begin{bmatrix} 1 \\ 0 \end{bmatrix}, \begin{bmatrix} 0 \\ 1 \end{bmatrix} \right\}$, suppose that the matrix equation

$$P[\alpha]_e = [\alpha]_b$$

or

$$\begin{bmatrix} 0 & -1 \\ 1 & 0 \end{bmatrix} \begin{bmatrix} 6 \\ 3 \end{bmatrix} = \begin{bmatrix} -3 \\ 6 \end{bmatrix}$$

represents a change in basis from b to e. Find e. *Hint:* See Examples 4 or 5.
16. In Theorem 5, prove that P is the only matrix for which $P[\alpha]_e = [\alpha]_b$.
17. Let b and e be two different bases for an n-dimensional vector space $V(R)$. Prove that the transition matrix from e to b is the inverse of the transition matrix from b to e.

8.5 The Effect of a Change of Basis on a Transformation Matrix

Now we consider the effect of a change of basis on a transformation matrix. Let $V(F)$ be an arbitrary vector space with basis $b = \{\beta_1, \beta_2, \ldots, \beta_n\}$, and let T be a linear operator over $V(F)$. Then

(8.5.1) $$[T]_b [\alpha]_b = [T(\alpha)]_b.$$

If we change the basis from b to $e = \{\varepsilon_1, \varepsilon_2, \ldots, \varepsilon_n\}$, then the coordinate matrices of α and $T(\alpha)$ change accordingly, as shown in the last section; that is, if P is the transition matrix from b to e, then

(8.5.2) $$P[\alpha]_e = [\alpha]_b \quad \text{and} \quad P[T(\alpha)]_e = [T(\alpha)]_b.$$

Substituting (8.5.2) into (8.5.1) we obtain

$$[T]_b P[\alpha]_e = P[T(\alpha)]_e.$$

Because P is invertible we can premultiply both sides by P^{-1}, and

$$(P^{-1}[T]_b P)[\alpha]_e = [T(\alpha)]_e.$$

The matrix $P^{-1}[T]_b P$ is the new transformation matrix with respect to the new basis; because $[T]_e$ is the *only* matrix for which $[T(\alpha)]_e = [T]_e[\alpha]_e$ (as proved in Exercise 24 of Section 8.3), then

$$[T]_e = P^{-1}[T]_b P.$$

Thus the change of basis effects a transformation of the original transformation matrix.

As we pointed out previously, one of the primary applications of linear algebra is to change the basis in such a way that the new transformation matrix, $P^{-1}[T]_b P$, will assume a desired form; one such desired form is a diagonal matrix.

Example 1. For a transformation represented by

$$Y = [T]_b X \quad \text{where} \quad [T]_b = \begin{bmatrix} 1 & 2 \\ -1 & 4 \end{bmatrix},$$

it is possible to change the basis by a transition matrix $P = \begin{bmatrix} 2 & -3 \\ 1 & -3 \end{bmatrix}$ in such a way that the new transformation matrix

$$[T]_e = P^{-1}[T]_b P$$

is a diagonal matrix. (A method of calculating such a P, when one exists, will be discussed in a later section.)

If $P = \begin{bmatrix} 2 & -3 \\ 1 & -3 \end{bmatrix}$, then $P^{-1} = \begin{bmatrix} 1 & -1 \\ \frac{1}{3} & -\frac{2}{3} \end{bmatrix}$, and

$$[T]_e = P^{-1}[T]_b P = \begin{bmatrix} 1 & -1 \\ \frac{1}{3} & -\frac{2}{3} \end{bmatrix} \begin{bmatrix} 1 & 2 \\ -1 & 4 \end{bmatrix} \begin{bmatrix} 2 & -3 \\ 1 & -3 \end{bmatrix} = \begin{bmatrix} 2 & 0 \\ 0 & 3 \end{bmatrix}.$$

Thus

$$[Y]_b = \begin{bmatrix} 1 & 2 \\ -1 & 4 \end{bmatrix} [X]_b$$

becomes $[Y]_e = \begin{bmatrix} 2 & 0 \\ 0 & 3 \end{bmatrix} [X]_e$ upon the change of basis.

The importance of changing the form of a transformation matrix by means of a change in basis cannot be overemphasized. Frequently we do not care what the new basis is; we are simply interested in finding a transition matrix P which will change the transformation matrix to the desired form. When an operator A is related to another matrix operator B according to $B = P^{-1}AP$, we say that A is *similar* to B.

8.5 THE EFFECT OF A CHANGE OF BASIS

Definition 4. *Let A and B be n by n matrices over F. Matrix A is said to be **similar** to matrix B if there exists an invertible matrix P over F such that $B = P^{-1}AP$.*

Example 2. The matrix $\begin{bmatrix} 1 & 2 \\ -1 & 4 \end{bmatrix}$ is similar to $\begin{bmatrix} 2 & 0 \\ 0 & 3 \end{bmatrix}$ because there exists an invertible matrix $P = \begin{bmatrix} 2 & -3 \\ 1 & -3 \end{bmatrix}$ such that

$$\begin{bmatrix} 2 & 0 \\ 0 & 3 \end{bmatrix} = P^{-1} \begin{bmatrix} 1 & 2 \\ -1 & 4 \end{bmatrix} P.$$

In the last example we expressed a relation between two matrices; that relation was "is similar to." According to Definition 4 of Section 2.2 such a relation between pairs of elements of a set S is known as a *binary relation* and frequently is designated by R. In general aRb means "a is in relation R to b." A binary relation R on a set S is said to be an **equivalence relation** if aRa, $aRb \Rightarrow bRa$, and $(aRb$ and $bRc) \Rightarrow aRc$, where a, b, and c belong to S. Verbally this says that R is an equivalence relation if: (1) element a is in relation to itself; (2) element a is in relation to element b implies that b is in relation to a; (3) element a is in relation to element b and b is in relation to element c implies that a is in relation to c. For example "equals" is an equivalence relation on the set of real numbers, since for any a, b, c in the set, $a = a$, $(a = b) \Rightarrow (b = a)$, and $(a = b, b = c) \Rightarrow (a = c)$.

Theorem 6. *Similarity of matrices is an equivalence relation over the set of n by n matrices over F.*

The proof is left as Exercise 13.

It can be proved that any two matrix representations, with respect to different bases, of a single linear operator on a vector space are similar, and, conversely, that if two matrices are similar, then they represent the same linear operator with respect to different bases. In other words, similarity of two matrices means that the two matrices are matrix representations of the same linear operator with respect to different bases.

APPLICATIONS

★**Example 3.** A very common problem encountered in applied work is to find the solution of a system of linear differential equations of the form

(8.5.3) $$\dot{X} = AX + B,$$

where A is an n by n matrix of constants, $B = \begin{bmatrix} b_1(t) \\ \vdots \\ b_n(t) \end{bmatrix}$, $X = \begin{bmatrix} x_1(t) \\ \vdots \\ x_n(t) \end{bmatrix}$, and $\dot{X} = \begin{bmatrix} D_t x_1 \\ \vdots \\ D_t x_n \end{bmatrix}$; all four matrices are C-matrices, and the dot notation represents differentiation. One method of solution of such a system involves a transition $X = PY$ where P is a transition matrix that is selected in such a way that $P^{-1}AP = D$ assumes a desired form, preferably a diagonal matrix if that is possible. Equation (8.5.3) then becomes

$$P\dot{Y} = APY + B,$$

or

(8.5.4) $$\dot{Y} = (P^{-1}AP)Y + P^{-1}B.$$

If $P^{-1}AP$ is a diagonal matrix, then the system (8.5.4) can be solved by simply solving the decoupled equations

$$\dot{y}_i(t) = d_{ii}y_i(t) + b_i^*(t), \quad (i = 1, \ldots, n),$$

where the $b_i^*(t)$ are the entries of $P^{-1}B$ and the d_{ii} are main diagonal entries of $D = P^{-1}AP$. From elementary differential equations, solutions are

$$y_i(t) = e^{d_{ii}t}\left(y_i(0) + \int_0^t e^{-d_{ii}t}b_i^*(t)\,dt\right), \quad (i = 1, \ldots, n).$$

Even if A cannot be diagonalized, A is always similar to what is known as the Jordan Canonical form, and from this form a solution may be found by elementary techniques of differential equations. As an illustration of system (8.5.3) where A can be diagonalized, let

$$A = \begin{bmatrix} -1 & -2 & -3 \\ -2 & 1 & 8 \\ -1 & 1 & 3 \end{bmatrix} \quad \text{and} \quad B = \begin{bmatrix} 0 \\ 0 \\ 0 \end{bmatrix}.$$

By methods that we will learn later, a matrix $P = \begin{bmatrix} -1 & 7 & 5 \\ 2 & 2 & -3 \\ 1 & 1 & 2 \end{bmatrix}$ can be constructed so that $P^{-1}AP$ is the diagonal matrix $D = \begin{bmatrix} 6 & 0 & 0 \\ 0 & -2 & 0 \\ 0 & 0 & -1 \end{bmatrix}$.

Hence (8.5.4) is

$$\begin{bmatrix} \dot{y}_1 \\ \dot{y}_2 \\ \dot{y}_3 \end{bmatrix} = \begin{bmatrix} 6 & 0 & 0 \\ 0 & -2 & 0 \\ 0 & 0 & -1 \end{bmatrix}\begin{bmatrix} y_1 \\ y_2 \\ y_3 \end{bmatrix},$$

or

$$\begin{cases} \dot{y}_1 = 6y_1, \\ \dot{y}_2 = -2y_2, \\ \dot{y}_3 = -y_3, \end{cases}$$

8.5 THE EFFECT OF A CHANGE OF BASIS

from which it is easy to find $Y = \begin{bmatrix} y_1 \\ y_2 \\ y_3 \end{bmatrix} = \begin{bmatrix} y_1(0)e^{6t} \\ y_2(0)e^{-2t} \\ y_3(0)e^{-t} \end{bmatrix}.$

The solution of the original system can be found easily from $X = PY$.

★**Example 4.** Example 3 and the techniques of this section may be observed in an applied context by reading pages 42–43 and pages 93–108 of [8] (Elgerd); in fact this book has numerous other illustrations of elementary applications of linear algebra in the area of automatic controls. The particular problem referred to above, however, is concerned with controlling the inside temperature of a double-capacitance heat system as shown in Figure 8.5.1.

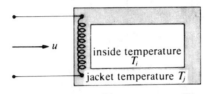

Figure 8.5.1

Let

u = rate of energy transfer to the system;
A_i, A_o = inside and outside jacket surface areas;
C_i, C_J = inside space and jacket heat capacities;
h_i, h_o = inside and outside surface film coefficients;
T_i, T_J, T_o = inside, jacket, and outside temperatures.

From a heat balance for the jacket and for the inside oven space, respectively, the following equations are obtained:

(8.5.5) $\quad \begin{cases} C_J \dot{T}_J = A_o h_o (T_o - T_J) + A_i h_i (T_i - T_J) + u, \\ C_i \dot{T}_i = A_i h_i (T_J - T_i). \end{cases}$

If we let x_1 and x_2 be the temperature differences $T_J - T_o$ and $T_i - T_o$, respectively, then the system of linear differential equations (8.5.5) becomes

(8.5.6) $\qquad\qquad\qquad \dot{X} = AX + Bu,$

where

$X = \begin{bmatrix} x_1 \\ x_2 \end{bmatrix}, \quad A = \begin{bmatrix} \dfrac{-1}{C_J}(A_o h_o + A_i h_i) & \dfrac{A_i h_i}{C_J} \\ \dfrac{A_i h_i}{C_i} & \dfrac{-A_i h_i}{C_i} \end{bmatrix}, \quad B = \begin{bmatrix} \dfrac{1}{C_J} \\ 0 \end{bmatrix},$

and

$$\dot{X} = \begin{bmatrix} D_t x_1 \\ D_t x_2 \end{bmatrix}.$$

Under a transition $X = PY$, equation (8.5.6) becomes

(8.5.7) $$\dot{Y} = (P^{-1}AP)Y + P^{-1}Bu,$$

where the new state variables are uncoupled if $P^{-1}AP$ is diagonal; system (8.5.7) can then be solved by the method shown in Example 3.

EXERCISES

In each of Exercises 1–4, let a change in basis of a vector space from b to e be made by use of a given transition matrix P. Find the new transformation matrix $[T]_e$.

1. $P = \begin{bmatrix} 2 & 1 \\ 1 & -2 \end{bmatrix}$; $[T]_b = \begin{bmatrix} 2 & -2 \\ -2 & 5 \end{bmatrix}$.

2. $P = \begin{bmatrix} 3 & 1 \\ 1 & 1 \end{bmatrix}$; $[T]_b = \begin{bmatrix} 1 & 3 \\ -1 & 5 \end{bmatrix}$.

3. $P = \begin{bmatrix} -1 & -3 \\ 1 & 1 \end{bmatrix}$; $[T]_b = \begin{bmatrix} 5 & 3 \\ -1 & 1 \end{bmatrix}$.

4. $P = \begin{bmatrix} 1 & -1 \\ 1 & 1 \end{bmatrix}$; $[T]_b = \begin{bmatrix} 0 & 2 \\ 2 & 0 \end{bmatrix}$.

5. For the vector space $V_2(R)$ let the basis be changed from $b = \{(1, 0), (0, 1)\}$ to $e = \{(1, 0), (1, 1)\}$. If $[T]_b = \begin{bmatrix} 3 & 2 \\ 1 & 4 \end{bmatrix}$, find $[T]_e$.

6. For the vector space $V(R)$ of all 2 by 1 R-matrices, let the basis be changed from $b = \left\{ \begin{bmatrix} 1 \\ 0 \end{bmatrix}, \begin{bmatrix} 0 \\ 1 \end{bmatrix} \right\}$ to $e = \left\{ \begin{bmatrix} 1 \\ 1 \end{bmatrix}, \begin{bmatrix} 0 \\ 2 \end{bmatrix} \right\}$. If $[T]_b = \begin{bmatrix} 1 & 0 \\ 3 & 2 \end{bmatrix}$, find $[T]_e$.

7–10. In each of Exercises 7–10 use the corresponding answers to Exercises 7–10 of Section 8.4 to find $[T]_e$ if T is defined by $ax + b \xmapsto{T} 2ax + b$.

11. Let the vector space be $V_2(R)$ with basis $b = \{(1, 0), (0, 1)\}$; let $T: V_2(R) \to V_2(R)$ be the shear defined by $(x, y) \xmapsto{T} (x + 5y, y)$, and let $\alpha = (2, 4)$.
   (a) Find $[\alpha]_b$ and $[T]_b$.
   (b) Find $[T(\alpha)]_b$ by using a matrix transformation.
   (c) If $e = \{(0, 1), (-1, 0)\}$, find the transition matrix P from b to e.
   (d) Find $[T]_e$.
   (e) Name three different ways of finding $[T(\alpha)]_e$, and use one of them.
   (f) Draw the geometric vectors corresponding to b, $[\alpha]_b$, $[T(\alpha)]_b$, e, $[\alpha]_e$, and $[T(\alpha)]_e$ on the same graph. Use one color to represent the basis b, and another color to represent the basis e.

★12. Let the vector space be $P_2(R)$ of polynomials with basis $b = \{1, x, x^2\}$; let $T: P_2(R) \to P_2(R)$ be the derivative D, and let $\alpha = x + 2x^2$.
   (a) Find $[\alpha]_b$ and $[T]_b$.
   (b) Find $[T(\alpha)]_b$ by using a matrix transformation.
   (c) If $e = \{1, -x^2, x\}$, find the transition matrix P from b to e.
   (d) Find $[T]_e$.
   (e) State three different ways of finding $[T(\alpha)]_e$, and use one of them.

13. Prove Theorem 6.

8.6 Linear Transformations of $V(F)$ into $U(F)$ (optional)[3]

So far we have dealt primarily with linear transformations of a vector space into itself. Now we consider the more general case of linear transformations from a vector space $V(F)$ into another vector space $U(F)$. As before, the coordinate matrix of an arbitrary element α of $V(F)$ depends upon a chosen basis $b = \{\beta_1, \beta_2, \ldots, \beta_n\}$ of $V(F)$. The *image* of α, however, will now belong to the vector space $U(F)$, and its coordinate matrix will depend upon a chosen basis $b' = \{\beta'_1, \beta'_2, \ldots, \beta'_m\}$ of that space. Notice that we let the dimension of $V(F)$ be n, and the dimension of $U(F)$ be m. As before, we list the images of the basis vectors of $V(F)$, but we must remember that these images, $T(\beta_i)$, belong to $U(F)$ and hence are expressed as linear combinations of the basis $b' = \{\beta'_1, \beta'_2, \ldots, \beta'_m\}$ of $U(F)$:

(8.6.1)
$$\begin{cases} T(\beta_1) = t_{11}\beta'_1 + t_{12}\beta'_2 + \cdots + t_{1m}\beta'_m, \\ T(\beta_2) = t_{21}\beta'_1 + t_{22}\beta'_2 + \cdots + t_{2m}\beta'_m, \\ \cdots \cdots \cdots \cdots \cdots \cdots \cdots \cdots \cdots \\ T(\beta_n) = t_{n1}\beta'_1 + t_{n2}\beta'_2 + \cdots + t_{nm}\beta'_m. \end{cases}$$

In a manner similar to that used in Section 8.2, we can justify the following generalization of Theorem 3.

Theorem 7. *For finite-dimensional vector spaces $V(F)$ and $U(F)$ with fixed bases $b = \{\beta_1, \beta_2, \ldots, \beta_n\}$ and $b' = \{\beta'_1, \beta'_2, \ldots, \beta'_m\}$, respectively, there is a one-to-one correspondence between the linear transformations T from $V(F)$ into $U(F)$ and the m by n matrices*

$$[T]_b^{b'} = [[T(\beta_1)]_{b'} \mid [T(\beta_2)]_{b'} \mid \cdots \mid [T(\beta_n)]_{b'}]$$

over F.

The proof is left as Exercise 13.

The matrix

$$[T]_b^{b'} = [[T(\beta_1)]_{b'} \mid [T(\beta_2)]_{b'} \mid \cdots \mid [T(\beta_n)]_{b'}]$$

$$= \begin{bmatrix} t_{11} & t_{21} & & t_{n1} \\ t_{12} & t_{22} & \cdots & t_{n2} \\ \vdots & \vdots & & \vdots \\ t_{1m} & t_{2m} & & t_{nm} \end{bmatrix}$$

[3] This section, however, is a prerequisite of the optional Sections 8.7, and 11.2 through 11.10.

of Theorem 7 is the **matrix representation** of a linear transformation T from $V(F)$ into $U(F)$ with respect to the respective bases b and b'. As before, the matrix representation of T will be referred to as the **transformation matrix**.

★**Example 1.** Consider the transformation $\int_1^x (\ ) dx$ from the vector space $P_1(R)$ of polynomials into the vector space $P_2(R)$ of polynomials. Let the basis of $P_1(R)$ be $b = \{1, x\}$, and let the basis of $P_2(R)$ be $b' = \{1, x, x^2\}$. Notice that $n = 2$ and $m = 3$.

$$\begin{cases} T(1) = \int_1^x (1)\, dx = x - 1 = -1(1) + 1(x) + 0(x^2), \\ T(x) = \int_1^x x\, dx = \tfrac{1}{2}x^2 - \tfrac{1}{2} = -\tfrac{1}{2}(1) + 0(x) + \tfrac{1}{2}(x^2). \end{cases}$$

(8.6.2) $\begin{cases} T(\beta_1) = -1(\beta'_1) + 1(\beta'_2) + 0(\beta'_3), \\ T(\beta_2) = -\tfrac{1}{2}(\beta'_1) + 0(\beta'_2) + \tfrac{1}{2}(\beta'_3). \end{cases}$

Therefore,

$$[T(\beta_1)]_{b'} = \begin{bmatrix} -1 \\ 1 \\ 0 \end{bmatrix} \text{ and } [T(\beta_2)]_{b'} = \begin{bmatrix} -\tfrac{1}{2} \\ 0 \\ \tfrac{1}{2} \end{bmatrix}.$$

Hence, the transformation matrix is

$$[T]_b^{b'} = \begin{bmatrix} -1 & -\tfrac{1}{2} \\ 1 & 0 \\ 0 & \tfrac{1}{2} \end{bmatrix}.$$

Notice that, as before, the transformation matrix is the transpose of the matrix of coefficients of (8.6.2).

We now generalize Theorem 4 of Section 8.3 and thereby offer a means of effecting any linear transformation from one vector space into another by means of a matrix transformation.

Theorem 8. *Let $V(F)$ and $U(F)$ be vector spaces of dimensions n and m, respectively. If T is a linear transformation of $V(F)$ into $U(F)$, and if α is an element of $V(F)$, then, with respect to the respective bases $b = \{\beta_1, \ldots, \beta_n\}$ and $b' = \{\beta'_1, \ldots, \beta'_m\}$, the coordinate matrix of α premultiplied by the unique matrix representation of T will equal the coordinate matrix of the image of α. That is,*

$$[T]_b^{b'}[\alpha]_b = [T(\alpha)]_{b'}.$$

The proof is left as Exercise 14.

8.6 LINEAR TRANSFORMATIONS OF V(F) INTO U(F)

★**Example 2.** Reconsider Example 1. There we found

$$[T]_b^{b'} = \begin{bmatrix} -1 & -\frac{1}{2} \\ 1 & 0 \\ 0 & \frac{1}{2} \end{bmatrix}.$$

Let $\alpha = 6 + 8x$; hence $[\alpha]_b = \begin{bmatrix} 6 \\ 8 \end{bmatrix}$. By Theorem 8, the image $\int_1^x (6 + 8x)\, dx$ of α can be expressed as $[T]_b^{b'}[\alpha]_b = [T(\alpha)]_{b'}$, or

$$\begin{bmatrix} -1 & -\frac{1}{2} \\ 1 & 0 \\ 0 & \frac{1}{2} \end{bmatrix} \begin{bmatrix} 6 \\ 8 \end{bmatrix} = \begin{bmatrix} -10 \\ 6 \\ 4 \end{bmatrix}.$$

Therefore the image of $(6 + 8x)$ is $(-10 + 6x + 4x^2)$.

APPLICATIONS

Example 3. An alternative approach for calculating the change ΔU of Example 7 of Section 4.1 is as follows. List the transactions of the firm, in order, as a column matrix $X = [40\ \ 35\ \ 15\ \ 25\ \ 30\ \ 5]^T$. If there are s accounts, then, since an account cannot transact with itself, there can be as many as $s(s-1)$ types of transactions. Let p be the actual number of types of transactions that occur. The set of all p by 1 matrices X can be thought of as elements of a vector space $V(R)$, and the set of all s by 1 matrices ΔU can be thought of as elements of a vector space $U(R)$. An s by p transformation matrix T from $V(R)$ into $U(R)$ can be written with

$$t_{ij} = \begin{cases} +1 & \text{if the } j\text{th entry of } X \text{ indicates the flow of the transaction toward account } i, \\ -1 & \text{if the } j\text{th entry of } X \text{ indicates the flow of the transaction away from account } i, \\ 0 & \text{if the } j\text{th entry of } X \text{ indicates no change in account } i, \end{cases}$$

such that $\Delta U = TX$.

For Example 7 of Section 4.1, we have

$$T = \begin{bmatrix} 0 & -1 & -1 & +1 & +1 & 0 \\ +1 & 0 & 0 & 0 & -1 & -1 \\ -1 & +1 & 0 & 0 & 0 & +1 \\ 0 & 0 & +1 & -1 & 0 & 0 \end{bmatrix}.$$

Thus $\Delta U = TX = \begin{bmatrix} 5 \\ 5 \\ 0 \\ -10 \end{bmatrix}$. One advantage of this alternative approach is that a general transformation matrix T can be calculated once and for all for all different types of transaction that might occur; this can then be recorded in a computer, and at the end of each period it can be allowed to operate on the column matrix X of that period to obtain ΔU. As might be expected, associated with a certain s by p matrix T there is a unique flow chart for p transactions among s accounts, and conversely. The flow chart for the specific T stated above is given in Figure 8.6.1; for example, in the second transaction (second column of T) flow is from account 1 to account 3 (-1 in first entry and $+1$ in third entry). Pages 92–94 of [14] (Ijiri) provide a good reference for this problem.

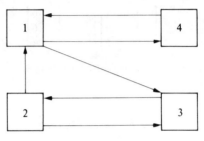

Figure 8.6.1

Example 4. The theory involved in the last example need not be confined to "accounts" of monetary value. A matrix may be used to provide for an accounting of people in a labor market, of student registration in certain courses, of registered voters within certain political units, of rental agencies (such as U-haul vehicles or automobiles), etc.

EXERCISES

In each of Exercises 1–4, consider the vector space $V_2(R)$ with basis $b = \{(1, 0), (0, 1)\}$, the vector space $V_3(R)$ with basis $b' = \{(1, 0, 0), (0, 1, 0), (0, 0, 1)\}$, and the transformation $T: V_2(R) \to V_3(R)$. For the given $[T]_b^{b'}$ and α, find $[T(\alpha)]_{b'}$. On two separate graphs, represent α in $V_2(R)$ and $T(\alpha)$ in $V_3(R)$; also represent the respective bases.

1. $[T]_b^{b'} = \begin{bmatrix} 1 & 0 \\ 0 & 1 \\ 1 & 1 \end{bmatrix}$; $\alpha = (3, 4)$.

2. $[T]_b^{b'} = \begin{bmatrix} 1 & 0 \\ 1 & 1 \\ 0 & 1 \end{bmatrix}$; $\alpha = (5, 6)$.

3. $[T]_b^{b'} = \begin{bmatrix} 0 & 1 \\ 1 & 0 \\ 0 & 2 \end{bmatrix}$; $\alpha = (9, 2)$.

4. $[T]_b^{b'} = \begin{bmatrix} 1 & 1 \\ 1 & -1 \\ -1 & 0 \end{bmatrix}; \quad \alpha = (2, 3).$

In each of Exercises 5–8, let the vector spaces and bases be as defined in Exercises 1–4. Find $[T]_b^{b'}$ for the linear transformation $T: V_2(R) \to V_3(R)$ defined in each exercise.

5. $(x, y) \xmapsto{T} (x, y, x + y)$.   **6.** $(x, y) \xmapsto{T} (x, x + y, y)$.

7. $(x, y) \xmapsto{T} (y, x, 2y)$.   **8.** $(x, y) \xmapsto{T} (x + y, x - y, -x)$.

In each of Exercises 9–12, consider the vector space $P_2(R)$ of polynomials with basis $b = \{1, x, x^2\}$, the vector space $P_3(R)$ of polynomials with basis $b' = \{1, x, x^2, x^3\}$, and a linear transformation $T: P_2(R) \to P_3(R)$. Find $[T]_b^{b'}$, and then find $[T(\alpha)]_{b'}$ by using a matrix transformation.

★**9.** $\alpha \xmapsto{T} \int_0^x \alpha \, dx; \quad \alpha = 1 + 2x + 5x^2$.

★**10.** $\alpha \xmapsto{T} \int_1^x \alpha \, dx; \quad \alpha = x^2$.

★**11.** $\alpha \xmapsto{T} \alpha + \int_1^x \alpha \, dx; \quad \alpha = 3 + x^2$.

★**12.** $\alpha \xmapsto{T} D_x \alpha + \int_0^x \alpha \, dx; \quad \alpha = 1 + x + x^2$.

13. Prove Theorem 7. *Hint*: Similar to proof of Theorem 3.

14. Prove Theorem 8. *Hint*: Similar to proof of Theorem 4.

8.7 Change of Basis for $V(F) \xrightarrow{T} U(F)$ (optional)[4]

We have learned that we can express any element α of a vector space $V(F)$ with respect to a chosen basis $b = \{\beta_1, \ldots, \beta_n\}$ as a coordinate matrix $[\alpha]_b$, and that if we change the basis to $e = \{\varepsilon_1, \ldots, \varepsilon_n\}$, then there exists an invertible transition matrix P such that

(8.7.1) $$[\alpha]_b = P[\alpha]_e.$$

Now consider a linear transformation T from $V(F)$ into $U(F)$. As before, the image of α, called $T(\alpha)$, which is in $U(F)$, can be expressed as the coordinate matrix $[T(\alpha)]_{b'}$ with respect to a basis b' of $U(F)$. If we choose, we can also change the basis of $U(F)$ to e'; if we do, we know that there exists an invertible transition matrix Q such that

(8.7.2) $$[T(\alpha)]_{b'} = Q[T(\alpha)]_{e'}.$$

From the last section we know that, corresponding to any linear transformation $T: V(F) \to U(F)$, there exists a transformation matrix $[T]_b^{b'}$ such that

(8.7.3) $$[T]_b^{b'}[\alpha]_b = [T(\alpha)]_{b'}.$$

[4] This section, however, is a prerequisite of the optional Sections 11.2 through 11.10.

If we substitute both (8.7.1) and (8.7.2) into (8.7.3), we get

$$[T]_b^{b'}(P[\alpha]_e) = Q[T(\alpha)]_{e'},$$

or, after premultiplying by Q^{-1},

$$(Q^{-1}[T]_b^{b'}P)[\alpha]_e = [T(\alpha)]_{e'}.$$

Therefore, after a change of basis in both $V(F)$ and $U(F)$, the matrix representation of T is

$$[T]_e^{e'} = Q^{-1}[T]_b^{b'}P.$$

If we change the basis only in $V(F)$ or only in $U(F)$, we get the matrix transformations

$$[T]_e^{b'} = [T]_b^{b'}P, \quad \text{or} \quad [T]_b^{e'} = Q^{-1}[T]_b^{b'},$$

respectively.

Example 1. Consider the vector space $V(R)$ of 3 by 1 R-matrices, generated by the basis

$$b = \{\beta_1, \beta_2, \beta_3\} = \left\{ \begin{bmatrix} 1 \\ 0 \\ 1 \end{bmatrix}, \begin{bmatrix} 0 \\ 1 \\ 0 \end{bmatrix}, \begin{bmatrix} 0 \\ 0 \\ 1 \end{bmatrix} \right\},$$

the vector space $U(R)$ of 2 by 1 R-matrices generated by the basis

$$b' = \{\beta_1', \beta_2'\} = \left\{ \begin{bmatrix} 1 \\ 0 \end{bmatrix}, \begin{bmatrix} 0 \\ 1 \end{bmatrix} \right\},$$

and the linear transformation $T: V(R) \to U(R)$ represented by

$$[T]_b^{b'} = \begin{bmatrix} 2 & 1 & 1 \\ 1 & 4 & 0 \end{bmatrix}.$$

The image $T(\alpha)$ of the vector α, where $[\alpha]_b = \begin{bmatrix} 0 \\ 3 \\ 4 \end{bmatrix}$, is given by

$$[T(\alpha)]_{b'} = [T]_b^{b'}[\alpha]_b = \begin{bmatrix} 2 & 1 & 1 \\ 1 & 4 & 0 \end{bmatrix} \begin{bmatrix} 0 \\ 3 \\ 4 \end{bmatrix} = \begin{bmatrix} 7 \\ 12 \end{bmatrix}.$$

Now suppose that we leave the basis of $U(R)$ unchanged but change the basis of $V(R)$ to

$$e = \{\varepsilon_1, \varepsilon_2, \varepsilon_3\} = \left\{ \begin{bmatrix} 1 \\ 0 \\ 0 \end{bmatrix}, \begin{bmatrix} 3 \\ 4 \\ 5 \end{bmatrix}, \begin{bmatrix} 0 \\ 0 \\ 1 \end{bmatrix} \right\}.$$

8.7 CHANGE OF BASIS FOR $V(F) \xrightarrow{T} U(F)$ (OPTIONAL)

The transition matrix from b to e is $P = \begin{bmatrix} 1 & 3 & 0 \\ 0 & 4 & 0 \\ -1 & 2 & 1 \end{bmatrix}$ (Example 3, Section 8.4). Therefore the transformation matrix with respect to the change in basis is

$$[T]_e^{b'} = [T]_b^{b'} P = \begin{bmatrix} 2 & 1 & 1 \\ 1 & 4 & 0 \end{bmatrix} \begin{bmatrix} 1 & 3 & 0 \\ 0 & 4 & 0 \\ -1 & 2 & 1 \end{bmatrix} = \begin{bmatrix} 1 & 12 & 1 \\ 1 & 19 & 0 \end{bmatrix}.$$

Of course, the coordinate matrix representing α will change because of the new basis. The reader should investigate what effect, if any, this change of basis has on the image $[T(\alpha)]_{b'}$, since the basis of $U(R)$ did not change.

APPLICATIONS

Example 2. Linear programming is a relatively new branch of mathematics and has numerous direct applications, some of which are described on pages 11–18 of [11] (Gass); a comprehensive bibliography may be found in [30] (Riley). The material of this section is fundamental in the study of the simplex method which is the primary method of solution of linear programming problems. Exercises 13 through 16 of this section are designed to give a brief insight into how the material of this section is used in the simplex method.

EXERCISES

In each of Exercises 1–8, let $V(F)$ be the vector space $V_2(R)$ with basis $b = \{(1, 0), (0, 1)\}$, and let $U(F)$ be the vector space $V_3(R)$ with basis $b' = \{(1, 0, 0), (0, 1, 0), (0, 0, 1)\}$. Assume that T is a linear transformation from $V(F)$ into $U(F)$.

1. Find $[T]_e^{e'}$, if $[T]_b^{b'} = \begin{bmatrix} 1 & 0 \\ 0 & 2 \\ 3 & 1 \end{bmatrix}$, if the transition matrix from b to e in $V(F)$ is

$\begin{bmatrix} 3 & 2 \\ 1 & 4 \end{bmatrix}$, and if the transition matrix from b' to e' in $U(F)$ is $\begin{bmatrix} 3 & 0 & 0 \\ 0 & 1 & 0 \\ 0 & 0 & 2 \end{bmatrix}$.

2. Find $[T]_e^{e'}$, if $[T]_b^{b'} = \begin{bmatrix} 3 & 9 \\ 4 & 2 \\ 0 & 1 \end{bmatrix}$, if the transition matrix from b to e in $V(F)$ is

$\begin{bmatrix} 4 & 0 \\ 8 & 2 \end{bmatrix}$, and if the transition matrix from b' to e' in $U(F)$ is $\begin{bmatrix} 1 & 0 & 0 \\ 0 & 3 & 0 \\ 0 & 0 & 4 \end{bmatrix}$.

3. Find $[T]_e^{b'}$, if $[T]_b^{b'} = \begin{bmatrix} 4 & 2 \\ 3 & 9 \\ 0 & 6 \end{bmatrix}$ and the transition matrix from b to e in $V(F)$ is $\begin{bmatrix} 4 & 2 \\ 8 & 2 \end{bmatrix}$.

4. Find $[T]_b^{e'}$, if $[T]_b^{b'} = \begin{bmatrix} 4 & 0 \\ 9 & 3 \\ 2 & 0 \end{bmatrix}$ and the transition matrix from b' to e' in $U(F)$ is $\begin{bmatrix} 1 & 0 & 0 \\ 0 & 3 & 0 \\ 0 & 0 & 4 \end{bmatrix}$.

5. Find $[T(\alpha)]_{e'}$ by means of a matrix transformation if $[\alpha]_b = \begin{bmatrix} 1 \\ 2 \end{bmatrix}$ in Exercise 1.

6. Find $[T(\alpha)]_{e'}$ by means of a matrix transformation if $[\alpha]_b = \begin{bmatrix} 3 \\ 4 \end{bmatrix}$ in Exercise 2.

7. Find $[T(\alpha)]_{e'}$ by means of a matrix transformation if $\alpha = (3, 1)$, $e = \{(0, 1), (-1, 0)\}$, $e' = b'$, and T is defined by

$$(x, y) \xmapsto{T} (x, 0, x + y).$$

8. Find $[T(\alpha)]_{e'}$ by means of a matrix transformation if $\alpha = (5, 4)$, $b = e$, $e' = \{(1, 0, 0), (0, 2, 0), (0, 0, 3)\}$, and T is defined by

$$(x, y) \xmapsto{T} (y, x, x - y).$$

In each of Exercises 9–12, consider the vector space $P_2(R)$ of polynomials, the vector space $P_3(R)$ of polynomials, and a linear transformation $T: P_2(R) \to P_3(R)$. Let the bases be $b = \{1, x, x^2\}$ and $b' = \{1, x, x^2, x^3\}$, respectively. Let $\alpha = 2 + x + 3x^2$.

★9. Find $[T]_b^{b'}$, $[T]_e^{e'}$, and then $[T(\alpha)]_{e'}$ if the transition matrices from b to e and from b' to e' are

$$\begin{bmatrix} 1 & 0 & 2 \\ 0 & 1 & 0 \\ 0 & 0 & 2 \end{bmatrix} \text{ and } \begin{bmatrix} 2 & 0 & 0 & 0 \\ 0 & 2 & 0 & 0 \\ 0 & 0 & 2 & 0 \\ 0 & 0 & 0 & 2 \end{bmatrix},$$

respectively, and if T is defined by

$$\alpha \xmapsto{T} \int_0^x \alpha \, dx.$$

★10. Find $[T]_b^{b'}$, $[T]_e^{e'}$, and then $[T(\alpha)]_{e'}$ if the transition matrices from b to e and from b' to e' are

$$\begin{bmatrix} 2 & 0 & 3 \\ 0 & 1 & 0 \\ 0 & 0 & 1 \end{bmatrix} \text{ and } \begin{bmatrix} 4 & 0 & 0 & 0 \\ 0 & 3 & 0 & 0 \\ 0 & 0 & 1 & 0 \\ 0 & 0 & 0 & 2 \end{bmatrix},$$

respectively, and if T is defined by
$$\alpha \overset{T}{\mapsto} \alpha + \int_1^x \alpha \, dx.$$

★11. Find $[T]_b^{b'}$, $[T]_e^{e'}$, and then $[T(\alpha)]_{e'}$ if $e = \{1, x + x^2, x^2\}$, $e' = \{x^3, x^2, x, 1\}$, and if T is defined by
$$\alpha \overset{T}{\mapsto} \int_1^x \alpha \, dx.$$

★12. Repeat Exercise 11 if $\alpha \overset{T}{\mapsto} \alpha + \int_0^x \alpha \, dx$.

13. Consider the matrix equation
$$\begin{bmatrix}4\\5\end{bmatrix}x_1 + \begin{bmatrix}1\\2\end{bmatrix}x_2 + \begin{bmatrix}1\\0\end{bmatrix}x_3 + \begin{bmatrix}0\\1\end{bmatrix}x_4 = \begin{bmatrix}p_{10}\\p_{20}\end{bmatrix},$$
which can be rewritten as $AX = P_0$, where
$$A = \begin{bmatrix}4 & 1 & 1 & 0\\5 & 2 & 0 & 1\end{bmatrix}, \quad X = \begin{bmatrix}x_1\\x_2\\x_3\\x_4\end{bmatrix}, \quad \text{and } P_0 = \begin{bmatrix}p_{10}\\p_{20}\end{bmatrix}.$$

Suppose that $AX = P_0$ represents a linear transformation from the vector space $V_4(R)$ of vectors (x_1, x_2, x_3, x_4) into the vector space $U(R)$ of all vectors generated by the columns of A (column space of A). Further suppose that we let the basis b of $V_4(R)$ remain unchanged, namely
$$\{(1, 0, 0, 0), (0, 1, 0, 0), (0, 0, 1, 0), (0, 0, 0, 1)\},$$
but we change the basis of $U(R)$ from the third and fourth columns of A (basis b') to the second and fourth columns of A (basis e').
(a) What is the transition matrix Q from basis b' to basis e'? What is Q^{-1}?
(b) What is the new matrix representation of the transformation? That is, what is the transformation matrix with respect to the new basis?

14. Repeat Exercise 13 if the basis of $U(R)$ is changed from the third and fourth columns of A to the third and first columns of A.

15. Repeat Exercise 13 if the basis of $U(R)$ is changed from the third and fourth columns of A to the first and fourth columns of A.

16. Repeat Exercise 13 if the basis of $U(R)$ is changed from the third and fourth columns of A to the third and second columns of A.

8.8 Linear Algebra of Linear Transformations (optional)

The purpose of this section is to show that the relationship between matrices and linear transformations extends beyond what we have developed so far. First, however, we must state more definitions and theorems concerning linear transformations.

Definition 5. *Let k be an element of a field F, let S and T be linear transformations over a finite-dimensional vector space V(F), and let α be an element of V(F), then:*

(i) *$S + T$ is a transformation over $V(F)$ such that*
$$(S + T)(\alpha) = S(\alpha) + T(\alpha);$$

(ii) *kS is a transformation over $V(F)$ such that*
$$(kS)(\alpha) = k(S(\alpha));$$

(iii) *ST is a transformation over $V(F)$ such that*
$$(ST)(\alpha) = S(T(\alpha)).$$

The third part of Definition 5 is referred to as the **composite** or **product** of two transformations and is illustrated in Figure 8.8.1; notice that T acts first, and then S acts on the image $T(\alpha)$.

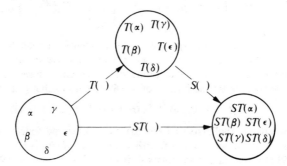

Figure 8.8.1

Theorem 9. *The transformations $S + T$, kS, and ST of Definition 5 are linear transformations.*

The proof is left as Exercise 17.

Theorem 9 assures us that the set of linear transformations over a vector space is closed under the operations defined in Definition 5; the fact that this set of linear transformations over a vector space forms a linear algebra under these operations can be established[5] by the rather lengthy but straightforward demonstration that each of the postulates of a linear algebra is satisfied.

Theorem 10. *With respect to the operations of Definition 5, the set of linear transformations over a vector space $V(F)$ forms a linear algebra.*

The proof is left as Exercise 18.

[5] McCoy, N. H., *Introduction to Modern Algebra*, Allyn and Bacon, Boston, 1960, page 273.

Example 1. Let $V(F)$ be the vector space $P_1(R)$ of polynomials. Let $\mathscr{L}$ represent the set of all linear transformations over $P_1(R)$. Theorem 10 assures us that the set $\mathscr{L}$ over the field R forms a linear algebra with respect to the operations of Definition 5 of this section. According to the definition of a linear algebra given in Section 6.5 this means that $\mathscr{L}$ forms a vector space over R with respect to addition and scalar multiplication (*i* and *ii* of Definition 5); also the following postulates must be valid: Let R, S, and T be elements of $\mathscr{L}(R)$, and let k be an element of R; then

$$R(S+T) = RS + RT \quad \text{(left distributive)},$$
$$(R+S)T = RT + ST \quad \text{(right distributive)},$$
$$k(ST) = (kS)T = S(kT).$$

In passing, we point out that in this example the additive identity **0** of $\mathscr{L}(R)$ is the transformation that maps any element of $P_1(R)$ into the zero polynomial since

$$(S + \mathbf{0})(\alpha) = S(\alpha) + \mathbf{0}(\alpha) = S(\alpha).$$

Also notice that the additive inverse of any transformation T of $\mathscr{L}(R)$ is the transformation $-T$ since

$$(T+(-T))(\alpha) = T(\alpha) + (-T(\alpha)) = T(\alpha) - T(\alpha) = \mathbf{0}(\alpha).$$

For example, if T is defined by

$$(ax+b) \stackrel{T}{\mapsto} (-bx+a),$$

then $-T$ is

$$(ax+b) \stackrel{-T}{\mapsto} (bx-a),$$

because

$$(-bx+a) + (bx-a) = 0x + 0.$$

We reemphasize that Theorem 10 implies that the set of linear transformations itself forms a vector space; this observation offers a springboard for considerable further study.

Finally we state one of the most important theorems in linear algebra. In Exercise 15 of Section 6.5 we found that the set of all n by n matrices over F formed a linear algebra. Theorem 10 assures us that the set of linear transformations over a vector space $V(F)$ of dimension n also forms a linear algebra. Then, since we recall that there is a one-to-one correspondence between n by n matrices over F and linear transformations over an n-dimensional vector space $V(F)$, we are led to investigate whether this one-to-one correspondence is preserved under the operations of Definition 5; it can be proved[6] that the one-to-one correspondence is preserved. Hence we have the

---

[6] Finkbeiner, D. T., *Introduction to Matrices, Vectors, and Linear Transformations*, Freeman, San Francisco, 1966, pages 85–86.

following theorem, with the understanding that an isomorphism between two algebras means that there is an isomorphism between the vector spaces plus the preservation of the one-to-one correspondence under the additional operation of "multiplication" of the linear algebras.

Theorem 11. *The linear algebra of n by n matrices over a field F is isomorphic to the linear algebra of linear transformations over an n-dimensional vector space $V(F)$.*

The proof is left as Exercise 19.

Theorem 11 is extremely important. In essence, a study of the linear algebra of linear transformations over a vector space of n dimensions may be conducted by a study of the linear algebra of n by n matrices, and vice versa. This approach greatly assists in the study of both algebras.

APPLICATION

Example 2. A simple illustration of products of transformations may be found on pages 19–21 of [4] (Buerger), where it is reported that three different sets of axes were assigned to the triclinic crystal, aximite, in the years 1892, 1897, and 1926. The matrices representing transformations from the first set of axes to the axes developed later are given as:

$$R = \begin{bmatrix} 1 & 1 & 0 \\ -2 & 0 & 0 \\ 0 & 0 & 2 \end{bmatrix}^{1892-1897}, \quad S = \begin{bmatrix} -1 & -1 & -1 \\ 1 & 0 & 0 \\ 0 & 0 & 1 \end{bmatrix}^{1897-1926}, \quad T = \begin{bmatrix} 1 & -1 & -2 \\ 1 & 1 & 0 \\ 0 & 0 & 2 \end{bmatrix}^{1892-1926}.$$

It can be verified that

$$T = SR.$$

EXERCISES

1. Consider the vector space $V_2(R)$ with basis $b = \{(1, 0), (0, 1)\}$. Let T and S be linear operators over $V_2(R)$ defined according to
$$(x, y) \overset{T}{\mapsto} (x, x + y),$$
$$(x, y) \overset{S}{\mapsto} (y, -x).$$

Find the image of (x, y) under $T + S$, TS, ST, kS, kT, and SS where k is in R.

8.8   LINEAR ALGEBRA OF LINEAR TRANSFORMATIONS

2. Repeat Exercise 1 except let T and S be defined according to
$$(x, y) \overset{T}{\mapsto} (-y, x),$$
$$(x, y) \overset{S}{\mapsto} (x + 2y, -y).$$

3. In Exercise 1, prove that $T + S$, TS, and kS are linear.
4. In Exercise 2, prove that $S + T$, ST, and kT are linear.
5. In Exercise 1, find the matrix representations of S and T, and, using these and Theorem 11, find the matrix representations of $S + T$, ST, and $3T$.
6. In Exercise 2, find the matrix representations of S and T, and, using these and Theorem 11, find the matrix representations of $S + T$, ST, and $3T$.
7. In Exercise 1, write the additive and multiplicative identities for the set of all linear transformations over $V_2(R)$. Also write the additive and multiplicative inverses of T if possible.
8. In Exercise 2, write the additive and multiplicative identities for the set of all linear transformations over $V_2(R)$. Also write the additive and multiplicative inverses of S if possible.
9. Let $V(R)$ be the vector space of all two-dimensional geometric vectors. On a graph, describe the linear transformation over $V(R)$ corresponding to the scalar matrix kI_2.
10. Let $V(R)$ be the vector space of all two-dimensional geometric vectors. On a graph, describe the linear transformation over $V(R)$ corresponding to the matrix $\begin{bmatrix} 0 & 1 \\ -1 & 0 \end{bmatrix}$.

In each of Exercises 11–14, let the vector space be $P_3(R)$ of polynomials. Let k belong to R, let T be the first derivative, and let S be the second derivative over $P_3(R)$.

★11. Find the image of $\alpha = ax^3 + bx^2 + cx + d$ under $T + S$, TS, ST, kS, kT, and TT.

★12. Prove that $T + S$, TS, and kS are linear.

★13. Find the matrix representations of S and T with respect to the basis $b = \{1, x, x^2, x^3\}$; using these representations and Theorem 11 find the matrix representations of $S + T$, ST, and $3T$.

★14. Write the additive and multiplicative identities for the set of all linear transformations over $P_3(R)$. Also write the additive and multiplicative inverses of T, if possible.

15. Prove that the set of all linear transformations from a vector space $V(F)$ into a vector space $U(F)$ forms a commutative group under addition.
16. Prove that the set of all linear transformations from a vector space $V(F)$ into a vector space $U(F)$ forms a vector space with respect to addition. *Hint:* Use the result of Exercise 15.
17. Prove Theorem 9.
18. Prove Theorem 10. *Hint:* Reread the paragraph preceding the theorem and use the reference in the footnote if necessary.
19. Prove Theorem 11. *Hint:* Reread the paragraph preceding the theorem and use the reference in the footnote if necessary.

NEW VOCABULARY

linear transformation   8.1
matrix transformation   8.1
matrix representation of a linear transformation (or transformation matrix)   8.2
linear operator   8.3
transition matrix from one basis to another   8.4
transition   8.4
similar matrices   8.5
equivalence relation   8.5
composite or product of two transformations   8.8

9

CHARACTERISTIC VALUES AND VECTORS

9.1 Basic Definitions

Consider (1) an arbitrary, nonzero element α of a vector space $V(F)$, (2) a scalar λ of F, and (3) a linear operator $T: V(F) \to V(F)$. Ordinarily the image $T(\alpha)$ cannot be expected to be a scalar multiple of α; if, however, the image $T(\alpha)$ *is* a scalar multiple, say $\lambda\alpha$, of some particular nonzero vector α, then λ is called a ***characteristic value of the linear operator T***, and the corresponding nonzero α is a ***characteristic vector of T***. Frequently, characteristic values are called ***eigenvalues*** or ***characteristic roots*** or ***latent roots***, and characteristic vectors are called ***eigenvectors***. *Eigen* is the German word for characteristic.

Example 1. Consider the vector space $V_2(R)$, and let T be the transformation of "reflection through the x-axis," defined according to

$$(x, y) \overset{T}{\mapsto} (x, -y).$$

Any nonzero vector of the form $\alpha = (0, a)$ will have an image $(0, -a)$ and the image is a scalar multiple, namely $(-1)\alpha$, of α. That is, if $\alpha = (0, a)$ then $T(\alpha) = (-1)\alpha$; therefore, $\lambda = -1$ is a characteristic value of T, and any nonzero vector of the form $(0, a)$ is a characteristic vector of T.

Example 2. Consider the vector space $V_2(R)$ of all two-dimensional real coordinate vectors, and let $T: V_2(R) \to V_2(R)$ be the linear transformation defined by

$$(x, y) \stackrel{T}{\mapsto} (2y, 2x).$$

Any nonzero vector α with equal components will have an image which is twice α; that is, if $\alpha = (a, a)$, then $T(\alpha) = 2\alpha$ (see Figure 9.1.1). Therefore, $\lambda = 2$ is a characteristic value of T, and any nonzero vector of the form (a, a) is a characteristic vector of T. The reader should verify that nonzero $\alpha = (a, -a)$ is also a characteristic vector of T, with $\lambda = -2$ as the corresponding characteristic value (see Figure 9.1.2).

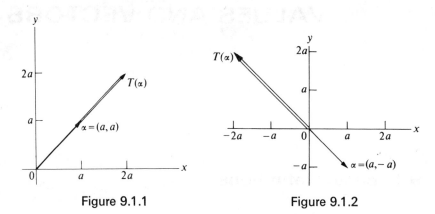

Figure 9.1.1   Figure 9.1.2

The linear transformation of Example 1 had one characteristic value, and the linear transformation of Example 2 had two characteristic values; we now consider a linear transformation that has no characteristic values.

Example 3. Consider the transformation over the vector space $V_2(R)$ that produces a "90° rotation." After some thought we can see that there is no nonzero vector in $V_2(R)$ which, when rotated 90°, will have an image that is some real multiple of itself.

As might be expected, the concepts of characteristic value and characteristic vector are also defined for the matrix representation of a linear transformation. With respect to some basis b of a vector space $V(F)$, let $[T]_b$ be the matrix representation of some linear operator $T: V(F) \to V(F)$, and let $[\alpha]_b$ be the coordinate matrix of a nonzero element α of $V(F)$; then finding characteristic values and characteristic vectors of T becomes a matter of solving the matrix equation

$$[T]_b[\alpha]_b = \lambda[\alpha]_b$$

9.1 BASIC DEFINITIONS

for λ and nonzero $[\alpha]_b$. Suppose we let $[T]_b = A$ and the unknown coordinate matrix $[\alpha]_b = X$. Then the problem can be restated: Solve $AX = \lambda X$ for λ and nonzero X.

Definition 1. *Let A be an n by n matrix over a field F. If X is an n by 1 nonzero matrix with entries from F, and if λ is an element of F such that*

$$AX = \lambda X,$$

*then X is said to be a **characteristic vector of the matrix A** corresponding to the **characteristic value λ of the matrix A**.*

If A is an n by n matrix over F, the matrix equation

$$AX = \lambda X$$

can be restated as

$$(A - \lambda I_n)X = 0,$$

which is the matrix representation of a system of homogeneous linear equations over F having a nonzero solution in F if and only if for λ in F the rank of $(A - \lambda I_n)$ is less than n; that is, if and only if, for λ in F, $\det(A - \lambda I_n) = 0$. The polynomial equation of nth degree, $\det(A - \lambda I_n) = 0$, is known as the **characteristic equation**, and its roots in F are the desired characteristic values.

Example 4. For the R-matrix $A = \begin{bmatrix} 1 & 2 \\ -1 & 4 \end{bmatrix}$, $AX = \lambda X$ is

$$\begin{bmatrix} 1 & 2 \\ -1 & 4 \end{bmatrix} \begin{bmatrix} x_1 \\ x_2 \end{bmatrix} = \lambda \begin{bmatrix} x_1 \\ x_2 \end{bmatrix},$$

or

$$\left(\begin{bmatrix} 1 & 2 \\ -1 & 4 \end{bmatrix} - \lambda \begin{bmatrix} 1 & 0 \\ 0 & 1 \end{bmatrix} \right) \begin{bmatrix} x_1 \\ x_2 \end{bmatrix} = \begin{bmatrix} 0 \\ 0 \end{bmatrix}.$$

Now, a solution $X \neq 0$ exists for the above homogeneous system if and only if, for λ in R,

$$\det\left(\begin{bmatrix} 1 & 2 \\ -1 & 4 \end{bmatrix} - \lambda \begin{bmatrix} 1 & 0 \\ 0 & 1 \end{bmatrix} \right) = 0, \quad \text{or} \quad \det \begin{bmatrix} 1-\lambda & 2 \\ -1 & 4-\lambda \end{bmatrix} = 0.$$

The expansion of this determinant yields the characteristic equation

$$\lambda^2 - 5\lambda + 6 = 0,$$

whose roots are $\lambda_1 = 2$ and $\lambda_2 = 3$. These roots are the characteristic values of the original matrix A.

Once the characteristic values have been found, we can then solve the homogeneous system

$$(A - \lambda I_n)X = 0$$

for nonzero X. [Characteristic value(s) ↑   ↑ characteristic vector(s)]

Example 5. In the last example we found $\lambda_1 = 2$, and $\lambda_2 = 3$. The problem now is to find the characteristic vectors that correspond to the respective characteristic values; that is, we must find nonzero solutions of

$$(A - 2I_2)X_1 = 0 \quad \text{and} \quad (A - 3I_2)X_2 = 0,$$

or

$$\begin{cases} -x_1 + 2x_2 = 0, \\ -x_1 + 2x_2 = 0, \end{cases} \quad \text{and} \quad \begin{cases} -2x_1 + 2x_2 = 0, \\ -x_1 + x_2 = 0. \end{cases}$$

The respective augmented matrices are

$$\begin{bmatrix} -1 & 2 & | & 0 \\ -1 & 2 & | & 0 \end{bmatrix} \quad \text{and} \quad \begin{bmatrix} -2 & 2 & | & 0 \\ -1 & 1 & | & 0 \end{bmatrix}.$$

The Gauss-Jordan elimination method produces

$$\begin{bmatrix} 1 & -2 & | & 0 \\ 0 & 0 & | & 0 \end{bmatrix} \quad \text{and} \quad \begin{bmatrix} 1 & -1 & | & 0 \\ 0 & 0 & | & 0 \end{bmatrix}.$$

Therefore, the respective complete solutions, with x_2 arbitrary, are

$$x_1 = 2x_2 \quad \text{and} \quad x_1 = x_2.$$

Hence, for $\lambda_1 = 2$,

$$X_1 = \begin{bmatrix} x_1 \\ x_2 \end{bmatrix} = \begin{bmatrix} 2x_2 \\ x_2 \end{bmatrix},$$

and, for $\lambda_2 = 3$,

$$X_2 = \begin{bmatrix} x_1 \\ x_2 \end{bmatrix} = \begin{bmatrix} x_2 \\ x_2 \end{bmatrix},$$

where the x_2 of each characteristic vector is an arbitrary, nonzero, real parameter. Suppose we arbitrarily let the parameter be 1 in X_1 and be -3 in X_2. We then obtain $X_1 = \begin{bmatrix} 2 \\ 1 \end{bmatrix}$ and $X_2 = \begin{bmatrix} -3 \\ -3 \end{bmatrix}$. Notice that neither X_1 nor X_2 is unique. If these characteristic vectors are normalized, we obtain

$$\begin{bmatrix} 2/\sqrt{5} \\ 1/\sqrt{5} \end{bmatrix} \quad \text{and} \quad \begin{bmatrix} -1/\sqrt{2} \\ -1/\sqrt{2} \end{bmatrix}.$$

Observe that *normalized characteristic vectors* are unique except for a possible change in sign of each vector.

9.1 BASIC DEFINITIONS

Example 6. We observed in Example 3 that no characteristic vectors exist in the vector space $V_2(R)$ for the transformation of rotation through 90°. Let us consider the corresponding matrix equation

$$\begin{bmatrix} 0 & -1 \\ 1 & 0 \end{bmatrix} \begin{bmatrix} x_1 \\ x_2 \end{bmatrix} = \lambda \begin{bmatrix} x_1 \\ x_2 \end{bmatrix},$$

or

$$\begin{bmatrix} -\lambda & -1 \\ 1 & -\lambda \end{bmatrix} \begin{bmatrix} x_1 \\ x_2 \end{bmatrix} = \begin{bmatrix} 0 \\ 0 \end{bmatrix}.$$

But

$$\det \begin{bmatrix} -\lambda & -1 \\ 1 & -\lambda \end{bmatrix} = \lambda^2 + 1 = 0$$

has no real roots, as we expected from Example 3. We have here an example of an R-matrix that has no real characteristic values and no real characteristic vectors. If, however, we do not limit ourselves to the field of real numbers, we find that there *are* complex characteristic values and vectors of the C-matrix $\begin{bmatrix} 0 & -1 \\ 1 & 0 \end{bmatrix}$. This example illustrates the fact that an underlying field *must* be designated when considering the existence of characteristic values and vectors of a particular matrix A.

The left-hand side of the characteristic equation, namely, $\det(A - \lambda I_n)$, is known as the **characteristic function** or the **characteristic polynomial of A**. The following well-known theorem establishes a relationship between A and its characteristic function.

Theorem 1. *Cayley-Hamilton Theorem.*[1] *Every square matrix A over F of order n satisfies its characteristic equation: that is, for the characteristic equation*

$$\lambda^n + b_{n-1}\lambda^{n-1} + \cdots + b_1\lambda + b_0 = 0,$$

we have

$$A^n + b_{n-1}A^{n-1} + \cdots + b_1 A + b_0 I_n = \mathbf{0}.$$

Example 7. The characteristic equation for Example 4 was found to be $\lambda^2 - 5\lambda + 6 = 0$. According to the Cayley-Hamilton theorem, $A^2 - 5A + 6I_2$ should be $\mathbf{0}$. To check this we have

$$\begin{bmatrix} 1 & 2 \\ -1 & 4 \end{bmatrix}^2 - 5\begin{bmatrix} 1 & 2 \\ -1 & 4 \end{bmatrix} + 6I_2 = \begin{bmatrix} -1 & 10 \\ -5 & 14 \end{bmatrix} - \begin{bmatrix} 5 & 10 \\ -5 & 20 \end{bmatrix} + \begin{bmatrix} 6 & 0 \\ 0 & 6 \end{bmatrix} = \begin{bmatrix} 0 & 0 \\ 0 & 0 \end{bmatrix}.$$

---

[1] For proofs see Eves, Howard, *Elementary Matrix Theory*, Allyn & Bacon, Boston, 1966, pages 200–202, or Lightstone, A. H., *Linear Algebra*, Appleton-Century-Crofts, New York, 1969, pages 295–298.

APPLICATIONS

Example 8. Consider the arbitrary linear transformation of an object of shape *abcd* as shown in Figure 9.1.3. After the transformation, which preserves the area of the object, the shape is *a'b'c'd'*. Vector Oc' becomes a scalar multiple of Oc, and vector Oe' becomes a scalar multiple of Oe; thus Oc and Oe are characteristic vectors according to the definition of a characteristic vector; the ratios $\dfrac{\|Oc'\|}{\|Oc\|}$ and $\dfrac{\|Oe'\|}{\|Oe\|}$ are the corresponding characteristic values under the transformation because they are the scalar multipliers, that is, $Oc' = \dfrac{\|Oc'\|}{\|Oc\|} Oc$ and $Oe' = \dfrac{\|Oe'\|}{\|Oe\|} Oe$, respectively.

Next consider a second transformation of an object of shape *abcd* as shown in Figure 9.1.4; this transformation also preserves the area of the object, and the

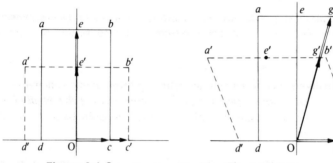

Figure 9.1.3   Figure 9.1.4

vector Oc' is a scalar multiple of Oc. Thus, Oc remains a characteristic vector—but notice that Oe no longer is. Figure 9.1.4 illustrates that under this transformation there exists a point g with an image g' such that Og' is a scalar multiple of Og, and hence Og is the other characteristic vector; the two characteristic vectors are no longer orthogonal (perpendicular in two dimensions).

Now, in the general two-dimensional case let T be a linear transformation with normalized, noncollinear, characteristic vectors ε_1 and ε_2 having corresponding characteristic values λ_1 and λ_2. Prior to the transformation T, any vector α can be expressed as

$$\alpha = k_1 \varepsilon_1 + k_2 \varepsilon_2.$$

Because the transformation is linear, the image of α is

$$\alpha' = T(\alpha) = k_1 T(\varepsilon_1) + k_2 T(\varepsilon_2)$$
$$= k_1(\lambda_1 \varepsilon_1) + k_2(\lambda_2 \varepsilon_2)$$
$$= \lambda_1(k_1 \varepsilon_1) + \lambda_2(k_2 \varepsilon_2),$$

9.1 BASIC DEFINITIONS

and hence α' can be expressed as a linear combination of $k_1 \varepsilon_1$ and $k_2 \varepsilon_2$. This is a rather useful idea and is restated in different words for emphasis: An arbitrary vector α may be resolved into vector components $k_1 \varepsilon_1$ and $k_2 \varepsilon_2$; the image of α under T may be considered as a linear combination of those same components, and, moreover, the coefficients of combination are the characteristic values λ_1 and λ_2. This idea serves as an underlying reason for changing the basis of $V_2(R)$ from the standard basis $\{(1, 0), (0, 1)\}$ to $\{\varepsilon_1, \varepsilon_2\}$ when applying the linear transformation T.

Example 9. Certain studies concerning a population of objects can be made by the procedure shown below.

A population matrix equation concerning certain living objects in a given environment can be constructed in the following way:

Let $x_{i,t}$ = the number of living objects of age i at time t.
Let p_i = the probability that an object of age i at time t will survive to time $t+1$.
Let f_i = number of new objects created per old object of age i.
Let n = maximum possible age.

Then a column matrix representing the expected number of living objects at time $t+1$ can be determined by the following matrix equation

(8.1.1)
$$\begin{bmatrix} x_{0,t+1} \\ x_{1,t+1} \\ x_{2,t+1} \\ \vdots \\ x_{n,t+1} \end{bmatrix} = \begin{bmatrix} f_0 & f_1 & f_2 & \cdots & f_n \\ p_0 & 0 & 0 & \cdots & 0 \\ 0 & p_1 & 0 & \cdots & 0 \\ \vdots & & \ddots & & \vdots \\ 0 & \cdots & 0 & p_{n-1} & 0 \end{bmatrix} \begin{bmatrix} x_{0,t} \\ x_{1,t} \\ x_{2,t} \\ \vdots \\ x_{n,t} \end{bmatrix}.$$

Let the respective matrices in Equation (8.1.1) be designated X_{t+1}, M, and X_t so that the matrix equation becomes

$$X_{t+1} = MX_t.$$

Suppose we wish to find the population at time t that will yield a proportional population at time $t+1$. That is, what population will produce a stability of age groups? Mathematically, taking t as the independent variable we are asked to find X_t such that X_{t+1} is a scalar multiple, λX_t, of X_t. If $X_{t+1} = \lambda X_t$, then the equation $X_{t+1} = MX_t$ becomes

$$\lambda X_t = MX_t,$$

which is the characteristic value problem. For biological implications of the procedure just described, see [19] (Leslie) and [3] (Bernadelli). The latter reference concerns a kind of hypothetical beetle whose M matrix is

$$\begin{bmatrix} 0 & 0 & 6 \\ \frac{1}{2} & 0 & 0 \\ 0 & \frac{1}{3} & 0 \end{bmatrix}.$$

One characteristic vector of this matrix is $\begin{bmatrix} 6 \\ 3 \\ 1 \end{bmatrix}$, hence any population of these beetles with an age distribution proportional to $6:3:1$ should keep a stable age distribution.

★**Example 10.** This example presupposes some knowledge of calculus and physics. Consider the mechanical system shown in Figure 9.1.5 where the down-

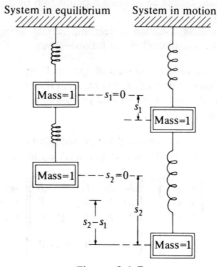

Figure 9.1.5

ward direction is considered to be the positive direction. The spring constant of the upper spring is k_1, and the spring constant of the lower spring is k_2; s_1 is the displacement of the upper mass, and s_2 is that of the lower one. The change in length of the first spring is s_1, and the change in length of the second spring is $(s_2 - s_1)$. The displacements s_1 and s_2 shown in the second diagram are due to a force F, which has been exerted on the system. The well-known physics formulas $F = ms''$ and $F = ks$ allow us to write $ms'' = ks$, where s'' is the acceleration equal to the second derivative of s with respect to t, and m is the mass of the object. Because the mass of each object is one unit in this example, and if the damping and the mass of the springs are neglected, the vertical motion of the system is governed by the equations

$$\begin{cases} s_1'' = -k_1 s_1 + k_2(s_2 - s_1), \\ s_2'' = -k_2(s_2 - s_1), \end{cases}$$

or

$$\begin{bmatrix} s_1'' \\ s_2'' \end{bmatrix} = \begin{bmatrix} -(k_1 + k_2) & k_2 \\ k_2 & -k_2 \end{bmatrix} \begin{bmatrix} s_1 \\ s_2 \end{bmatrix}.$$

9.1 BASIC DEFINITIONS

In the first equation the right-hand member is the sum of the force $-k_1 s_1$ exerted on the first body by the first spring (negative if $s_1 > 0$) and the force $k_2(s_2 - s_1)$ exerted on the first body by the second spring (positive if $s_2 - s_1 > 0$). In the second equation we use the force $-k_2(s_2 - s_1)$ exerted on the second body (negative if $s_2 - s_1 > 0$). In order to solve this system we assume the solution that is obtained by calculus,

$$\begin{bmatrix} s_1 \\ s_2 \end{bmatrix} = \begin{bmatrix} c_1 \\ c_2 \end{bmatrix} e^{wt},$$

where c_1 and c_2 are arbitrary constants which are to be determined. After differentiating twice we obtain

$$\begin{bmatrix} s_1'' \\ s_2'' \end{bmatrix} = w^2 \begin{bmatrix} c_1 \\ c_2 \end{bmatrix} e^{wt}.$$

Upon substitution in both sides of the original matrix equation we obtain

$$w^2 \begin{bmatrix} c_1 \\ c_2 \end{bmatrix} e^{wt} = \begin{bmatrix} -(k_1 + k_2) & k_2 \\ k_2 & -k_2 \end{bmatrix} \begin{bmatrix} c_1 \\ c_2 \end{bmatrix} e^{wt},$$

or

$$w^2 \begin{bmatrix} c_1 \\ c_2 \end{bmatrix} = \begin{bmatrix} -(k_1 + k_2) & k_2 \\ k_2 & -k_2 \end{bmatrix} \begin{bmatrix} c_1 \\ c_2 \end{bmatrix},$$

which is a characteristic value problem. A solution then is a characteristic vector multiplied by e^{wt}, where w^2 is the corresponding characteristic value. Notice that the matrix is symmetric; hence w^2 is real as we will prove in Theorem 2, page 279.

Example 11. Consider some organism or other object which reproduces itself every month, and suppose that the new offspring must wait 2 months before it begins to reproduce. If none of the objects die, this situation can be represented by the *difference equation*

$$x_n = x_{n-1} + x_{n-2} \quad (n = 2, 3, \ldots),$$

where x_n represents the number of objects after n months. Assume that $x_0 = 1$ and $x_1 = 1$; then $x_2 = 2$, $x_3 = 3$, $x_4 = 5$, etc. We wish to find a general expression for x_n. Such equations in which the independent variable n must assume certain integral values are called difference equations; they appear in many types of problems dealing with electrical theory, investment problems, the social sciences, statistics, and mechanics. A solution of the given difference equation can be found by use of matrices as follows:

Let $y_{n-1} = x_{n-2}$ or $y_n = x_{n-1}$ be introduced so that we obtain the system of equations

$$\begin{cases} x_n = x_{n-1} + y_{n-1}, \\ y_n = x_{n-1}, \end{cases} \quad \text{or} \quad \begin{bmatrix} x_n \\ y_n \end{bmatrix} = \begin{bmatrix} 1 & 1 \\ 1 & 0 \end{bmatrix} \begin{bmatrix} x_{n-1} \\ y_{n-1} \end{bmatrix}.$$

If $n = 2$,

$$\begin{bmatrix} x_2 \\ y_2 \end{bmatrix} = A \begin{bmatrix} x_1 \\ y_1 \end{bmatrix}, \text{ where } A = \begin{bmatrix} 1 & 1 \\ 1 & 0 \end{bmatrix}.$$

If $n = 3$,
$$\begin{bmatrix} x_3 \\ y_3 \end{bmatrix} = A \begin{bmatrix} x_2 \\ y_2 \end{bmatrix} = A^2 \begin{bmatrix} x_1 \\ y_1 \end{bmatrix}.$$

In general,
$$\begin{bmatrix} x_n \\ y_n \end{bmatrix} = A^{n-1} \begin{bmatrix} x_1 \\ y_1 \end{bmatrix} \quad \text{for} \quad n = 2, 3, \ldots.$$

Since we know $x_1 = 1$ and $y_1 = x_0 = 1$, all we need is A^{n-1}.

A formula for finding A^k, where A is a 2 by 2 C-matrix, is

$$A^k = \frac{\lambda_2 \lambda_1^k - \lambda_1 \lambda_2^k}{\lambda_2 - \lambda_1} I_2 + \frac{\lambda_2 k - \lambda_1 k}{\lambda_2 - \lambda_1} A,$$

if λ_1 and λ_2 are distinct characteristic values of A.[2] In this example, $\lambda_1 = \frac{1}{2}(1 + \sqrt{5})$ and $\lambda_2 = \frac{1}{2}(1 - \sqrt{5})$, and after considerable manipulation we find that

$$x_n = \frac{1}{\sqrt{5}} \left[\left(\frac{1 + \sqrt{5}}{2} \right)^{n+1} - \left(\frac{1 - \sqrt{5}}{2} \right)^{n+1} \right].$$

EXERCISES

1. By inspection, determine the characteristic vectors and characteristic values, if any, for the shearing transformation defined by

$$(x, y) \xmapsto{T} (x + 3y, y)$$

over the vector space $V_2(R)$.

2. By inspection, determine the characteristic vectors and characteristic values, if any, for the transformation "projection on the x-axis" defined by

$$(x, y) \xmapsto{T} (x, 0)$$

over the vector space $V_2(R)$.

In each of Exercises 3–8, find the characteristic equation, characteristic values, and normalized characteristic vectors for each of the following R-matrices.

3. $\begin{bmatrix} 2 & -2 \\ -2 & 5 \end{bmatrix}$.

4. $\begin{bmatrix} 1 & 3 \\ -1 & 5 \end{bmatrix}$.

5. $\begin{bmatrix} 5 & 3 \\ -1 & 1 \end{bmatrix}$.

6. $\begin{bmatrix} 4 & 3 \\ 1 & 2 \end{bmatrix}$.

7. $\begin{bmatrix} 2 & 1 & 1 \\ 1 & 1 & 0 \\ 1 & 0 & 1 \end{bmatrix}$.

8. $\begin{bmatrix} 2 & 0 & 0 \\ 0 & 3 & 1 \\ 0 & 1 & 3 \end{bmatrix}$.

[2] A derivation of this formula (which makes use of the Cayley-Hamilton theorem) can be found in P. J. Davis, *The Mathematics of Matrices*, Blaisdell, New York, 1965, pp. 274–275.

In each of Exercises 9–12, find a nonzero real solution of the system $AX = \lambda X$.

9. $A = \begin{bmatrix} 5 & 1 \\ 4 & 8 \end{bmatrix}$.

10. $A = \begin{bmatrix} 0 & 3 \\ 1 & 2 \end{bmatrix}$.

11. $A = \begin{bmatrix} 0 & 0 & 2 \\ 0 & 1 & 0 \\ 2 & 0 & 0 \end{bmatrix}$.

12. $A = \begin{bmatrix} 2 & -2 & 0 \\ 0 & 4 & 0 \\ 2 & 5 & 1 \end{bmatrix}$.

In each of Exercises 13–16, find a nonzero normalized real column vector X such that the image of the matrix transformation defined by $X \overset{T}{\mapsto} AX$ is a scalar multiple of X. Also find the scalar multiple.

13. $A = \begin{bmatrix} 2 & 6 \\ 6 & -3 \end{bmatrix}$.

14. $A = \begin{bmatrix} 0 & 2 \\ 2 & 0 \end{bmatrix}$.

15. $A = \begin{bmatrix} 3 & 1 & 2 \\ 0 & 1 & 0 \\ 0 & 0 & 4 \end{bmatrix}$.

16. $A = \begin{bmatrix} 1 & 0 & 0 \\ 1 & 3 & 0 \\ 1 & 1 & 4 \end{bmatrix}$.

17. What is the number of characteristic values of an n by n C-matrix?

18. Find the characteristic values of

(a) $\begin{bmatrix} 3 & 1 \\ 0 & -4 \end{bmatrix}$;

(b) $\begin{bmatrix} a & c \\ 0 & b \end{bmatrix}$;

(c) Generalize the result found in (b) to nth-order triangular C-matrices.

19. Illustrate the Cayley-Hamilton Theorem for the matrix of Exercise 3.
20. Illustrate the Cayley-Hamilton Theorem for the matrix of Exercise 8.
21. In Figure 9.1.3 of Example 8 verify that, with respect to a basis of unit vectors along the coordinate axes, $T = \begin{bmatrix} \frac{3}{2} & 0 \\ 0 & \frac{2}{3} \end{bmatrix}$ is the matrix representation of the transformation if $a = (-2, 6)$, $b = (2, 6)$, $a' = (-3, 4)$, and $b' = (3, 4)$; find the characteristic values.

22. In Figure 9.1.4 of Example 8, verify that, with respect to a basis of unit vectors along the coordinate axes, $T = \begin{bmatrix} \frac{3}{2} & -\frac{1}{4} \\ 0 & \frac{2}{3} \end{bmatrix}$ is the matrix representation of the transformation if $a = (-2, 6)$, $b = (2, 6)$, $a' = (-9/2, 4)$, and $b' = (3/2, 4)$. Find the characteristic vectors Oc and Og.

9.2 Theorems

We now consider a few important theorems concerning characteristic values and vectors of matrices. For reasons that will become apparent later we will pay particular attention to symmetric matrices.

Theorem 2. *Over C, the characteristic values of a symmetric matrix with real entries are real.*

Proof. Let A be a symmetric matrix with real entries. The complex conjugate of $AX = \lambda X$ is $\overline{AX} = \overline{\lambda X}$ or $A\overline{X} = \overline{\lambda}\overline{X}$, since A has real entries. Premultiplying the first equation by $\overline{X}^T$ and the third equation by X^T we obtain

$$\overline{X}^T AX = \lambda \overline{X}^T X \quad \text{and} \quad X^T A\overline{X} = \overline{\lambda} X^T \overline{X}.$$

Subtraction yields

$$\overline{X}^T AX - X^T A\overline{X} = \lambda \overline{X}^T X - \overline{\lambda} X^T \overline{X}.$$

But the left-hand side is the zero matrix because

$$\overline{X}^T AX = (\overline{X}^T AX)^T = X^T A^T \overline{X} = X^T A\overline{X}.$$

The first step is valid because $\overline{X}^T AX$ is a 1 by 1 matrix, and in the last step we use the fact that A is symmetric. Therefore,

$$0 = \lambda \overline{X}^T X - \overline{\lambda} X^T \overline{X}.$$

But $\overline{X}^T X = X^T \overline{X}$ which is real and positive (since $(a + bi)(a - bi) = a^2 + b^2$); hence,

$$0 = (\lambda - \overline{\lambda}) X^T \overline{X},$$

which implies $\lambda - \overline{\lambda} = 0$ and $\lambda = \overline{\lambda}$. Therefore λ must be real. □

Actually the last theorem is a special case of the following theorem since a symmetric matrix with real entries is Hermitian.

Theorem 3. *Over C, the characteristic values of a Hermitian matrix are real.*

The proof parallels that given for Theorem 2 and is left as Exercise 15 of this section.

We hasten to add that over C, nonsymmetric matrices with real entries, or non-Hermitian matrices, may have some characteristic values that are not real, as Exercises 1 and 2 of this section illustrate.

Theorem 4. *If X_i and X_j are characteristic vectors corresponding to two distinct characteristic values λ_i, λ_j of a symmetric R-matrix, then $X_j^T X_i = 0$.*

Proof. Let A be a symmetric R-matrix. Also, we are given that $AX_i = \lambda_i X_i$ and $AX_j = \lambda_j X_j$. Hence, if we premultiply the first equation by X_j^T and postmultiply the second equation by X_i after taking the transpose of both sides of the second equation, we obtain

$$X_j^T(AX_i) = X_j^T(\lambda_i X_i) \quad \text{and} \quad (X_j^T A)X_i = (X_j^T \lambda_j)X_i.$$

9.2 THEOREMS

Subtracting, we get

$$X_j^T A X_i - X_j^T A X_i = \lambda_i X_j^T X_i - \lambda_j X_j^T X_i,$$

or

$$0 = (\lambda_i - \lambda_j)(X_j^T X_i),$$

and since by hypothesis $\lambda_i \neq \lambda_j$, then $X_j^T X_i = 0$. □

The last theorem is a special case of the following theorem.

Theorem 5. *If X_i and X_j are characteristic vectors corresponding to two distinct characteristic values λ_i, λ_j of a Hermitian C-matrix, then $X_j^* X_i = 0$.*

The proof parallels that given for Theorem 4 and is left as Exercise 16.

It follows from Theorems 4 and 5 that any pair of characteristic vectors corresponding to distinct characteristic values of a symmetric R-matrix or of a Hermitian C-matrix are orthogonal with respect to the standard inner product.

Example 1. The symmetric R-matrix

$$A = \begin{bmatrix} 2 & \sqrt{6} \\ \sqrt{6} & 1 \end{bmatrix}$$

has real characteristic values $\lambda_1 = -1$, $\lambda_2 = 4$, and corresponding characteristic vectors $X_1 = \begin{bmatrix} \sqrt{6} \\ -3 \end{bmatrix}$, $X_2 = \begin{bmatrix} \sqrt{6} \\ 2 \end{bmatrix}$. It can be verified easily that X_1 and X_2 are orthogonal (that is, their standard inner product is zero). Each of these characteristic vectors can be normalized by dividing each component by the magnitude of the vector. Thus an orthonormal set of characteristic vectors is

$$\begin{bmatrix} \sqrt{6}/\sqrt{15} \\ -3/\sqrt{15} \end{bmatrix} \quad \text{and} \quad \begin{bmatrix} \sqrt{6}/\sqrt{10} \\ 2/\sqrt{10} \end{bmatrix}.$$

Pairs of characteristic vectors of nonsymmetric R-matrices or of non-Hermitian C-matrices need not be orthogonal, as Exercises 12 and 13 illustrate.

Definition 2. *Let A be an n by n matrix over a field F with n characteristic vectors $X_1, X_2, \ldots, X_n$ corresponding to the n characteristic values $\lambda_1, \lambda_2, \ldots, \lambda_n$, respectively. The matrix*

$$P = [X_1 \mid X_2 \mid \cdots \mid X_n]$$

is called a modal matrix of A. (The characteristic values of A may or may not be distinct.)

Example 2. In Example 1, $X_1 = \begin{bmatrix} \sqrt{6} \\ -3 \end{bmatrix}$ and $X_2 = \begin{bmatrix} \sqrt{6} \\ 2 \end{bmatrix}$ were a pair of characteristic vectors. According to Definition 2 the matrix

$$P = [X_1 \mid X_2] = \begin{bmatrix} \sqrt{6} & \sqrt{6} \\ -3 & 2 \end{bmatrix}$$

is a modal matrix of A.

Notice that a modal matrix of A is not unique. Also notice that an n by n matrix A has an *invertible* modal matrix if and only if A has a set of n linearly independent characteristic vectors.

We now recall from Theorem 13 of Chapter 4 that an R-matrix is an orthogonal matrix (that is, $A^T = A^{-1}$) if and only if its columns form an orthonormal set. Using this fact and also Theorem 4 of the present chapter, we can justify the following theorem.

Theorem 6. *If a modal matrix P of an n by n symmetric R-matrix, with n distinct characteristic values, has normalized columns, then P is an orthogonal matrix.*

The proof is left as Exercise 17.

Using Theorem 15 of Chapter 4 and Theorem 5 of this chapter we have the following generalization of Theorem 6.

Theorem 7. *If a modal matrix P of an n by n Hermitian C-matrix with n distinct characteristic values has normalized columns, then P is a unitary matrix.*

The proof is left as Exercise 18.

Further insight into the relationships among linear transformations, characteristic values, characteristic vectors, and modal matrices can be obtained by considering the special situation where a set of characteristic vectors $b = \{\beta_1, \ldots, \beta_n\}$ of a given linear operator T are linearly independent and hence can be chosen as the basis of the vector space over which T is defined. We then have

$$\begin{cases} T(\beta_1) = \lambda_1 \beta_1, \\ T(\beta_2) = \lambda_2 \beta_2, \\ \vdots \\ T(\beta_n) = \lambda_n \beta_n. \end{cases}$$

In Chapter 8 we learned that the matrix representation of T with respect to this basis is therefore

$$[T]_b = \begin{bmatrix} \lambda_1 & & 0 \\ & \ddots & \\ 0 & & \lambda_n \end{bmatrix}.$$

Conversely, we observe that if a matrix representation of a linear operator has been diagonalized, then the diagonal entries are the characteristic values and the basis is a set of characteristic vectors of the operator. Since it is often desirable to have a diagonal transformation matrix, one of the fundamental problems of linear algebra is, if possible, to find a transition matrix that will change the basis to one consisting of the characteristic vectors. In Section 10.2 we shall prove that when such a transition matrix exists it will be a special modal matrix.

APPLICATIONS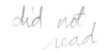

Example 3. Some of the concepts mentioned in this section are useful in the study of various stability problems in economics. One problem is to consider what is called the *stability* of a system of difference equations, such as

$$\begin{cases} x_n = a_{11}x_{n-1} + a_{12}y_{n-1}, \\ y_n = a_{21}x_{n-1} + a_{22}y_{n-1}. \end{cases}$$

Perhaps x_n and y_n represent changes in certain economic indices after n intervals of time, and the question is whether or not these changes are finite as n gets arbitrarily large. We rewrite the system using matrix notation,

$$\begin{bmatrix} x_n \\ y_n \end{bmatrix} = A \begin{bmatrix} x_{n-1} \\ y_{n-1} \end{bmatrix}, \quad \text{or} \quad X_n = AX_{n-1},$$

and, as we found in Example 11 of Section 9.1, the latter statement can be written as

$$X_n = A^{n-1}X_1.$$

Now, if a similarity transformation matrix P exists such that $P^{-1}AP = D$ is a diagonal matrix, then

$$A = PDP^{-1},$$
$$A^2 = (PDP^{-1})(PDP^{-1}) = PD^2P^{-1},$$
$$\vdots$$
$$A^{n-1} = PD^{n-1}P^{-1}.$$

Therefore,

$$X_n = A^{n-1}X_1 = (PD^{n-1}P^{-1})X_1$$
$$= P \begin{bmatrix} \lambda_1^{n-1} & 0 \\ 0 & \lambda_2^{n-1} \end{bmatrix} P^{-1}X_1,$$

where the λ_i are the characteristic values of A, and P is a modal matrix of A. The matrix X_1 represents the initial conditions. (The last equation offers an alternative method of solving Example 11 of Section 9.1.) From the last equation we can see that the solutions are of the form

$$\begin{cases} x_n = c_1 \lambda_1^{n-1} + c_2 \lambda_2^{n-1}, \\ y_n = c_3 \lambda_1^{n-1} + c_4 \lambda_2^{n-1}, \end{cases}$$

where the c_i are known constants. If $|\lambda_i| < 1$, then x_n and y_n must converge to zero (achieve stability) as $n \to \infty$.

Example 4. Characteristic values and vectors are also useful in an extension of Example 9 of Section 9.1. Let X_0 represent the original population, then the population after one time interval is

$$X_1 = MX_0;$$

after two time intervals

$$X_2 = MX_1 = M(MX_0) = M^2 X_0.$$

Then

$$X_3 = MX_2 = M(M^2 X_0) = M^3 X_0.$$

Eventually after k time intervals we have

$$X_k = M^k X_0,$$

which allows a projection of population after k intervals of time based on the initial population. As explained in Example 3, one way of calculating powers of matrices that can be diagonalized is given by the formula

$$M^k = P \begin{bmatrix} \lambda_1^k & \cdot & \cdot & \cdot & 0 \\ \cdot & & & & \cdot \\ \cdot & & & & \cdot \\ \cdot & & & & \cdot \\ 0 & \cdot & \cdot & \cdot & \lambda_n^k \end{bmatrix} P^{-1},$$

where P is a modal matrix of M.

★**Example 5.** The transition matrix P in the very important application given in Examples 3 and 4 of Section 8.5 is a *modal matrix*; a justification of this assertion will be given in Section 10.2.

★**Example 6.** In control systems theory, it can be shown that a linear time-invariant system described by the C-matrix equation $\dot{X} = AX + uB$ is stable (that is, $X \to 0$ as $t \to \infty$) if and only if the n eigenvalues of A have negative real parts. See Example 4 of Section 8.5 for a concrete illustration of such a system. A general discussion of the stability of a linear time-invariant system may be found on pages 71–75 of [8] (Elgerd).

9.2 THEOREMS

★**Example 7.** For a simple treatment of transients in an electrical circuit see [23] (Noble), pages 102–110. The treatment uses matrix methods to solve a typical system of linear differential equations that arise from either a certain electrical circuit or mechanical system. The uses of characteristic values, characteristic vectors, basis, and change of basis are demonstrated.

EXERCISES

1. Over C, find a matrix with real entries whose characteristic values are not real.
2. Over C, find a matrix with at least one nonreal entry whose characteristic values are not real.
3. Find a symmetric C-matrix whose characteristic values are not real.
4. Is it possible to find a Hermitian C-matrix whose characteristic values are not real?
5. Calculate the discriminant of the quadratic characteristic equation of the R-matrix $\begin{bmatrix} a & b \\ c & d \end{bmatrix}$. Show that real characteristic values exist that are unequal if b and c have the same sign (that is, $bc > 0$). Under what condition are the characteristic values equal?
6. Use Theorem 4 to say what you can about the orthogonality of pairs of characteristic vectors of the R-matrices of Exercises 4, 6, and 8 of Section 9.1. Use the answers of those exercises to verify your results.
7. Use Theorem 4 to say what you can about the orthogonality of pairs of characteristic vectors of the R-matrices of Exercises 3, 5, and 7 of Section 9.1. Use the answers of those exercises to verify your results.
8. Using the answers to Exercises 3, 5, and 7 of Section 9.1, construct a modal matrix of each of the given R-matrices in those exercises. Which, if any, of the modal matrices are orthogonal according to Theorem 6?
9. Using the answers to Exercises 9 and 11 of Section 9.1, construct a modal matrix of each of the given R-matrices in those exercises.
10. Using the answers to Exercises 13 and 15 of Section 9.1, construct a modal matrix of each of the given R-matrices in those exercises.
11. Determine a modal matrix M of
$$A = \begin{bmatrix} 3 & 1 & 1 \\ 0 & 2 & 1 \\ 0 & 0 & 4 \end{bmatrix}.$$
After normalizing the columns of M, determine whether the resulting modal matrix is orthogonal.
12. Verify that the characteristic vectors of the nonsymmetric R-matrix $A = \begin{bmatrix} 1 & 2 \\ 4 & 3 \end{bmatrix}$ are not orthogonal. Write a modal matrix of A.

13. Verify that the characteristic vectors of the non-Hermitian C-matrix
$$A = \begin{bmatrix} 3 & -2i \\ 4i & 1 \end{bmatrix}$$ are not orthogonal. Write a modal matrix of A.

14. Show that the C-matrix $A = \begin{bmatrix} 1 & i \\ -i & 1 \end{bmatrix}$ is Hermitian. Find a modal matrix of A after having normalized columns, and verify that this modal matrix is a unitary matrix as Theorem 7 predicts.

15. Prove Theorem 3. *Hint:* The proof is very similar to the proof of Theorem 2.
16. Prove Theorem 5. *Hint:* The proof is very similar to the proof of Theorem 4.
17. Prove Theorem 6.
18. Prove Theorem 7.
19. Determine whether or not the following system has stability. Give reasons for your answer.
$$\begin{cases} x_n = x_{n-1} - \tfrac{1}{6} y_{n-1}, \\ y_n = x_{n-1} + \tfrac{1}{6} y_{n-1}. \end{cases}$$

Hint: See Example 3.

9.3 Diagonalization of Real Quadratic Forms[3]

Consider the equation of an ellipse,
$$2x^2 - 4xy + 7y^2 = 6,$$
whose graph is shown in Figure 9.3.1. The expression on the left side of this equation involves only 2nd degree terms and is an example of a *real quadratic form*. In general, a **real quadratic form** is an algebraic expression involving only 2nd degree terms with real coefficients and which, if there are n variables, is of the form

$$\begin{aligned}
c_{11} x_1^2 &+ 2c_{12} x_1 x_2 + 2c_{13} x_1 x_3 + \cdots + 2c_{1n} x_1 x_n \\
&+ c_{22} x_2^2 \quad + 2c_{23} x_2 x_3 + \cdots + 2c_{2n} x_2 x_n \\
&\qquad\qquad + c_{33} x_3^2 \quad + \cdots + 2c_{3n} x_3 x_n \\
&\qquad\qquad\qquad\qquad\qquad\qquad \vdots \\
&\qquad\qquad\qquad\qquad\qquad\qquad + c_{nn} x_n^2.
\end{aligned}$$

The quadratic form can be represented by the matrix expression
$$[x_1 x_2 \cdots x_n] \begin{bmatrix} c_{11} & c_{12} & \cdots & c_{1n} \\ c_{12} & c_{22} & \cdots & c_{2n} \\ \vdots & \vdots & & \vdots \\ c_{1n} & c_{2n} & \cdots & c_{nn} \end{bmatrix} \begin{bmatrix} x_1 \\ x_2 \\ \vdots \\ x_n \end{bmatrix},$$

[3] This section should be omitted if Chapter 10 is to be studied.

9.3 DIAGONALIZATION OF REAL QUADRATIC FORMS

or

$$X^T C X.$$

The symmetric R-matrix C is called the *matrix of the real quadratic form*.

Example 1. In the plane, the left side of the equation

$$2x^2 - 4xy + 7y^2 = 6$$

is a real quadratic form, and the graph of the equation is the ellipse shown in Figure 9.3.1. After rewriting the equation as

$$2x^2 + (-2)xy + (-2)xy + 7y^2 = 6,$$

we see that the quadratic equation can be expressed as the matrix equation

$$[x \ y] \begin{bmatrix} 2 & -2 \\ -2 & 7 \end{bmatrix} \begin{bmatrix} x \\ y \end{bmatrix} = [6],$$

or

$$X^T C X = 6.^4$$

The matrix of the form is $\begin{bmatrix} 2 & -2 \\ -2 & 7 \end{bmatrix}$.

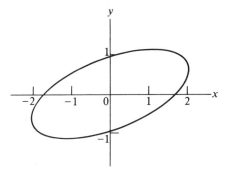

Figure 9.3.1

Let α be an element of an n-dimensional vector space $V(R)$, let the coordinate matrix of α be $X = \begin{bmatrix} x_1 \\ \vdots \\ x_n \end{bmatrix}$ with respect to a given basis, and let C be a symmetric R-matrix; the real quadratic form $q(X) = X^T C X$ can be thought of as a transformation from $V(R)$ into R. The occurrence of quadratic forms is fairly frequent in applied work, and consequently matrix methods for the

---

[4] The scalar 6 is often used here rather than the matrix [6]; this is permissible because the field R of real numbers is isomorphic to the field of 1 by 1 R-matrices, and hence they are algebraically equivalent.

manipulations of real quadratic forms are desirable. We will be particularly concerned with diagonalizing the matrix of a quadratic form. Under a change of basis of $V(R)$ accomplished by $X = PU$, the real quadratic form

$$X^T C X$$

becomes

$$(PU)^T C(PU),$$

or

$$U^T(P^T C P)U.$$

Hence the matrix of the real quadratic form has been transformed to the matrix

$$P^T C P$$

by a change in basis of the vector space; such a transformation of matrices is an example of a *congruence transformation*.

Definition 3. *Let A and B be n by n matrices over a field F. Matrix A is said to be **congruent** to matrix B if there exists an invertible matrix P over F such that $B = P^T A P$.*

Example 2. The matrix $\begin{bmatrix} 1 & 4 \\ 2 & 5 \end{bmatrix}$ is congruent to $\begin{bmatrix} 1 & 2 \\ 0 & -3 \end{bmatrix}$ because there exists an invertible matrix $P = \begin{bmatrix} 1 & 2 \\ 0 & 1 \end{bmatrix}$ such that

$$\begin{bmatrix} 1 & 2 \\ 0 & -3 \end{bmatrix} = P^T \begin{bmatrix} 1 & 4 \\ 2 & 5 \end{bmatrix} P.$$

For the symmetric matrix C of a real quadratic form, the following definition will be particularly significant.

Definition 4. *An n by n R-matrix A is said to be **orthogonally congruent** (or **orthogonally similar**) to an R-matrix B if there exists an orthogonal R-matrix P such that $B = P^T A P = P^{-1} A P$.*

Example 3. The matrix $\begin{bmatrix} 1 & 2 \\ 2 & 1 \end{bmatrix}$ is orthogonally congruent to $\begin{bmatrix} -1 & 0 \\ 0 & 3 \end{bmatrix}$ because there exists an orthogonal matrix

$$P = \begin{bmatrix} 1/\sqrt{2} & 1/\sqrt{2} \\ -1/\sqrt{2} & 1/\sqrt{2} \end{bmatrix} \text{ such that } \begin{bmatrix} -1 & 0 \\ 0 & 3 \end{bmatrix} = P^T \begin{bmatrix} 1 & 2 \\ 2 & 1 \end{bmatrix} P = P^{-1} \begin{bmatrix} 1 & 2 \\ 2 & 1 \end{bmatrix} P.$$

The following theorem points the way toward diagonalizing the matrix of a real quadratic form, which then enables us to express the quadratic form in terms involving squared variables only.

9.3 DIAGONALIZATION OF REAL QUADRATIC FORMS

Theorem 8. *If A is an n by n symmetric R-matrix with n distinct characteristic values λ_i and corresponding normalized characteristic vectors X_i, then there exists a modal matrix P of A where*

$$P = [X_1 \mid X_2 \mid \cdots \mid X_n]$$

such that

$$P^T A P = \begin{bmatrix} \lambda_1 & 0 & \cdots & 0 \\ 0 & \lambda_2 & \cdots & 0 \\ \vdots & \vdots & \ddots & \vdots \\ 0 & 0 & \cdots & \lambda_n \end{bmatrix}.$$

Proof. (Most of the reasons are left as Exercise 15.)

$$\begin{aligned}
AP &= A[X_1 \mid X_2 \mid \cdots \mid X_n] \\
&= [AX_1 \mid AX_2 \mid \cdots \mid AX_n] \\
&= [\lambda_1 X_1 \mid \lambda_2 X_2 \mid \cdots \mid \lambda_n X_n] \\
&= [X_1 \lambda_1 \mid X_2 \lambda_2 \mid \cdots \mid X_n \lambda_n] \\
&= [X_1 \mid X_2 \mid \cdots \mid X_n] \begin{bmatrix} \lambda_1 & 0 & \cdots & 0 \\ 0 & \lambda_2 & \cdots & 0 \\ \vdots & \vdots & \ddots & \vdots \\ 0 & 0 & \cdots & \lambda_n \end{bmatrix} \\
&= P \begin{bmatrix} \lambda_1 & 0 & \cdots & 0 \\ 0 & \lambda_2 & \cdots & 0 \\ \vdots & \vdots & \ddots & \vdots \\ 0 & 0 & \cdots & \lambda_n \end{bmatrix}.
\end{aligned}$$

By Theorem 6, P is an orthogonal matrix; hence

$$P^{-1} A P = P^T A P = \begin{bmatrix} \lambda_1 & 0 & \cdots & 0 \\ 0 & \lambda_2 & \cdots & 0 \\ \vdots & \vdots & \ddots & \vdots \\ 0 & 0 & \cdots & \lambda_n \end{bmatrix}. \quad \square$$

Example 4. Consider the equation of a certain surface:

$$g(x, y, z) = 5x^2 + 3y^2 + 3z^2 - 2xy + 2yz - 2xz = 1,$$

or

$$[x \; y \; z] \begin{bmatrix} 5 & -1 & -1 \\ -1 & 3 & 1 \\ -1 & 1 & 3 \end{bmatrix} \begin{bmatrix} x \\ y \\ z \end{bmatrix} = 1,$$

or

$$g(x, y, z) = X^T A X = 1.$$

The characteristic values of the matrix of the real quadratic form are found to be 2, 3, and 6, with the corresponding normalized characteristic vectors

$$X_1 = \begin{bmatrix} 0 \\ 1/\sqrt{2} \\ -1/\sqrt{2} \end{bmatrix}, \quad X_2 = \begin{bmatrix} 1/\sqrt{3} \\ 1/\sqrt{3} \\ 1/\sqrt{3} \end{bmatrix}, \quad X_3 = \begin{bmatrix} -2/\sqrt{6} \\ 1/\sqrt{6} \\ 1/\sqrt{6} \end{bmatrix}.$$

Thus, under the transition matrix

$$P = \begin{bmatrix} 0 & 1/\sqrt{3} & -2/\sqrt{6} \\ 1/\sqrt{2} & 1/\sqrt{3} & 1/\sqrt{6} \\ -1/\sqrt{2} & 1/\sqrt{3} & 1/\sqrt{6} \end{bmatrix},$$

$$P^T A P = \begin{bmatrix} 2 & 0 & 0 \\ 0 & 3 & 0 \\ 0 & 0 & 6 \end{bmatrix}.$$

Under the transition $X = PU$, where $U = [u \ v \ w]^T$,

$$X^T A X = 1$$

becomes

$$(PU)^T A (PU) = 1,$$

or

$$U^T (P^T A P) U = 1,$$

which is

$$2u^2 + 3v^2 + 6w^2 = 1.$$

The reader should note that neither P nor the diagonalized matrix is unique, because the characteristic values may be listed in any order. (However, for a rigid rotation, det $P = +1$.) The reader should also note that the hypothesis of Theorem 8 required that the matrix be symmetric and that the characteristic values be distinct. In Chapter 10 it is learned that the requirement of distinct characteristic values can be removed and the conclusion of Theorem 8 will follow. The proof, however, is considerably more involved than it is in the special case. The next example will illustrate the computational procedure when the characteristic values are *not* all distinct.

Example 5. For the quadratic form

$$f(x, y, z) = X^T A X = X^T \begin{bmatrix} 2 & 0 & 1 \\ 0 & 3 & 0 \\ 1 & 0 & 2 \end{bmatrix} X,$$

the characteristic values of A are $\lambda_1 = 1$, $\lambda_2 = \lambda_3 = 3$. A normalized characteristic vector corresponding to $\lambda_1 = 1$ is found to be $X_1 = \begin{bmatrix} 1/\sqrt{2} \\ 0 \\ -1/\sqrt{2} \end{bmatrix}$. The characteristic

9.3 DIAGONALIZATION OF REAL QUADRATIC FORMS

values $\lambda_2 = \lambda_3 = 3$ yield

$$(A - 3I_3)X = 0 \quad \text{or} \quad \begin{bmatrix} -1 & 0 & 1 \\ 0 & 0 & 0 \\ 1 & 0 & -1 \end{bmatrix} \begin{bmatrix} x \\ y \\ z \end{bmatrix} = \begin{bmatrix} 0 \\ 0 \\ 0 \end{bmatrix},$$

whose complete solution is

$$\begin{cases} x = z, \\ y \text{ arbitrary}. \end{cases}$$

Therefore we must choose two vectors subject to these conditions such that they are orthogonal to X_1 and to each other. The vectors $\begin{bmatrix} 1 \\ 1 \\ 1 \end{bmatrix}$ and $\begin{bmatrix} 1 \\ -2 \\ 1 \end{bmatrix}$ suffice; notice that they are not unique. Normalized, these vectors become

$$X_2 = \begin{bmatrix} 1/\sqrt{3} \\ 1/\sqrt{3} \\ 1/\sqrt{3} \end{bmatrix} \quad \text{and} \quad X_3 = \begin{bmatrix} 1/\sqrt{6} \\ -2/\sqrt{6} \\ 1/\sqrt{6} \end{bmatrix}.$$

Therefore, $P = \begin{bmatrix} 1/\sqrt{2} & 1/\sqrt{3} & 1/\sqrt{6} \\ 0 & 1/\sqrt{3} & -2/\sqrt{6} \\ -1/\sqrt{2} & 1/\sqrt{3} & 1/\sqrt{6} \end{bmatrix}$ and $P^T A P = \begin{bmatrix} 1 & 0 & 0 \\ 0 & 3 & 0 \\ 0 & 0 & 3 \end{bmatrix}$.

So far in this section we have been concerned primarily with R-matrices because of our interest in real quadratic forms. It may be of further interest, however, to state the following extensions of Definition 4 and Theorem 8 for C-matrices.

Definition 5. *An n by n C-matrix A is said to be **unitarily congruent** (or **unitarily similar**) to a matrix B if there exists a unitary C-matrix P such that $B = P^* A P = P^{-1} A P$.*

Theorem 9. *If A is an n by n Hermitian C-matrix with n distinct characteristic values λ_i and corresponding normalized characteristic vectors X_i, then there exists a modal matrix P of A where*

$$P = [X_1 \mid X_2 \mid \cdots \mid X_n]$$

such that

$$P^* A P = \begin{bmatrix} \lambda_1 & 0 & \cdots & 0 \\ 0 & \lambda_2 & \cdots & 0 \\ \vdots & \vdots & \ddots & \vdots \\ 0 & 0 & \cdots & \lambda_n \end{bmatrix}.$$

The proof of this theorem parallels the proof of Theorem 8 and is left as Exercise 16.

In the hypothesis of Theorem 9, the requirement that the characteristic values of A be distinct can be removed and the conclusion will still follow; the proof, however, is considerably more involved.

APPLICATIONS

Applications making use of the material covered in this section are numerous, but many of them are quite sophisticated; the textbook *Matrix Methods for Engineering* by L. A. Pipes [29] offers scores of illustrations of this statement. Applications of this material are not limited to engineering, however.

Example 6. Consider the symmetric correlation matrix in Example 9 of Section 1.2; such matrices are of considerable importance in the branch of statistics known as factor analysis. One of the primary problems of factor analysis involves the diagonalization of matrices of correlation.[5] Although psychologists did much of the original work in this area, there is considerable potential for its usefulness in other areas of the social sciences such as sociology and political science.[6]

Example 7. Consider a rigid body in motion with respect to an X-coordinate system, and let a moving coordinate system Y be fixed in the body with the origin at the center of mass. We may regard the rigid body with mass m as a set of particles whose relative distances from each other are constant. The symmetric matrix

$$J_y = \begin{bmatrix} m(y_2^2 + y_3^2) & -my_1 y_2 & -my_1 y_3 \\ -my_1 y_2 & m(y_1^2 + y_3^2) & -my_2 y_3 \\ -my_1 y_3 & -my_2 y_3 & m(y_1^2 + y_2^2) \end{bmatrix}$$

is called the *inertia matrix* of the rigid body. The entries on the main diagonal represent the moments of inertia with respect to the coordinate axes y_1, y_2, and y_3. The other entries represent the products of inertia of the body. It is desirable to introduce a new set of axes z_1, z_2, and z_3 by a rotation transformation such that the products of inertia vanish. The transformation matrix R (such that $Y = RZ$) has as its columns the normalized characteristic vectors of J_y. The resulting inertia matrix with respect to the Z-coordinate system is

$$J_z = \begin{bmatrix} m(z_2^2 + z_3^2) & 0 & 0 \\ 0 & m(z_1^2 + z_3^2) & 0 \\ 0 & 0 & m(z_1^2 + z_2^2) \end{bmatrix}.$$

[5] L. L. Thurstone, *Multiple Factor Analysis*, University of Chicago Press, Chicago, 1947, Chap. 20. A later reference is D. N. Lawley and A. E. Maxwell, *Factor Analysis as a Statistical Method*, Butterworths, London, 1963.

[6] For instance, see H. Alker, Jr., "Dimensions of Conflict in the General Assembly," *American Political Science Review*, Vol. 58 (1964), pp. 642–657.

9.3 DIAGONALIZATION OF REAL QUADRATIC FORMS

The entries on the main diagonal are the characteristic values of J_y and are called the principal moments of inertia. The axes z_1, z_2, and z_3 are called the principal axes of inertia.

Example 8. Chapter 12 of [23] (Noble) entitled "The Structure and Analysis of Linear Chemical Reaction Systems" illustrates the application of many of the concepts discussed in this book, and particularly, in this section. The chapter is based on a paper by J. Wei and C. D. Prater, The Structure and Analysis of Complex Reaction Systems, *Advances in Catalysis*, Vol. 13, Academic Press, New York, 1962, pages 203–392. A summary article by the same authors is a New Approach to First-Order Chemical Reaction Systems, *Am. Inst. Chem. Eng., J.* 9, pages 17–81 (1963). A perusal of any of this material should convince the reader of the relevancy of linear algebra to chemical research.

EXERCISES

In each of Exercises 1–9, eliminate the product terms (that is, the terms involving xy, xz, or yz) by finding and using an orthogonal transition matrix P.

1. $2x^2 - 4xy + 5y^2 = 6$. Use answer to Exercise 3 of Section 9.1.
2. $2x^2 + 12xy - 3y^2 = 4$. Use answer to Exercise 13 of Section 9.1.
3. $x^2 + 4xy + y^2 = 1$.
4. $x^2 - 4xy - 2y^2 = 9$.
5. $2x^2 + 2xy + y^2 + z^2 + 2xz = 9$. Use answer to Exercise 7 of Section 9.1.
6. $y^2 + 4xz = 8$. Use answer to Exercise 11 of Section 9.1.
7. $3x^2 + 2xy + 3y^2 + 2z^2 = 4$.
8. $x^2 + 4y^2 + 3z^2 - 4xy = 5$.
9. $xy = 4$.
10. A primary problem in analytic geometry is to rotate the axes so that the equation of a given conic contains no terms that include the product of two different variables such as xy or yz or xz. Demonstrate that this is what was accomplished in Exercise 9, by graphing the given equation and the resulting equation in the answer.
11. Find a transition matrix P such that $P^T A P$ is equal to a diagonal matrix, and then find that diagonal matrix.

$$A = \begin{bmatrix} 2 & -1 & 1 \\ -1 & 2 & -1 \\ 1 & -1 & 2 \end{bmatrix}.$$

Hint: See Example 5 of this section.

12. *Conjecture:* An R-matrix A with distinct characteristic values is similar to a diagonal matrix.
    (a) How does this conjecture differ from Theorem 8?
    (b) Prove or disprove this conjecture if we are given that A has an invertible modal matrix P.
13. Prove: If an R-matrix A is orthogonally similar to a diagonal matrix, then A is symmetric.

14. Prove: If a C-matrix A is unitarily similar to a diagonal R-matrix, then A is Hermitian.
15. Give the reasons for the statements in the proof of Theorem 8.
16. Prove Theorem 9.
17. Apply Theorem 8 to Example 1 of this Section. Graph the new quadratic equation and compare with Figure 9.3.1.

NEW VOCABULARY

characteristic value 9.1
characteristic vector 9.1
eigenvalue 9.1
characteristic roots 9.1
latent roots 9.1
eigenvectors 9.1
characteristic equation 9.1
normalized characteristic vector 9.1
characteristic function 9.1
characteristic polynomial 9.1

modal matrix 9.2
quadratic form 9.3
real quadratic form 9.3
matrix of a real quadratic form 9.3
congruent matrices 9.3
orthogonally congruent matrices 9.3
orthogonally similar matrices 9.3
unitarily congruent 9.3
unitarily similar 9.3

SPECIAL PROJECTS

I. An elementary discussion of series of matrices appears in the book by J. T. Schwartz entitled *Matrices and Vectors*, McGraw-Hill, New York, 1961, pages 152–160. A discussion of sequences and series of matrices, along with matrices of functions, the derivative and integral of such matrices, and their application to the study of differential equations, may be found in Chapter 8 of Cullen, C. G., *Matrices and Linear Transformations*, Addison-Wesley, Reading, Mass., 1966; study these topics in the given references or other books,[7] and write a short paper reflecting your study.

II. The so-called lambda matrices are discussed in many books on matrix theory or linear algebra. Two such discussions of an elementary nature may be found in Chapter 4 of Eves, Howard, *Elementary Matrix Theory*, Allyn & Bacon, Boston, 1966, and in Chapter 23 of Ayres, Frank Jr., *Matrices*, Schaum Publishing Co., New York, 1962. After studying this subject in one or more books, write a short expository paper on lambda matrices.

---

[7] Frazer, R. A., Duncan, W. J., Collar, A. R., *Elementary Matrices*, 1950, Chapter 2. Pipes, L. A., *Matrix Methods for Engineering*, Prentice-Hall, Englewood Cliffs, N. J., 1963, Chapter 4.

10

TRANSFORMATIONS OF MATRICES (optional)

10.1 Introduction

So far in this book we have seen that matrices can be used for a variety of purposes. For example, a system of linear algebraic equations may be represented by a unique augmented matrix, and, as we studied more recently, a matrix offers a unique representation of a linear transformation from one vector space to another. In both of these examples *a fundamental problem of linear algebra is to transform a matrix representation to a desired form.* The purpose of this chapter is to introduce some of the transformations of matrices that are commonly used and some of their applications. In other words we shall examine a few transformations defined over a vector space of matrices; these transformations are listed in Table 10.1. When a matrix A has been transformed to B by the indicated transformation, we say that A is related to B according to the relationship indicated in Table 10.1; it can be proved[1] that *each transformation is linear* and that *each relation is an equivalence relation.*

The first of these transformations, defined by $A \stackrel{T}{\mapsto} PA$ where P is invertible, originally appeared in this book when the augmented matrix A of

---

[1] This is accomplished in exercises throughout this book.

a system of linear equations was transformed by a succession of elementary row operations into a reduced echelon matrix, PA. A reduced echelon matrix was desired because from it a solution of the corresponding system could be obtained easily. This transformation was discussed in Sections 2.2 and 2.3.

The second transformation, defined by $A \xmapsto{T} PAQ$ where P and Q are invertible, was first used in Section 3.4 to effect a succession of elementary row and column operations on A. In Section 8.7 the transformation was in essence used to represent a change in the matrix representation of a linear transformation from one vector space to another under a change in basis of both vector spaces.

TABLE 10.1

$A \xmapsto{T} PA = B$;	P is invertible.	A is *row equivalent* to B.
$A \xmapsto{T} PAQ = B$;	P and Q are invertible.	A is *equivalent* to B.
$A \xmapsto{T} P^{-1}AP = B$;	P is invertible.	A is *similar* to B.
$A \xmapsto{T} P^T AP = B$;	P is invertible.	A is *congruent* to B.
$A \xmapsto{T} P^T AP = B$;	$P^T = P^{-1}$. A is real.	A is *orthogonally congruent* or *orthogonally similar* to B.
$A \xmapsto{T} P^* AP = B$;	$P^* = P^{-1}$.	A is *unitarily congruent* or *unitarily similar* to B.

As we have shown, the third transformation, called the *similarity transformation* and defined by $A \xmapsto{T} P^{-1}AP$ where P is invertible, arises naturally from the alteration of a matrix representation of a linear operator over a single vector space due to a change in basis of that vector space. It was pointed out in Section 8.5 that similar matrices can be thought of as different matrix representations of the same linear operator with respect to different bases of the given vector space. Properties of this transformation will be pursued further in the next section.

The remainder of the transformations of Table 10.1 will be introduced in later sections of this chapter.

As we study these particular linear transformations of matrices, it is interesting and useful to observe that certain properties of a given matrix will remain unchanged (or are *invariant*) under a certain transformation. For example we proved in Theorem 11 of Chapter 5 that the ranks of equivalent C-matrices are the same, and, since all of the other relations listed in Table 10.1 are special cases of the relation of matrix equivalence, then the rank of a matrix remains unchanged or invariant under any of the transformations listed. Other invariants will be pointed out throughout this chapter.

10.2 Similarity

As we have stated previously, similarity is a special case of matrix equivalence and, as shown in Section 8.5, arises naturally from matrix representations of one linear operator with respect to different bases. In this section we first wish to establish a few invariants under the similarity transformation; we shall then determine conditions under which a matrix can be diagonalized.

Theorem 1. *If A is similar to B, then*
  (i) *A and B have the same characteristic values;*
  (ii) tr A = tr B;
  (iii) det A = det B.

The proof is left as Exercises 14, 15, and 16.

In applied work it is often very helpful to be able to diagonalize a matrix under a similarity transformation, but the fact is that this cannot always be done; hence we should state the conditions that will determine whether or not a matrix is similar to a diagonal matrix.

Theorem 2. *An n by n matrix A over a field F is similar to a diagonal matrix D if and only if there exists an invertible modal matrix of A. Moreover, the diagonal entries of D are the characteristic values of A.*

Proof. PART I. Assume that there exists an invertible modal matrix M of A. (The reasons are left for Exercise 17.)

$$\begin{aligned}
AM &= A[M_1 \mid \cdots \mid M_n] \\
&= [AM_1 \mid \cdots \mid AM_n] \\
&= [\lambda_1 M_1 \mid \cdots \mid \lambda_n M_n] \\
&= [M_1 \lambda_1 \mid \cdots \mid M_n \lambda_n] \\
&= MD, \text{ where } D = \begin{bmatrix} \lambda_1 & & 0 \\ & \ddots & \\ 0 & & \lambda_n \end{bmatrix}.
\end{aligned}$$

Therefore, $A = MDM^{-1}$, which implies that A is similar to D.

PART II. Assume that A is similar to a diagonal matrix D. There exists an invertible matrix P such that $P^{-1}AP = D$, hence

$$AP = PD,$$

or

$$[AP_1 \mid \cdots \mid AP_n] = [P_1 \lambda_1 \mid \cdots \mid P_n \lambda_n].$$

Therefore
$$AP_1 = \lambda_1 P_1, \quad \ldots, \quad AP_n = \lambda_n P_n,$$
and consequently $P_1, \ldots, P_n$ are characteristic vectors. Matrix P must be a modal matrix. ∎

Example 1. The R-matrix $A = \begin{bmatrix} 2 & 0 & 0 \\ 0 & 2 & 2 \\ 0 & 0 & 4 \end{bmatrix}$ has the characteristic values $\lambda_1 = 2$, $\lambda_2 = 2$, and $\lambda_3 = 4$ with corresponding characteristic vectors

$$\begin{bmatrix} a_1 \\ a_2 \\ 0 \end{bmatrix}, \quad \begin{bmatrix} b_1 \\ b_2 \\ 0 \end{bmatrix}, \quad \text{and} \quad \begin{bmatrix} 0 \\ c \\ c \end{bmatrix},$$

where a_1, a_2, b_1, b_2, and c are arbitrary real numbers. If these arbitrary numbers are chosen so that $c \neq 0$ and $a_1 b_2 - a_2 b_1 \neq 0$, then the modal matrix $P = \begin{bmatrix} a_1 & b_1 & 0 \\ a_2 & b_2 & c \\ 0 & 0 & c \end{bmatrix}$ is invertible and A is similar to a diagonal matrix. On the other hand, the matrix

$$B = \begin{bmatrix} 2 & 2 & 0 \\ 0 & 2 & 0 \\ 0 & 0 & 4 \end{bmatrix}$$

has the same characteristic values $\lambda_1 = 2$, $\lambda_2 = 2$, and $\lambda_3 = 4$ with corresponding characteristic vectors

$$\begin{bmatrix} a \\ 0 \\ 0 \end{bmatrix}, \quad \begin{bmatrix} b \\ 0 \\ 0 \end{bmatrix}, \quad \text{and} \quad \begin{bmatrix} 0 \\ 0 \\ c \end{bmatrix},$$

where a, b, and c are arbitrary real numbers. However, no matter what values are assigned to a, b, and c, the modal matrix

$$P = \begin{bmatrix} a & b & 0 \\ 0 & 0 & 0 \\ 0 & 0 & c \end{bmatrix}$$

is not invertible, and hence B cannot be similar to a diagonal matrix.

Since a square matrix over F is invertible if and only if its columns are linearly independent, and since n coordinate vectors span $V_n(F)$ if and only if they are linearly independent (by Exercises 20 and 21 of Section 7.1), we have the following Corollaries to Theorem 2.

Corollary. *An n by n matrix A over F is similar to a diagonal matrix D if and only if A has n linearly independent characteristic vectors.*

10.2 SIMILARITY

Corollary. *An n by n matrix A over F is similar to a diagonal matrix D if and only if the characteristic vectors of A span the vector space of all n by 1 matrices over F.*

From Chapter 8, recall that a matrix representation of a linear operator T over a vector space $V(F)$ is diagonal if and only if there exists a basis $e = \{\varepsilon_1, \ldots, \varepsilon_n\}$ such that

(10.2.1)
$$\begin{cases} T(\varepsilon_1) = t_{11}\varepsilon_1, \\ T(\varepsilon_2) = \phantom{t_{11}\varepsilon_1,}t_{22}\varepsilon_2, \\ \vdots \\ T(\varepsilon_n) = \phantom{t_{11}\varepsilon_1,\ t_{22}\varepsilon_2,}t_{nn}\varepsilon_n. \end{cases}$$

Notice, then, that in (10.2.1) each ε_i is a characteristic vector and each t_{ii} is a characteristic value. In other words, a linear operator T has a diagonal matrix representation if and only if there exists a set of n characteristic vectors that will serve as a basis of $V(F)$. To find such a set, if one exists, becomes a matter of finding a transition matrix which will change a given basis b to $e = \{\varepsilon_1, \ldots, \varepsilon_n\}$. In the proof of Theorem 2 it is apparent that such a transition matrix, if one exists, will be an invertible modal matrix of $[T]_b$.

Example 2. For the vector space $V_2(R)$, let $T: V_2(R) \to V_2(R)$ be defined by $(x, y) \overset{T}{\mapsto} (2y, 2x)$, and let the initial basis be $b = \{(1, 0), (0, 1)\}$. By the methods of Chapter 8 we find

$$[T]_b = \begin{bmatrix} 0 & 2 \\ 2 & 0 \end{bmatrix}.$$

Now, if possible, we change the basis to some $e = \{\varepsilon_1, \varepsilon_2\}$ so that $[T]_e$ is a diagonal matrix. According to Theorem 2 we seek an invertible modal matrix P of $\begin{bmatrix} 0 & 2 \\ 2 & 0 \end{bmatrix}$ such that $P^{-1}[T]_b P = \begin{bmatrix} \lambda_1 & 0 \\ 0 & \lambda_2 \end{bmatrix}$, where λ_1 and λ_2 are characteristic values of $[T]_b$. For the matrix $\begin{bmatrix} 0 & 2 \\ 2 & 0 \end{bmatrix}$ we find $\lambda_1 = 2$ and $\lambda_2 = -2$ and $X_1 = \begin{bmatrix} 1 \\ 1 \end{bmatrix}$ and $X_2 = \begin{bmatrix} 1 \\ -1 \end{bmatrix}$. Since $[X_1 \vdots X_2]$ is invertible, then the modal matrix

$$P = [X_1 \vdots X_2] = \begin{bmatrix} 1 & 1 \\ 1 & -1 \end{bmatrix}$$

is a transition matrix that will change the basis to $e = \{(1, 1), (1, -1)\}$, so that the matrix representation of T will be the diagonal matrix

$$[T]_e = \begin{bmatrix} 2 & 0 \\ 0 & -2 \end{bmatrix}.$$

The reader may verify that $P^{-1}[T]_b P = [T]_e$, that is, that

$$\begin{bmatrix} 1 & 1 \\ 1 & -1 \end{bmatrix}^{-1} \begin{bmatrix} 0 & 2 \\ 2 & 0 \end{bmatrix} \begin{bmatrix} 1 & 1 \\ 1 & -1 \end{bmatrix} = \begin{bmatrix} 2 & 0 \\ 0 & -2 \end{bmatrix}.$$

Next, we determine certain special conditions under which A must be similar to a diagonal matrix. Note, however, that the converses of the following theorems are not necessarily true.

Theorem 3. *Any n by n R-matrix that is symmetric is similar to a diagonal matrix.*

Theorem 3 is a corollary to Theorem 13 (of this chapter), which will be proved later.

Theorem 4. *An n by n C-matrix that is Hermitian is similar to a diagonal matrix.*

Theorem 4 is a corollary to Theorem 14 (of this chapter), which will be proved later.

Theorem 5. *An n by n C-matrix with distinct characteristic values is similar to a diagonal matrix.*

The proof is left as Exercise 18. *Hint:* See proof of Theorem 8 of Chapter 9.

As we have seen, not every matrix can be transformed to a diagonal matrix by means of the similarity transformation; we can, however, always transform a given matrix with complex entries to a triangular matrix, as the next theorem guarantees.

Theorem 6. *Any n by n C-matrix A is similar to an upper triangular matrix whose diagonal entries are the characteristic values of A.*

See Theorem A.10.6, on page A5, for the proof.

There are other useful forms into which a transformation matrix can be transformed by means of a similarity transformation. These forms, such as the Jordan canonical form and the rational canonical form, may be studied in more advanced books.

APPLICATIONS

Example 3. Applications that require integral powers of matrices are numerous. For instance, any application that calls for the solution of the matrix equation $X_k = AX_{k-1}$, where X is a column vector of entries determined by the discrete variable k, may make use of A^k since

$$X_k = AX_{k-1} = A(AX_{k-2}) = \cdots = A^k X_0 \quad (\text{for } k = 1, 2, \ldots).$$

(See Example 11 of Section 9.1 or Example 4 of Section 9.2.) Also, any application that makes use of a stochastic matrix P may make use of P^k, since the ijth entry of P^k represents the probability of transition from state i to state j after k successive stages of transition. (See Examples 10 and 11 of Section 3.1.) If A is similar to a diagonal matrix D, then there exists an invertible matrix P such that $P^{-1}AP = D$; hence

$$A = PDP^{-1},$$
$$A^2 = (PDP^{-1})(PDP^{-1}) = PD^2P^{-1},$$
$$\vdots$$
$$A^k = PD^kP^{-1}.$$

As an illustration, let $A = \begin{bmatrix} 1 & 2 \\ -1 & 4 \end{bmatrix}$. An invertible modal matrix is found to be $P = \begin{bmatrix} 2 & -3 \\ 1 & -3 \end{bmatrix}$; then, by Theorem 2, $P^{-1}AP$ is the diagonal matrix of characteristic values, $D = \begin{bmatrix} 2 & 0 \\ 0 & 3 \end{bmatrix}$. Therefore $A^8 = PD^8P^{-1} = P\begin{bmatrix} 2^8 & 0 \\ 0 & 3^8 \end{bmatrix}P^{-1}$.

Example 4. An illustration of a biological use of integral powers of matrices, characteristic vectors, and characteristic values may be found on page 172 of Searle [31]. The illustration calculates the probability of a certain genotype of type T_i having a descendant of type T_j after k generations of selfing; reference is made to page 107 of [21] (Li).

Example 5. This example will illustrate and justify a method for finding roots of an R-matrix. In Example 3 it was proved that if A is similar to a diagonal matrix D then there exists an invertible matrix P such that

$$A^k = PD^kP^{-1},$$

where k is a positive integer. This result may be extended for R-matrices to allow k to be any rational number, but first we must define what is meant by a matrix raised to a *rational power*.

Let A be a p by p R-matrix. Let m and n be integers, where n is positive. $A^{1/n}$ is defined to be a p by p R-matrix B, if one exists, such that $B^n = A$. (B may not be unique.) If B exists,

$$A^{m/n} = (A^{1/n})^m = B^m.$$

We now prove that if A is similar to a diagonal matrix D and if $A^{m/n}$ exists, then

$$A^{m/n} = PD^{m/n}P^{-1}.$$

This will follow from the definition of negative powers of matrices (when $m < 0$) and from Example 3 if we can show that

$$A^{1/n} = PD^{1/n}P^{-1}.$$

If A is similar to a diagonal matrix D, then there exists an invertible matrix P such that

$$A = PDP^{-1}.$$

Now, if possible, define a new matrix B such that

$$PD^{1/n}P^{-1} = B,$$

and then show that B must equal $A^{1/n}$.

$$(PD^{1/n}P^{-1} = B) \Rightarrow (PDP^{-1} = B^n) \Rightarrow (A = B^n) \Rightarrow (A^{1/n} = B).$$

For example, find $A^{1/2} = \begin{bmatrix} 1 & 2 \\ -1 & 4 \end{bmatrix}^{1/2}$. We observed in Example 3 that for $A = \begin{bmatrix} 1 & 2 \\ -1 & 4 \end{bmatrix}$ there exists an invertible modal matrix $P = \begin{bmatrix} 2 & -3 \\ 1 & -3 \end{bmatrix}$ such that $P^{-1}AP = D = \begin{bmatrix} 2 & 0 \\ 0 & 3 \end{bmatrix}$. Therefore if we choose $D^{1/2} = \begin{bmatrix} \sqrt{2} & 0 \\ 0 & \sqrt{3} \end{bmatrix}$, then

$$A^{1/2} = P \begin{bmatrix} \sqrt{2} & 0 \\ 0 & \sqrt{3} \end{bmatrix} P^{-1}.$$

Notice that $D^{1/2}$ could have been chosen to be

$$\begin{bmatrix} -\sqrt{2} & 0 \\ 0 & \sqrt{3} \end{bmatrix}, \quad \text{or} \quad \begin{bmatrix} \sqrt{2} & 0 \\ 0 & -\sqrt{3} \end{bmatrix}, \quad \text{or} \quad \begin{bmatrix} -\sqrt{2} & 0 \\ 0 & -\sqrt{3} \end{bmatrix}.$$

Also notice that if the entries of $A^{1/2}$ must be rational then $A^{1/2}$ does not exist.

Example 6. From Theorem 1 of this section we know that the determinant of a matrix and the trace of a matrix are invariant under a similarity transformation. We can use this knowledge in the following way: According to Theorem 2, certain matrices are similar to a diagonal matrix D, where the diagonal elements of D are the characteristic values; therefore, for such an n by n matrix A

(1) $\qquad (|\lambda_i| < 1 \text{ for all } i) \Rightarrow \left(\left| \sum_{i=1}^{n} \lambda_i \right| < n \right)$

$\Rightarrow (|\text{tr } D| < n)$

$\Rightarrow (|\text{tr } A| < n) \Rightarrow (-n < \text{tr } A < n).$

10.2 SIMILARITY

(2) $(\lambda_i < 0 \text{ for all } i) \Rightarrow \left(\sum_{i=1}^{n} \lambda_i < 0\right)$
$\Rightarrow (\text{tr } D < 0) \Rightarrow (\text{tr } A < 0).$

(3) $(|\lambda_i| < 1 \text{ for all } i) \Rightarrow (|\det D| < 1)$
$\Rightarrow (|\det A| < 1) \Rightarrow (-1 < \det A < 1).$

(4) $(\lambda_i < 0 \text{ for all } i) \Rightarrow \begin{cases} \det D < 0 \text{ if } n \text{ is odd,} \\ \det D > 0 \text{ if } n \text{ is even.} \end{cases}$
$\Rightarrow \begin{cases} \det A < 0 \text{ if } n \text{ is odd,} \\ \det A > 0 \text{ if } n \text{ is even.} \end{cases}$

These implications and the contrapositives of these implications are sometimes useful.

EXERCISES

1. Use Theorem 1 to prove that the following matrices are not similar:
$$\begin{bmatrix} 3 & 2 & 0 \\ 4 & 1 & 0 \\ 6 & 9 & 2 \end{bmatrix}, \begin{bmatrix} 6 & 4 & 1 \\ 0 & 0 & 2 \\ 9 & 2 & 3 \end{bmatrix}.$$

2. Which theorems of this section guarantee that each of the following matrices is similar to a diagonal matrix?
   (a) $\begin{bmatrix} 1 & 2 \\ 2 & 1 \end{bmatrix}$; (b) $\begin{bmatrix} 1 & 4 \\ 2 & 3 \end{bmatrix}$; (c) $\begin{bmatrix} 2 & -i \\ i & 1 \end{bmatrix}$.

In each of Exercises 3–8, find an invertible modal matrix P of a given transformation matrix $[T]_b$ such that $P^{-1}[T]_b P$ is a diagonal matrix, and verify that the resulting diagonal matrix has the characteristic values of $[T]_b$ on the main diagonal.

3. $[T]_b = \begin{bmatrix} 3 & 3 \\ 4 & 2 \end{bmatrix}.$

4. $[T]_b = \begin{bmatrix} 2 & 4 \\ 5 & 1 \end{bmatrix}.$

5. $[T]_b = \begin{bmatrix} 3 & 4 \\ 1 & 0 \end{bmatrix}.$

6. $[T]_b = \begin{bmatrix} 3 & 0 \\ 5 & 2 \end{bmatrix}.$

7. $[T]_b = \begin{bmatrix} 3 & 1 & 0 \\ 0 & 0 & 4 \\ 0 & 0 & 2 \end{bmatrix}.$

8. $[T]_b = \begin{bmatrix} 2 & -2 & 0 \\ 0 & 4 & 0 \\ 2 & 5 & 1 \end{bmatrix}.$

In each of Exercises 9–12, show that an invertible modal matrix of $[T]_b$ does not exist for the given R-matrices, and hence, by Theorem 2, that $[T]_b$ is not similar to a diagonal R-matrix.

9. $[T]_b = \begin{bmatrix} 1 & -1 \\ 1 & 3 \end{bmatrix}.$

10. $[T]_b = \begin{bmatrix} 4 & -1 \\ 1 & 2 \end{bmatrix}.$

11. $[T]_b = \begin{bmatrix} 2 & 1 & 0 \\ 0 & 2 & 0 \\ 0 & 0 & 1 \end{bmatrix}.$

12. $[T]_b = \begin{bmatrix} 1 & 0 & 0 \\ 0 & 2 & 0 \\ 1 & 0 & 1 \end{bmatrix}.$

13. Do matrices A and B of Example 1 (*not* Exercise 1) of this section furnish a counterexample to the converse of each part of Theorem 1?
14. Prove part *i* of Theorem 1. *Hint:* Show that $|B - \lambda I_n| = |A - \lambda I_n|$, and let $I_n = P^{-1} I_n P$.
15. Prove part *ii* of Theorem 1.
16. Prove part *iii* of Theorem 1.
17. Give reasons for the statements given in the proof of Theorem 2.
18. Prove Theorem 5.
19. In Example 3 of this section use the method presented there to calculate A^4. (Matrix A is given in Example 3.)

10.3 Congruence

The so-called congruence transformation of a square matrix arises naturally from a study of quadratic forms. Consider the equation of an ellipse

$$2x^2 - 4xy + 7y^2 = 6,$$

whose graph is shown in Figure 10.3.1. The expression on the left side of this equation involves only 2nd degree terms and is an example of a *quadratic form*.

Definition 1. *If F is a field where $1 + 1 \neq 0$, and if $X = (x_1, x_2, \ldots, x_n)$ is an element of $V_n(F)$, then an expression, with coefficients in F, of the form*

$$\begin{aligned} f(X) = &c_{11} x_1^2 + 2c_{12} x_1 x_2 + 2c_{13} x_1 x_3 + \cdots + 2c_{1n} x_1 x_n \\ &+ c_{22} x_2^2 + 2c_{23} x_2 x_3 + \cdots + 2c_{2n} x_2 x_n \\ &\phantom{+ c_{22} x_2^2} + c_{33} x_3^2 + \cdots + 2c_{3n} x_3 x_n \\ &\phantom{+ c_{22} x_2^2 + c_{33} x_3^2 + \cdots} \vdots \\ &\phantom{+ c_{22} x_2^2 + c_{33} x_3^2 + \cdots} + c_{nn} x_n^2, \end{aligned}$$

is called a **quadratic form** *and can be represented by the matrix expression*

$$[x_1 \ x_2 \ \cdots \ x_n] \begin{bmatrix} c_{11} & c_{12} & \cdots & c_{1n} \\ c_{12} & c_{22} & \cdots & c_{2n} \\ \vdots & \vdots & & \vdots \\ c_{1n} & c_{2n} & \cdots & c_{nn} \end{bmatrix} \begin{bmatrix} x_1 \\ x_2 \\ \vdots \\ x_n \end{bmatrix}$$

or

$$X^T C X.$$

The symmetric matrix C is called the **matrix of the quadratic form**.[2]

[2] We require that $1 + 1 \neq 0$ because the nondiagonal entries of the matrix of the quadratic form are found by multiplying the corresponding entries of the quadratic form by the multiplicative inverse of $1 + 1$; however, a field where $1 + 1 = 0$ does not contain a multiplicative inverse of $1 + 1$.

10.3 CONGRUENCE

Notice from Definition 1 that a quadratic form is simply a transformation $T: V_n(F) \to F$. In this book we are primarily concerned with the case where F is the field of real numbers; a quadratic form for which F is the field of real numbers will be called a **real quadratic form**. The occurrence of quadratic forms is fairly frequent in applied work, and, consequently, matrix methods for the manipulations of quadratic forms are desirable.

Example 1. In the plane, the left side of the equation

$$2x^2 - 4xy + 7y^2 = 6$$

is a real quadratic form. The graph of the equation is the ellipse shown in Figure 10.3.1.

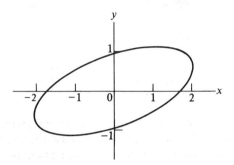

Figure 10.3.1

After rewriting the equation as

$$2x^2 + (-2)xy + (-2)yx + 7y^2 = 6,$$

we see that the quadratic equation can be expressed as the matrix equation

$$[x \ y] \begin{bmatrix} 2 & -2 \\ -2 & 7 \end{bmatrix} \begin{bmatrix} x \\ y \end{bmatrix} = [6],$$

or

$$X^T C X = 6.^3$$

Let $V_n(F)$ be the vector space. Under a transformation $U = P^{-1}X$ or $X = PU$, the quadratic form

$$X^T C X$$

becomes

$$(PU)^T C (PU)$$

---

[3] The scalar 6 is often used here rather than the matrix [6]; this is permissible because the field R of real numbers is isomorphic to the field of 1 by 1 R-matrices.

or
$$U^T(P^TCP)U.$$

Hence the matrix C of the quadratic form has been transformed to the matrix P^TCP.

Definition 2. *An n by n matrix A over a field F is said to be **congruent** to matrix B if there exists an invertible matrix P over F such that $B = P^TAP$. A transformation T of A such that $A \xmapsto{T} P^TAP$ is called a **congruence transformation** of A.*

Example 2. $\begin{bmatrix} 1 & 4 \\ 2 & 5 \end{bmatrix}$ is congruent to $\begin{bmatrix} 1 & 2 \\ 0 & -3 \end{bmatrix}$ because there exists an invertible matrix $P = \begin{bmatrix} 1 & 2 \\ 0 & 1 \end{bmatrix}$ such that

$$\begin{bmatrix} 1 & 2 \\ 0 & -3 \end{bmatrix} = P^T \begin{bmatrix} 1 & 4 \\ 2 & 5 \end{bmatrix} P.$$

Because much of the known theory of the congruence transformation applies only to symmetric matrices, and because the matrix of a quadratic form is a symmetric matrix, we will confine our attention to the congruence transformation of symmetric matrices. We will find: (*i*) that for any field F in which $1 + 1 \neq 0$, any symmetric matrix over F can be transformed by a congruence transformation to a diagonal matrix; (*ii*) in particular, if F is the field of complex numbers, then any n by n symmetric C-matrix A is congruent to $\begin{bmatrix} I_r & 0 \\ \hline 0 & 0 \end{bmatrix}$ where r is the rank of A; (*iii*) if, however, F is the field of real numbers, then any symmetric R-matrix A is congruent to

$$\begin{bmatrix} I_p & 0 & 0 \\ \hline 0 & -I_{r-p} & 0 \\ \hline 0 & 0 & 0 \end{bmatrix},$$

where p is an integer that is uniquely determined by A; notice that if $p = 0$ or if $p = r$ we will obtain the notation I_0, which will mean that I_p or $-I_{r-p}$ is nonexistent. First, however, we need to establish that the symmetry of a matrix is unchanged under a congruence transformation.

Theorem 7. *If A is a symmetric matrix over F, and if A is congruent to B, then B is symmetric.*

The proof is left as Exercise 16.

Theorem 8. *If A is a symmetric n by n matrix over a field F in which $1 + 1 \neq 0$, then A is congruent to an n by n diagonal matrix in which the number of nonzero entries on the diagonal is equal to the rank of A.*

Proof. The theorem is obvious if $n = 1$; assume $n > 1$.

(1) The following algorithm (or set of procedures described for Cases I, II, III, IV below) reduces any matrix A over F to the form

$$\begin{bmatrix} a_{11} & 0 \\ \hline 0 & B \end{bmatrix}.$$

CASE I. If $a_{11} \neq 0$, perform the elementary operations

$$\frac{-a_{i1}}{a_{11}} R_1 + R_i \qquad (i = 2, \ldots, n)$$

and then

$$\frac{-a_{1i}}{a_{11}} C_1 + C_i \qquad (i = 2, \ldots, n).$$

CASE II. If $a_{11} = 0$ and some $a_{ii} \neq 0$, perform the elementary operations

$$R_1 \leftrightarrow R_i \quad \text{and} \quad C_1 \leftrightarrow C_i$$

and Case I will apply.

CASE III. If $a_{ii} = 0$ for all i, but there exists some entry $a_{ij} \neq 0$, perform the elementary operations

$$R_i + R_j \quad \text{and} \quad C_i + C_j;$$

Case II or Case I will then apply.

CASE IV. If $a_{ij} = 0$ for all i and j, the matrix is already diagonalized.

(2) Because the elementary operations used in (1) can be accomplished by postmultiplying A by a product of elementary matrices P and by premultiplying the result by P^T, and because of Theorem 7, we know that B is symmetric.

(3) By repeated use of the algorithm up to at most $n - 1$ times, the matrix can be diagonalized.

(4) Since only elementary operations have been used, the rank is invariant, and hence the number of nonzero entries in the resulting diagonal matrix must equal the rank of A. ☐

Theorem 9. *If A is a symmetric n by n C-matrix, then A is congruent to*

$$\begin{bmatrix} I_r & 0 \\ \hline 0 & 0 \end{bmatrix}_{(n,n)}$$

where r is the rank of A.

The proof is left as Exercise 17.

Example 3. The matrix $A = \begin{bmatrix} 1 & 2 & 0 \\ 2 & 4 & 2 \\ 0 & 2 & 4 \end{bmatrix}$ can be diagonalized by applying the algorithm given in Part (1) of the proof of Theorem 8. In order to calculate the product of elementary row transformation matrices P^T, we will perform the same elementary row (but not column) operations on I_3 that we perform on A. We show the operation on A at the left and on I_3 at the right.

$$\begin{bmatrix} 1 & 2 & 0 \\ 2 & 4 & 2 \\ 0 & 2 & 4 \end{bmatrix} \qquad \begin{bmatrix} 1 & 0 & 0 \\ 0 & 1 & 0 \\ 0 & 0 & 1 \end{bmatrix}$$

$$\underset{\sim}{\overset{-2R_1+R_2}{-2C_1+C_2}} \begin{bmatrix} 1 & 0 & 0 \\ 0 & 0 & 2 \\ 0 & 2 & 4 \end{bmatrix} \qquad \underset{\sim}{\overset{-2R_1+R_2}{}} \begin{bmatrix} 1 & 0 & 0 \\ -2 & 1 & 0 \\ 0 & 0 & 1 \end{bmatrix}$$

$$\underset{\sim}{\overset{R_2\leftrightarrow R_3}{C_2\leftrightarrow C_3}} \begin{bmatrix} 1 & 0 & 0 \\ 0 & 4 & 2 \\ 0 & 2 & 0 \end{bmatrix} \qquad \underset{\sim}{\overset{R_2\leftrightarrow R_3}{}} \begin{bmatrix} 1 & 0 & 0 \\ 0 & 0 & 1 \\ -2 & 1 & 0 \end{bmatrix}$$

$$\underset{\sim}{\overset{-\frac{1}{2}R_2+R_3}{-\frac{1}{2}C_2+C_3}} \begin{bmatrix} 1 & 0 & 0 \\ 0 & 4 & 0 \\ 0 & 0 & -1 \end{bmatrix}. \qquad \underset{\sim}{\overset{-\frac{1}{2}R_2+R_3}{}} \begin{bmatrix} 1 & 0 & 0 \\ 0 & 0 & 1 \\ -2 & 1 & -\frac{1}{2} \end{bmatrix}.$$

According to the proof of Theorem 9 (see answer to Exercise 17), the diagonal matrix that we have just found can be reduced further by the elementary operations $\frac{1}{2}R_2, \frac{1}{2}C_2, i^{-1}R_3$, and $i^{-1}C_3$.

$$\begin{bmatrix} 1 & 0 & 0 \\ 0 & 1 & 0 \\ 0 & 0 & 1 \end{bmatrix} = D. \qquad \begin{bmatrix} 1 & 0 & 0 \\ 0 & 0 & \frac{1}{2} \\ -2i^{-1} & i^{-1} & -\frac{1}{2}i^{-1} \end{bmatrix} = P^T.$$

Therefore $P^T A P = D$ is

$$\begin{bmatrix} 1 & 0 & 0 \\ 0 & 0 & \frac{1}{2} \\ -2i^{-1} & i^{-1} & -\frac{1}{2}i^{-1} \end{bmatrix} \begin{bmatrix} 1 & 2 & 0 \\ 2 & 4 & 2 \\ 0 & 2 & 4 \end{bmatrix} \begin{bmatrix} 1 & 0 & -2i^{-1} \\ 0 & 0 & i^{-1} \\ 0 & \frac{1}{2} & -\frac{1}{2}i^{-1} \end{bmatrix} = \begin{bmatrix} 1 & 0 & 0 \\ 0 & 1 & 0 \\ 0 & 0 & 1 \end{bmatrix}.$$

Theorem 10. *If A is a symmetric n by n R-matrix, then A is congruent to*

$$\begin{bmatrix} I_p & 0 & 0 \\ 0 & -I_{r-p} & 0 \\ 0 & 0 & 0 \end{bmatrix}_{(n,n)},$$

*where r is the rank of A and p is an integer that is uniquely determined by A. The integer p is called the **index** of A.*

Proof. The first part of the theorem follows easily from Theorem 8 by performing the necessary transformations to reduce the diagonal entries to ± 1 (see Example 3 and the answer to Exercise 17). To show that p is a unique integer, the reader is referred to Theorem 5.5.5 on page 239 of Eves, *Elementary Matrix Theory*, Allyn & Bacon, Boston, 1966. □

Example 4. Consider the equation of the ellipse of Example 1,

$$2x^2 - 4xy + 7y^2 = 6,$$

which can be written as

$$X^T \begin{bmatrix} 2 & -2 \\ -2 & 7 \end{bmatrix} X = 6.$$

By Theorem 10 there exists a transformation $U = P^{-1}X$ or $X = PU$ such that the new equation is

$$U^T DU = 6,$$

where D is a diagonal matrix of the form given in that theorem and is found by

$$D = P^T \begin{bmatrix} 2 & -2 \\ -2 & 7 \end{bmatrix} P.$$

By the methods of Example 3,

$$\begin{bmatrix} 2 & -2 \\ -2 & 7 \end{bmatrix} \begin{smallmatrix} R_1 + R_2 \\ C_1 + C_2 \\ \sim \end{smallmatrix} \begin{bmatrix} 2 & 0 \\ 0 & 5 \end{bmatrix}$$

$$\begin{smallmatrix} \frac{1}{2}\sqrt{2}R_1 \\ \frac{1}{2}\sqrt{2}C_1 \\ \sim \end{smallmatrix} \begin{bmatrix} 1 & 0 \\ 0 & 5 \end{bmatrix}$$

$$\begin{smallmatrix} \frac{1}{5}\sqrt{5}R_2 \\ \frac{1}{5}\sqrt{5}C_2 \\ \sim \end{smallmatrix} \begin{bmatrix} 1 & 0 \\ 0 & 1 \end{bmatrix} = D.$$

$$\begin{bmatrix} 1 & 0 \\ 0 & 1 \end{bmatrix} \begin{smallmatrix} R_1 + R_2 \\ \sim \end{smallmatrix} \begin{bmatrix} 1 & 0 \\ 1 & 1 \end{bmatrix}$$

$$\begin{smallmatrix} \frac{1}{2}\sqrt{2}R_1 \\ \sim \end{smallmatrix} \begin{bmatrix} \frac{1}{2}\sqrt{2} & 0 \\ 1 & 1 \end{bmatrix}$$

$$\begin{smallmatrix} \frac{1}{5}\sqrt{5}R_2 \\ \sim \end{smallmatrix} \begin{bmatrix} \frac{1}{2}\sqrt{2} & 0 \\ \frac{1}{5}\sqrt{5} & \frac{1}{5}\sqrt{5} \end{bmatrix} = P^T.$$

Hence $P = \begin{bmatrix} \frac{1}{2}\sqrt{2} & \frac{1}{5}\sqrt{5} \\ 0 & \frac{1}{5}\sqrt{5} \end{bmatrix}$, and the image of the original ellipse under the linear transformation T defined by

$$X \xmapsto{T} P^{-1}X \quad \text{or} \quad U = P^{-1}X$$

is

$$u^2 + v^2 = 6,$$

which is a circle of radius $\sqrt{6}$. This transformation also can be thought of as a change in basis, and what was an ellipse in the original xy-coordinate system is a circle in the new uv-coordinate system. As shown in Figure 10.3.2, the graphical representation undergoes a considerable distortion. By dividing the members of the original quadratic equation by 6, and by altering P as needed, we can transform the original equation into

$$u^2 + v^2 = 1.$$

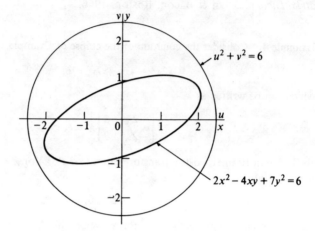

Figure 10.3.2

Example 5. Consider the equation

$$x^2 + 6xy + 5y^2 = 1,$$

or

$$X^T \begin{bmatrix} 1 & 3 \\ 3 & 5 \end{bmatrix} X = 1.$$

Perform the real transformation $U = P^{-1}X$ or $X = PU$ in such a way that the matrix of the quadratic form is diagonalized to the form given in Theorem 10. Using the methods of Example 3, we get

$$\begin{bmatrix} 1 & 3 \\ 3 & 5 \end{bmatrix} \underset{\sim}{\overset{-3R_1+R_2}{-3C_1+C_2}} \begin{bmatrix} 1 & 0 \\ 0 & -4 \end{bmatrix} \quad \vdots \quad \begin{bmatrix} 1 & 0 \\ 0 & 1 \end{bmatrix} \underset{\sim}{\overset{-3R_1+R_2}{}} \begin{bmatrix} 1 & 0 \\ -3 & 1 \end{bmatrix}$$

$$\underset{\sim}{\overset{\frac{1}{2}R_2}{\frac{1}{2}C_2}} \begin{bmatrix} 1 & 0 \\ 0 & -1 \end{bmatrix} = D. \quad \vdots \quad \underset{\sim}{\overset{\frac{1}{2}R_2}{}} \begin{bmatrix} 1 & 0 \\ -\frac{3}{2} & \frac{1}{2} \end{bmatrix} = P^T.$$

Therefore $P = \begin{bmatrix} 1 & -\frac{3}{2} \\ 0 & \frac{1}{2} \end{bmatrix}$.

Hence the original equation can be transformed by $U = P^{-1}X$ or $X = PU$ to the equation of an equilateral hyperbola

$$u^2 - v^2 = 1.$$

The graphs of the original hyperbola and its image are shown in Figure 10.3.3. Again, notice the distortion of the graph under the transformation.

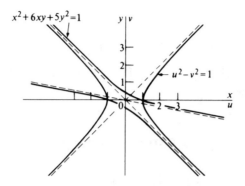

Figure 10.3.3

As the last two examples suggest, any real quadratic equation of the form

$$ax^2 + bxy + cy^2 = d$$

can be reduced, by a congruence transformation of the matrix of the quadratic form, to one of the following real quadratic forms:

$$\begin{cases} u^2 + v^2 = 1 & \text{unit circle,} \\ u^2 - v^2 = 1 & \text{equilateral hyperbola,} \\ -u^2 - v^2 = 1 & \text{no locus,} \\ u^2 = 1 & \text{two parallel lines,} \\ u^2 = -1 & \text{no locus.} \end{cases} \quad \text{if } d \neq 0,$$

or

$$\begin{cases} u^2 + v^2 = 0 & \text{point,} \\ u^2 - v^2 = 0 & \text{two intersecting lines,} \\ -u^2 - v^2 = 0 & \text{point,} \\ u^2 = 0 & \text{one line.} \end{cases} \quad \text{if } d = 0.$$

Somewhat analogous to the study of the congruence transformation of symmetric R-matrices is the study of the *Hermitian congruence* (or *conjunctive*) *transformations* of Hermitian C-matrices. An n by n C-matrix A is said to be **Hermitely congruent** (or **conjunctive**) to a matrix B if there exists an invertible C-matrix P such that

$$B = P^*AP.$$

If A is Hermitian, the analogue of Theorem 10 follows, namely, that A is Hermitely congruent to

$$\begin{bmatrix} I_p & 0 & 0 \\ 0 & -I_{r-p} & 0 \\ 0 & 0 & 0 \end{bmatrix}_{(n,\,n)}$$

where r is the rank of A, and where p is an integer that is uniquely determined by A.

APPLICATIONS

Example 6. In Example 2–17 on pages 126–128 of [13] (Hollingsworth), a procedure is given for determining the *optical axes* of a crystal. The procedure involves a change of the underlying coordinate system of a quadratic form. Suppose that the crystal has three indices of refraction α, β, γ such that $\gamma > \beta > \alpha$. The indicatrix ellipsoid is given by

$$X^T A X = 1,$$

where

$$A = \begin{bmatrix} \alpha^{-2} & 0 & 0 \\ 0 & \beta^{-2} & 0 \\ 0 & 0 & \gamma^{-2} \end{bmatrix}, \quad \text{and} \quad X = \begin{bmatrix} x_1 \\ x_2 \\ x_3 \end{bmatrix}.$$

A rotation of the coordinate system is made about the x_2-axis through an angle V in such a way that in the new $y_1 y_2 y_3$-coordinate system when $y_3 = 0$, then

$$\alpha^{-2} y_1^2 + \beta^{-2} y_2^2 = 1,$$

and hence the new y_3-axis is along the optic axis. Such a transformation can be accomplished by the rotation transformation $Y = R^{-1} X$ or $X = RY$ where

$$R = \begin{bmatrix} \cos V & 0 & \sin V \\ 0 & 1 & 0 \\ -\sin V & 0 & \cos V \end{bmatrix},$$

and $\sin V$ and $\cos V$ are determined by the equations

$$\sin V = \frac{\gamma}{\beta} \sqrt{\frac{\beta^2 - \alpha^2}{\gamma^2 - \alpha^2}}$$

and

$$\cos V = \pm \frac{\alpha^2}{\beta^2} \sqrt{\frac{\beta^2 - \gamma^2}{\alpha^2 - \gamma^2}}.$$

Because of the two values for cos V, there are two optic axes. Under the change of coordinates the equation

$$X^T A X = 1$$

becomes

$$Y^T(R^T A R) Y = 1.$$

EXERCISES

In each of Exercises 1–4, write the quadratic equation as a matrix equation.
1. $x^2 + y^2 - 3z^2 + xy + 6yz + 4xz = 6$.
2. $2x^2 - 3z^2 + 5xy - 8xz = 10$.
3. $w^2 + y^2 + z^2 + 6xy - 4yz + 8wx = 7$.
4. $4w^2 - x^2 + y^2 - 2xz + 4wy - 8wx = 10$.

In each of Exercises 5–8, illustrate Theorem 10 by finding an invertible matrix P such that $P^T C P$ is a diagonal matrix of the form specified in that theorem.

5. $C = \begin{bmatrix} 2 & 2 \\ 2 & 4 \end{bmatrix}$.

6. $C = \begin{bmatrix} 1 & 3 \\ 3 & 6 \end{bmatrix}$.

7. $C = \begin{bmatrix} 1 & 0 & 2 \\ 0 & 1 & 0 \\ 2 & 0 & 3 \end{bmatrix}$.

8. $C = \begin{bmatrix} 1 & 3 & 0 \\ 3 & 9 & 0 \\ 0 & 0 & 4 \end{bmatrix}$.

In each of Exercises 9–12, illustrate Theorem 9 by finding an invertible matrix P such that $P^T C P$ is a diagonal matrix of the form specified in that theorem.

9. $C = \begin{bmatrix} 3 & 9i \\ 9i & 2 \end{bmatrix}$.

10. $C = \begin{bmatrix} 2 & 8 \\ 8 & 7 \end{bmatrix}$.

11. $C = \begin{bmatrix} 1 & 2 & 0 \\ 2 & 4 & 0 \\ 0 & 0 & -9 \end{bmatrix}$.

12. $C = \begin{bmatrix} 1 & 0 & 2 \\ 0 & i & 0 \\ 2 & 0 & 3 \end{bmatrix}$.

13. Graph the ellipse whose equation is

$$x^2 + 8xy + 20y^2 = 80$$

by plotting the intercepts and the points where $x = y$ and $x = -y$. Determine the matrix P such that the transformation defined by $X = PU$ or $U = P^{-1}X$ will map the ellipse into a circle with radius $\sqrt{80}$.

14. Graph the ellipse whose equation is

$$x^2 + 4xy + 8y^2 = 117$$

by plotting the intercepts and the points where $x = y$ and $x = -y$. Determine the invertible matrix P such that the transformation defined by $X = PU$ or $U = P^{-1}X$ will map the ellipse into a circle with radius $\sqrt{117}$.

15. Let X be an arbitrary element of the vector space $V_3(R)$, let
$$C = \begin{bmatrix} c_{11} & c_{12} & c_{13} \\ c_{12} & c_{22} & c_{23} \\ c_{13} & c_{23} & c_{33} \end{bmatrix}$$
be the matrix of a real quadratic form $f(x_1, x_2, x_3) = X^T C X$, and let P designate an invertible matrix such that $P^T C P = D$ where D is a diagonal matrix. Write $f(x_1, x_2, x_3)$ and, without calculating P, write the transformation over $V_3(R)$ that will cause f to be mapped into a real quadratic form without the product terms (that is, terms involving $x_1 x_2$, $x_2 x_3$, and $x_1 x_3$); if the rank of C is 3, what is the image of f?

16. Prove Theorem 7.

17. Prove Theorem 9. (Answer is provided.)

10.4 Orthogonal Transformations

In Section 10.3 we found that linear transformations exist that will always diagonalize the matrix of a quadratic form. In the plane, we saw that such transformations may produce considerable distortions in the graph of a quadratic equation. For instance, in Example 4 of Section 10.3 we mapped an ellipse into a circle as shown in Figure 10.3.2. In Figure 10.4.1, two geometric vectors and their images are shown under the same transformation. Obviously neither the lengths of these vectors nor the angle between them is preserved under the transformation. The question naturally arises as to whether it is possible to identify a transformation in the plane for which there is no dis-

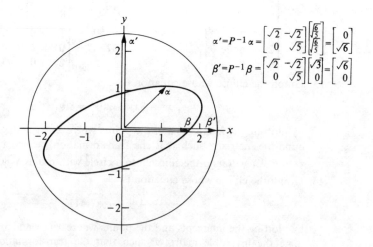

Figure 10.4.1

tortion in the lengths of vectors, or the angle between two vectors, or the graph of a quadratic equation. Recall from Section 7.5 that in general the length of a vector, the angle between two vectors, and the distance between two vectors were defined in terms of inner products over a vector space; consequently any linear transformation that preserves the inner product will also preserve lengths, angles, and distances. This motivates the following important definition. From Section 7.4 we recall that an inner product of two vectors α and β is denoted by $\langle \alpha, \beta \rangle$.

Definition 3. *A linear operator T over a real inner product space (Euclidean space) $V(R)$ is **orthogonal** if for every α and β in $V(R)$*

$$\langle T(\alpha), T(\beta) \rangle \equiv \langle \alpha, \beta \rangle.$$

Example 1. Let $V_2(R)$ designate the inner product space of all two-dimensional, real, coordinate vectors with the standard inner product. The linear transformation of reflection, T, over $V_2(R)$, defined by

$$(x, y) \xmapsto{T} (x, -y)$$

is an example of an orthogonal transformation T over $V_2(R)$ because

$$\begin{aligned}
\langle T(x_1, y_1), T(x_2, y_2) \rangle &= (x_1, -y_1) \cdot (x_2, -y_2) \\
&= x_1 x_2 + (-y_1)(-y_2) \\
&= x_1 x_2 + y_1 y_2 \\
&= (x_1, y_1) \cdot (x_2, y_2) \\
&= \langle (x_1, y_1), (x_2, y_2) \rangle.
\end{aligned}$$

Example 2. Let $V_2(R)$ be defined as it was in Example 1. The linear transformation of rotation, T, over $V_2(R)$, defined by

$$(x, y) \xmapsto{T} (\{x \cos \theta - y \sin \theta\}, \{x \sin \theta + y \cos \theta\})$$

is an example of an orthogonal transformation over $V_2(R)$ because we can show that

$$(x_1, y_1) \cdot (x_2, y_2) = \{T(x_1, y_1)\} \cdot \{T(x_2, y_2)\}.$$

The reader is asked to verify this fact in Exercise 17.

In Section 8.2 we learned that there is a one-to-one correspondence between the set of all linear transformations over an n-dimensional vector space $V(F)$ and the set of all n by n matrices over F. Now it seems logical that we try to identify the subset of n by n matrices over F that corresponds to the subset of orthogonal linear transformations over $V(F)$.

Recall that an orthogonal matrix was defined in Section 4.4 as an n by n R-matrix P such that $P^T = P^{-1}$. Although the reader probably did not realize it at that time, this definition was chosen in such a way that P is the matrix representation of a unique orthogonal transformation over an inner product space $V(R)$, *provided that the basis of $V(R)$ is an orthonormal set of vectors*. This fact is justified by the following theorem.

Theorem 11. *Let $V(R)$ be a real inner product space with an orthonormal basis $b = \{\beta_1, \ldots, \beta_n\}$, and let T be a linear transformation defined over $V(R)$. T is an orthogonal linear transformation if and only if $[T]_b^T = [T]_b^{-1}$.*

For the proof of this, see Theorem A.10.11, in Appendix B, page A6.

Example 3. The matrix $A = \begin{bmatrix} 0 & 0 & 1 \\ \frac{1}{2}\sqrt{2} & -\frac{1}{2}\sqrt{2} & 0 \\ \frac{1}{2}\sqrt{2} & \frac{1}{2}\sqrt{2} & 0 \end{bmatrix}$ has the property that

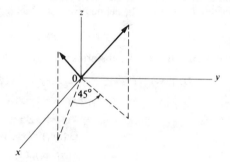

Figure 10.4.2

$A^T = A^{-1}$; hence, by Theorem 11, A is the matrix representation of an orthogonal transformation T over a three-dimensional inner product space with an orthonormal basis. For the inner product space of all three-dimensional geometric vectors with a basis consisting of unit vectors along the axes, A is the matrix representation of the orthogonal transformation that rotates a vector through 45° about the z-axis, as shown in Figure 10.4.2. Under such a transformation, lengths of vectors and angles between pairs of vectors are preserved.

So far in this section we have been concerned only with linear operators over a real inner product space $V(R)$. For a complex inner product space we have the following generalizations of Definition 3 and Theorem 11.

Definition 4. *A linear operator T over a complex inner product space (unitary space) $V(C)$ is **unitary** if for every α and β in $V(C)$*

$$\langle T(\alpha), T(\beta) \rangle \equiv \langle \alpha, \beta \rangle.$$

Theorem 12. *Let $V(C)$ be a complex inner product space with an orthonormal basis $b = \{\beta_1, \ldots, \beta_n\}$, and let T be a linear transformation defined over $V(C)$. T is a unitary linear transformation if and only if*

$$[T]_b^* = [T]_b^{-1}.$$

The proof is left as Exercise 20. *Hint:* The proof is analogous to the proof of Theorem A.10.11.

EXERCISES

In each of Exercises 1–4, use Definition 3 to determine whether or not the given linear operator T is orthogonal. Let T be defined over the inner product space $V_2(R)$ of all two-dimensional real coordinate vectors with standard inner product. Graph (3, 1) and (1, 0) and their images; what do you observe about the preservation of angles, lengths, and distances under the transformation?

1. T is defined by $(x, y) \xmapsto{T} (y, x)$.
2. T is defined by $(x, y) \xmapsto{T} (-y, x)$.
3. T is defined by $(x, y) \xmapsto{T} (2y, 2x)$.
4. T is defined by $(x, y) \xmapsto{T} (x + y, y)$.

In each of Exercises 5 and 6, use Definition 3 to determine whether or not the given linear operator T is orthogonal. Let T be defined over the inner product space $V(R)$ of all 2 by 2 R-matrices with inner product $\langle A, B \rangle = \text{tr}(B^T A)$.

5. T is defined by $A \xmapsto{T} P^{-1}AP$ where P is invertible.
6. T is defined by $A \xmapsto{T} P^T A P$ where P is invertible.

In each of Exercises 7 and 8, use Definition 3 to determine whether or not the given linear operator T is orthogonal. Let T be defined over the inner product space $P_1(R)$ consisting of all polynomials of degree 1 or less over the real numbers and the zero polynomial. The inner product is defined according to

$$\langle p(x), q(x) \rangle = \int_0^1 p(x)\, q(x)\, dx.$$

★7. T is the derivative operator.

★8. T is defined according to $p(x) \xmapsto{T} k\, p(x)$ where k is a real number and $k \neq 0$.

9. Let $b = \{\beta_1, \beta_2, \beta_3\}$ be an orthonormal basis of an inner product space $V(R)$ over which a linear transformation T is defined according to

$$\begin{cases} T(\beta_1) = 2\beta_1 + 3\beta_2, \\ T(\beta_2) = \phantom{2\beta_1 + {}} 4\beta_2 + \beta_3, \\ T(\beta_3) = \beta_1 \phantom{{} + 4\beta_2} + 4\beta_3. \end{cases}$$

Use Theorem 11 to determine whether or not T is an orthogonal transformation. *Hint:* Use Theorem 12 or Theorem 13 of Chapter 4.

10. Repeat Exercise 9 if
$$\begin{cases} T(\beta_1) = \beta_1 + \beta_2, \\ T(\beta_2) = \beta_2 - \beta_3, \\ T(\beta_3) = \beta_1 + \beta_2 + \beta_3. \end{cases}$$

11. Repeat Exercise 9 if
$$\begin{cases} T(\beta_1) = \tfrac{1}{3}\sqrt{3}\,\beta_1 + \tfrac{1}{3}\sqrt{3}\,\beta_2 + \tfrac{1}{3}\sqrt{3}\,\beta_3, \\ T(\beta_2) = \tfrac{1}{6}\sqrt{6}\,\beta_1 - \tfrac{1}{3}\sqrt{6}\,\beta_2 + \tfrac{1}{6}\sqrt{6}\,\beta_3, \\ T(\beta_3) = -\tfrac{1}{2}\sqrt{2}\,\beta_1 \phantom{+ \tfrac{1}{3}\sqrt{6}\,\beta_2} + \tfrac{1}{2}\sqrt{2}\,\beta_3. \end{cases}$$

12. Repeat Exercise 9 if
$$\begin{cases} T(\beta_1) = \tfrac{4}{5}\beta_1 + \tfrac{3}{5}\beta_2, \\ T(\beta_2) = -\tfrac{3}{5}\beta_1 + \tfrac{4}{5}\beta_2, \\ T(\beta_3) = \phantom{-\tfrac{3}{5}\beta_1 + \tfrac{4}{5}} \beta_3. \end{cases}$$

13. Let $b = \{\beta_1, \beta_2\}$ be the basis of the inner product space $V_2(R)$ over which a linear transformation T is defined according to
$$\begin{cases} T(\beta_1) = \tfrac{1}{2}\sqrt{2}\,\beta_1 + \tfrac{1}{2}\sqrt{2}\,\beta_2, \\ T(\beta_2) = -\tfrac{1}{2}\sqrt{2}\,\beta_1 + \tfrac{1}{2}\sqrt{2}\,\beta_2. \end{cases}$$

The inner product of $V_2(R)$ is $\langle \alpha, \gamma \rangle = \alpha \cdot \gamma$; if $\beta_1 = (1, 0)$, and $\beta_2 = (1, 1)$, can we conclude from Theorem 11 that T is orthogonal, and why?

14. Repeat Exercise 13 if $\beta_1 = (1, 0)$, $\beta_2 = (0, 1)$, and
$$\begin{cases} T(\beta_1) = \tfrac{1}{2}\sqrt{3}\,\beta_1 + \tfrac{1}{2}\beta_2, \\ T(\beta_2) = -\tfrac{1}{2}\beta_1 + \tfrac{1}{2}\sqrt{3}\,\beta_2. \end{cases}$$

★15. Repeat Exercise 13 if the inner product space is $P_1(R)$ with
$$\langle p(x), q(x) \rangle = \int_0^1 p(x)q(x)\,dx,$$
and $\beta_1 = 1$ and $\beta_2 = x - \tfrac{1}{2}$. Let T be defined as it was in Exercise 13.

16. Repeat Exercise 13 if the inner product space is $V(R)$ consisting of the set of all 2 by 2 R-matrices where $\langle A, B \rangle = \operatorname{tr}(B^T A)$, and where
$$\beta_1 = \begin{bmatrix} 1 & 0 \\ 0 & 0 \end{bmatrix}, \quad \beta_2 = \begin{bmatrix} 0 & 1 \\ 0 & 0 \end{bmatrix}, \quad \beta_3 = \begin{bmatrix} 0 & 0 \\ 1 & 0 \end{bmatrix}, \quad \text{and} \quad \beta_4 = \begin{bmatrix} 0 & 0 \\ 0 & 1 \end{bmatrix}.$$
Let T be defined according to
$$\begin{cases} T(\beta_1) = \tfrac{1}{2}\sqrt{2}\,\beta_1 + \tfrac{1}{2}\sqrt{2}\,\beta_3, \\ T(\beta_2) = \beta_4, \\ T(\beta_3) = -\tfrac{1}{2}\sqrt{2}\,\beta_1 + \tfrac{1}{2}\sqrt{2}\,\beta_3, \\ T(\beta_4) = \beta_2. \end{cases}$$

17. In Example 2 verify that
$$(x_1, y_1) \cdot (x_2, y_2) = \{T(x_1, y_1)\} \cdot \{T(x_2, y_2)\}.$$

10.5 ORTHOGONAL CONGRUENCE 319

18. Use Definition 4 to determine whether or not the given linear operator T is unitary. Let T be defined by
$$(x, y) \xmapsto{T} (-x, y)$$
over the inner product space $V_2(C)$ with standard inner product.

19. Use Theorem 12 to determine whether or not the linear operator T defined by
$$\begin{cases} T(\beta_1) = \tfrac{1}{10}i\sqrt{10}\,\beta_1 - \tfrac{3}{10}\sqrt{10}\,\beta_2, \\ T(\beta_2) = \tfrac{3}{10}\sqrt{10}\,\beta_1 - \tfrac{1}{10}i\sqrt{10}\,\beta_2, \end{cases}$$
is a unitary transformation over a complex inner product space with an orthonormal basis $b = \{\beta_1, \beta_2\}$.

20. Prove Theorem 12.
21. Prove: An n by n R-matrix P is orthogonal if and only if $P^T P = I_n$.
22. Prove: An n by n C-matrix P is unitary if and only if $P^* P = I_n$.
23. Prove: The product of two n by n orthogonal matrices is an n by n orthogonal matrix.
24. Prove: The product of two n by n unitary matrices is an n by n unitary matrix.
25. Prove: If A is an orthogonal matrix then:
    (i) A^{-1} is orthogonal;   (ii) A^T is orthogonal.
26. Prove: If A is a unitary matrix, then:
    (i) A^{-1} is unitary;   (ii) A^T is unitary;   (iii) $\bar{A}$ is unitary;
    (iv) A^* is unitary.
27. Prove: The set of all n by n orthogonal matrices forms a group under Cayley multiplication.
28. Prove: The set of all n by n unitary matrices forms a group under Cayley multiplication.
29. Prove: The determinant of an orthogonal matrix has absolute value 1.
30. Prove: The determinant of a unitary matrix is a number which has modulus 1.

10.5 Orthogonal Congruence

Notice that under $X = PU$, where $P^T = P^{-1}$, the real quadratic form
$$X^T A X$$
becomes
$$U^T(P^T A P)U \quad \text{or} \quad U^T(P^{-1} A P)U.$$
The resulting transformation of the R-matrix of the quadratic form
$$P^T A P \quad \text{or} \quad P^{-1} A P, \quad \text{where } P^T = P^{-1},$$
is called the **orthogonal congruence** (or **orthogonal similarity**) **transformation** of A.

Definition 5. *An n by n R-matrix A is said to be **orthogonally congruent** (or **orthogonally similar**) to an R-matrix B if there exists an orthogonal R-matrix P such that*

$$B = P^{-1}AP = P^T AP.$$

Example 1. The matrix $A = \begin{bmatrix} 1 & 2 \\ 2 & 1 \end{bmatrix}$ is orthogonally congruent to $B = \begin{bmatrix} -1 & 0 \\ 0 & 3 \end{bmatrix}$ because there exists a matrix $P = \begin{bmatrix} 1/\sqrt{2} & 1/\sqrt{2} \\ -1/\sqrt{2} & 1/\sqrt{2} \end{bmatrix}$ such that

$$\begin{bmatrix} -1 & 0 \\ 0 & 3 \end{bmatrix} = P^{-1} \begin{bmatrix} 1 & 2 \\ 2 & 1 \end{bmatrix} P = P^T \begin{bmatrix} 1 & 2 \\ 2 & 1 \end{bmatrix} P.$$

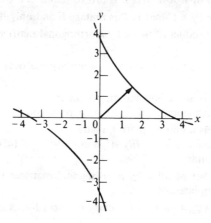

Figure 10.5.1

Example 2. Consider the inner product space $V_2(R)$ with standard inner product and with basis $b = \{(1, 0), (0, 1)\}$. The orthogonal matrix P in $X = PU$ can be viewed in two ways. First, it can be thought of as the inverse of an orthogonal transformation matrix $[T]_b$ which preserves lengths of vectors and angles between pairs of vectors with respect to the same basis. Second, P can be thought of as a transition matrix that effects a change in basis such that the orthogonality of the basis is preserved under the change.

We illustrate these two interpretations with the orthogonal matrix

$$P = \begin{bmatrix} 1/\sqrt{2} & 1/\sqrt{2} \\ -1/\sqrt{2} & 1/\sqrt{2} \end{bmatrix}$$

and the quadratic equation $x^2 + 4xy + y^2 = 12$, or $X^T \begin{bmatrix} 1 & 2 \\ 2 & 1 \end{bmatrix} X = 12$. The vectors X that satisfy this equation can be described geometrically as originating at the origin and terminating on the hyperbola shown in Figure 10.5.1.

10.5 ORTHOGONAL CONGRUENCE

First of all, from $X = PU$ we obtain the transformation $U = P^{-1}X$, or

$$\begin{bmatrix} u \\ v \end{bmatrix} = \begin{bmatrix} 1/\sqrt{2} & -1/\sqrt{2} \\ 1/\sqrt{2} & 1/\sqrt{2} \end{bmatrix} \begin{bmatrix} x \\ y \end{bmatrix}.$$

This transformation will transform the vectors terminating on the hyperbola $x^2 + 4xy + y^2 = 12$ shown in Figure 10.5.1 to corresponding vectors terminating on the hyperbola $-u^2 + 3v^2 = 12$ shown in Figure 10.5.2. In Figure 10.5.2 notice that the coordinate system (or basis) has not moved but that the hyperbola itself has undergone a rigid 45° rotation. The length of any particular vector and the angle between any pair of vectors have been preserved.

Secondly, however, from $X = PU$ we can think of orthogonal P as a transition matrix from basis b to a new basis e. Hence the quadratic equation

$$-u^2 + 3v^2 = 12$$

can be viewed as the unchanged hyperbola expressed in terms of a new orthogonal coordinate system as shown in Figure 10.5.3.

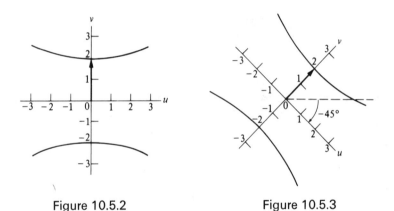

Figure 10.5.2   Figure 10.5.3

In either interpretation, the matrix A of the quadratic form is transformed to its image $P^T A P$ where $P^T = P^{-1}$, and this transformation of matrix A is an example of an *orthogonal similarity* or an *orthogonal congruence* transformation.

The following theorem is one of the most important in linear algebra. Among other things it will assure us that we can always transform the matrix of a real quadratic form to a diagonal matrix of characteristic values.

Theorem 13. *If A is an n by n symmetric R-matrix, then A is orthogonally congruent (or orthogonally similar) to a diagonal matrix with the n diagonal entries equal to the n characteristic values of A.*

Proof. If $n = 1$, the Theorem is obvious. If $n > 1$, we will show that there exists an orthogonal matrix P such that

$$P^T A P = \begin{bmatrix} \lambda_1 & & 0 \\ & \ddots & \\ 0 & & \lambda_n \end{bmatrix}.$$

(1) Let λ_1 be a distinct or multiple characteristic root of A. By Theorem 2 of Chapter 9, λ_1 exists and is real, and hence there is a normalized real characteristic vector $X_1 \neq \mathbf{0}$ such that

$$A X_1 = \lambda_1 X_1.$$

(2) With X_1 written as the first column, an orthogonal matrix Q can be built according to the Gram-Schmidt procedure of Section 7.6.

(3) Form the product AQ or $A[X_1 \mid Q_2 \mid \cdots \mid Q_n]$. The first column of AQ is AX_1 which by Statement (1) is $\lambda_1 X_1$.

(4) Now form the product $M = Q^T(AQ)$ or

$$M = \begin{bmatrix} X_1^T \\ \hline Q_2^T \\ \hline \vdots \\ \hline Q_n^T \end{bmatrix} [\lambda_1 X_1 \mid AQ_2 \mid \cdots \mid AQ_n].$$

The first entry in the first column of M is $X_1^T(\lambda_1 X_1)$, which is λ_1 since X_1 is normal. The remaining entries in the first column of M are $Q_i^T(\lambda_1 X_1) = 0$ because the rows of Q^T (columns of Q) were constructed so as to be orthogonal to X_1. Hence M has the form

$$Q^T A Q = \begin{bmatrix} \lambda_1 & F \\ \hline 0 & B \end{bmatrix},$$

where F and B will be discussed next. (An alternate argument here is that since $Q^{-1} = Q^T$, the first column of $Q^T A Q$ is $\lambda_1 Q^{-1} X_1$, which is the same as λ_1 times the first column of $Q^{-1} Q$, which is λ_1 times the first column of I_n.)

(5) Because A is symmetric, $Q^T A Q$ is symmetric according to Theorem 7; therefore $F = \mathbf{0}$, and

$$Q^T A Q = \begin{bmatrix} \lambda_1 & 0 \\ \hline 0 & B \end{bmatrix},$$

where B is an $n - 1$ by $n - 1$ symmetric R-matrix.

(6) Let λ_2 be a characteristic value of A. By Theorem 1, similar matrices have the same characteristic values, and therefore λ_2 is a characteristic value of B (regardless of whether λ_1 equals λ_2 or not).

10.5 ORTHOGONAL CONGRUENCE

(7) If $n > 2$, then by a repetition of Statements (1)–(5) we can show that there exists an orthogonal matrix R such that

$$R^T B R = \begin{bmatrix} \lambda_2 & 0 \\ \hline 0 & C \end{bmatrix},$$

where C is an $n - 2$ by $n - 2$ symmetric R-matrix with characteristic values $\lambda_3, \ldots, \lambda_n$, which are the remaining characteristic values of A.

(8) Because R is orthogonal, $S = \begin{bmatrix} 1 & 0 \\ \hline 0 & R \end{bmatrix}$ is orthogonal, and consequently QS is orthogonal. Thus we have

$$(QS)^T A (QS) = S^T(Q^T A Q)S = \begin{bmatrix} 1 & 0 \\ \hline 0 & R^T \end{bmatrix} \begin{bmatrix} \lambda_1 & 0 \\ \hline 0 & B \end{bmatrix} \begin{bmatrix} 1 & 0 \\ \hline 0 & R \end{bmatrix}$$

$$= \begin{bmatrix} \lambda_1 & 0 \\ \hline 0 & R^T B R \end{bmatrix} = \begin{bmatrix} \lambda_1 & 0 & 0 \\ 0 & \lambda_2 & 0 \\ \hline 0 & & C \end{bmatrix}.$$

(9) The remaining stages can be accomplished in the same manner as outlined above, and we will obtain a product of n orthogonal matrices which we define to be P. Hence there exists an orthogonal matrix P such that

$$P^{-1} A P = P^T A P = \begin{bmatrix} \lambda_1 & & 0 \\ & \ddots & \\ 0 & & \lambda_n \end{bmatrix}. \quad \square$$

The proof of the previous theorem does not provide us with an efficient method of constructing the orthogonal matrix P such that $P^T A P = P^{-1} A P$ is a diagonal matrix with the characteristic values on the main diagonal. This can be done easily, however, because the proof of Theorem 2 assures us that P must be a modal matrix, and because Theorem 13 of Chapter 4 states that a matrix will be orthogonal if its columns form an orthonormal set. Hence we must find a modal matrix with orthonormal columns. If the characteristic values are distinct, it follows from Theorem 4 of Chapter 9 that the columns of a modal matrix of a symmetric matrix must be orthogonal, and, of course, they can be normalized. If, however, the characteristic values are not distinct, then the orthogonal characteristic vectors corresponding to a multiple characteristic value can be constructed as shown in the next example; Theorem 13 assures us that this can be done.

Example 3. For the quadratic form

$$f(x, y, z) = X^T A X = X^T \begin{bmatrix} 2 & 0 & 1 \\ 0 & 3 & 0 \\ 1 & 0 & 2 \end{bmatrix} X,$$

the characteristic values of A are $\lambda_1 = 1$, $\lambda_2 = \lambda_3 = 3$. A normalized characteristic vector corresponding to $\lambda_1 = 1$ is found to be $X_1 = \begin{bmatrix} 1/\sqrt{2} \\ 0 \\ -1/\sqrt{2} \end{bmatrix}$. The characteristic values $\lambda_2 = \lambda_3 = 3$ yield

$$(A - 3I_3)X = 0 \quad \text{or} \quad \begin{bmatrix} -1 & 0 & 1 \\ 0 & 0 & 0 \\ 1 & 0 & -1 \end{bmatrix} \begin{bmatrix} x \\ y \\ z \end{bmatrix} = \begin{bmatrix} 0 \\ 0 \\ 0 \end{bmatrix}$$

whose complete solution is

$$\begin{cases} x = z, \\ y \quad \text{arbitrary}; \end{cases}$$

therefore we must choose two vectors subject to these conditions and orthogonal to X_1 and to each other. The vectors $\begin{bmatrix} 1 \\ 1 \\ 1 \end{bmatrix}$ and $\begin{bmatrix} 1 \\ -2 \\ 1 \end{bmatrix}$ suffice; notice that they are not unique. Normalized, these vectors become $X_2 = \begin{bmatrix} 1/\sqrt{3} \\ 1/\sqrt{3} \\ 1/\sqrt{3} \end{bmatrix}$ and $X_3 = \begin{bmatrix} 1/\sqrt{6} \\ -2/\sqrt{6} \\ 1/\sqrt{6} \end{bmatrix}$.

Therefore $P = \begin{bmatrix} 1/\sqrt{2} & 1/\sqrt{3} & 1/\sqrt{6} \\ 0 & 1/\sqrt{3} & -2/\sqrt{6} \\ -1/\sqrt{2} & 1/\sqrt{3} & 1/\sqrt{6} \end{bmatrix}$ and $P^T A P = \begin{bmatrix} 1 & 0 & 0 \\ 0 & 3 & 0 \\ 0 & 0 & 3 \end{bmatrix}$.

Example 4. The characteristic values of the matrix

$$A = \begin{bmatrix} 5 & -1 & -1 \\ -1 & 3 & 1 \\ -1 & 1 & 3 \end{bmatrix}$$

are 2, 3, and 6. Characteristic vectors corresponding to these values are

$$\begin{bmatrix} 0 \\ 1 \\ -1 \end{bmatrix}, \begin{bmatrix} 1 \\ 1 \\ 1 \end{bmatrix}, \text{ and } \begin{bmatrix} -2 \\ 1 \\ 1 \end{bmatrix}.$$

Because A is symmetric and the characteristic values are distinct, the characteristic vectors must be orthogonal; it can be verified that they are in fact orthogonal. If these vectors are normalized, then the following orthogonal matrix is obtained:

$$P = \begin{bmatrix} 0 & 1/\sqrt{3} & -2/\sqrt{6} \\ 1/\sqrt{2} & 1/\sqrt{3} & 1/\sqrt{6} \\ -1/\sqrt{2} & 1/\sqrt{3} & 1/\sqrt{6} \end{bmatrix}.$$

It can be verified that

$$P^T A P = P^{-1} A P = \begin{bmatrix} 2 & 0 & 0 \\ 0 & 3 & 0 \\ 0 & 0 & 6 \end{bmatrix}.$$

Therefore the quadratic equation
$$5x^2 + 3y^2 + 3z^2 - 2xy + 2yz - 2xz = 1$$
becomes
$$2u^2 + 3v^2 + 6w^2 = 1$$
under $X = PU$.

The reader should note that neither P nor P^TAP is unique, because the characteristic values may be listed in any order.

So far in this section we have dealt exclusively with R-matrices. Definition 5 and Theorem 13 can be generalized to deal with C-matrices.

Definition 6. *A C-matrix A is said to be **unitarily congruent** (or **unitarily similar**) to a C-matrix B if there exists a unitary C-matrix P such that*
$$B = P^{-1}AP = P^*AP.$$

Theorem 14. *If A is an n by n Hermitian C-matrix, then A is unitarily congruent (or unitarily similar) to a diagonal matrix with the n diagonal entries equal to the n characteristic values of A.*

The proof parallels that of Theorem 13 and is left as Exercise 19.

In Theorem 14 the reader should notice that, since Hermitian matrices have real characteristic values, the unitarily congruent diagonal matrix will be real. Hence we are led to the still more general theorem where a C-matrix is unitarily congruent to a diagonal matrix that may not be real.

Theorem 15. *An n by n C-matrix A is unitarily congruent (or unitarily similar) to a diagonal matrix if and only if A is normal (that is, $A^*A = AA^*$).*

Proof. Part I of the proof is left for Exercise 18. For the other part, the reader is referred to page 81 of Lancaster, *Theory of Matrices*, Academic Press, New York 1969. □

Because the characteristic values are invariant or unchanged under a similarity transformation, the diagonal elements of the diagonal matrix of the last theorem must be the characteristic values of A.

APPLICATIONS

For applications of the material of this section refer to Examples 6, 7, and 8 of Section 9.3.

EXERCISES

In each of Exercises 1–9, eliminate the product terms (that is, the terms involving xy, xz, or yz) by finding and using an orthogonal transition matrix P.

1. $2x^2 - 4xy + 5y^2 = 6$. Use answer to Exercise 3 of Section 9.1.
2. $2x^2 + 12xy - 3y^2 = 4$. Use answer to Exercise 13 of Section 9.1.
3. $x^2 + 4xy + y^2 = 1$.
4. $x^2 - 4xy - 2y^2 = 9$.
5. $2x^2 + 2xy + y^2 + z^2 + 2xz = 9$. Use answer to Exercise 7 of Section 9.1.
6. $y^2 + 4xz = 8$. Use answer to Exercise 11 of Section 9.1.
7. $3x^2 + 2xy + 3y^2 + 2z^2 = 4$.
8. $x^2 + 4y^2 + 3z^2 - 4xy = 5$.
9. $xy = 4$.
10. A primary problem in analytic geometry is to rotate the axes so that the equation of a given conic does not have terms that include the product of two different variables such as xy, or yz, or xz. Demonstrate that this is what was accomplished in Exercise 9, by graphing the given equation and the resulting equation in the answer.
11. Find a transition matrix P such that $P^T A P$ is equal to a diagonal matrix, and then find that diagonal matrix, if

$$A = \begin{bmatrix} 2 & -1 & 1 \\ -1 & 2 & -1 \\ 1 & -1 & 2 \end{bmatrix}.$$

Hint: See Example 3 of this section.

12. In Example 5 of Section 10.3 a *nonorthogonal* transition matrix P was found and graphs were made of a given hyperbola before and after the transformation T defined by $X \xmapsto{T} P^{-1}X$. According to the methods of this present section find the new equation of the hyperbola, and graph the hyperbola after an orthogonal transformation. Then compare your graph with that of Example 5 of Section 10.3. The equation of the hyperbola is

$$x^2 + 6xy + 5y^2 = 1.$$

13. In Example 4 of Section 10.3 a *nonorthogonal* transition matrix P was found and graphs were made of a given ellipse before and after the transformation T defined by $X \xmapsto{T} P^{-1}X$. According to the methods of this present section find the new equation of the ellipse, and graph the ellipse after an orthogonal transformation. Then compare your graph with that of Example 4 of Section 10.3. The equation of the ellipse is

$$2x^2 - 4xy + 7y^2 = 6.$$

14. Find a diagonal matrix which is unitarily congruent to $A = \begin{bmatrix} 1 & -2i \\ 2i & 1 \end{bmatrix}$. Is A normal (that is, does $A^*A = AA^*$)? Is A Hermitian?

15. Find a diagonal matrix which is unitarily congruent to $A = \begin{bmatrix} 1 & 2i \\ 2i & 1 \end{bmatrix}$. Is A normal (that is, does $A^*A = AA^*$)? Is A Hermitian?

16. Prove: If an R-matrix A is orthogonally congruent to a diagonal matrix, then A is symmetric.

17. Prove: If a C-matrix A is unitarily congruent to a diagonal R-matrix, then A is Hermitian. (Notice here that the diagonal matrix is real. See Exercise 14.)

18. Prove: If an n by n C-matrix A is unitarily congruent to a diagonal matrix (not necessarily real), then A is normal (that is, $A^*A = AA^*$).

19. Prove Theorem 14.

10.6 Definite and Semi-Definite Forms

It is sometimes convenient to classify real quadratic forms with n variables $x_1, x_2, \ldots, x_n$ into the five categories of Definition 7.

Definition 7. Let $X = (x_1, x_2, \ldots, x_n)$. A real quadratic form $f(X)$ is defined to be:

(i) **positive definite** if $f(X) > 0$ when $X \neq 0$; (In other words, for $X \neq 0$, $f(X)$ is definitely positive),

(ii) **positive semidefinite** if $f(X) \geq 0$ and there exists some $X \neq 0$ for which $f(X) = 0$; (In other words, $f(X)$ is not positive definite and yet $f(X)$ is never negative),

(iii) **negative definite** if $f(X) < 0$ when $X \neq 0$; (In other words, for $X \neq 0$, $f(X)$ is definitely negative),

(iv) **negative semidefinite** if $f(X) \leq 0$ and there exists some $X \neq 0$ for which $f(X) = 0$; (In other words, $f(X)$ is not negative definite and yet $f(X)$ is never positive),

(v) **indefinite** for all other cases.

Example 1. The real quadratic form $f(X) = x_1^2 - 6x_1x_2 + 9x_2^2$ is positive or zero because

$$f(X) = (x_1 - 3x_2)^2$$

can never be negative for real values of x_1 and x_2; if however $X = (3, 1)$ we see that $f(X) = 0$, and hence $f(X)$ is not positive definite; therefore $f(X)$ is positive semidefinite.

Example 2. Given the real quadratic form

$$f(X) = -x_1^2 - x_2^2;$$

if $X \neq 0$, then $f(X)$ must always be negative for real values of x_1 and x_2. We then say that $f(X)$ is negative definite.

Example 3. The real quadratic form

$$f(X) = x_1^2 - 3x_1x_2 + x_2^2$$

is negative for $X = (1, 1)$ whereas $f(X)$ is positive for $X = (5, 1)$, therefore we say that $f(X)$ is indefinite.

Definition 8. *A nonzero symmetric R-matrix A is **positive definite**, **positive semidefinite**, **negative definite**, **negative semidefinite**, or **indefinite** according to the classification of the associated real quadratic form of A.*

We now state several theorems involving the concepts just defined. Some of these theorems can be useful in classifying a particular quadratic form.

Theorem 16. *If A is a nonzero n by n symmetric R-matrix having rank r and index p (see Theorem 10 for the meaning of p) then A is:*
  (i) *positive definite if and only if $p = r = n$,*
  (ii) *positive semidefinite if and only if $p = r < n$,*
  (iii) *negative definite if and only if $p = 0$ and $r = n$,*
  (iv) *negative semidefinite if and only if $p = 0$ and $r < n$.*

Proof. A proof of the first two parts of this theorem can be found in Perlis, S., *Theory of Matrices*, Addison Wesley, Reading, Mass., 1952, pages 93–94. The proof of the last two parts is similar to the proof of the first two parts. ☐

Theorem 17. *If A is a nonzero n by n symmetric R-matrix having rank r, then A is:*
  (i) *positive definite if and only if A is congruent to I_n,*
  (ii) *positive semidefinite if and only if A is congruent to*

$$\left[\begin{array}{c|c} I_r & 0 \\ \hline 0 & 0 \end{array}\right]_{(n,n)}, \quad n \neq r,$$

  (iii) *negative definite if and only if A is congruent to $-I_n$,*
  (iv) *negative semidefinite if and only if A is congruent to*

$$\left[\begin{array}{c|c} -I_r & 0 \\ \hline 0 & 0 \end{array}\right]_{(n,n)}, \quad n \neq r.$$

The proof follows almost immediately from Theorem 16 and is left as Exercises 15 and 16.

Theorem 18. *If A is a nonzero n by n symmetric R-matrix having rank r, then A is:*

(i) *positive definite if and only if there exists an invertible R-matrix P such that $A = P^T P$,*

(ii) *positive semidefinite if and only if there exists a noninvertible R-matrix P of rank r such that $A = P^T P$,*

(iii) *negative definite if and only if there exists an invertible R-matrix P such that $A = -P^T P$,*

(iv) *negative semidefinite if and only if there exists a noninvertible R-matrix P of rank r such that $A = -P^T P$.*

The proof follows easily from Theorem 17 and is left as Exercises 17 and 18.

Theorem 19. *If A is a nonzero n by n symmetric R-matrix having rank r, then A is:*

(i) *positive definite if and only if all of the characteristic values of A are positive,*

(ii) *positive semidefinite if and only if r characteristic values of A are positive and $n - r$ characteristic values of A are zero,*

(iii) *negative definite if and only if all of the characteristic values of A are negative,*

(iv) *negative semidefinite if and only if r of the characteristic values of A are negative and $n - r$ characteristic values of A are zero.*

Proof. An outline of a proof of the first part follows, and a formal proof of the entire theorem is left as Exercise 20. Because A is symmetric it is orthogonally congruent, and hence congruent, to a diagonal R-matrix D with characteristic values of A on the main diagonal. By Theorem 17, A is positive definite if and only if A is congruent to I_n and hence if and only if D is congruent to I_n. But D is congruent to I_n if and only if all entries on the main diagonal of D are positive. □

For the convenience of the reader the definitions and theorems of this section are summarized in Table 10.6.1.

APPLICATIONS

★**Example 4.** Consider a real function $f(x, y)$ having continuous second partial derivatives at a point p_0 and first partial derivatives for which

$$\left.\frac{\partial f}{\partial x}\right|_{p_0} = 0 \quad \text{and} \quad \left.\frac{\partial f}{\partial y}\right|_{p_0} = 0.$$

TABLE 10.6.1

Quadratic forms	By Definitions	By Theorem 16	By Theorem 17	By Theorem 18	By Theorem 19
Positive Definite R-matrix A	Quadratic form > 0 for all $X \neq \mathbf{0}$.	$p = r = n$.	$A \stackrel{c}{=} I_n$, where $\stackrel{c}{=}$ designates "is congruent to."	There exists an invertible R-matrix P such that $P^T P = A$.	All characteristic values of A are positive.
Positive Semidefinite R-matrix A	Quadratic form ≥ 0 for all $X \neq \mathbf{0}$ and $= 0$ for some $X \neq \mathbf{0}$.	$p = r < n$.	$A \stackrel{c}{=} \begin{bmatrix} I_r & 0 \\ \hline 0 & 0 \end{bmatrix}_{(n,n)}$. $r < n$.	There exists a noninvertible R-matrix P of rank r such that $P^T P = A$.	r characteristic values of A are positive and $n - r$ of them are 0.
Indefinite R-matrix A	Quadratic form > 0 for some X and < 0 for some X.	$p \neq r$. $p \neq 0$.	$A \stackrel{c}{=} \begin{bmatrix} I_p & 0 & 0 \\ 0 & -I_{r-p} & 0 \\ 0 & 0 & 0 \end{bmatrix}_{(n,n)}$. $p \neq r$.		There exist some positive and some negative characteristic values.
Negative Semidefinite R-matrix A	Quadratic form ≤ 0 for all $X \neq \mathbf{0}$, and $= 0$ for some $X \neq \mathbf{0}$.	$p = 0$. $r < n$.	$A \stackrel{c}{=} \begin{bmatrix} -I_r & 0 \\ \hline 0 & 0 \end{bmatrix}_{(n,n)}$. $r < n$.	There exists a noninvertible R-matrix P of rank r such that $-P^T P = A$.	r characteristic values of A are negative and $n - r$ of them are 0.
Negative Definite R-matrix A	Quadratic form < 0 for all $X \neq \mathbf{0}$.	$p = 0$. $r = n$.	$A \stackrel{c}{=} -I_n$.	There exists an invertible R-matrix P such that $-P^T P = A$.	All characteristic values of A are negative.

10.6 DEFINITE AND SEMI-DEFINITE FORMS

Form the matrix

$$M_0 = \begin{bmatrix} \dfrac{\partial^2 f}{\partial x^2}\bigg|_{p_0} & \dfrac{\partial^2 f}{\partial y \partial x}\bigg|_{p_0} \\ \dfrac{\partial^2 f}{\partial x \partial y}\bigg|_{p_0} & \dfrac{\partial^2 f}{\partial y^2}\bigg|_{p_0} \end{bmatrix}.$$

It can be proved[4] that:

If M_0 is positive definite, f has a minimum value at p_0.
If M_0 is negative definite, f has a maximum value at p_0.
If M_0 is indefinite, f has neither a maximum nor a minimum value at p_0.

The other alternative, M_0 is semidefinite, simply implies that the test fails. As an illustration consider the function

$$f(x, y) = x^2 + xy + y^2 - 5x - 4y + 1.$$

$$\frac{\partial f}{\partial x} = 2x + y - 5,$$

$$\frac{\partial f}{\partial y} = x + 2y - 4,$$

and

$$\frac{\partial f}{\partial x} = \frac{\partial f}{\partial y} = 0 \quad \text{at} \quad p_0 = (2, 1).$$

Since

$$M_0 = \begin{bmatrix} 2 & 1 \\ 1 & 2 \end{bmatrix}$$

has the characteristic values 1 and 3, both of which are positive, then M_0 is positive definite. Therefore f has a minimum value at $(2, 1)$. Procedures similar to those used in this example are valid for appropriate functions of three or more variables. The reader may also find it instructive to show that the procedure of this example is what is commonly called the second derivative test in the optimization of functions of a single variable.

★**Example 5.** One use of the material of this section and of Section 10.5 may be found in electrical and mechanical systems having two degrees of freedom. In this example, we will limit our considerations to the frictionless mechanical system pictured in Figure 10.6.1 in which two masses m_1 and m_2 and three springs k, k_1, and k_2 are connected to two fixed walls; the variables x_1 and x_2 measure the displacement of the masses from the position of equilibrium. The electrical analogue is shown in Figure 10.6.2. (A more comprehensive treatment of this

---

[4] Birkoff and MacLane, *A Survey of Modern Algebra*, 3rd Edition, MacMillan, New York, 1965, page 256.

traditional problem may be found on pages 103–108 of Franklin, J. N., *Matrix Theory*, Prentice-Hall, Englewood Cliffs, N.J., 1968, and on pages 218–226 of Pipes, L. A., *Matrix Methods for Engineering*, Prentice Hall, Englewood Cliffs, N.J., 1963.) It can be determined that the equation of motion is

(10.6.1) $$M\ddot{X} + KX = 0,$$

where

$$M = \begin{bmatrix} m_1 & 0 \\ 0 & m_2 \end{bmatrix}, \quad K = \begin{bmatrix} (k_1 + k) & -k \\ -k & (k_2 + k) \end{bmatrix}, \quad \text{and} \quad X = \begin{bmatrix} x_1(t) \\ x_2(t) \end{bmatrix}.$$

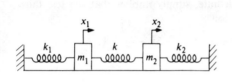
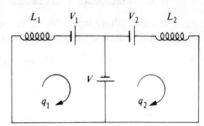

Figure 10.6.1   Figure 10.6.2

Notice that M and K are symmetric. It can also be determined that expressions for the kinetic energy T and potential energy V are the quadratic forms

$$T = \tfrac{1}{2}\dot{X}^T M \dot{X}$$

and

$$V = \tfrac{1}{2} X^T K X,$$

respectively. Moreover, it is easy to show by the definition that both of these quadratic forms are positive definite (Exercise 26). Consequently we employ the following theorem:

Theorem 20. *If M and K are two n by n symmetric R-matrices where M is positive definite, then there exists an invertible R-matrix P such that*

$$P^T M P = I_n \text{ and } P^T K P = D,$$

where D is a diagonal matrix whose diagonal entries d_{ii} are roots of the equation

$$|K - \lambda M| = 0.$$

Moreover, if K is positive definite, then all d_{ii} are positive.

Proof. See Franklin, J. N., *Matrix Theory*, Prentice Hall, Englewood Cliffs, N.J., 1968, page 106. □

10.6 DEFINITE AND SEMI-DEFINITE FORMS

Theorem 20 is used to verify that, under the substitution $X = PY$ and pre-multiplication by P^T, the original equation (10.6.1) becomes

$$(P^T M P)\ddot{Y} + (P^T K P) Y = 0,$$

or

$$\ddot{Y} + DY = 0,$$

where $d_{ii} > 0$. Since $d_{ii} > 0$, we introduce a new diagonal matrix $\Omega = \begin{bmatrix} \omega_1 & 0 \\ 0 & \omega_2 \end{bmatrix}$ such that $D = \Omega^2$. This gives $d_{ii} = \omega_i^2$, and we have

$$\ddot{Y} + \Omega^2 Y = 0,$$

or

$$\begin{cases} y_1''(t) + \omega_1^2 y_1(t) = 0, \\ y_2''(t) + \omega_2^2 y_2(t) = 0. \end{cases}$$

Therefore

$$Y = \begin{bmatrix} c_1 \cos \omega_1 t + c_2 \sin \omega_1 t \\ c_3 \cos \omega_2 t + c_4 \sin \omega_2 t \end{bmatrix},$$

and

$$X = PY.$$

EXERCISES

1. Use Definition 7 to classify $f(x_1, x_2) = x_1^2 - x_2^2$.
2. Use Definition 7 to classify $f(x_1, x_2) = 2x_1^2 + 4x_1 x_2 + 2x_2^2$.
3. Use Definition 8 and Theorem 16 to classify $f(x, y) = 2x^2 + 2xy + 2y^2$.
4. Use Definition 8 and Theorem 17 to classify $f(x, y) = 2x^2 + 2xy + 2y^2$.
5. For the matrix $A = \begin{bmatrix} 2 & 1 \\ 1 & 2 \end{bmatrix}$, find the matrix P of Theorem 18 and then classify A according to that theorem.
6. Use Theorem 19 to classify $A = \begin{bmatrix} 2 & 1 \\ 1 & 2 \end{bmatrix}$.

In each of Exercises 7–14, classify each of the following symmetric R-matrices according to whether they are positive definite, negative definite, positive semi-definite, negative semidefinite, or indefinite.

7. $\begin{bmatrix} 2 & -2 \\ -2 & 5 \end{bmatrix}$. 8. $\begin{bmatrix} 0 & 1 \\ 1 & 0 \end{bmatrix}$. 9. $\begin{bmatrix} 1 & 2 \\ 2 & 1 \end{bmatrix}$. 10. $\begin{bmatrix} 2 & 6 \\ 6 & -3 \end{bmatrix}$.

11. $\begin{bmatrix} 2 & 1 & 1 \\ 1 & 1 & 0 \\ 1 & 0 & 1 \end{bmatrix}$. 12. $\begin{bmatrix} 1 & -2 & 0 \\ -2 & 4 & 0 \\ 0 & 0 & 3 \end{bmatrix}$.

13. $\begin{bmatrix} 1 & 0 & 1 \\ 0 & 1 & 2 \\ 1 & 2 & 5 \end{bmatrix}$. 14. $\begin{bmatrix} 2 & -1 & 1 \\ -1 & 2 & -1 \\ 1 & -1 & 2 \end{bmatrix}$.

15. Prove parts (*i*) and (*ii*) of Theorem 17.

16. Prove parts (*iii*) and (*iv*) of Theorem 17.
17. Prove parts (*i*) and (*ii*) of Theorem 18.
18. Prove parts (*iii*) and (*iv*) of Theorem 18.
19. Give an outline of the proof of the second part of Theorem 19.
20. Prove Theorem 19.

In each of Exercises 21–24, test for maximum and minimum values of the given function.

21. $f(x, y) = x^3 + y^3 - 6xy$.
22. $f(x, y) = -x^4 - x^2y^2 - y^4 + 9x^2 + 6y^2 + 8$.
23. $f(x, y, z) = 4 - y^4 - x^2z^2 - x^2 + y^2 + 2z^2 + 2xyz$; test only at the origin.
24. $f(x, y, z) = 6 - 2y^3 + 5xz^2 - 2x^2 - y^2 - z^2 - 3xyz$.
25. Use the function

$$f(x, y) = -2x^4 + y^2$$

to show that the following conjecture is false: in Example 4, if f has a maximum at a point X_0 then M_0 is positive definite.

26. Prove that matrix K of Example 5 is positive definite. *Hint:* Show that the quadratic form $X^T K X$ is positive definite.
27. Use the results of Example 5 to solve a spring vibration problem,

$$M\ddot{X} + KX = 0,$$

with $m_1 = 2$, $m_2 = 0$, $k = 4$, $k_1 = 2$, $k_2 = 1$.

NEW VOCABULARY

similarity transformation   10.1
quadratic form   10.1
matrix of a quadratic form   10.3
real quadratic form   10.3
congruent matrices   10.3
congruence transformation   10.3
index of a symmetric R-matrix   10.3
Hermitian congruence transformations   10.3
Hermitely congruent (conjunctive) matrices   10.3
orthogonal linear operator   10.4
unitary linear operator   10.4
orthogonal congruence transformation   10.5

orthogonal similarity transformation   10.5
orthogonally congruent matrices   10.5
orthogonally similar matrices   10.5
unitarily congruent matrices   10.5
unitarily similar matrices   10.5
positive definite real quadratic form   10.6
positive semidefinite real quadratic form   10.6
negative definite real quadratic form   10.6
negative semidefinite real quadratic form   10.6
indefinite real quadratic form   10.6

11

LINEAR PROGRAMMING (optional)

11.1 Introduction

We now turn our attention to a special type of optimization problem known as a *linear programming problem*. We start with the following example of such a problem.

Example 1. Maximize the *linear* function

$$f(x_1, x_2, x_3, x_4) = 4x_1 + 2x_2 + 6x_3 + 0x_4$$

subject to the conditions or *constraints* that all variables must be nonnegative and that

$$\begin{cases} x_1 + x_2 + x_3 & \leq 3, \\ x_1 + 2x_2 + x_3 + x_4 = 5, \\ -3x_1 + x_3 \geq -6. \end{cases}$$

Notice that all constraints are linear equations or inequalities.

In general a linear programming problem seeks to find values of nonnegative real variables that will optimize a linear function of those variables subject to constraints in the form of linear equations or inequalities. Problems of this type, which are frequently concerned with decision making, have attracted considerable attention in recent years. After making the following important definition we shall give a very simple problem and its solution.

Definition 1. *A real coordinate vector that simultaneously satisfies all of the constraints of a linear programming problem is called a **feasible solution** of the problem. The set of all feasible solutions is called the **feasible set** of the problem.*

Example 2. Two factories manufacture three different grades of paper, and there is some demand for each grade. The company that controls the factories has contracts to supply 16 tons of low-grade, 5 tons of medium-grade, and 20 tons of high-grade paper. It costs $1,000 per day to operate the first factory and $2,000 per day to operate the second factory. When it operates, Factory 1 produces 8 tons of low-grade, 1 ton of medium-grade, and 2 tons of high-grade paper per day of operation. Factory 2 produces 2 tons of low-grade, 1 ton of medium-grade, and 7 tons of high-grade paper per day of operation. How many days should each factory be operated in order to fill the orders most economically?

	Factory 1	Factory 2	Tons Needed
Low-Grade	8 tons per day	2 tons per day	16
Medium-Grade	1 ton per day	1 ton per day	5
High-Grade	2 tons per day	7 tons per day	20
Daily Cost	$1,000	$2,000	

The table shows the information stated in the problem. The entries in the table indicate the production of each grade of paper by each factory, the minimum requirements to fill the orders, and the daily cost of operating each factory. Let x_1 and x_2 be the number of days that Factories 1 and 2, respectively, operate in order to produce the required amounts.

(1) $8x_1 + 2x_2 \geq 16$.   (At least 16 tons of low-grade paper are required.)
(2) $x_1 + x_2 \geq 5$.   (At least 5 tons of medium-grade paper are required.)
(3) $2x_1 + 7x_2 \geq 20$.   (At least 20 tons of high-grade paper are required.)
(4) $x_1 \geq 0$, (5) $x_2 \geq 0$.   (Number of days operated must be nonnegative.)

The five inequalities above represent five restrictions on our variables. Subject to these restrictions, or constraints, we wish to minimize the total operating cost of the factories, namely,

$$f(x_1, x_2) = 1{,}000 x_1 + 2{,}000 x_2.$$

Geometric Solution. First we will give a geometric solution of the problem. Consider $8x_1 + 2x_2 \geq 16$ or equivalently $4x_1 + x_2 \geq 8$; this represents all points on and above the line

(1) $4x_1 + x_2 = 8.$

Likewise, graph the other inequalities. All points in the shaded region in Figure 11.1.1 satisfy the five inequalities and represent the feasible set.

The cost function is $f(x_1, x_2) = 1{,}000x_1 + 2{,}000x_2$ or $f = 1{,}000x_1 + 2{,}000x_2$. If f is assigned arbitrary values, the equation in each case represents a straight line of slope $-\tfrac{1}{2}$. With f as a parameter, $1{,}000x_1 + 2{,}000x_2 = f$ is the equation of a family of parallel lines. The lines correspond to increasing values of f as they go up from the origin. From the figure it appears that the line corresponding to $f = 7{,}000$ is the "lowest" line that has a point in common with the shaded region. Any lower line does not intersect the feasible set, and higher lines represent values of f larger than the minimum. Hence the point $(3, 2)$, which is the intersection of (2) $x_1 + x_2 = 5$ and (3) $2x_1 + 7x_2 = 20$, gives the minimum f, namely $(1{,}000)(3) + (2{,}000)(2) = 7{,}000$. Thus if Factory 1 operates three days, and Factory 2 operates two days, the requirements will be met at less cost than with any other combination.

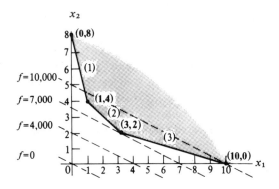

Figure 11.1.1

Notice that if $x_1 = 3$, and if $x_2 = 2$, then $8x_1 + 2x_2 = 28$, or 28 tons of low-grade paper are produced. This is more than the minimum requirement, but any other combination of operating days (including fractions) will result in a greater cost. At first it seems strange that a surplus is obtained at minimum cost.

It is possible for the feasible set of a linear programming problem to be empty; procedures for identifying this case will be developed in Section 11.9.

An R-matrix A is "**greater than or equal to**" ($\geq$) an R-matrix B of the same order when each entry of A is greater than or equal to the corresponding entry of B ($<$, $>$, or $\leq$ can be substituted for $\geq$ with the corresponding changes in meaning). The linear programming problem of Example 2 then can be stated using matrix notation as follows:

minimize CX, subject to

$$\begin{cases} AX \geq P_0, \\ X \geq 0, \end{cases}$$

where

$$C = [1000 \quad 2000], \quad X = \begin{bmatrix} x_1 \\ x_2 \end{bmatrix}, \quad A = \begin{bmatrix} 8 & 2 \\ 1 & 1 \\ 2 & 7 \end{bmatrix}, \quad \text{and} \quad P_0 = \begin{bmatrix} 16 \\ 5 \\ 20 \end{bmatrix}.$$

In linear programming problems it is customary to require that all unknowns be nonnegative, and these imposed constraints are appropriately known as the **nonnegativity constraints**; all other constraints are called **structural constraints**. The linear function that is to be optimized in a linear programming problem is called the **objective function**. The reader should observe that there is no loss in generality in requiring all of the unknowns to be nonnegative, because if in a certain problem there exist possible negative unknowns u_i, then u_i can always be expressed as $u_i = x_i - y_i$ where both x_i and y_i are required to be nonnegative.

Linear programming is a relatively new field in mathematics. Most of the basic work was done in the 1940s by such men as F. L. Hitchcock, T. C. Koopmans, and G. B. Dantzig, although some basic ideas were suggested by L. Kantorovitch in 1939. Much of this basic work was spurred by the economic theories of J. Von Neumann and W. Leontief in the 1930s. In 1947 Dantzig, who was a member of a group studying allocation problems for the U.S. Air Force, formulated the general linear programming problem and developed the simplex method of solution. Since that time the subject has received widespread attention in such diverse fields as nutrition, engineering, economics, agriculture, and many others.

APPLICATIONS

To give the reader some idea of how linear programming can be applied to various fields, the basic mathematical model of Example 2 will be given four different interpretations.

Example 3. The minimum requirements of the amount of certain chemicals necessary to grow a certain crop successfully have been found. Knowing the content and cost of two special types of commercial fertilizers, packaged in 100 pound bags, a grower wants to know how many bags of each type should be applied to his crop to assure proper growth at minimum cost.

	Fertilizer 1	Fertilizer 2	Units needed
Chemical A	8 units per bag	2 units per bag	16
Chemical B	1 unit per bag	1 unit per bag	5
Chemical C	2 units per bag	7 units per bag	20
Cost Per Bag	1,000¢	2,000¢	

Let x_1 be the number of bags of Fertilizer 1 that are applied.
Let x_2 be the number of bags of Fertilizer 2 that are applied.

11.1 INTRODUCTION

Example 4. A nutritionist working for a space craft designer is faced with the problem of minimizing weight subject to certain nutritional requirements. He is considering two foods which are packaged in tubes.

	Food 1	Food 2	Requirements
Carbohydrates	8 units per tube	2 units per tube	16
Fats	1 unit per tube	1 unit per tube	5
Proteins	2 units per tube	7 units per tube	20
Wt. (grams) per tube	1,000	2,000	

Let x_1 be the number of tubes of Food 1 sent on the space craft.
Let x_2 be the number of tubes of Food 2 sent on the space craft.

Example 5. An office manager must assign his two groups of employees in such a way that each of three tasks is performed a certain number of times and the time to do them is held to a minimum.

	Group 1	Group 2	Times needed
Task A	8 times per assignment	2 times per assignment	16
Task B	1 time per assignment	1 time per assignment	5
Task C	2 times per assignment	7 times per assignment	20
Minutes taken per assignment	1,000	2,000	

Let x_1 be the number of times employee Group 1 is assigned to work.
Let x_2 be the number of times employee Group 2 is assigned to work.

Example 6. A fruit buyer needs a certain number of units of various varieties of fruit. Two suppliers can supply his needs but will sell only in full truckloads consisting of a specified number of units of each variety (let one unit equal 10 boxes). How many loads should the buyer order from each supplier in order to save time and money by holding the total shipping distance to a minimum?

	Supplier 1	Supplier 2	Units needed
Variety A	8 units per load	2 units per load	16
Variety B	1 unit per load	1 unit per load	5
Variety C	2 units per load	7 units per load	20
Distance in miles	1,000	2,000	

Let x_1 be the number of truckloads ordered from Supplier 1.
Let x_2 be the number of truckloads ordered from Supplier 2.

The student should observe that in Examples 4 and 6 the solution vector had to have integral components. When applied problems of this type are encountered, and if such a solution vector is not obtained by our methods, then techniques beyond the scope of this text must be employed. The study of such problems is a very interesting branch of linear programming known as *integer programming*.

The petroleum industry has found linear programming useful in refining procedures. Agriculture may use it to secure proper mixes of feed, fertilizer, seed, and the like. The transportation industry is interested in the problem of minimizing shipping costs while supplying many consumers with specified requirements. Some of these problems are more complicated than those treated in this text, and the interested reader should refer to other books. There are, of course, many other possible applications of linear programming, many yet unknown.

EXERCISES

Solve Exercises 1–5 by the geometric method taught in this section.

1. Minimize $x_1 + x_2$, subject to
$$\begin{cases} x_1 + 2x_2 \geq 12, \\ x_1 \geq 6, \\ x_2 \geq 0. \end{cases}$$

2. Minimize $2x_1 + x_2$, subject to
$$\begin{cases} x_1 + x_2 \geq 1, \\ x_1 - x_2 \geq -1, \\ x_1 + 2x_2 \leq 4, \end{cases} \text{ and } \begin{cases} x_1 \geq 0, \\ x_2 \geq 0. \end{cases}$$

3. Maximize $2x_1 + x_2$, subject to the same restrictions as in Exercise 2.

4. Maximize $x_1 - 3x_2$, subject to
$$\begin{bmatrix} -1 & 1 \\ 1 & 1 \end{bmatrix} \begin{bmatrix} x_1 \\ x_2 \end{bmatrix} \leq \begin{bmatrix} 1 \\ 1 \end{bmatrix} \text{ and } \begin{bmatrix} x_1 \\ x_2 \end{bmatrix} \geq \begin{bmatrix} 0 \\ 0 \end{bmatrix}.$$

5. Minimize $x_1 + 2x_2$, subject to
$$\begin{cases} 2x_1 + x_2 \geq 8, \\ x_1 + x_2 \geq 6, \\ x_1 + 2x_2 \geq 9, \end{cases} \text{ and } \begin{cases} x_1 \geq 0, \\ x_2 \geq 0. \end{cases}$$

6. Let $f = 8x_1 + 6x_2$, and let the feasible set be defined by
$$\begin{cases} 4x_1 + 3x_2 \geq 18, \\ 2x_1 + 5x_2 \geq 16, \end{cases} \text{ and } \begin{cases} x_1 \geq 0, \\ x_2 \geq 0. \end{cases}$$

Show that more than one feasible solution will yield minimum f, and give a geometric explanation of that fact.

7. Let $f = 1{,}000x_1 + 1{,}000x_2$, and let the feasible set be that shown in Figure 11.1.1. Show that there is more than one solution that will yield minimum f, and give a geometric explanation of that fact.

8. Show that $f = x_1 - 2x_2$ has no maximum or minimum values over the feasible set shown in Figure 11.1.1.

11.1 INTRODUCTION

9. Show that $g = 4x_1 - 5x_2$ has no maximum or minimum when defined over the feasible set of Exercise 6.

10. Show by a graph that the feasible set of the constraints

$$\begin{cases} x_1 + x_2 \leq 1, \\ x_1 + 2x_2 \geq 4, \\ x_1 \geq 0, \\ x_2 \geq 0, \end{cases}$$

is empty.

11. Two oil refineries produce three grades of gasoline, A, B, and C. At each refinery, the various grades of gasoline are produced in a single operation so that they are in fixed proportions. Assume that one operation at Refinery 1 produces 1 unit of A, 3 units of B, and 1 unit of C. One operation at Refinery 2 produces 1 unit of A, 4 units of B, and 5 units of C. Refinery 1 charges \$300 for the production of one operation, and Refinery 2 charges \$500 for the production of one operation. A consumer needs 100 units of A, 340 units of B, and 150 units of C. How should the orders be placed if the consumer is to meet his needs most economically?

12. Suppose a certain company has two methods, M_1 and M_2, of manufacturing three automobile gadgets, G_1, G_2, and G_3. The first method will produce one of each gadget in 3 hours. The second method will produce three G_1's and one G_3 in 4 hours. The company has an order for six G_1's, two G_2's, and four G_3's. How many times should it employ each method to fill the order and minimize the time spent in production?

13. A trucking company owns 2 types of trucks. Type A has 20 cubic yards of refrigerated space and 30 cubic yards of nonrefrigerated space. Type B has 20 cubic yards of refrigerated space and 10 cubic yards of nonrefrigerated space. A customer wants to haul some produce a certain distance and will require 160 cubic yards of refrigerated space and 120 cubic yards of nonrefrigerated space. The trucking company figures it will take 300 gallons of gas for the type A truck to make the trip and 200 gallons of gas for the type B truck. Find the number of trucks of each type that the company should allow for the job in order to minimize gas consumption.

14. A nutritionist in a large institution wishes to serve food that provides the necessary vitamins and minerals for the inhabitants. Foods F_1 and F_2 contain the following amounts of vitamins and minerals per pound of food eaten.

	F_1	F_2
Vitamins	2 units	4 units
Minerals	5 units	2 units

At least 80 units of vitamins and at least 60 units of minerals must be provided. If the costs of F_1 and F_2 are \$1 and \$.80 per pound, respectively, how many pounds of each food should be ordered to meet minimum diet requirements while also minimizing the total cost of the foods purchased?

15. A local television network is faced with the following problem. At certain times each week it has been found that Program A with 20 minutes of

music and 1 minute of advertisement draws 30,000 viewers while Program B with 10 minutes of music and 1 minute of advertisement draws 10,000 viewers. Within one week the advertiser insists that at least 6 minutes be devoted to his advertisement and the network can afford no more than 80 minutes of music. How many times per week should each program be given in order to obtain the maximum number of viewers?

16. A fruit dealer ships 800 boxes of fruit north on a certain truck. If he must ship at least 200 boxes of oranges at 20¢ a box profit, at least 100 boxes of grapefruit at 10¢ a box profit, and at most 200 boxes of tangerines at 30¢ a box profit, how should he load his truck for maximum profit?

In each of Exercises 17–20, express the linear programming problems using matrix notation, and specify the structural and nonnegativity constraints.

17. Exercise 5 of this section.
18. Exercise 2 of this section.
19. Exercise 15 of this section.
20. Exercise 11 of this section.

11.2 Basic Definitions

When the number of variables of a linear programming problem exceeds three, a geometric method of solution as presented in Section 11.1 is insufficient. Therefore, one must develop an algebraic approach that can be applied to any linear programming problem.

To begin with, we must understand that any structural constraint can be expressed as an equation; an example will illustrate this.

Example 1. Consider the constraints

$$\begin{cases} x_1 + 2x_2 \leq 4, \\ 2x_1 + 3x_2 \geq 3, \quad \text{and} \quad X \geq 0. \\ x_1 + x_2 = 3, \end{cases}$$

Throughout this book the column vector X represents all of the unknowns currently used in the structural constraints. If $x_1 + 2x_2 \leq 4$, then there exists a nonnegative unknown, say x_3, such that

$$(x_1 + 2x_2) + x_3 = 4.$$

For the second constraint, $2x_1 + 3x_2 \geq 3$, the left side is greater than or equal to 3; therefore, there exists some nonnegative unknown, say x_4, such that

$$(2x_1 + 3x_2) - x_4 = 3.$$

The new unknowns, x_3 and x_4, are called *slack variables*; notice that they must be nonnegative. The last constraint is already an equation; therefore the system of constraints can now be written

$$\begin{cases} x_1 + 2x_2 + x_3 = 4, \\ 2x_1 + 3x_2 - x_4 = 3, \quad \text{and} \quad X \geq 0, \\ x_1 + x_2 = 3, \end{cases}$$

or

$$AX = P_0 \quad \text{and} \quad X \geq 0,$$

where A is the matrix of coefficients, $P_0 = \begin{bmatrix} 4 \\ 3 \\ 3 \end{bmatrix}$, and X is now the column matrix of the four current unknowns.

Definition 2. *A slack variable is a nonnegative variable x_q which changes an inequality structural constraint*

$$a_{i1}x_1 + \cdots + a_{ip}x_p \leq p_{i0},$$

or

$$a_{i1}x_1 + \cdots + a_{ip}x_p \geq p_{i0},$$

to an equation

$$a_{i1}x_1 + \cdots + a_{ip}x_p + x_q = p_{i0},$$

or

$$a_{i1}x_1 + \cdots + a_{ip}x_p - x_q = p_{i0},$$

respectively.

The presence of slack variables in the constraints need have no effect on the linear objective function, because the coefficients of the slack variables can be assigned the value zero:

$$\begin{aligned} f &= c_1 x_1 + \cdots + c_p x_p \\ &= c_1 x_1 + \cdots + c_p x_p + 0 x_{p+1} + \cdots + 0 x_n. \end{aligned}$$

Thus the structural constraints can always be stated in the form $AX = P_0$ where X now includes the slack variables as well as the original variables; the nonnegativity constraints are extended to include the nonnegativity of the slack variables.

The matrix equation representing the structural constraints

$$AX = P_0$$

or

$$\begin{bmatrix} a_{11} & a_{12} & \cdots & a_{1n} \\ a_{21} & a_{22} & \cdots & a_{2n} \\ \vdots & \vdots & & \vdots \\ a_{m1} & a_{m2} & \cdots & a_{mn} \end{bmatrix} \begin{bmatrix} x_1 \\ x_2 \\ \vdots \\ x_n \end{bmatrix} = \begin{bmatrix} p_{10} \\ p_{20} \\ \vdots \\ p_{m0} \end{bmatrix}$$

can be thought of as a linear transformation from the vector space of all n by 1 column R-matrices into the vector space generated by the columns of A, that is, the column space of A.

Example 2. Consider the structural constraints

$$\begin{cases} 2x_1 + 3x_2 + x_3 & = p_{10}, \\ x_1 + 4x_2 & + x_4 = p_{20}, \end{cases}$$

or

$$\begin{bmatrix} 2 & 3 & 1 & 0 \\ 1 & 4 & 0 & 1 \end{bmatrix} \begin{bmatrix} x_1 \\ x_2 \\ x_3 \\ x_4 \end{bmatrix} = \begin{bmatrix} p_{10} \\ p_{20} \end{bmatrix}.$$

This matrix equation can be considered as a linear transformation of $V(R)$ into $U(R)$, where $V(R)$ is the vector space of all 4 by 1 column R-matrices and $U(R)$ is the column space of $\begin{bmatrix} 2 & 3 & 1 & 0 \\ 1 & 4 & 0 & 1 \end{bmatrix}$; that is, in this case, the set of all real column vectors of the form $\begin{bmatrix} p_{10} \\ p_{20} \end{bmatrix}$. In this problem, $A = \begin{bmatrix} 2 & 3 & 1 & 0 \\ 1 & 4 & 0 & 1 \end{bmatrix}$ serves as a transformation matrix from $V(R)$ into $U(R)$ with respect to the bases

$$b = \left\{ \begin{bmatrix} 1 \\ 0 \\ 0 \\ 0 \end{bmatrix}, \begin{bmatrix} 0 \\ 1 \\ 0 \\ 0 \end{bmatrix}, \begin{bmatrix} 0 \\ 0 \\ 1 \\ 0 \end{bmatrix}, \begin{bmatrix} 0 \\ 0 \\ 0 \\ 1 \end{bmatrix} \right\} \text{ of } V(R) \text{ and } b' = \left\{ \begin{bmatrix} 1 \\ 0 \end{bmatrix}, \begin{bmatrix} 0 \\ 1 \end{bmatrix} \right\} \text{ of } U(R).$$

Frequently it is convenient to express the structural constraints $AX = P_0$ as a linear combination of the columns of A.

$$x_1 \begin{bmatrix} a_{11} \\ \vdots \\ a_{m1} \end{bmatrix} + x_2 \begin{bmatrix} a_{12} \\ \vdots \\ a_{m2} \end{bmatrix} + \cdots + x_n \begin{bmatrix} a_{1n} \\ \vdots \\ a_{mn} \end{bmatrix} = P_0,$$

or

$$x_1 A_1 + x_2 A_2 + \cdots + x_n A_n = P_0,$$

where the notation A_i represents the ith column of A. We shall require that $n \geq m$;[1] then if the rank[2] of A is m, there will exist m linearly independent columns of A which will serve as a basis of the column space of the matrix A. This means that P_0 can be expressed as a linear combination of the basis column vectors of A; such an expression implies that the coefficients of the nonbasis vectors are zero. This observation leads us to a very important definition.

Definition 3. *Consider the structural constraints $AX = P_0$ or*

$$x_1 \begin{bmatrix} a_{11} \\ \vdots \\ a_{m1} \end{bmatrix} + x_2 \begin{bmatrix} a_{12} \\ \vdots \\ a_{m2} \end{bmatrix} + \cdots + x_n \begin{bmatrix} a_{1n} \\ \vdots \\ a_{mn} \end{bmatrix} = P_0,$$

where $n \geq m$ and where a certain m column vectors of A constitute a basis of the column space of A; the unknown coefficients of the vectors selected to be in the basis are called **basic variables,** *and the rest of the unknowns are called* **nonbasic variables.** *A* **basic solution** *of $AX = P_0$ is a solution in which the nonbasic variables are zero.*

Definition 4. *A* **basic feasible solution** *is a basic solution that is feasible; the associated basis of the column space of A (in Definition 3) is called* **a feasible basis.**

Definition 5. *A* **degenerate basic solution** *is a basic solution in which at least one basic variable is zero; if every basic variable is nonzero, then the basic solution is* **nondegenerate.**

Definition 6. *An* **optimal solution** *is a feasible solution at which the linear objective function is optimum.*

Example 3. For the constraints

$$\begin{bmatrix} 2 & 3 & 1 & 0 \\ 1 & 4 & 0 & 1 \end{bmatrix} \begin{bmatrix} x_1 \\ x_2 \\ x_3 \\ x_4 \end{bmatrix} = \begin{bmatrix} 8 \\ 7 \end{bmatrix},$$

or

$$x_1 \begin{bmatrix} 2 \\ 1 \end{bmatrix} + x_2 \begin{bmatrix} 3 \\ 4 \end{bmatrix} + x_3 \begin{bmatrix} 1 \\ 0 \end{bmatrix} + x_4 \begin{bmatrix} 0 \\ 1 \end{bmatrix} = \begin{bmatrix} 8 \\ 7 \end{bmatrix},$$

---

[1] No generality is lost here, because if $m > n$ then $m - n$ structural constraints are redundant in a consistent system.

[2] The method of solution that we will employ requires that the rank of A be m. If initially the rank of A is not m then the procedure to be followed will be explained in Section 11.9.

we may choose any pair of linearly independent columns to serve as the basis of the vector space $U(R)$ generated by the columns of the matrix. Suppose we choose $\left\{ \begin{bmatrix} 1 \\ 0 \end{bmatrix}, \begin{bmatrix} 0 \\ 1 \end{bmatrix} \right\}$; if the coefficients of the other two columns are zero (that is, $x_1 = 0$, $x_2 = 0$), then the resulting solution $X = \begin{bmatrix} 0 \\ 0 \\ 8 \\ 7 \end{bmatrix}$ is a basic solution. Moreover, since all coordinates are nonnegative, we have a basic feasible solution, and, since the basis vectors had nonzero coefficients, the basic feasible solution is nondegenerate. Of course there are at most 6 (that is, $C(4, 2) = 4!/2!2!$) basic solutions to be obtained by choosing pairs of column vectors to serve as bases; not all, however, will be feasible. Suppose we choose $\begin{bmatrix} 2 \\ 1 \end{bmatrix}$ and $\begin{bmatrix} 1 \\ 0 \end{bmatrix}$ to be the basis; then $x_2 = 0$, $x_4 = 0$, and consequently $x_1 = 7, x_3 = -6$. Hence another basic solution is $\begin{bmatrix} 7 \\ 0 \\ -6 \\ 0 \end{bmatrix}$, but this solution is not feasible, because the third coordinate is negative and the nonnegativity constraints are not satisfied; the basic solution is nondegenerate because the basic variables are nonzero.

The basic feasible solutions are important because we shall find that an optimal solution, if one exists, must be among the basic feasible solutions.

EXERCISES

In each of Exercises 1–4, express the structural constraints as a single matrix equation, write the new nonnegativity constraints, and find any two basic solutions of the matrix equation. Determine whether the two basic solutions are basic feasible solutions and whether they are degenerate.

1. $\begin{cases} -x_1 - x_2 + x_3 \geq 2, \\ -2x_1 + x_2 + x_3 \geq 1, \end{cases}$ and $X \geq 0$.

2. $\begin{cases} x_1 + x_2 \leq 4, \\ x_1 + 4x_2 \leq 7, \\ 2x_2 \leq 3, \end{cases}$ and $X \geq 0$.

3. $\begin{cases} x_1 + 2x_2 + x_3 \geq 8, \\ x_1 + x_2 \leq 5, \\ x_1 + x_2 + x_3 = 2, \end{cases}$ and $X \geq 0$.

4. $\begin{cases} 2x_1 - x_2 + x_4 \leq 1, \\ x_1 + 2x_2 + x_4 \geq 3, \\ x_1 - x_2 - x_3 = 2, \end{cases}$ and $X \geq 0$.

11.3 A FUNDAMENTAL THEOREM

5. (a) Represent graphically the constraints
$$\begin{cases} z_1 + 2z_2 \leq 5, \\ z_1 + 3z_2 \leq 7, \end{cases} \text{ and } Z \geq 0.$$

(b) The resulting feasible set has four bounding lines; find the six points each of which is common to some two of these bounding lines.

(c) Rewrite the structural constraints as a matrix equation, find the six basic solutions, and show that they correspond to the six points of intersection found in part (b).

(d) Which of the basic solutions are basic feasible solutions?

(e) Are any of the basic feasible solutions degenerate?

6. Repeat Exercise 5 if the constraints are
$$\begin{cases} x_1 + 2x_2 \leq 4, \\ 3x_1 + x_2 \leq 6, \end{cases} \text{ and } X \geq 0.$$

7. (a) Rewrite the following linear programming problem in matrix notation, expressing the structural constraints as a matrix equation.

Maximize $4x_1 - 6x_2 + 5x_3$, subject to
$$\begin{cases} -x_1 + x_2 & \leq p_{10}, \\ x_2 + 2x_3 & \leq p_{20}, \\ 2x_1 \quad + x_3 & \leq p_{30}, \\ 2x_2 + x_3 & \leq p_{40}, \end{cases} \text{ and } X \geq 0.$$

(b) The matrix equation $AX = P_0$ of structural constraints may be thought of as a linear transformation T from vector space $V(R)$ into another vector space $U(R)$. Describe the two vector spaces in (a), and state their dimension.

(c) Write a basis for $V(R)$.

(d) Write two sets of bases for $U(R)$.

(e) If $P_0 = \begin{bmatrix} 1 \\ 4 \\ 6 \\ 1 \end{bmatrix}$, and if $X \xmapsto{T} AX$, write a vector $X = X_0$ in $V(R)$ whose image in $U(R)$ is P_0.

8. Repeat the first four parts of Exercise 7 if the problem is maximize $2x_1 + x_2 + 6x_3 + x_4$, subject to
$$\begin{cases} x_1 + 3x_2 + x_3 + x_4 \leq 4, \\ x_1 \qquad + x_3 + 2x_4 \leq 5, \\ x_2 + x_3 \qquad \leq 2, \end{cases} \text{ and } X \geq 0.$$

11.3 A Fundamental Theorem

We now state a theorem that is a cornerstone in the theory of linear programming because it assures us that an optimal solution, if one exists, can be found among the basic feasible solutions. The significant point here is that there are a finite number of basic feasible solutions, hence the problem of finding an optimal solution is manageable.

Theorem 1. *If there is an optimal solution to a given problem in linear programming, then an optimal solution can be found among the basic feasible solutions. Moreover, there may be more than one optimal solution.*

A proof of this theorem is outlined on page 373.

We now illustrate how this theorem can be used *under the assumption that an optimal solution exists.*

Example 1. For the linear programming problem, maximize $x_1 + x_2$, subject to

$$\begin{cases} x_1 + 2x_2 \leq 6, \\ 3x_1 + 4x_2 \leq 12, \end{cases} \text{ and } X \geq 0,$$

we add two nonnegative slack variables x_3 and x_4 to change the structural constraints into equations. The objective function then can be written $x_1 + x_2 + 0x_3 + 0x_4$. The reader can verify that the basic solutions of

$$\begin{cases} x_1 + 2x_2 + x_3 = 6, \\ 3x_1 + 4x_2 + x_4 = 12, \end{cases}$$

are $(0, 0, 6, 12)$, $(4, 0, 2, 0)$, $(6, 0, 0, -6)$, and $(0, 3, 0, 0)$. Notice that the last degenerate basic feasible solution results from three different bases. Of the basic solutions, $(6, 0, 0, -6)$ is not feasible, because the fourth coordinate is negative and hence does not satisfy the nonnegativity constraints. According to Theorem 1, an optimal solution, if one exists, can be found among the basic feasible solutions; hence, if a maximum exists, we can evaluate the objective function at each of the basic feasible solutions and choose the solution that yields maximum CX.

$$CX_1 = [1 \ \ 1 \ \ 0 \ \ 0] \begin{bmatrix} 0 \\ 0 \\ 6 \\ 12 \end{bmatrix} = 0;$$

$$CX_2 = [1 \ \ 1 \ \ 0 \ \ 0] \begin{bmatrix} 4 \\ 0 \\ 2 \\ 0 \end{bmatrix} = 4;$$

$$CX_3 = [1 \ \ 1 \ \ 0 \ \ 0] \begin{bmatrix} 0 \\ 3 \\ 0 \\ 0 \end{bmatrix} = 3.$$

Therefore, if a maximum exists, then

$$\max CX = 4 \text{ at } X = \begin{bmatrix} 4 \\ 0 \\ 2 \\ 0 \end{bmatrix}.$$

11.3 A FUNDAMENTAL THEOREM

Without the slack variables the solution of the original problem is $\begin{bmatrix} 4 \\ 0 \end{bmatrix}$. We reemphasize that the basic feasible solution $(0, 3, 0, 0)$ is degenerate because the number of nonzero coordinates is less than $m = 2$, that is, one of the basic variables is zero.

The linear programming problem of Example 1 could have been solved by the geometric method as illustrated in Figure 11.3.1.

Notice the one-to-one correspondence between the basic feasible solutions and the intersections of the bounding lines of the feasible set as shown in Figure 11.3.1.

$$(0, 0, 6, 12) \leftrightarrow (0, 0).$$
$$(4, 0, 2, 0) \leftrightarrow (4, 0).$$
$$(0, 3, 0, 0) \leftrightarrow (0, 3).$$

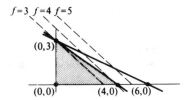

Figure 11.3.1

Each of the two nondegenerate basic feasible solutions corresponds to a specific basis of the column space of A. The degenerate basic feasible solution corresponds to three different bases. The nonfeasible basic solution $(6, 0, 0, -6)$ corresponds to the sixth and remaining basis of the column space of A.

We now illustrate that it is possible to have more than one optimal solution for a given problem.

Example 2. Maximize $3x_1 + x_2$, subject to

$$\begin{cases} x_1 + 2x_2 + x_3 = 4, \\ 3x_1 + x_2 + x_4 = 3, \end{cases} \text{ and } X \geq 0.$$

The basic feasible solutions are $(0, 0, 4, 3)$, $(1, 0, 3, 0)$, $(0, 2, 0, 1)$, and $(\frac{2}{5}, \frac{9}{5}, 0, 0)$. Corresponding values of the objective function are:

$$CX_1 = [3 \ 1 \ 0 \ 0] \begin{bmatrix} 0 \\ 0 \\ 4 \\ 3 \end{bmatrix} = 0; \quad CX_2 = [3 \ 1 \ 0 \ 0] \begin{bmatrix} 1 \\ 0 \\ 3 \\ 0 \end{bmatrix} = 3;$$

$$CX_3 = \begin{bmatrix} 3 & 1 & 0 & 0 \end{bmatrix} \begin{bmatrix} 0 \\ 2 \\ 0 \\ 1 \end{bmatrix} = 2; \qquad CX_4 = \begin{bmatrix} 3 & 1 & 0 & 0 \end{bmatrix} \begin{bmatrix} \frac{2}{5} \\ \frac{9}{5} \\ 0 \\ 0 \end{bmatrix} = 3.$$

Therefore maximum $CX = 3$ at both X_2 and X_4.

A *convex combination* of n vectors $X_1, \ldots, X_n$ is defined to be

$$Y = t_1 X_1 + \cdots + t_n X_n \quad \text{where} \quad 0 \leq t_i \leq 1 \quad \text{and} \quad \sum_{i=1}^{n} t_i = 1.$$

In two-space a convex combination of two vectors may be interpreted geometrically as a point on the line segment between two points. In this example, the objective function evaluated at any convex combination of X_2 and X_4 is also 3. Hence we have an example of a linear programming problem with an infinite number of optimal solutions.

APPLICATIONS

Example 3. On pages 11–18 of [11], Gass gives a survey of linear programming applications. This survey covers the following areas: agriculture, contract awards, economics, military practice, personnel assignment, production scheduling, inventory control, structural design, traffic analysis, transportation, and network theory. Industrial applications from the following kinds of industries are also discussed: chemical, coal, commercial aviation, communication, iron and steel, paper, petroleum, and railroad. On pages 325–337 of the same book a bibliography is given for applications in each of the categories listed above. The reader is strongly encouraged at least to glance through the material in [11] that was just referred to in order to appreciate properly the role of linear programming in modern applications of mathematics in our society; such an appreciation should help to motivate the student to study the simplex method, which is to be studied next, and which is the primary method of solution of linear programming problems.

EXERCISES

In each of Exercises 1–4, use Theorem 1 (page 348) to maximize $3x_1 + x_2$ and also to minimize $3x_1 + x_2$ subject to the given constraints. It is given that both the maximum and minimum do exist.

1. $\begin{cases} x_1 + 2x_2 + x_3 \quad\quad\; = 4, \\ 4x_1 + 3x_2 \quad\quad + x_4 = 9, \end{cases}$ and $X \geq 0$.

2. $\begin{cases} x_1 + x_2 \leq 3, \\ \quad\;\; x_2 \leq 2, \end{cases}$ and $X \geq 0$.

3. $\begin{cases} x_1 + x_2 + x_3 = 4, \\ x_2 + x_3 \geq 2, \end{cases}$ and $X \geq 0$.

4. $\begin{cases} x_1 + x_3 \leq 5, \\ x_1 + 2x_2 + x_3 = 6, \end{cases}$ and $X \geq 0$.

5. Use Theorem 1 to show that more than one solution will minimize $8x_1 + 6x_2$ subject to the following constraints.

$$\begin{cases} 4x_1 + 3x_2 \geq 18, \\ 2x_1 + 5x_2 \geq 10, \end{cases} \text{ and } X \geq 0.$$

Illustrate by solving the problem geometrically.

6. Use Theorem 1 to solve the problem: maximize $x_1 + x_2 + x_3 + x_4$, subject to

$$\begin{cases} 2x_1 + 4x_2 + x_3 + x_4 = 6, \\ x_1 + 3x_2 \phantom{{}+ x_3} + x_4 = 4, \end{cases} \text{ and } X \geq 0.$$

Assume that an optimal solution does exist. Also

minimize $x_1 + x_2 + x_3 + x_4$.

7. (a) Find the basic solutions of the system $\begin{cases} 2x_1 + x_2 + x_3 \phantom{{}+ x_4} = 4, \\ x_1 + x_2 \phantom{{}+ x_3} + x_4 = 2. \end{cases}$

(b) Which of the basic solutions are feasible, and which are degenerate?
(c) Identify the basis or bases that correspond to each basic solution.
(d) Write the basic feasible solution and corresponding basis that produces maximum $x_1 + 2x_2$. Assume that a maximum exists.

11.4 Introduction to the Simplex Method

So far we have found that linear programming problems are a class of optimization problems for which one seeks a set of nonnegative values for the unknowns that will produce the optimum of a linear function subject to linear constraints. We have presented a geometrical method of solution which is obviously insufficient when the number of unknowns is greater than three. Hence there is a need to develop a general method for solving linear programming problems. Some progress was made toward this goal in the last section when it was stated that an optimal solution, if one exists, is to be found among the basic feasible solutions. This progress does not eliminate all of our difficulties, however, because for large linear programming problems the task of finding all of the basic feasible solutions and evaluating the objective function for each basic feasible solution can be very time consuming—even for a computer. Also one is confronted with determining whether or not the problem has an optimal solution.

The general method of solution that is commonly used to solve a linear programming problem is called the *simplex method*; it was developed[3] by George Dantzig in 1947. We give a general description of this method in this section and then discuss the implementation of the method in the remaining sections.

We are seeking a basic feasible solution $X = \begin{bmatrix} x_1 \\ x_2 \\ \vdots \\ x_n \end{bmatrix}$ belonging to an n-dimensional vector space $V(R)$ such that P_0 is the image under the matrix transformation $AX = P_0$ and such that CX is optimal; the column vector, P_0, is an element of the m-dimensional[4] column space $U(R)$ of the matrix A. Now, because each basic solution X is associated with a basis of column vectors of $U(R)$, it is our task to locate such a basis of column vectors of A that optimizes CX. The so-called simplex method is an iterative procedure which begins with an initial basis, then replaces the columns in the basis one at a time in such a way that the objective function is at least as large (for the maximum problem) with each change of basis. This succession of changes of basis continues until it becomes apparent that no optimal solution exists, or that no improvement is possible, in which case the optimum has been attained.

In practice, the simplex method can be thought of as consisting of three stages, and it will be helpful to the reader if he keeps these stages in mind in order to view the material of the next few sections in proper perspective. The implementation of these three stages is the subject of the next few sections.

Stage 1. *Determine whether either the current basic feasible solution is optimal or no optimal solution exists. If one of these two situations is the case, then the simplex method is terminated; otherwise proceed with the next two stages, which are concerned with a replacement of a vector in the current basis.*

Stage 2. *To introduce a new vector into the basis, select a nonbasic column of A that will improve, or at least maintain, the value of the objective function.*

Stage 3. *If a vector enters (is introduced into) the basis then some vector must be chosen to leave the basis so that the new basic solution is feasible. Then return to Stage 1.*

In order to facilitate the implementation of the three stages of the simplex method it is customary to limit our attention to problems of the type:

$$\text{Optimize } CX$$

---

[3] For a brief history of the development of linear programming see: Dantzig, George, *Linear Programming and Extensions*, Princeton University Press, Princeton, N.J., 1963, pages 12–31.

[4] We are assuming, of course, that the rank of A is m.

11.4 INTRODUCTION TO THE SIMPLEX METHOD

subject to $AX = P_0$ and $X \geq 0$ where $P_0 \geq 0$ and A is an m by n matrix having m columns equal to the m columns of the identity matrix I_m. Let $I_{(i)}$ represent the ith column of I_m and let A_i represent the ith column of A; then (by reordering the unknowns if necessary) A can be expressed as $[A_1 \mid \cdots \mid A_{n-m} \mid I_{(1)} \mid \cdots \mid I_{(m)}]$. The case where these restrictions are not satisfied will be presented in Section 11.9. The theory developed in the meantime, however, will be general, because we show later that each linear programming problem can be solved by considering an associated problem that does satisfy these restrictions. We should point out that, with these restrictions imposed, it is impossible for the feasible set to be empty. Further discussion of the empty feasible set can be found in Section 11.9.

For emphasis we discuss these restrictions by means of two examples.

Example 1. For the problem, maximize $3x_1 + 2x_2 + 6x_3$, subject to

$$\begin{cases} -3x_1 - x_2 - x_3 = -6, \\ 4x_1 + 3x_2 \leq 2, \\ -x_1 - 9x_2 \geq -1, \end{cases} \text{ and } X \geq 0,$$

notice that one structural constraint is an equation, one is a "less than or equal to" inequality, and one is a "greater than or equal to" inequality. By adding a slack variable in the second constraint, by subtracting a slack variable in the third constraint, and then by multiplying both the first and third constraints by -1, we get the following form of the original problem:

maximize $3x_1 + 2x_2 + 6x_3 + 0x_4 + 0x_5$, subject to

$$\begin{cases} 3x_1 + x_2 + x_3 = 6, \\ 4x_1 + 3x_2 + x_4 = 2, \\ x_1 + 9x_2 + x_5 = 1, \end{cases} \text{ and } X \geq 0.$$

In its latter form the problem does meet our restrictions that the structural constraints be $AX = P_0$ where $P_0 \geq 0$ and the unit column vectors

$$I_{(1)} = \begin{bmatrix} 1 \\ 0 \\ 0 \end{bmatrix}, \quad I_{(2)} = \begin{bmatrix} 0 \\ 1 \\ 0 \end{bmatrix}, \text{ and } I_{(3)} = \begin{bmatrix} 0 \\ 0 \\ 1 \end{bmatrix}$$

are columns of A.

Example 2. Now consider a problem that does not meet the imposed restrictions. Minimize $x_1 + 2x_2$, subject to

$$\begin{cases} x_1 + x_2 \geq 3, \\ -2x_1 + 3x_2 = -6, \end{cases} \text{ and } X \geq 0;$$

this can be rewritten as:

minimize $x_1 + 2x_2$ subject to

$$\begin{cases} x_1 + x_2 - x_3 = 3, \\ 2x_1 - 3x_2 = 6, \end{cases} \text{ and } X \geq 0.$$

The constraints in their present form, however, do not meet the restriction that the unit column vectors

$$I_{(1)} = \begin{bmatrix} 1 \\ 0 \end{bmatrix} \quad \text{and} \quad I_{(2)} = \begin{bmatrix} 0 \\ 1 \end{bmatrix}$$

are columns of the coefficient matrix. There is a way to circumvent this apparent difficulty but we will delay its presentation until Section 11.9. Thus for the next four sections we will recognize that problems of this type are temporarily beyond our capabilities.

APPLICATIONS

Example 3. The simplex method also is a very important tool in solving certain game theory problems. Game theory is a relatively new branch of mathematics, and one that shows considerable promise in some of the social sciences. An application of the simplex method for this purpose may be found in Part 5 of [12] (Glicksman) and on pages 448–453 of [17] (Kemeny et. al.). Two examples of game theory applied to economic theory are given on pages 436–446 of [17]; one example is a model of an expanding economy, and the other is concerned with the existence of an economic equilibrium.

EXERCISES

1. Which of the following sets of constraints meet the temporary restrictions placed in this section? If they do not meet the restrictions, state why they do not.

    (a) $\begin{cases} 2x_1 + x_2 \leq 5, \\ x_1 + 3x_2 \geq 6, \end{cases}$ and $X \geq 0$.

    (b) $\begin{cases} 2x_1 + x_2 + x_3 = 4, \\ x_1 + x_2 \leq 3, \\ x_2 + x_3 \leq 5, \end{cases}$ and $X \geq 0$.

    (c) $\begin{cases} x_1 + 5x_2 \geq -3, \\ 3x_1 - 2x_2 \leq 6, \\ x_1 + x_2 + x_3 = 7, \end{cases}$ and $X \geq 0$.

2. One of the following sets of constraints meets the temporary restrictions of this section. Do not renumber the unknowns, but, if possible, write A, A_1, A_2, A_3, $I_{(1)}$, $I_{(2)}$, $I_{(3)}$, and P_0. State why the other set does not meet the temporary restrictions of this section.

    (a) $\begin{cases} 4x_1 + x_2 + x_3 \leq 4, \\ x_1 - x_2 + x_4 = 6, \\ x_1 + 2x_2 + x_3 \geq -2, \end{cases}$ and $X \geq 0$.

(b) $\begin{cases} x_1 + 6x_2 + 3x_3 = 4, \\ x_1 + x_2 + x_3 \leq 5, \text{ and } X \geq 0. \\ x_2 + x_3 \geq 9, \end{cases}$

3. For the structural constraints $\begin{cases} x_1 + x_2 + x_3 & = 4, \\ 3x_1 + x_2 & + x_4 & = 2, \\ 2x_2 & & + x_5 = 3, \end{cases}$ find the

following: $A_1, A_2, A_4, I_{(3)}, P_0, \dfrac{p_{20}}{a_{2k}}$ if $k=1$, $\dfrac{p_{10}}{a_{1k}}$ if $k=2$.

4. State two reasons why procedures beyond Theorem 1 are needed to solve linear programming problems.

11.5 Selection of the Vector to Leave the Basis

For theoretical reasons we will consider the three stages of the simplex method in reverse order.

For the constraints $AX = P_0$ and $X \geq 0$, where $P_0 \geq 0$ and

$$A = [A_1 \vdots A_2 \vdots \cdots \vdots A_{n-m} \vdots I_{(1)} \vdots I_{(2)} \vdots \cdots \vdots I_{(m)}],$$

an initial basic feasible solution may be obtained by letting the columns $\{I_{(1)}, I_{(2)}, \ldots, I_{(m)}\}$ serve as the initial basis. Then some particular nonbasic vector A_k is selected to enter the basis (the method of selecting A_k is the subject of the next section). The immediate problem of this section is to determine which vector in the basis is to be replaced if the new basic solution is to be feasible. Remember that the notation p_{i0} represents the ith entry of P_0 and that a_{ik} represents the ith entry of A_k.

Theorem 2. *Consider a linear programming problem with constraints*

$$[A_1 \vdots \cdots \vdots A_{n-m} \vdots I_{(1)} \vdots \cdots \vdots I_{(m)}]X = P_0, \text{ and } X \geq 0,$$

where $P_0 \geq 0$; if a column vector A_k with at least one positive component replaces $I_{(j)}$ in the basis $\{I_{(1)}, \ldots, I_{(m)}\}$, and if j is determined to satisfy

$$\frac{p_{j0}}{a_{jk}} = \min\left\{\frac{p_{i0}}{a_{ik}} \text{ where } i = 1, 2, \ldots, m, \text{ and } a_{ik} > 0\right\},$$

then $\{I_{(1)}, \ldots, I_{(j-1)}, A_k, I_{(j+1)}, \ldots, I_{(m)}\}$ is a new basis, and the corresponding basic solution is feasible.

The proof of Theorem 2 will follow Example 1.

Example 1. Consider the structural constraints

$$\begin{bmatrix} 1 & 2 & 2 & 1 & 0 & 0 \\ 1 & -1 & 1 & 0 & 1 & 0 \\ 2 & 0 & 4 & 0 & 0 & 1 \end{bmatrix} \begin{bmatrix} x_1 \\ x_2 \\ x_3 \\ x_4 \\ x_5 \\ x_6 \end{bmatrix} = \begin{bmatrix} 7 \\ 3 \\ 8 \end{bmatrix}.$$

If $\{I_{(1)}, I_{(2)}, I_{(3)}\}$ is chosen as the initial basis, and if A_1 is chosen to enter the basis, then we form the ratios p_{i0}/a_{ik} where $k=1$ and i varies from 1 to 3.

$$\frac{p_{10}}{a_{11}} = \frac{7}{1}, \quad \frac{p_{20}}{a_{21}} = \frac{3}{1}, \quad \frac{p_{30}}{a_{31}} = \frac{8}{2}.$$

The minimum of the ratios is p_{20}/a_{21}, hence we have $j=2$ and $I_{(j)} = \begin{bmatrix} 0 \\ 1 \\ 0 \end{bmatrix}$.

Therefore Theorem 2 assures us that if A_1 replaces $I_{(2)}$ in the basis, then the new basic solution will be feasible. If A_2 had been chosen to enter the basis, then $I_{(1)}$ would have to leave because p_{10}/a_{12} is the only ratio where $a_{i2} > 0$.

Proof of Theorem 2. Express A_k and P_0 as a linear combination of the initial basis.

$$A_k = \begin{bmatrix} a_{1k} \\ a_{2k} \\ \vdots \\ a_{mk} \end{bmatrix} = a_{1k} \begin{bmatrix} 1 \\ 0 \\ 0 \\ \vdots \\ 0 \\ 0 \end{bmatrix} + a_{2k} \begin{bmatrix} 0 \\ 1 \\ 0 \\ \vdots \\ 0 \\ 0 \end{bmatrix} + \cdots + a_{mk} \begin{bmatrix} 0 \\ 0 \\ 0 \\ \vdots \\ 0 \\ 1 \end{bmatrix},$$

and

$$P_0 = \begin{bmatrix} p_{10} \\ p_{20} \\ \vdots \\ p_{m0} \end{bmatrix} = p_{10} \begin{bmatrix} 1 \\ 0 \\ 0 \\ \vdots \\ 0 \\ 0 \end{bmatrix} + p_{20} \begin{bmatrix} 0 \\ 1 \\ 0 \\ \vdots \\ 0 \\ 0 \end{bmatrix} + \cdots + p_{m0} \begin{bmatrix} 0 \\ 0 \\ 0 \\ \vdots \\ 0 \\ 1 \end{bmatrix};$$

that is,

$$\begin{cases} A_k = a_{1k} I_{(1)} + a_{2k} I_{(2)} + \cdots + a_{mk} I_{(m)}, \\ P_0 = p_{10} I_{(1)} + p_{20} I_{(2)} + \cdots + p_{m0} I_{(m)}. \end{cases}$$

Multiply the members of the first equation by a real number $-t$, and add

11.5 SELECTION OF THE VECTOR TO LEAVE THE BASIS

the members of the two equations. After rearranging terms, we obtain the very important expression of P_0 as a linear combination of A_k and the original basis vectors:

(11.5.1) $$P_0 = tA_k + (p_{10} - ta_{1k})I_{(1)} \\ + (p_{20} - ta_{2k})I_{(2)} + \cdots + (p_{m0} - ta_{mk})I_{(m)}.$$

If we choose t so that one of the coefficients, say $(p_{j0} - ta_{jk})$, is zero, and if $a_{jk} \neq 0$ so that A_k can replace $I_{(j)}$ in the basis (see Exercise 14), then we have a new basic solution. If we also require that the other coefficients be nonnegative, then the new basic solution will be feasible; we now show that both can be accomplished if t is the minimum of the nonnegative ratios[5] formed by dividing each positive component of A_k into the corresponding component of P_0. These nonnegative ratios are among

$$\frac{p_{10}}{a_{1k}}, \quad \frac{p_{20}}{a_{2k}}, \quad \ldots, \quad \frac{p_{m0}}{a_{mk}};$$

suppose that $t = (p_{j0}/a_{jk})$ is the minimum of the nonnegative ratios. The coefficient of an arbitrary basis vector $I_{(i)}$ in (11.5.1) is $(p_{i0} - ta_{ik})$; if $a_{ik} \leq 0$, then $(p_{i0} - ta_{ik}) \geq 0$ because $P_0 \geq 0$, or, if $a_{ik} > 0$, then we have

$$p_{i0} - ta_{ik} = a_{ik}\left(\frac{p_{i0}}{a_{ik}} - t\right) \geq 0$$

because t was chosen so that $(p_{i0}/a_{ik}) \geq t$. Thus we have feasibility. Moreover since $(p_{i0}/a_{ik}) = t$ when $i = j$, then $I_{(j)}$ will have a zero coefficient in (11.5.1). Therefore we have a new basic solution and it is feasible. ☐

Example 2. Maximize $16x_1 + 5x_2 + 20x_3$, subject to

$$x_1\begin{bmatrix}8\\2\end{bmatrix} + x_2\begin{bmatrix}1\\1\end{bmatrix} + x_3\begin{bmatrix}2\\7\end{bmatrix} + x_4\begin{bmatrix}1\\0\end{bmatrix} + x_5\begin{bmatrix}0\\1\end{bmatrix} = \begin{bmatrix}10\\20\end{bmatrix}, \text{ and } X \geq 0.$$

Choose $\begin{bmatrix}1\\0\end{bmatrix}$ and $\begin{bmatrix}0\\1\end{bmatrix}$ as the initial basis; hence $(0, 0, 0, 10, 20)$ is the initial basic feasible solution. If $A_1 = \begin{bmatrix}8\\2\end{bmatrix}$ is chosen to enter the basis, determine which vector should leave the basis, and determine the new basic feasible solution. From (11.5.1),

$$P_0 = t\begin{bmatrix}8\\2\end{bmatrix} + (10 - 8t)\begin{bmatrix}1\\0\end{bmatrix} + (20 - 2t)\begin{bmatrix}0\\1\end{bmatrix}.$$

---

[5] Notice that t must be nonnegative because it is to be the value of the new basic variable x_k. Because $P_0 \geq 0$, the denominators of the nonnegative ratios must be positive; that is, $a_{ik} > 0$.

Of the ratios 10/8 and 20/2, the first is the smaller; hence $t = \frac{10}{8}$, and $I_{(1)}$ is removed from the basis. Now from the last equation,

$$P_0 = \tfrac{10}{8}\begin{bmatrix}8\\2\end{bmatrix} + 0\begin{bmatrix}1\\0\end{bmatrix} + \tfrac{3.5}{2}\begin{bmatrix}0\\1\end{bmatrix} = \tfrac{5}{4}A_1 + \tfrac{3.5}{2}I_{(2)},$$

and the new basic feasible solution is $(\tfrac{5}{4}, 0, 0, 0, \tfrac{3.5}{2})$. Notice that if $a_{ik} > 0$ and we replace a vector $I_{(i)}$ that is *not* associated with the minimum ratio, then a nonfeasible basic solution is determined.

In practice, the work of the preceding example is accomplished more efficiently by observing that a change of basis with associated changes in the transformation matrix A and image, P_0, can be accomplished by applying to $[A \mid P_0]$ the same elementary row transformations that are necessary to transform A_k to $I_{(j)}$.

Let Q represent the transition matrix that changes the basis $b' = \{I_{(1)}, \ldots, I_{(m)}\}$ of the column space of A in such a way that A_k replaces $I_{(j)}$. Call the new basis e'. The original transformation

$$AX = P_0$$

(notice that $P_0 = [P_0]_{b'}$), under the transition $P_0 = Q[P_0]_{e'}$, becomes

$$AX = Q[P_0]_{e'},$$

or

$$(Q^{-1}A)X = [P_0]_{e'}.$$

Now express A_k as a linear combination of the new basis e'.

$$A_k = 0I_{(1)} + \cdots + 0I_{(j-1)} + 1A_k + 0I_{(j+1)} + \cdots + 0I_{(m)}.$$

Therefore the coordinate matrix of A_k with respect to the new basis is

$$[A_k]_{e'} = I_{(j)}.$$

Hence

$$A_k = Q[A_k]_{e'} = QI_{(j)},$$

or

$$Q^{-1}A_k = I_{(j)}.$$

Because Q^{-1} is invertible, and consequently can be viewed as a product of elementary row transformation matrices, we may conclude that the sequence of elementary row operations necessary to reduce A_k to $I_{(j)}$ is exactly the sequence of operations needed to effect the desired change of basis for

$$AX = P_0.$$

In other words, if Q is the transition matrix that effects a change in

11.5 SELECTION OF THE VECTOR TO LEAVE THE BASIS

basis from b' to e', then the transformation matrix A becomes $Q^{-1}A$ and P_0 becomes $Q^{-1}P_0$ with respect to the basis e'. But we have shown that $Q^{-1}A_k = I_{(j)}$, therefore the sequence of elementary row operations that transforms A_k to $I_{(j)}$ will also transform both A and P_0 to new representations with respect to the new basis e'. In practice, we treat the augmented matrix $[A \mid P_0]$ of the matrix transformation $AX = P_0$.

Example 3. The structural constraints of Example 2 were

$$\begin{cases} 8x_1 + x_2 + 2x_3 + x_4 = 10, \\ 2x_1 + x_2 + 7x_3 + x_5 = 20. \end{cases}$$

The augmented matrix $[A \mid P_0]$ then is

$$\begin{bmatrix} 8 & 1 & 2 & 1 & 0 & \vdots & 10 \\ 2 & 1 & 7 & 0 & 1 & \vdots & 20 \end{bmatrix}.$$

If A_1 is chosen to enter the basis, the ratios p_{10}/a_{11} and p_{20}/a_{21} can be formed from the augmented matrix as shown in Figure 11.5.1.

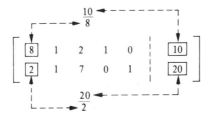

Figure 11.5.1

Since $p_{10}/a_{11} = 10/8$ is the smaller of these two ratios, $A_4 = I_{(1)}$ will leave the basis, and the row operations necessary to effect this change will be the same elementary row operations that transform $\begin{bmatrix} 8 \\ 2 \end{bmatrix}$ to $\begin{bmatrix} 1 \\ 0 \end{bmatrix}$.

$$[A \mid P_0] \;\widetilde{R_1/8}\; \begin{bmatrix} 1 & 1/8 & 1/4 & 1/8 & 0 & \vdots & 5/4 \\ 2 & 1 & 7 & 0 & 1 & \vdots & 20 \end{bmatrix}$$

$$\widetilde{-2R_1+R_2}\; \begin{bmatrix} 1 & 1/8 & 1/4 & 1/8 & 0 & \vdots & 5/4 \\ 0 & 3/4 & 13/2 & -1/4 & 1 & \vdots & 35/2 \end{bmatrix}$$

$$= [Q^{-1}A \mid Q^{-1}P_0].$$

It is obvious that $(\frac{5}{4}, 0, 0, 0, \frac{35}{2})$ is the new basic feasible solution. Moreover we have a "new" set of constraints

$$A'X = P'_0$$

(primes denote that a change of basis has been made) of the same type that we started with. Hence we are ready to repeat the process if necessary.

In the event that there is a "tie" for the minimum nonnegative ratio, there is a choice of vectors to be removed from the basis at that stage. The resulting basic feasible solution will be degenerate; at least one of the basis vectors will have the coefficient zero, and consequently more than one set of basis vectors are associated with that particular basic solution. There is a very small chance that a "cycle" could develop among degenerate basic solutions with no resulting improvement of the objective function. In other words, a certain sequence of bases will keep recurring and the process will not terminate. The probability of cycling is so small that ordinarily a computer program will incorporate some simple rule for resolving the tie with no consideration for cycling. A more comprehensive discussion of degeneracy may be found in many books[6] on linear programming.

In Theorem 2, notice that p_{i0} can be zero; this can occur in connection with a degenerate solution.

Example 4. The structural constraints of a certain linear programming problem are

$$x_1 \begin{bmatrix} 1 \\ 2 \end{bmatrix} + x_2 \begin{bmatrix} 1 \\ 3 \end{bmatrix} + x_3 \begin{bmatrix} 1 \\ 0 \end{bmatrix} + x_4 \begin{bmatrix} 0 \\ 1 \end{bmatrix} = \begin{bmatrix} 3 \\ 6 \end{bmatrix}.$$

The augmented matrix $[A \mid P_0]$ is

$$\begin{bmatrix} 1 & 1 & 1 & 0 & \vdots & 3 \\ 2 & 3 & 0 & 1 & \vdots & 6 \end{bmatrix},$$

and the initial basic feasible solution is $(0, 0, 3, 6)$. If A_1 is chosen to enter the basis, then the ratios p_{10}/a_{11} and p_{20}/a_{21} can be formed and are $3/1$ and $6/2$, respectively. Because of the "tie" we may eliminate either A_3 or A_4 from the basis. Suppose we choose to eliminate A_3; then the elementary row operation $-2R_1 + R_2$ will transform A_1 to $I_{(1)}$, and $[A \mid P_0]$ with respect to the new basis is

$$[A' \mid P'_0] = \begin{bmatrix} 1 & 1 & 1 & 0 & \vdots & 3 \\ 0 & 1 & -2 & 1 & \vdots & 0 \end{bmatrix}.$$

The new basic feasible solution is $(3, 0, 0, 0)$. If we continue the replacement process by bringing A'_2 into the basis, then the ratios p'_{10}/a'_{12} and p'_{20}/a'_{22} are found to be $3/1$ and $0/1$ respectively. The latter is smaller, hence A'_2 replaces A'_4 in the basis. The elementary row operation $-R_2 + R_1$ transforms A'_2 to $I_{(2)}$, and $[A' \mid P'_0]$ with respect to the new basis is

$$[A'' \mid P''_0] = \begin{bmatrix} 1 & 0 & 3 & -1 & \vdots & 3 \\ 0 & 1 & -2 & 1 & \vdots & 0 \end{bmatrix}.$$

Notice that, although the basis changed, the basic feasible solution remains $(3, 0, 0, 0)$. This is because this solution is degenerate.

---

[6] Gass, S., *Linear Programming*, 3rd Edition, McGraw-Hill, New York, 1969, Chapter 7, is one such book.

11.5 SELECTION OF THE VECTOR TO LEAVE THE BASIS

EXERCISES

1. The constraints of a certain linear programming problem are
$$\begin{cases} 2x_1 + x_2 \leq 6, \\ x_1 + x_2 \leq 5, \end{cases} \text{ and } X \geq 0.$$

   (a) Express the structural constraints as a matrix equation $AX = P_0$, and find the initial basic solution of $AX = P_0$ if the initial basis is $b = \{I_{(1)}, I_{(2)}\}$.
   (b) Bring A_1 into the basis and find the corresponding basic feasible solution of $AX = P_0$.
   (c) Will A_1 and A_3 serve as a basis for the column space of A? Is the corresponding basic solution feasible? How does Theorem 2 reflect this?
   (d) By continued use of Theorem 2 find the remaining two basic feasible solutions of $AX = P_0$.
   (e) Graph the constraints as originally given, and determine the points on the graph that correspond to each basic feasible solution of $AX = P_0$.

2. Repeat Exercise 1 for the constraints
$$\begin{cases} x_1 + 3x_2 \leq 6, \\ x_1 - x_2 \leq 2, \end{cases} \text{ and } X \geq 0.$$

3. If $[A \mid P_0] = \begin{bmatrix} 3 & 2 & 2 & 1 & 0 & 0 & | & 8 \\ 1 & -2 & 0 & 0 & 1 & 0 & | & 4 \\ 1 & 1 & -1 & 0 & 0 & 1 & | & 2 \end{bmatrix}$, let $b = \{I_{(1)}, I_{(2)}, I_{(3)}\}$

   be the initial basis of the column space of A. Use Theorem 2 to determine a new feasible basis if
   (a) A_1 is to enter the basis b;
   (b) A_2 is to enter the basis b;
   (c) A_3 is to enter the basis b.

4. If $[A \mid P_0] = \begin{bmatrix} 2 & 3 & 1 & 0 & 0 & | & 2 \\ 1 & 1 & 0 & 1 & 0 & | & 3 \\ 1 & 2 & 0 & 0 & 1 & | & 4 \end{bmatrix}$, let $b = \{I_{(1)}, I_{(2)}, I_{(3)}\}$ be the initial basis of the column space of A. Use Theorem 2 to determine a new feasible basis if
   (a) A_1 is to enter the basis b;
   (b) A_2 is to enter the basis b.

5. In part (a) of Exercise 3,
   (a) find the initial basic solution;
   (b) using elementary row operations find the new transformation matrix A' (that is, $Q^{-1}A$) with respect to the new basis if A_1 enters;
   (c) using elementary row operations find the new image P_0' (that is, $Q^{-1}P_0$) of the transformation with respect to the new basis if A_1 enters;
   (d) find the new basic feasible solution;

(e) write the elementary matrices corresponding to each elementary row transformation used to obtain A'; multiply these matrices to get Q^{-1}. (Be careful to get the elementary matrices in the right order when multiplying to obtain Q^{-1}.)

6. Repeat Exercise 5 for part (a) of Exercise 4.
7. Repeat Exercise 5 for part (b) of Exercise 3.
8. Repeat Exercise 5 for part (b) of Exercise 4.
9. For $[A \mid P_0] = \begin{bmatrix} 4 & 0 & 1 & 0 & 0 & | & 9 \\ 2 & -1 & 0 & 1 & 0 & | & 4 \\ 1 & 2 & 0 & 0 & 1 & | & 2 \end{bmatrix}$, let $b = \{I_{(1)}, I_{(2)}, I_{(3)}\}$ be the basis of the column space of A.

   (a) What pair of new feasible bases could result if Theorem 2 is used to bring A_1 into the basis?
   (b) Write the basic feasible solutions corresponding to each of the two new bases that include A_1.
   (c) What special name is given to a basic feasible solution of the form given in answer to part (b)?

10. The constraints of a linear programming problem are

$$\begin{bmatrix} 1 & 2 & 1 & 1 & 0 & 0 \\ 2 & a_{22} & 0 & 0 & 1 & 0 \\ 3 & 3 & 2 & 0 & 0 & 1 \end{bmatrix} \begin{bmatrix} x_1 \\ x_2 \\ x_3 \\ x_4 \\ x_5 \\ x_6 \end{bmatrix} = \begin{bmatrix} 8 \\ 3 \\ 9 \end{bmatrix}, \text{ and } X \geq 0.$$

   (a) Determine the value of a_{22} that would cause a degenerate basic feasible solution if A_2 is introduced into the basis.
   (b) What is the degenerate basic feasible solution?

11. Consider a linear programming problem with constraints

$$x_1 A_1 + x_2 A_2 + x_3 I_{(1)} + x_4 I_{(2)} + x_5 I_{(3)} = \begin{bmatrix} 2 \\ 1 \\ 6 \end{bmatrix}, \text{ and } X \geq 0.$$

   (a) What is the initial basic feasible solution if the initial basis is
   $$b = \{I_{(1)}, I_{(2)}, I_{(3)}\}?$$
   (b) Express $P_0 = \begin{bmatrix} 2 \\ 1 \\ 6 \end{bmatrix}$ as a linear combination of the basis b.
   (c) Express $A_2 = \begin{bmatrix} a_{12} \\ a_{22} \\ a_{32} \end{bmatrix}$ as a linear combination of the basis b.
   (d) Use the answers to (b) and (c) to express P_0 as a linear combination of A_2 and the initial basis where the coefficient of combination of A_2 is t.
   (e) If A_2 enters the basis and A_5 is removed, what is t?
   (f) What is the new basic solution if A_2 replaces A_5 in the basis?
   (g) Under what condition is the new basic solution of part (f) feasible?

11.6 SELECTION OF THE VECTOR TO ENTER THE BASIS

12. Let $A = [A_1 | \cdots | A_{n-m} | I_{(1)} | \cdots | I_{(m)}]$ and $P_0 \geq 0$. For the transformation $AX = P_0$, from the vector space $V(R)$ of all n by 1 R-matrices into the column space $U(R)$ of A, let Q represent a transition matrix that effects the replacement of $I_{(j)}$ by A_k in the basis $b = \{I_{(1)}, \ldots, I_{(m)}\}$ of $U(R)$. Prove or disprove the following conjecture: Q is the inverse of the product of elementary matrices (in proper order) which transforms A_k into $I_{(j)}$.

13. Let m of the columns of A be the columns of I_m. The structural constraints, $AX = P_0$, represent a transformation from an n-dimensional vector space $V(R)$ (to which X belongs) into an m-dimensional vector space $U(R)$ (which is the column space of A).
    (a) What is a basis of $V(R)$, and does the basis of $V(R)$ change in the course of applying the simplex method?
    (b) What is a basis of $U(R)$?
    (c) Is the transition matrix that is used for a certain basis change of $U(R)$ a product of elementary matrices?
    (d) After a basis change has been made, is the old transformation matrix A row equivalent to the new transformation matrix A'? Give reasons for your answer.

14. Suppose that the set $b = \{I_{(1)}, \ldots, I_{(m)}\}$ is a basis of the column space $U(R)$ of an m by n matrix A. Prove that if $a_{jk} \neq 0$, then a new set, formed from b by replacing $I_{(j)}$ with A_k, is a basis of $U(R)$.

15. Let $A = \begin{bmatrix} A_1 & I_{(1)} & I_{(2)} \\ 2 & 1 & 0 \\ 0 & 0 & 1 \end{bmatrix}$ and $P_0 = \begin{bmatrix} 4 \\ 0 \end{bmatrix}$. Then

$$P_0 = tA_1 + (4 - t \cdot 2)I_{(1)} + (0 - t \cdot 0)I_{(2)}$$

for the initial basis $\{I_{(1)}, I_{(2)}\}$. Let $t = 1$, then the coefficient of $I_{(2)}$ is 0 and $P_0 = A_1 + (4 - 1 \cdot 2)I_{(1)}$. But $\{I_{(1)}, A_1\}$ is not a basis, and hence $(1, 2, 0)$ is not a basic solution. Is this a contradiction of Theorem 2? Give reasons for your answer.

11.6 Selection of the Vector to Enter the Basis

We now turn our attention to a method of selecting a vector to enter the basis which is Stage 2 of the simplex method. We will select a vector which, if it enters into the basis, will increase or leave unchanged the value of the objective function of a maximum problem and will decrease or leave unchanged the value of the objective function of a minimum problem.

First define a row vector Z with n components such that the kth component is

$$z_k = c_{(1)}a_{1k} + c_{(2)}a_{2k} + \cdots + c_{(m)}a_{mk},$$

where $c_{(i)}$ are the coefficients of the current basic variables in the objective function. *Notice that z_k is the dot product of the column vector, A_k, and the row vector $\hat{C}$ consisting of the coefficients of the basic variables in the objective function.*

Example 1. Consider the problem: maximize $3x_1 + 2x_2 + 6x_3 + 0x_4 + 0x_5$,

subject to $x_1 \begin{bmatrix} 3 \\ 4 \\ 1 \end{bmatrix} + x_2 \begin{bmatrix} 1 \\ 3 \\ 9 \end{bmatrix} + x_3 \begin{bmatrix} 1 \\ 0 \\ 0 \end{bmatrix} + x_4 \begin{bmatrix} 0 \\ 1 \\ 0 \end{bmatrix} + x_5 \begin{bmatrix} 0 \\ 0 \\ 1 \end{bmatrix} = \begin{bmatrix} 6 \\ 2 \\ 1 \end{bmatrix}$, and $X \geq 0$. If we choose $\{I_{(1)}, I_{(2)}, I_{(3)}\}$ to be the basis, then x_3, x_4, and x_5 are the basic variables; the coefficients of these basic variables in the objective function are 6, 0, 0, respectively. Therefore $\hat{C} = [6 \ 0 \ 0]$, and

$$z_1 = \hat{C} \cdot A_1 = [6 \ 0 \ 0] \begin{bmatrix} 3 \\ 4 \\ 1 \end{bmatrix} = 18,$$

$$z_2 = \hat{C} \cdot A_2 = [6 \ 0 \ 0] \begin{bmatrix} 1 \\ 3 \\ 9 \end{bmatrix} = 6,$$

$$z_3 = \hat{C} \cdot A_3 = [6 \ 0 \ 0] \begin{bmatrix} 1 \\ 0 \\ 0 \end{bmatrix} = 6,$$

$$z_4 = \hat{C} \cdot A_4 = [6 \ 0 \ 0] \begin{bmatrix} 0 \\ 1 \\ 0 \end{bmatrix} = 0,$$

$$z_5 = \hat{C} \cdot A_5 = [6 \ 0 \ 0] \begin{bmatrix} 0 \\ 0 \\ 1 \end{bmatrix} = 0.$$

Hence

$$Z = [z_1 \ z_2 \ z_3 \ z_4 \ z_5] = [18 \ 6 \ 6 \ 0 \ 0].$$

Notice that

$$\begin{aligned} Z &= [z_1 \cdots z_n] \\ &= [\hat{C}A_1 \cdots \hat{C}A_n] \\ &= \hat{C}[A_1 \vdots \cdots \vdots A_n] \\ &= \hat{C}A. \end{aligned}$$

The fact that $Z = \hat{C}A$ is important, and it will be used again.

11.6 SELECTION OF THE VECTOR TO ENTER THE BASIS

The reason for defining Z will become apparent in the proof of Theorem 3, where we will develop a formula for the change in f (change in CX) if A_k is brought into the basis. The formula is

$$(\text{change in } f) = t(c_k - z_k)$$

where nonnegative t is defined as the minimum ratio in Theorem 2. If t is positive, the change in f will be positive or negative according as $c_k - z_k$ is positive or negative; if $t = 0$, the change in f is zero. Thus we see that as long as there is a positive $c_k - z_k$ for some nonbasic vector A_k, then f can be made as large or larger by introducing A_k into the basis. The real number $c_k - z_k$ is sometimes called the **evaluator** of A_k or x_k with respect to a given basis.

Theorem 3. *Consider the linear programming problem with m structural constraints:*

$$\text{maximize } CX, \text{ subject to } AX = P_0, \text{ and } X \geq 0,$$

where $P_0 \geq 0$, and all of the columns of I_m are columns of A: if there exists some column of A, say A_k, such that $A_k \not\leq 0$ and $c_k - z_k$ is positive, then CX at the current basic feasible solution is less than or equal to CX at the basic feasible solution obtained by introducing A_k into the basis. If however, all $c_k - z_k$ are nonpositive, then maximum CX occurs at the current basic feasible solution.

A proof of Theorem 3 will follow Example 2.

A corresponding theorem can be stated and proved for a minimum problem (Exercises 11 and 12).

We hasten to add that the maximum may also occur at more than one basic feasible solution; this situation can be discovered by simply searching for those $A_k \not\leq 0$ that are not in the basis for which $c_k - z_k = 0$ or $t = 0$.

Example 2. Maximize $4x_1 + 2x_2 - 6x_3 + x_5$, subject to

$$x_1 \begin{bmatrix} 1 \\ 1 \\ 3 \end{bmatrix} + x_2 \begin{bmatrix} 1 \\ 2 \\ 0 \end{bmatrix} + x_3 \begin{bmatrix} 1 \\ 1 \\ 1 \end{bmatrix} + x_4 \begin{bmatrix} 1 \\ 0 \\ 0 \end{bmatrix} + x_5 \begin{bmatrix} 0 \\ 1 \\ 0 \end{bmatrix} + x_6 \begin{bmatrix} 0 \\ 0 \\ 1 \end{bmatrix} = \begin{bmatrix} 3 \\ 5 \\ 2 \end{bmatrix}, \text{ and } X \geq 0.$$

If we choose the initial basis to be $\{I_{(1)}, I_{(2)}, I_{(3)}\}$, then the coefficients of the basic variables in the objective function are 0, 1, and 0, respectively; hence $\hat{C} = [0\ 1\ 0]$. Now calculate Z and then $C - Z$.

$$\left.\begin{array}{l} z_1 = \hat{C} \cdot A_1 = 1, \\ z_2 = \hat{C} \cdot A_2 = 2, \\ z_3 = \hat{C} \cdot A_3 = 1, \\ z_4 = \hat{C} \cdot I_{(1)} = 0, \\ z_5 = \hat{C} \cdot I_{(2)} = 1, \\ z_6 = \hat{C} \cdot I_{(3)} = 0, \end{array}\right\} Z = [1\ 2\ 1\ 0\ 1\ 0].$$

$$\left.\begin{array}{l} c_1 - z_1 = 4 - 1 = 3, \\ c_2 - z_2 = 2 - 2 = 0, \\ c_3 - z_3 = -6 - 1 = -7, \\ c_4 - z_4 = 0 - 0 = 0, \\ c_5 - z_5 = 1 - 1 = 0, \\ c_6 - z_6 = 0 - 0 = 0, \end{array}\right\} \quad C - Z = [3 \quad 0 \quad -7 \quad 0 \quad 0 \quad 0].$$

The quantity $c_i - z_i$ represents the gain in the objective function per unit of A_i if A_i is introduced into the basis and if $t \neq 0$. In this example, clearly, we choose A_1 to enter the basis because $c_1 - z_1$ is the only positive term among the $c_i - z_i$, and consequently the only vector that will improve the value of the function. Notice that $c_2 - z_2 = 0$, and therefore the entrance of A_2 into the basis at this point will leave the value of f unchanged. The $c_i - z_i$ for the basis vectors, naturally, will always be zero (Exercise 14).

Proof of Theorem 3. PART I. Let f_1 be the value of the objective function for the initial basic feasible solution, and let f_2 be the value of the objective function if some nonbasic variable x_k is allowed to be nonzero and all other nonbasic variables remain equal to zero. We will derive a formula for $f_2 - f_1$ which is the change in f if $x_k \neq 0$.

(1) Initially the vectors $\{I_{(1)}, \ldots, I_{(m)}\}$ are chosen as a basis and

$$P_0 = p_{10} I_{(1)} + \cdots + p_{m0} I_{(m)}.$$

Hence, if $x_{(i)}$ represents the value of the basic variable associated with each $I_{(i)}$, then initially a basic feasible solution is

$$x_{(1)} = p_{10},$$
$$\vdots$$
$$x_{(m)} = p_{m0},$$

and all nonbasic variables are zero. Thus

$$f_1 = c_{(1)} p_{10} + \cdots + c_{(m)} p_{m0}.$$

(2) In the proof of Theorem 2 we developed the equation

$$P_0 = t A_k + (p_{10} - t a_{1k}) I_{(1)} + \cdots + (p_{m0} - t a_{mk}) I_{(m)}.$$

Therefore, if t is chosen so that all coefficients are nonnegative, a feasible solution is

$$x'_k = t,$$
$$x'_{(1)} = (p_{10} - t a_{1k}),$$
$$\vdots$$
$$x'_{(m)} = (p_{m0} - t a_{mk}),$$

11.6 SELECTION OF THE VECTOR TO ENTER THE BASIS

and all other variables have the value zero. Thus

$$f_2 = c_k t + c_{(1)}(p_{10} - ta_{1k}) + \cdots + c_{(m)}(p_{m0} - ta_{mk}).$$

(3) If we subtract f_1 from f_2, some terms cancel, and we have

$$f_2 - f_1 = c_k t - c_{(1)} ta_{1k} - \cdots - c_{(m)} ta_{mk},$$

or, by factoring,

(11.6.1) $$f_2 - f_1 = t(c_k - (c_{(1)} a_{1k} + \cdots + c_{(m)} a_{mk})).$$

(4) We recognize the expression in the inner parentheses of (11.6.1) as z_k from the definition of z_k at the beginning of this section. Hence (11.6.1) can be written as

(11.6.2) $$f_2 - f_1 = t(c_k - z_k).$$

(5) Because $A_k \not\leq 0$, then t can be determined as the minimum ratio of Theorem 2, and consequently f_2 is the value of f at the basic feasible solution obtained by introducing A_k into the basis. If $t = 0$ there is no change in f; if, however, $t \neq 0$, then $t > 0$, because $t = x_k$ and x_k must satisfy the nonnegativity constraints; thus f will increase if there exists a positive $c_k - z_k$ and a positive t, while f remains the same if there exists a positive $c_k - z_k$ and $t = 0$.

PART II. If, however, $c_k - z_k$ is nonpositive for all k, we now show that any other feasible solution (basic or otherwise) cannot improve the value of the objective function evaluated at the current basic solution. Let X_0 be the m by 1 column matrix of the values of all current basic variables, in order, and let $\hat{C}$ be the corresponding 1 by m row matrix of coefficients of the basic variables in the objective function; then $f_1 = \hat{C} X_0$. Let Y be the n by 1 column matrix of an arbitrary feasible solution; then $f = CY$. We will show that $f - f_1$ cannot be positive if $C - Z$ is nonpositive, that is, $c_k - z_k \leq 0$ for all k.

(1) $f - f_1 = CY - \hat{C} X_0$.

(2) From Statement (1) of Part I we have $X_0 = P_0$, and since Y is a feasible solution, then the structural constraints $AX = P_0$ become

(11.6.3) $$AY = X_0.$$

(3) Substitute (11.6.3) in Statement (1); then

$$\begin{aligned} f - f_1 &= CY - \hat{C}(AY) \\ &= (C - \hat{C}A)Y \\ &= (C - Z)Y \end{aligned}$$

(remember that $Z = \hat{C}A$, from page 364).

(4) Each entry of Y must be nonnegative, and if each entry of $C - Z$ is nonpositive, then $f - f_1$ must be nonpositive and f must have achieved its maximum value f_1 at the current basic solution. ∎

APPLICATIONS

Example 3.[7] Suppose that availability, A, is the probability that a missile battery will be functioning properly when called upon to perform a mission. Also suppose that it has been determined that the inclusion in the system or the mandatory stockage of one unit of a certain item Q_i will increase the availability of a missile battery by an amount a_i. Further, suppose that the cost of including one unit of item Q_i is c_i. The amount that can be spent cannot exceed E, and the amount spent for the most expensive item cannot exceed 80 percent of E. The storage requirement on each Q_i at the battery is v_i. The storage limitation at the battery is V. Let x_i be the number of units of Q_i that are stocked. Determine the amount of mandatory stockage, x_i, for each item that will maximize the increase in availability, ΔA, subject to the given restrictions. Since availability, A, is a probability, it is dimensionless and is represented by a real number; thus ΔA is also dimensionless.

The problem is to maximize

$$\Delta A = a_1 x_1 + \cdots + a_n x_n,$$

subject to
$$\begin{cases} c_1 x_1 + \cdots + c_n x_n \leq E, \\ c_i x_i \leq .80E \quad (Q_i \text{ is the most expensive item}), \\ v_1 x_1 + \cdots + v_n x_n \leq V, \end{cases}$$

and $X \geq 0$.

As a numerical illustration, assume the following data. $E = 500$; $V = 180$ cubic feet; and

i	a_i	c_i	v_i
1	0.001	5	3
2	0.005	20	5
3	0.002	10	4

Thus the problem becomes

$$\text{maximize } 0.001 x_1 + 0.005 x_2 + 0.002 x_3,$$

subject to
$$\begin{cases} 5x_1 + 20x_2 + 10x_3 \leq 500, \\ x_2 \leq 20, \quad \text{and} \quad X \geq 0. \\ 3x_1 + 5x_2 + 4x_3 \leq 180, \end{cases}$$

Introducing slack variables, we get the structural constraints

$$\begin{bmatrix} 5 \\ 0 \\ 3 \end{bmatrix} x_1 + \begin{bmatrix} 20 \\ 1 \\ 5 \end{bmatrix} x_2 + \begin{bmatrix} 10 \\ 0 \\ 4 \end{bmatrix} x_3 + \begin{bmatrix} 1 \\ 0 \\ 0 \end{bmatrix} x_4 + \begin{bmatrix} 0 \\ 1 \\ 0 \end{bmatrix} x_5 + \begin{bmatrix} 0 \\ 0 \\ 1 \end{bmatrix} x_6 = \begin{bmatrix} 500 \\ 20 \\ 180 \end{bmatrix}.$$

[7] This example was contributed by Mr. Robert Ailor, when he was a graduate student at Virginia Polytechnic Institute and State University.

11.6 SELECTION OF THE VECTOR TO ENTER THE BASIS

Since $\hat{C} = 0$, then $Z = 0$, and $C - Z = [0.001 \quad 0.005 \quad 0.002 \quad 0 \quad 0 \quad 0]$. If A_2 is brought into the basis, A_5 will leave the basis and the new $C' - Z'$ (prime denotes that one iteration has been accomplished) is $[0.001 \quad 0 \quad 0.002 \quad 0 \quad -0.005 \quad 0]$. If A'_3 is now brought into the basis, then A'_4 is removed and the new $C'' - Z''$ is $[0 \quad 0 \quad 0 \quad -0.002 \quad -0.001 \quad 0]$ which is nonpositive, and Theorem 3 assures us that a maximum has been reached. Notice that A''_1 is not in the basis and yet $c''_1 - z''_1 = 0$; this means that if A''_1 is brought into the basis the value of the objective function will not change. In other words there is more than one basic solution of this problem that will yield max ΔA. By methods that will be explored later, one basic solution is $x_1 = 0$, $x_2 = 20$, $x_3 = 10$, and maximum $\Delta A = 0.120$.

EXERCISES

For each of the linear programming problems in Exercises 1–4 (a) write $\hat{C}$; (b) calculate $c_k - z_k$ for each nonbasic A_k; (c) state which columns may improve the objective function if entered in the basis.

1. Maximize $x_1 - 3x_2$, subject to

$$\begin{cases} x_1 - x_2 \geq -1, \\ x_1 + x_2 \leq 1, \end{cases} \text{ and } X \geq 0.$$

2. Maximize $x_1 + 2x_2$, subject to

$$\begin{cases} x_1 + x_2 \leq 4, \\ x_1 + 4x_2 \leq 7, \end{cases} \text{ and } X \geq 0.$$

3. Minimize $x_1 + x_2 + 2x_3$, subject to

$$\begin{cases} x_1 + 2x_2 + x_3 = 4, \\ 3x_1 + x_2 \leq 6, \\ x_1 \leq 10, \end{cases} \text{ and } X \geq 0.$$

4. Minimize $x_1 + x_2 + 2x_5$, subject to

$$\begin{cases} 2x_1 - 2x_2 + x_3 + x_4 = 4, \\ x_2 + 2x_3 \quad\quad + x_5 = 4, \\ -x_1 + 3x_2 - x_3 \quad\quad \leq 6, \end{cases} \text{ and } X \geq 0.$$

5. Consider the linear programming problem: maximize $f = x_1 + x_2 + 2x_5$, subject to

$$\begin{cases} 2x_1 - 2x_2 + x_3 + x_4 = 4, \\ x_2 + 2x_3 \quad\quad + x_5 = 4, \\ -x_1 + 3x_2 - x_3 \quad\quad + x_6 = 6, \end{cases} \text{ and } X \geq 0.$$

The structural constraints can be written $AX = P_0$ or

$$x_1 A_1 + x_2 A_2 + x_3 A_3 + x_4 A_4 + x_5 A_5 + x_6 A_6 = P_0,$$

where $A_4 = I_{(1)}$, $A_5 = I_{(2)}$, $A_6 = I_{(3)}$.

(a) Let $\{A_4, A_5, A_6\}$ be the basis of the column space of A. Find $x_{(1)}$, $x_{(2)}$, $x_{(3)}$, f_1, $\hat{C}$, C, Z, and $C - Z$. Which vectors are eligible to enter the basis for possible improvement in f?
(b) If it were a minimum problem instead of a maximum problem, which vectors would be eligible to enter the basis for possible improvement in f?
(c) If A_1 is selected to enter the original basis, which vector must leave?
(d) If A_2 is selected to enter the original basis, which vector must leave?
(e) If A_3 is selected to enter the original basis, which vector must leave?

6. If A_k is a nonbasic vector, and if $c_k - z_k = 0$, what is the effect on the objective function if A_k is brought into the basis?

7. The real number $c_k - z_k$ has been called the *evaluator* of A_k or x_k with respect to a given basis.
(a) State what the evaluator means with regard to the objective function.
(b) What is it used for in the simplex method?
(c) Define z_k in terms of a standard inner product.
(d) What does the row matrix $\hat{C}$ represent?
(e) What is c_k?
(f) Show how $c_k - z_k$ originates in the theory of the simplex method.

8. Consider the linear programming problem: maximize $2x_1 + 3x_2 + 5x_4$,

subject to $x_1 A_1 + x_2 A_2 + x_3 I_{(1)} + x_4 I_{(2)} + x_5 I_{(3)} = \begin{bmatrix} 2 \\ 1 \\ 6 \end{bmatrix}$ and $X \geq 0$.

(a) What is the initial value of the objective function if the initial basis is $b = \{I_{(1)}, I_{(2)}, I_{(3)}\}$?
(b) Calculate $c_k - z_k$ for each A_k not in the basis.
(c) What does $c_k - z_k$ represent?
(d) When will the introduction of some A_k into the basis improve the values of the objective function for this problem?
(e) Suppose A_2 is chosen to enter the basis. In the formula $f_2 - f_1 = t(c_2 - z_2)$, what is t?

9. Determine the solution of: maximize $2x_1 + x_4$, subject to
$\begin{cases} x_1 + x_2 + 2x_3 + x_4 = 2, \\ x_1 + 2x_2 + x_3 \leq 5, \end{cases}$ and $X \geq 0$.

10. Determine the solution of: minimize $3x_1 + x_3$, subject to
$\begin{cases} x_1 + x_2 + x_3 = 3, \\ x_1 + 2x_2 + x_4 = 7, \end{cases}$ and $X \geq 0$.

11. State a theorem corresponding to Theorem 3 for a minimum problem.

12. Prove the theorem of Exercise 11.

13. Consider the problem: maximize $x_1 - 3x_4 - 3x_5$, subject to
$\begin{cases} 3x_1 + x_2 + 2x_4 - x_5 = 0, \\ -x_1 + x_3 - x_4 + x_5 = 1, \end{cases}$ and $X \geq 0$.

The current basic feasible solution obviously is $(0, 0, 1, 0, 0)$. Show that there exists a positive $c_k - z_k$, and yet the optimal solution after one iteration is the same feasible solution $(0, 0, 1, 0, 0)$. Does this contradict Theorem 3? Give reasons for your answer.

14. Prove that if A_i is the unit vector $I_{(j)}$ in the basis, then $c_i - z_i = 0$.

11.7 Unbounded Feasible Set with no Optimal Solution

In this section we examine the linear programming problem with an unbounded feasible set and no optimal solution; in particular, we wish to determine how the simplex method will exhibit this situation.

Example 1. The problem, maximize $x_1 + x_2$ subject to

$$3x_1 - x_2 + x_3 = 1, \quad \text{and} \quad X \geq 0,$$

is graphed in Figure 11.7.1, where f represents the objective function. It can be seen easily that, for this problem, f is unbounded because the feasible set is unbounded. We hasten to add that even though the feasible set is unbounded it is possible to have an optimal solution in certain problems; for example, consider the same constraints but suppose we seek minimum f.

Theorem 4. *Consider the linear programming problem with m structural constraints:*

maximize CX, subject to $AX = P_0$ and $X \geq 0$,

where $P_0 \geq 0$, and all of the columns of I_m are columns of A; if there exists some column of A, say A_k, such that $A_k \leq 0$ and $c_k - z_k$ is positive, then CX is unbounded.

Proof. Recall that, in the last section, initially, with basis $\{I_{(1)}, \ldots, I_{(m)}\}$,

$$f_1 = c_{(1)}p_{10} + c_{(2)}p_{20} + \cdots + c_{(m)}p_{m0}.$$

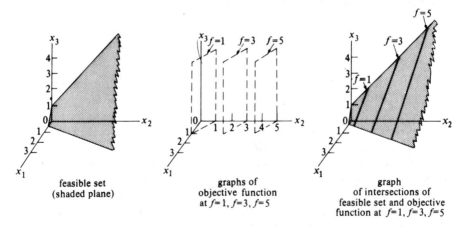

feasible set
(shaded plane)

graphs of
objective function
at $f=1, f=3, f=5$

graph
of intersections of
feasible set and objective
function at $f=1, f=3, f=5$

Figure 11.7.1

Also recall that a new value of the objective function can be expressed as

$$f_2 = c_k t + c_{(1)}(p_{10} - ta_{1k}) + \cdots + c_{(m)}(p_{m0} - ta_{mk})$$
$$= f_1 + t(c_k - z_k).$$

Remember that t and P_0 must be nonnegative; therefore if every component of an arbitrary A_k, namely, $a_{1k}, a_{2k}, \ldots, a_{mk}$, is nonpositive then each of

$$x'_k = t,$$
$$x'_{(1)} = p_{10} - ta_{1k},$$
$$x'_{(2)} = p_{20} - ta_{2k},$$
$$\vdots$$
$$x'_{(m)} = p_{m0} - ta_{mk},$$

must be nonnegative. Hence we have a feasible solution for any nonnegative t that is assigned regardless of its size. Thus we see that if $c_k - z_k > 0$, then $f_2 = f_1 + t(c_k - z_k)$ can be made arbitrarily large at nonbasic feasible solutions by choosing appropriate values for t. ☐

Example 2. Maximize $5x_1 - 2x_2$, subject to

$$\begin{cases} -x_1 + x_2 + x_3 = 3, \\ -2x_1 + x_2 + x_4 = 2, \end{cases} \text{ and } X \geq 0.$$

If we choose $I_{(1)}$ and $I_{(2)}$ as the initial basis, then the row vector of the coefficients of the basic variables in $5x_1 - 2x_2 + 0x_3 + 0x_4$, is $\hat{C} = [0 \ \ 0]$ and

$$(\hat{C} = [0 \ \ 0]) \Rightarrow \begin{cases} z_1 = 0, \\ z_2 = 0, \end{cases} \Rightarrow \begin{cases} c_1 - z_1 = 5, \\ c_2 - z_2 = -2. \end{cases}$$

Improvement in f can be made by bringing in A_1, but, because $A_1 \leq 0$, f has no maximum according to Theorem 4. The reader can illustrate this result by solving the following problem geometrically: maximize $5x_1 - 2x_2$, subject to

$$\begin{cases} -x_1 + x_2 \leq 3, \\ -2x_1 + x_2 \leq 2, \end{cases} \text{ and } X \geq 0.$$

The statement of a theorem corresponding to Theorem 4 for a minimum problem is left as Exercise 8. Particular attention is called to the fact that for a minimum problem where some $A_k \leq 0$ then $c_k - z_k$ must be *negative* for the guaranteed existence of an unbounded objective function (rather than positive as is the case for a maximum problem).

We are now in a position to consider an outline of a proof of Theorem 1 in Section 11.3 which stated: *If there is an optimal solution to a given linear programming problem, then an optimal solution can be found among the basic feasible solutions. Moreover, there may be more than one optimal solution.* Theorem 1 was stated first for instructional purposes and was not used in the proofs of Theorems 2, 3, or 4, hence the use of these theorems in this outlined proof is legitimate. Of course we must realize that for m structural constraints $AX = P_0$, Theorems 2, 3, and 4 required that $P_0 \geq 0$ and that $I_{(1)}, \ldots, I_{(m)}$ be among the columns of A. Hence, in any proof using these theorems we must make these same requirements. It turns out, however, that Theorems 5 and 6 in Section 11.9 will allow us to extend easily the following proof to those linear programming problems that do not meet the requirements specified above.

We assume a maximum problem. Evaluate the objective function f at each basic feasible solution and let f_1 denote the largest of these values. Calculate $C - Z$ for a basic feasible solution that produces f_1. We consider three cases.

CASE I. ($C - Z \leq 0$.) In this case, f_1 is larger than or equal to f at any other solution (basic or nonbasic) according to Part II of the proof of Theorem 3. Hence maximum f occurs at a basic feasible solution.

CASE II. ($C - Z \not\leq 0$ where, for some k, $c_k - z_k > 0$, and $A_k \leq 0$.) In this case f is unbounded according to Theorem 4, and this violates the hypothesis of Theorem 1 that f has an optimal solution.

CASE III. ($C - Z \not\leq 0$ where $A_k \not\leq 0$ for all k such that $c_k - z_k > 0$.) Recall from the proof of Theorem 3 that

$$(\text{change in } f) = t(c_k - z_k).$$

Using Part I of the proof of Theorem 3 we conclude that if $A_k \not\leq 0$ and t is chosen to maintain feasibility, and if t can be nonzero, then A_k can be brought into the basis with an increase in f. But we have assumed that f cannot be increased at another basic feasible solution, hence, by the contrapositive argument, t must equal zero. If t must be zero in order to maintain feasibility, then, although A_k comes into the basis, the (degenerate) basic feasible solution and f remain unchanged. We should then prove that a finite number of basis changes of this type will produce Case I, from which we could conclude that this (degenerate) basic feasible solution of Case III is indeed optimal: this, however, is beyond the scope of this book, and the reader is referred to more comprehensive books on linear programming.[8] □

EXERCISES

In Exercises 1 and 2, use Theorem 4 or the corresponding theorem for minimum problems to prove that the linear programming problem has feasible solutions such that the objective function is unbounded.

---

[8] See Theorem 3 on page 97 of Smythe, W. R., and Johnson, L. A., *Introduction to Linear Programming with Applications*, Prentice Hall, Englewood Cliffs, N.J., 1966, or see pages 183–186 of Hadley, G., *Linear Programming*, Addison-Wesley Publishing Co., Reading, Mass., 1962.

1. Minimize $x_2 - x_4$, subject to

$$\begin{cases} x_1 - 2x_2 & + x_4 = 2, \\ x_1 - x_2 + x_3 & \geq 3, \end{cases} \quad \text{and} \quad X \geq 0.$$

2. Maximize $3x_2 + 3x_3 + x_4$, subject to

$$\begin{cases} x_1 - x_2 + x_3 & = 3, \\ x_1 - 2x_2 & + x_4 = 4, \end{cases} \quad \text{and} \quad X \geq 0.$$

3. Given the problem: minimize $f = x_1 - 2x_2$, subject to

$$\begin{cases} x_1 - x_2 \leq 2, \\ 3x_1 - x_2 \geq -3, \end{cases} \quad \text{and} \quad X \geq 0.$$

(a) Show geometrically that f has no minimum.
(b) Write the structural constraints in the form $AX = P_0$, where $P_0 \geq 0$. Show that the initial basic feasible solution is (0, 0, 2, 3). Show that $f_1 = 0$.
(c) Find $C, \hat{C}, Z, C - Z$. Show that A_2 must enter the basis and that hence $I_{(2)}$ must leave the basis. Make this change of basis using elementary row operations. Show that the new basic feasible solution is (0, 3, 5, 0) and that $f_2 = -6$. (Note the improvement in the objective function).
(d) Now we are ready to change the basis again if necessary. At this point your new matrix should be $[A' \mid P_0'] = \begin{bmatrix} -2 & 0 & 1 & 1 & 5 \\ -3 & 1 & 0 & 1 & 3 \end{bmatrix}$. The new $\hat{C}$, which we call $\hat{C}'$, is $[0 \ -2]$. Why? Show that the new Z, which we call Z', is $[6 \ -2 \ 0 \ -2]$ and find $C - Z'$. Now apply Theorem 4 (with the necessary changes for a minimum problem) to show that f is unbounded below (which we already knew from part (a) above).

In Exercises 4 and 5, use previous theorems of this chapter to prove that the following linear programming problem has feasible solutions such that the objective function is unbounded.

4. Maximize $x_1 + x_2$, subject to

$$\begin{cases} -2x_1 + x_2 \leq 4, \\ x_1 - 3x_2 \leq 6, \end{cases} \quad \text{and} \quad X \geq 0.$$

5. Minimize $x_1 - x_2$, subject to

$$\begin{cases} 2x_1 - 5x_2 \leq 0, \\ -3x_1 + x_2 \leq 6, \end{cases} \quad \text{and} \quad X \geq 0.$$

6. Illustrate the proof of Theorem 4 by substituting the numbers of Exercise 1 for the letters in the proof.

7. Illustrate the proof of Theorem 4 by substituting the numbers of Exercise 2 for the letters in the proof.

8. State and prove a theorem corresponding to Theorem 4 for a minimum problem.

9. Determine whether the following problem has an optimal solution, then verify your conclusions graphically. (Use more than one iteration if necessary.)
Maximize $2z_1 + z_2$, subject to
$$\begin{cases} -z_1 + z_2 \leq 1, \\ z_1 - 2z_2 \leq 2, \end{cases} \text{ and } Z \geq 0.$$

11.8 The Simplex Method and Tableau

So far we have found that the structural constraints can be treated as the linear transformation $AX = P_0$, and each iteration of the simplex method is simply a change of basis of the image space (column space of A) of the transformation. The execution of each iteration of the simplex method is usually accomplished by means of elementary row operations on the so-called simplex tableau or simplex matrix. We now state the computational steps for the simplex method and illustrate the accompanying tableau or matrix for each step with the following example.

Example 1. Maximize $8x_1 + 6x_2 + 9x_3$, subject to
$$\begin{cases} 2x_1 + x_2 + x_3 \leq 1, \\ x_1 + x_2 + 2x_3 \leq 2, \end{cases} \text{ and } X \geq 0.$$

Step I: *Write the structural constraints as equations.* The purpose of this step is to express the system of structural constraints as a linear transformation from one vector space $V(R)$ into another vector space $U(R)$. For Example 1,
$$\begin{cases} 2x_1 + x_2 + x_3 + x_4 \quad\quad\; = 1, \\ x_1 + x_2 + 2x_3 \quad\quad + x_5 = 2, \end{cases} \text{ where } X \geq 0.$$

Step II: *Write the simplex tableau or simplex matrix*
$$\begin{bmatrix} A & | & P_0 \\ \hline C-Z & | & f-f_1 \end{bmatrix},$$
which is the augmented matrix of the system
$$\begin{cases} AX = P_0, \\ (C-Z)X = f - f_1. \end{cases}$$

Here f_1 represents the first value of f corresponding to the initial basic feasible solution. The right side of the last equation is $f - f_1$ because
$$(C-Z)X = CX - ZX$$
$$= CX - \hat{C}AX$$
$$= CX - \hat{C}P_0$$
$$= f - f_1.$$

(Remember that $Z = \hat{C}A$, from page 364). In our example

$$\left[\begin{array}{c|c} A & P_0 \\ \hline C-Z & f-f_1 \end{array}\right] = \left[\begin{array}{ccccc|c} 2 & 1 & 1 & 1 & 0 & 1 \\ 1 & 1 & 2 & 0 & 1 & 2 \\ \hline 8 & 6 & 9 & 0 & 0 & f-0 \end{array}\right].$$

The last row can be calculated by observing that $\hat{C} = [0 \ 0]$, hence $Z = 0$, and therefore $C - Z = C = [8 \ 6 \ 9 \ 0 \ 0]$. Also f_1, the initial value of f, is

$$8(0) + 6(0) + 9(0) + 0(1) + 0(2) = 0,$$

therefore $f - f_1$ is $f - 0$.

Step III: *Use Theorems 2, 3, and 4 to determine which column vectors should enter and which should leave the basis of $U(R)$ or to determine that the procedure has terminated.* In our example we could choose A_1, A_2, or A_3 to enter the basis. Suppose we choose A_2, which we mark with an arrow below the column. Then, according to Theorem 2, form the ratios

$$\frac{p_{10}}{a_{12}} = \frac{1}{1} \quad \text{and} \quad \frac{p_{20}}{a_{22}} = \frac{2}{1}.$$

The first ratio is the smaller, so we call a_{12} the **pivot** and mark this entry with an asterisk.

$$\left[\begin{array}{ccccc|c} 2 & 1^* & 1 & 1 & 0 & 1 \\ 1 & 1 & 2 & 0 & 1 & 2 \\ \hline 8 & 6 & 9 & 0 & 0 & f-0 \\ & \uparrow & & & & \end{array}\right]$$

Step IV: *Perform the following elementary row operations: Multiply the pivot row by a scalar to reduce the pivot to 1. Add multiples of the pivot row to the other rows to reduce all other entries of the pivot column to 0.* The purpose of this step is to effect the change of basis and thereby accomplish the following: (1) change the transformation matrix, A, with respect to the old basis to the new transformation matrix, $Q^{-1}A$ or A', with respect to the new basis; (2) change P_0 with respect to the old basis to $Q^{-1}P_0$ or P_0' with respect to the new basis; (3) change the last row so that the evaluator $c_k - z_k$ for the new basis vector A_k is 0; it can be proved[9] that the other entries of the transformed last row are the new $c_i - z_i'$ and $f - f_2$. (Here, f_2 is the value of the objective function corresponding to the second basis.) In our example, the elementary row transformations $-R_1 + R_2$ and $-6R_1 + R_3$ yield

$$\begin{array}{cccccc} A_1' & I_{(1)} & A_3' & A_4' & I_{(2)} & P_0' \end{array}$$
$$\left[\begin{array}{ccccc|c} 2 & 1 & 1 & 1 & 0 & 1 \\ -1 & 0 & 1^* & -1 & 1 & 1 \\ \hline -4 & 0 & 3 & -6 & 0 & f-6 \\ & & \uparrow & & & \end{array}\right].$$

---

[9] A proof is given in Appendix B (Theorem A.11.0, page A7).

Step V: *Determine whether further improvement of the objective function is possible. If it is possible, begin the second iteration. Repeat until the procedure terminates.*

From the last matrix observe that the only positive entry in the last row is $c_3 - z_3' = 3$. Therefore we choose A_3' to enter the basis and form the ratios

$$\frac{p_{10}'}{a_{13}'} = \frac{1}{1} \quad \text{and} \quad \frac{p_{20}'}{a_{23}'} = \frac{1}{1}$$

which are the same; therefore, we may choose either. Suppose we choose a_{23}' as the pivot; then the operations $-R_2 + R_1$ and $-3R_2 + R_3$ transform

$$\left[\begin{array}{c|c} A' & P_0' \\ \hline C - Z' & f - f_2 \end{array}\right]$$

to

$$\left[\begin{array}{ccccc|c} A_1'' & I_{(1)} & I_{(2)} & A_4'' & A_5'' & P_0'' \\ 3 & 1 & 0 & 2 & -1 & 0 \\ -1 & 0 & 1 & -1 & 1 & 1 \\ \hline -1 & 0 & 0 & -3 & -3 & f - 9 \end{array}\right].$$

Because no further improvement in the objective function is possible, the simplex procedure for this example is terminated. The last matrix may be thought of as an augmented matrix of the system

$$\begin{cases} 3x_1 + x_2 + 2x_4 - x_5 = 0, \\ -x_1 + x_3 - x_4 + x_5 = 1, \\ -x_1 - 3x_4 - 3x_5 = f - 9. \end{cases}$$

The optimal basic feasible solution is obviously $(0, 0, 1, 0, 0)$, and from the last equation $0 = f - 9$ or $f = 9$. Let f_c represent the current value of f; in Exercise 12 the reader is asked to prove that the left side of the last equation $(C - Z)X = f - f_c$ is always zero so that the value of f after each change of basis can be determined from the lower right-hand entry of the simplex matrix. Notice that the optimal solution in this example is degenerate; this is a consequence of the tie in the minimum ratios of the second iteration. If we had chosen the a_{13}' for the pivot rather than a_{23}' the final matrix would have been different, but the optimal basic feasible solution would still have been $(0, 0, 1, 0, 0)$; the degenerate solution is considered as one basic feasible solution even though the associated bases were the second and third columns in one case and the third and fifth columns in the other case.

Because the last two variables are slack variables, the answer to the original linear programming problem is $f = 9$ at $(0, 0, 1)$.

Example 2. Maximize $-2x_4 - x_5 + x_6$, subject to

$$\begin{cases} 2x_1 - x_2 + x_4 = 5, \\ -2x_1 + 2x_2 + x_3 + x_5 = 4, \text{ and } X \geq 0. \\ x_2 - x_3 + x_6 = 6, \end{cases}$$

Step I: The constraints are equations.

Step II: $\hat{C} = [-2 \quad -1 \quad 1]$.

$$\begin{cases} c_1 - z_1 = c_1 - (\hat{C} \cdot A_1) = 0 - (-2) = 2, \\ c_2 - z_2 = c_2 - (\hat{C} \cdot A_2) = 0 - 1 = -1, \\ c_3 - z_3 = c_3 - (\hat{C} \cdot A_3) = 0 - (-2) = 2, \end{cases}$$

and

$$\begin{aligned} f_1 &= c_{(1)}x_{(1)} + c_{(2)}x_{(2)} + c_{(3)}x_{(3)} \\ &= -2x_4 - x_5 + x_6 \\ &= -2(5) - (4) + (6) \\ &= -8. \end{aligned}$$

Therefore

$$\left[\begin{array}{c|c} A & P_0 \\ \hline C-Z & f-f_1 \end{array}\right] = \left[\begin{array}{cccccc|c} 2 & -1 & 0 & 1 & 0 & 0 & 5 \\ -2 & 2 & 1^* & 0 & 1 & 0 & 4 \\ 0 & 1 & -1 & 0 & 0 & 1 & 6 \\ \hline 2 & -1 & 2 & 0 & 0 & 0 & f+8 \end{array}\right].$$

Step III: Choose A_3 (we could choose A_1) to enter the basis. The entry a_{23} is the pivot, because p_{20}/a_{23} is the only nonnegative ratio for A_3.

Step IV: Perform the elementary row operations $R_2 + R_3$ and $-2R_2 + R_4$ to reduce $\left[\begin{array}{c} A_3 \\ \hline c_3 - z_3 \end{array}\right]$ to $\left[\begin{array}{c} I_{(2)} \\ \hline 0 \end{array}\right]$. The resulting matrix is

$$\left[\begin{array}{cccccc|c} 2^* & -1 & 0 & 1 & 0 & 0 & 5 \\ -2 & 2 & 1 & 0 & 1 & 0 & 4 \\ -2 & 3 & 0 & 0 & 1 & 1 & 10 \\ \hline 6 & -5 & 0 & 0 & -2 & 0 & f-0 \end{array}\right].$$

Step V: Next, we must choose A_1' to replace $I_{(1)}$. Perform $\tfrac{1}{2}R_1$, then $2R_1 + R_2$, $2R_1 + R_3$, $-6R_1 + R_4$, *or* perform $R_1 + R_2$, $R_1 + R_3$, $-3R_1 + R_4$, and lastly $\tfrac{1}{2}R_1$, to get

$$\left[\begin{array}{cccccc|c} 1 & -\tfrac{1}{2} & 0 & \tfrac{1}{2} & 0 & 0 & \tfrac{5}{2} \\ 0 & 1 & 1 & 1 & 1 & 0 & 9 \\ 0 & 2 & 0 & 1 & 1 & 1 & 15 \\ \hline 0 & -2 & 0 & -3 & -2 & 0 & f-15 \end{array}\right].$$

Because no further improvement is possible, we read the solution from the corresponding system

$$x_1 \begin{bmatrix} 1 \\ 0 \\ 0 \\ 0 \end{bmatrix} + 0\begin{bmatrix} -\tfrac{1}{2} \\ 1 \\ 2 \\ -2 \end{bmatrix} + x_3 \begin{bmatrix} 0 \\ 1 \\ 0 \\ 0 \end{bmatrix} + 0\begin{bmatrix} \tfrac{1}{2} \\ 1 \\ 1 \\ -3 \end{bmatrix} + 0\begin{bmatrix} 0 \\ 1 \\ 1 \\ -2 \end{bmatrix} + x_6 \begin{bmatrix} 0 \\ 0 \\ 1 \\ 0 \end{bmatrix} = \begin{bmatrix} \tfrac{5}{2} \\ 9 \\ 15 \\ f-15 \end{bmatrix}.$$

Therefore, $f = 15$ at $(\tfrac{5}{2}, 0, 9, 0, 0, 15)$.

11.8 THE SIMPLEX METHOD AND TABLEAU

A final question needs to be raised concerning the simplex method as applied so far to the restricted problems considered, in which the restrictions are that $P_0 \geq 0$ and all of the columns of $I_{(m)}$ are columns of A. Are we guaranteed that the simplex method applied to any such linear programming problem will terminate after a finite number of iterations? Indeed problems can occur (although fortunately rarely) in which a certain sequence of iterations will cycle. However, it can be proved [10] that although one sequence of iterations may cycle there is at least one sequence that will not cycle, hence the answer to our question is that we are guaranteed a finite number of iterations of the simplex method when applied to a linear programming problem of the type studied so far.

APPLICATIONS

Example 3.[11] An application of linear programming to a probability problem is described in this example.

Suppose two types of missiles (say, A and B) are available to fire at two types of targets, say, Target I and Target II. Let

x_1 = number of "A" missiles fired at Target I,
x_2 = number of "B" missiles fired at Target I,
x_3 = number of "A" missiles fired at Target II,
x_4 = number of "B" missiles fired at Target II,
$Q_1 = .10$ = probability that a single "A" missile will fail to destroy Target I,
$Q_2 = .20$ = probability that a single "B" missile will fail to destroy Target I,
$Q_3 = .15$ = probability that a single "A" missile will fail to destroy Target II,
$Q_4 = .25$ = probability that a single "B" missile will fail to destroy Target II,
$c_1 = \$20,000$ = cost of a single "A" missile,
$c_2 = \$15,000$ = cost of a single "B" missile,
$E = \$1,000,000$ = maximum amount that can be spent on a given engagement.

If no more than 20 missiles of each type, and no more than 30 missiles altogether can be expended on a given engagement, determine the number of missiles (quantities x_1, x_2, x_3, and x_4) that will minimize the overall probability of failure, which is given as

$$Q = Q_1^{x_1} Q_2^{x_2} Q_3^{x_3} Q_4^{x_4}.$$

Minimizing Q is equivalent to minimizing

$$\ln Q = x_1 \ln Q_1 + x_2 \ln Q_2 + x_3 \ln Q_3 + x_4 \ln Q_4,$$

[10] Smythe, W. R., Jr., Johnson, L. A., *Introduction to Linear Programming with Applications*, Prentice-Hall, Englewood Cliffs, N.J., 1966, pages 95–102.
Hadley, G., *Linear Programming*, Addison-Wesley Publishing Co., Reading, Mass., 1962, pages 183–186.
[11] This example was contributed by Mr. Robert Ailor, when he was a graduate student at Virginia Polytechnic Institute and State University.

because the iso-Q curves are straight lines. That is, for a constant probability of Q, say Q_0, then $\ln Q_0$ is a constant made up of a linear combination of the x_i's. Thus the problem is to minimize[12] $-2.3x_1 - 1.6x_2 - 1.9x_3 - 1.4x_4$, subject to

$$\begin{cases} 20{,}000x_1 + 15{,}000\,x_2 + 20{,}000\,x_3 + 15{,}000\,x_4 \leq 1{,}000{,}000, \\ \quad x_1 + \quad\quad x_2 + \quad\quad x_3 + \quad\quad x_4 \leq 30, \\ \quad x_1 \quad\quad\quad\quad + \quad\quad x_3 \quad\quad\quad\quad \leq 20, \\ \quad\quad\quad\quad x_2 \quad\quad\quad\quad + \quad\quad x_4 \leq 20, \end{cases}$$

and $X \geq 0$.

After dividing the first inequality by 1,000 and adding four slack variables, we obtain the initial tableau

$$\left[\begin{array}{cccccccc|c} 20 & 15 & 20 & 15 & 1 & 0 & 0 & 0 & 1000 \\ 1 & 1 & 1 & 1 & 0 & 1 & 0 & 0 & 30 \\ 1 & 0 & 1 & 0 & 0 & 0 & 1 & 0 & 20 \\ 0 & 1 & 0 & 1 & 0 & 0 & 0 & 1 & 20 \\ \hline -2.3 & -1.6 & -1.9 & -1.4 & 0 & 0 & 0 & 0 & f-0 \end{array}\right].$$

After three iterations it is possible to determine that the number of missiles that will minimize the overall probability of failure is

$$x_1 = 20, \quad x_2 = 10, \quad x_3 = 0, \quad \text{and} \quad x_4 = 0.$$

Example 4. The last example need not be restricted to missiles fired at targets. Actually, the model will serve any similar situation in which m "programs" are available for use to solve n "problems" where the probabilities of failure are known and where the constraints are linear. For example, one may have m chemicals available to attack n diseases, or m plans available to apply to n social problems.

Example 5. The usefulness of linear programming in management accounting is made quite evident by Ijiri in [14]. Any serious student of business should at least glance through [14] to observe first-hand the degree of application of linear algebra, in general, and linear programming, in particular, in accounting; other references illustrating the use of linear programming in accounting are:

Charnes, A., Cooper, W. W., Millier, M. H., *Application of Linear Programming to Financial Budgeting and the Costing of Funds*, Journal of Business of the University of Chicago 32, 1, Jan., 1959, pages 20–46.

Dorfman, R., *Application of Linear Programming to the Theory of the Firm, Including an Analysis of Monopolistic Firms by Non-linear Programming*, University of California Press, Berkeley, 1951.

Ijiri, Y., Levy, F. K., and Lyon, R. C., *A Linear Programming Model for Budgeting and Financial Planning*, Journal of Accounting Research, 1, 2, Autumn, 1963, pages 198–212.

---

[12] The coefficients of the objective function are obtained from a table of natural logarithms; that is,

$$\ln(.10) = -2.3; \quad \ln(.20) = -1.6; \quad \ln(.15) = -1.9; \quad \ln(.25) = -1.4.$$

EXERCISES

In each of Exercises 1–11, write the simplex matrix and then use the procedures of this chapter to solve the optimization problem.

1. Maximize $2x_1 + x_4$, subject to
$$\begin{cases} x_1 + x_2 + 2x_3 + x_4 = 2, \\ x_1 + 2x_2 + x_3 \le 5, \end{cases} \text{ and } X \ge 0.$$

2. Minimize $3x_1 + x_3$, subject to
$$\begin{cases} x_1 + x_2 + x_3 = 3, \\ x_1 + 2x_2 + x_4 = 7, \end{cases} \text{ and } X \ge 0.$$

3. Maximize $x_1 + 2x_2$, subject to
$$\begin{cases} x_1 + x_2 \le 4, \\ x_1 + 4x_2 \le 7, \end{cases} \text{ and } X \ge 0.$$

4. Maximize $x_1 - 3x_2$, subject to
$$\begin{cases} x_1 - x_2 \ge -1, \\ x_1 + x_2 \le 1, \end{cases} \text{ and } X \ge 0.$$

5. Maximize $2x_1 + x_2 + 6x_3 + x_4$, subject to
$$\begin{cases} x_1 + 3x_2 + x_3 + x_4 \le 4, \\ x_1 + x_3 + 2x_4 \le 5, \\ x_2 + x_3 \le 2, \end{cases} \text{ and } X \ge 0.$$

6. Maximize $4x_1 - 6x_2 + 5x_3$, subject to
$$\begin{cases} -x_1 + x_2 \le 1, \\ x_2 + 2x_3 \le 4, \\ 2x_1 + x_3 \le 6, \\ 2x_2 + x_3 \le 1, \end{cases} \text{ and } X \ge 0.$$

7. Minimize $x_1 + 2x_3$, subject to
$$\begin{cases} x_1 - 2x_2 + x_3 = 4, \\ x_1 + x_2 \le 6, \end{cases} \text{ and } X \ge 0.$$

8. Minimize $-x_2 + 4x_4$, subject to
$$\begin{cases} x_2 + x_3 + x_4 = 3, \\ x_1 - 2x_2 + x_4 = 6, \end{cases} \text{ and } X \ge 0.$$

9. Maximize $-x_1 + 2x_2 + x_4$, subject to
$$\begin{cases} 2x_1 - x_3 \le 4, \\ x_1 + x_2 + x_3 + x_4 \le 8, \\ -x_1 + 2x_2 - x_4 \le 2, \end{cases} \text{ and } X \ge 0.$$

10. Maximize $x_4 - x_5 - 2x_6$, subject to

$$\begin{cases} -x_1 & + x_6 = 5, \\ 2x_1 + x_2 - 2x_3 & + x_5 = 4, \\ x_1 - x_2 + x_4 = 6, \end{cases} \text{ and } X \geq 0.$$

11. Maximize $2x_1 - x_2 + x_3$, subject to

$$\begin{cases} -x_1 + 5x_2 - 2x_3 \leq 10, \\ 2x_1 + x_2 - x_3 \leq 5, \\ x_1 - x_2 + 2x_3 \leq 4, \end{cases} \text{ and } X \geq 0.$$

12. The original simplex matrix can be viewed as the augmented matrix of the system

$$\begin{cases} AX = P_0, \\ (C - Z)X = f - f_1. \end{cases}$$

Prove that the left side of the last equation must be 0. That is, prove that

$$(C - Z)X = 0.$$

(Answer is provided in booklet of Answers to Selected Even Numbered Exercises.)

13. Maximize $x_1 + x_2$, subject to

$$\begin{cases} -x_1 - x_2 + x_3 \leq 1, \\ -2x_1 - x_2 + x_3 \leq 1, \\ x_1 + x_2 - 8x_3 \leq 8, \end{cases} \text{ and } X \geq 0.$$

11.9 Artificial Variables

So far we have limited our discussion to problems with m structural constraints of the form $AX = P_0$, where $P_0 \geq 0$ and where m columns of A are $I_{(1)}, I_{(2)}, \ldots, I_{(m)}$ which served as an initial basis. Of course not all problems meet these restrictions, as the next examples will demonstrate; in these examples a method will be presented whereby any problem that does not meet these restrictions will be solved by considering an *augmented problem* that does meet them. We shall also determine whether or not there are any feasible solutions of the original problem.

Example 1. Minimize $x_1 + x_2 + 2x_3$, subject to

(11.9.1) $$\begin{cases} x_1 + 2x_2 + x_3 \leq 16, \\ 2x_1 + x_2 + x_3 = 10, \\ x_1 + x_3 \geq 2, \end{cases} \text{ and } X \geq 0.$$

11.9 ARTIFICIAL VARIABLES 383

After adding and subtracting the slack variables, we have as the constraints

$$\begin{cases} x_1 + 2x_2 + x_3 + x_5 = 16, \\ 2x_1 + x_2 + x_3 = 10, \\ x_1 + x_3 - x_4 = 2, \end{cases} \text{ and } X \geq 0.$$

Notice that $I_{(1)}$ is a column of the coefficient matrix, A, but not $I_{(2)}$ nor $I_{(3)}$. To alter this situation, we create an augmented problem by adding two nonnegative "artificial variables" x_6 and x_7 to the last two structural constraints; these artificial variables certainly must eventually become zero if the constraints of the augmented problem are to be consistent with the constraints of the original problem (11.9.1).

(11.9.2)
$$\begin{cases} x_1 + 2x_2 + x_3 + x_5 = 16, \\ 2x_1 + x_2 + x_3 + x_6 = 10, \\ x_1 + x_3 - x_4 + x_7 = 2. \end{cases}$$

We will show later that if (11.9.1) is a consistent system, then we can force x_6 and x_7 in (11.9.2) to be zero by making the coefficients of x_6 and x_7 in the objective function so large that the minimum cannot be attained as long as x_6 and x_7 are different from zero. Suppose we let M be an arbitrarily large number and create the objective function

$$g = x_1 + x_2 + 2x_3 + 0x_4 + 0x_5 + Mx_6 + Mx_7$$

for the augmented problem. The original objective function is $f = x_1 + x_2 + 2x_3 + 0x_4 + 0x_5$. Obviously minimum f will not equal minimum g unless x_6 and x_7 are zero; this can be accomplished if $I_{(2)}$ and $I_{(3)}$ are removed from the basis. To set up the simplex tableau of the augmented problem, notice that $\hat{C} = [0 \; M \; M]$, and consequently,

$$\begin{cases} c_1 - z_1 = c_1 - \hat{C} \cdot A_1 = 1 - 3M, \\ c_2 - z_2 = c_2 - \hat{C} \cdot A_2 = 1 - 1M, \\ c_3 - z_3 = c_3 - \hat{C} \cdot A_3 = 2 - 2M, \\ c_4 - z_4 = c_4 - \hat{C} \cdot A_4 = 0 - (-M). \end{cases}$$

Remember that, for the basis vectors, $c_i - z_i = 0$; therefore,

$$c_5 - z_5 = c_6 - z_6 = c_7 - z_7 = 0,$$

and

$$\begin{aligned} g_1 &= (0) + (0) + 2(0) + 0(0) + 0(x_5) + M(x_6) + M(x_7) \\ &= 0(16) + M(10) + M(2) \\ &= 12M. \end{aligned}$$

The simplex matrix is

$$\begin{bmatrix} A & \vline & P_0 \\ \hline C-Z & \vline & g-g_1 \end{bmatrix} = \left[\begin{array}{cccccccc|c} 1 & 2 & 1 & 0 & 1 & 0 & 0 & & 16 \\ 2 & 1 & 1 & 0 & 0 & 1 & 0 & & 10 \\ 1^* & 0 & 1 & -1 & 0 & 0 & 1 & & 2 \\ \hline 1-3M & 1-M & 2-2M & M & 0 & 0 & 0 & & g-12M \\ \uparrow & & & & & & & & \end{array} \right].$$

Because this is a minimum problem we look for negative entries of the last row. Introducing any one of the first three columns into the basis will improve g. Arbitrarily we pick A_1, and after forming the ratios p_{i0}/a_{i1} ($i = 1, 2, 3$), we find that a_{31} is the pivot and $I_{(3)}$ must leave the basis. If the row operations $-R_3 + R_1$, $-2R_3 + R_2$, and $-(1 - 3M)R_3 + R_4$ are performed, the next simplex matrix is

$$\begin{bmatrix} 0 & 2 & 0 & 1 & 1 & 0 & -1 & 14 \\ 0 & 1^* & -1 & 2 & 0 & 1 & -2 & 6 \\ 1 & 0 & 1 & -1 & 0 & 0 & 1 & 2 \\ \hline 0 & 1-M & 1+M & 1-2M & 0 & 0 & 3M-1 & g-(6M+2) \\ & \uparrow & & & & & & \end{bmatrix}.$$

Continue with the next iteration: We choose the second column to enter and replace $I_{(2)}$. Perform $-2R_2 + R_1$ and $-(1 - M)R_2 + R_4$; the result is

$$\begin{bmatrix} 0 & 0 & 2 & -3 & 1 & -2 & 3 & 2 \\ 0 & 1 & -1 & 2^* & 0 & 1 & -2 & 6 \\ 1 & 0 & 1 & -1 & 0 & 0 & 1 & 2 \\ \hline 0 & 0 & 2 & -1 & 0 & M-1 & M+1 & g-8 \\ & & & \uparrow & & & & \end{bmatrix}.$$

One more iteration ($\frac{3}{2}R_2 + R_1$, $\frac{1}{2}R_2 + R_3$, $\frac{1}{2}R_2 + R_4$, followed by $\frac{1}{2}R_2$) produces

$$\begin{bmatrix} 0 & \frac{3}{2} & \frac{1}{2} & 0 & 1 & -\frac{1}{2} & 0 & 11 \\ 0 & \frac{1}{2} & -\frac{1}{2} & 1 & 0 & \frac{1}{2} & -1 & 3 \\ 1 & \frac{1}{2} & \frac{1}{2} & 0 & 0 & \frac{1}{2} & 0 & 5 \\ \hline 0 & \frac{1}{2} & \frac{3}{2} & 0 & 0 & M-\frac{1}{2} & M & g-5 \end{bmatrix}.$$

Because the entries on the last row are all nonnegative, the objective function can no longer be improved. Therefore the minimum is $g = 5$ at

$$X = [5 \quad 0 \quad 0 \quad 3 \quad 11 \quad 0 \quad 0]^T.$$

Later, in Theorem 6, we will prove that because $x_6 = x_7 = 0$,

$$\text{minimum } f = \text{minimum } g = 5 \text{ at } X = [5 \quad 0 \quad 0 \quad 3 \quad 11]^T.$$

Example 2. Maximize $x_1 - 2x_2$, subject to

$$\begin{cases} x_1 - x_2 \leq 2, \\ -x_1 - x_2 \leq -4, \end{cases} \text{ and } X \geq 0,$$

or subject to

$$\begin{cases} x_1 - x_2 + x_3 = 2, \\ -x_1 - x_2 + x_4 = -4, \end{cases} \text{ and } X \geq 0.$$

Notice that P_0 is not nonnegative, but this can be changed by multiplying the last equation by (-1).

$$\begin{cases} x_1 - x_2 + x_3 = 2, \\ x_1 + x_2 - x_4 = 4. \end{cases}$$

However, $I_{(2)}$ is not now a column of the coefficient matrix. Thus we create an augmented problem by adding a nonnegative artificial variable x_5 to the last constraint and add $(-Mx_5)$ to the objective function, where M is an arbitrarily large number. The purpose of the artificial variable is to provide an initial basis $\{I_{(1)}, I_{(2)}\}$, which is needed to begin the iterative procedure as we have studied it. The new objective function is

$$g = x_1 - 2x_2 - Mx_5,$$

and the augmented problem is to maximize g, subject to

$$\begin{bmatrix}1\\1\end{bmatrix}x_1 + \begin{bmatrix}-1\\1\end{bmatrix}x_2 + \begin{bmatrix}1\\0\end{bmatrix}x_3 + \begin{bmatrix}0\\-1\end{bmatrix}x_4 + \begin{bmatrix}0\\1\end{bmatrix}x_5 = \begin{bmatrix}2\\4\end{bmatrix}, \text{ and } X \geq 0.$$

If the original problem has a feasible solution, say (s_1, s_2, s_3, s_4), then the augmented problem has a feasible solution $(s_1, s_2, s_3, s_4, 0)$, and, as long as x_5 is nonzero, the term $(-Mx_5)$ is small enough to require

$$(x_1 - 2x_2 - Mx_5) < (s_1 - 2s_2 - M \cdot 0).$$

Hence g has not achieved its maximum at $(x_1, x_2, x_3, x_4, x_5)$; by Theorem 3, there exists some $c_k - z_k > 0$, and the simplex procedure continues. Thus if an optimum of the augmented problem is ever to be obtained, x_5 must equal 0.

When a linear programming problem is approached by means of an augmented problem as was done in the last two examples, certain questions should be raised.

First. Under what conditions can we be assured that the artificial variables will be zero in an optimal solution? This question will be answered by the following theorem;

Theorem 5. *If the feasible set of a given linear programming problem is nonempty, then the artificial variables in an optimal solution of the augmented problem will be zero.*

The proof is left as Exercise 16. *Hint:* Generalize the latter portion of Example 2.

At this point, observe that because of degeneracy it is possible for all of the artificial variables to be zero and yet have the corresponding *artificial* vectors in the basis; in other words, the attainment of zero values for all artificial variables does not always imply that all artificial vectors have been removed from the basis.

Second. Even if the *augmented* problem has been optimized and all of the artificial variables are zero, how do we know that the *original* problem has been optimized, and what is its solution? Theorem 6 will answer this question.

Theorem 6. *If all of the artificial variables are zero in an optimal solution of an augmented linear programming problem, then the values of the non-artificial variables form an optimal solution of the original linear programming problem.*

Proof. Let the original linear programming problem be expressed as "optimize $f(X)$ subject to $AX = P_0$ and $X \geq 0$, where $P_0 \geq 0$," and let the augmented problem be expressed as "optimize $g(Y)$ subject to $[A \mid B]Y = P_0$ and $Y \geq 0$ where Y consists of all of the components of X and the artificial variables, and B is the matrix of coefficients of the artificial variables."

(1) We have assumed that an optimum of the augmented problem exists; Theorem 1 assures us that the optimum will be attained at a basic feasible solution, say, at $Y = Y_0$; hence there is an invertible transition matrix Q such that

$$Q^{-1}[A \mid B]Y_0 = Q^{-1}P_0,$$

and $g(Y_0)$ is optimum.

(2) If the artificial variables in Y_0 are zero (as we assumed), then $Y_0 = \begin{bmatrix} X_0 \\ \hline 0 \end{bmatrix}$ and $Q^{-1}AX_0 = Q^{-1}P_0$ and $g(Y_0) = f(X_0)$.

(3) If all artificial vectors can be eliminated from an optimal basis, then the evaluators $c_k - z_k$ of the nonartificial variables are the same in both the original and augmented problems; hence $f(X_0) = g(Y_0)$ is optimum for the original problem.

(4) If, however, an artificial vector (with corresponding zero-valued artificial variable) cannot be eliminated from an optimal basis, then it can be shown[13] that certain of the original constraints were redundant and can be removed; the problem will then be of the type treated in Statement (3) of this proof. ☐

Third. What are all of the possible outcomes when the simplex method is applied to an augmented problem, and what is the significance of each outcome? There are four possibilities:

(1) No artificial vectors are in the basis and yet an optimum has been reached. Theorem 6 then applies. Examples 1 and 2 have illustrated this case.

(2) No artificial vectors are in the basis and the objective function is unbounded for feasible solutions. Theorem 4 will apply in this case. It may be possible for Theorem 4 to apply prior to the elimination of all artificial vectors, but the artificial vectors can be eliminated first if that is desired. See Exercise 17.

(3) Artificial vectors are in the basis with corresponding nonzero artificial variables and the augmented problem has been optimized. The contrapositive of Theorem 5 can be applied to show that in this case there are no feasible solutions to the original problem. This can mean that the original

---

[13] Hadley, G., *Linear Programming*, Addison-Wesley Publishing Co., Reading, Mass., 1962, pages 122–123.

structural constraint equations either were inconsistent or had only non-feasible solutions.

(4) Because of degeneracy, artificial vectors are in the basis, but with corresponding *zero-valued* artificial variables, and an optimum has been reached. Theorem 6 applies but redundancy may exist in the structural constraints of the original problem;[14] that is, the rank of A in the original problem may be less than m.

EXERCISES

In each of Exercises 1–4, set up the simplex matrix and decide which A_k could enter the basis and improve the objective function.

1. Minimize $x_1 + x_2 + 4x_3$, subject to
$$\begin{cases} -x_1 - x_2 + x_3 \geq 2, \\ -x_1 + x_2 + 2x_3 \geq 1, \end{cases} \text{ and } X \geq 0.$$

2. Minimize $2x_1 + x_2$, subject to
$$\begin{cases} x_1 + x_2 \geq 1, \\ x_1 - x_2 \geq -1, \end{cases} \text{ and } X \geq 0.$$

3. Minimize $3x_1 + x_2 - x_3$, subject to
$$\begin{cases} -3x_1 - x_2 + 5x_3 \geq 18, \\ -x_1 - x_2 + 2x_3 \geq 5, \\ -x_1 + 2x_2 + x_3 \leq 6, \end{cases} \text{ and } X \geq 0.$$

4. Minimize $3x_1 + 5x_2 + 2x_3$, subject to
$$\begin{cases} x_1 + x_2 + 3x_3 \geq 4, \\ x_1 + 2x_2 \geq 2, \\ x_1 + x_2 + x_3 \geq 6, \end{cases} \text{ and } X \geq 0.$$

In each of Exercises 5–8, which refer to the preceding Exercises 1–4, perform one iteration of the simplex method by entering the column with the smallest evaluator $(c_k - z_k)$.

5. Exercise 1.   6. Exercise 2.   7. Exercise 3.   8. Exercise 4.

9. Solve by the simplex method.
Minimize $x_1 + x_2 + 4x_3 + x_4$, subject to
$$\begin{cases} -x_1 - x_2 + x_3 + x_4 = 2, \\ -x_1 + x_2 + 2x_3 \geq 1, \end{cases} \text{ and } X \geq 0.$$

---

[14] A more thorough discussion of item (4) in particular and the other items in general may be found on pages 121–124 of Hadley, G., *Linear Programming*, Addison-Wesley Publishing Co., Reading, Mass., 1962.

10. Solve by the simplex method.
    Minimize $x_2 + 2x_3$, subject to
    $$\begin{cases} x_1 + x_2 + x_3 + x_4 = 3, \\ x_1 + x_2 + 2x_4 = 1, \end{cases} \text{ and } X \geq 0.$$

11. Solve by the simplex method.
    Minimize $2x_1 + x_2$, subject to
    $$\begin{cases} 3x_1 + x_2 \geq 6, \\ x_1 + x_2 \geq 4, \end{cases} \text{ and } X \geq 0.$$

12. Solve by the simplex method.
    Minimize $x_1 + x_2$, subject to
    $$\begin{cases} 2x_1 + x_2 = 4, \\ x_1 + 2x_2 = 4, \end{cases} \text{ and } X \geq 0.$$

13. Solve by the simplex method.
    Maximize $5x_2 - 8x_3$, subject to
    $$\begin{cases} x_1 + x_2 + x_3 = 6, \\ x_1 + x_3 \geq 3, \\ x_1 + x_2 \leq 4, \end{cases} \text{ and } X \geq 0.$$

    Remember that the coefficient of an artificial variable in the objective function of a maximum problem is $(-M)$.

14. Solve by the simplex method.
    Maximize $2x_2 + x_3$, subject to
    $$\begin{cases} x_1 + x_3 \geq 2, \\ 2x_1 + x_2 + x_3 = 8, \\ x_2 \leq 5, \end{cases} \text{ and } X \geq 0.$$

15. (a) How is an unbounded feasible set with no optimal solution recognized in the simplex method?
    (b) How is a feasible set with an infinite number of optimal solutions recognized in the simplex method?
    (c) How is an empty feasible set recognized in the simplex method?

16. Prove Theorem 5.

17. Apply Theorem 4 to the following problem before and after the artificial vector has been removed from the basis. Also solve graphically.
    Maximize $2x_1$, subject to
    $$\begin{cases} x_2 = 2, \\ -x_1 + x_2 \leq 1, \end{cases} \text{ and } X \geq 0.$$

18. Use the contrapositive statement of Theorem 5 to prove that the following problem has an empty feasible set.
    Maximize $2x_1 + x_2$, subject to
    $$\begin{cases} x_1 + x_2 \leq 1, \\ x_1 + 2x_2 \geq 4, \end{cases} \text{ and } X \geq 0.$$

11.10 The Revised Simplex Method

It is possible to streamline the standard simplex method to a more efficient procedure. Recall that the original simplex matrix is

$$M_1 = \left[\begin{array}{c|c} A & P_0 \\ \hline C-Z & f-f_1 \end{array}\right].$$

Assume an initial basis $\{I_{(1)}, \ldots, I_{(m)}\}$, and examine $C-Z$ and $f-f_1$.

$$\begin{aligned} C - Z &= C - [z_1 \mid z_2 \mid \cdots \mid z_n] \\ &= C - [\hat{C}A_1 \mid \hat{C}A_2 \mid \cdots \mid \hat{C}A_n] \\ &= C - \hat{C}[A_1 \mid A_2 \mid \cdots \mid A_n] \\ &= C - \hat{C}A. \end{aligned}$$

Next,

$$\begin{aligned} f - f_1 &= f - \{c_{(1)} x_{(1)} + c_{(2)} x_{(2)} + \cdots + c_{(m)} x_{(m)}\} \\ &= f - \hat{C} P_0, \end{aligned}$$

since initially $x_{(i)} = p_{i0}$. Therefore M_1 can be expressed as

$$M_1 = \left[\begin{array}{c|c} A & P_0 \\ \hline C - \hat{C}A & f - \hat{C}P_0 \end{array}\right],$$

which in turn can be expressed as the product of two matrices

$$M_1 = \left[\begin{array}{c|c} I_m & 0 \\ \hline -\hat{C} & 1 \end{array}\right] \left[\begin{array}{c|c} A & P_0 \\ \hline C & f \end{array}\right];$$

the order of each submatrix in this product is given below

$$\left[\begin{array}{c|c} m \text{ by } n & m \text{ by } 1 \\ \hline 1 \text{ by } n & 1 \text{ by } 1 \end{array}\right] = \left[\begin{array}{c|c} m \text{ by } m & m \text{ by } 1 \\ \hline 1 \text{ by } m & 1 \text{ by } 1 \end{array}\right] \left[\begin{array}{c|c} m \text{ by } n & m \text{ by } 1 \\ \hline 1 \text{ by } n & 1 \text{ by } 1 \end{array}\right].$$

For future reference these two matrices will be designated as H_1 and M_0, respectively; that is,

$$H_1 = \left[\begin{array}{c|c} I_m & 0 \\ \hline -\hat{C} & 1 \end{array}\right], \quad M_0 = \left[\begin{array}{c|c} A & P_0 \\ \hline C & f \end{array}\right],$$

and the first simplex tableau is

$$M_1 = H_1 M_0.$$

It should be obvious that M_0 is simply the augmented matrix of the system consisting of the structural constraints and the objective function

$$\begin{cases} AX = P_0, \\ CX = f, \end{cases}$$

and thus is very easy to write.

It has become apparent in preceding sections that one iteration of the simplex method can be accomplished by elementary row operations on the existing simplex matrix. Such elementary row operations can be performed by the premultiplication of elementary matrices. Suppose we let E_1 represent the product of the elementary matrices used to execute the first iteration; then the second simplex matrix M_2 can be expressed as

$$M_2 = E_1 M_1 = E_1(H_1 M_0) = (E_1 H_1)M_0 = H_2 M_0.$$

Now let E_2 represent the product of elementary matrices used to execute the second iteration; then the third simplex matrix M_3 can be expressed as

$$M_3 = E_2 M_2 = E_2(H_2 M_0) = (E_2 H_2)M_0 = H_3 M_0.$$

Generalizing we have

$$M_k = E_{k-1} M_{k-1} = E_{k-1}(H_{k-1} M_0) = (E_{k-1} H_{k-1})M_0 = H_k M_0.$$

The advantage in using these forms is that by operating on H_i, rather than on the entire tableau M_i, a considerable number of calculations can be by-passed, as will be demonstrated in the next example. Another advantage is the ease with which $C - Z$ is found.

Example 1. Maximize $-2x_4 - x_5 + x_6$, subject to

$$\begin{cases} -x_2 + 2x_3 + x_4 & = 5, \\ x_1 + 2x_2 - 2x_3 + x_5 & = 4, \text{ and } X \geq 0. \\ -x_1 + x_2 + x_6 & = 6, \end{cases}$$

We have seen that

$$H_1 = \begin{bmatrix} I_m & 0 \\ \hline -\hat{C} & 1 \end{bmatrix} \quad \text{and} \quad M_0 = \begin{bmatrix} A & P_0 \\ \hline C & f \end{bmatrix}.$$

$$M_1 = H_1 M_0 = \begin{bmatrix} 1 & 0 & 0 & 0 \\ 0 & 1 & 0 & 0 \\ 0 & 0 & 1 & 0 \\ \hline +2 & +1 & -1 & 1 \end{bmatrix} \begin{bmatrix} 0 & -1 & 2 & 1 & 0 & 0 & 5 \\ 1 & 2 & -2 & 0 & 1 & 0 & 4 \\ -1 & 1 & 0 & 0 & 0 & 1 & 6 \\ \hline 0 & 0 & 0 & -2 & -1 & 1 & f \end{bmatrix}$$

11.10 THE REVISED SIMPLEX METHOD

$$= \begin{bmatrix} 0 & & & & & 5 \\ 1^* & & & & & 4 \\ -1 & & & & & 6 \\ \hline 2 & & & & & f+8 \end{bmatrix}.$$

We calculate only enough of the last row of M_1 to find a suitable A_k to enter the basis; then we also write A_k and P_0 in order to determine the pivot and to determine the row operations that we must perform on H_1. In this example, we must apply $R_2 + R_3$ and $-2R_2 + R_4$ to H_1 to obtain H_2.

$$H_2 = \begin{bmatrix} 1 & 0 & 0 & 0 \\ 0 & 1 & 0 & 0 \\ 0 & 1 & 1 & 0 \\ \hline 2 & -1 & -1 & 1 \end{bmatrix}.$$

Repeat the process.

$$M_2 = H_2 M_0 = \begin{bmatrix} 1 & 0 & 0 & 0 \\ 0 & 1 & 0 & 0 \\ 0 & 1 & 1 & 0 \\ \hline 2 & -1 & -1 & 1 \end{bmatrix} \begin{bmatrix} 0 & -1 & 2 & 1 & 0 & 0 & 5 \\ 1 & 2 & -2 & 0 & 1 & 0 & 4 \\ -1 & 1 & 0 & 0 & 0 & 1 & 6 \\ \hline 0 & 0 & 0 & -2 & -1 & 1 & f \end{bmatrix}$$

$$= \begin{bmatrix} & 2^* & & & & & 5 \\ & -2 & & & & & 4 \\ & -2 & & & & & 10 \\ \hline 0 & -5 & 6 & & & & f-0 \end{bmatrix}.$$

Thus A_3' should enter, a_{13}' is the pivot, and the row operations are $R_1 + R_2$, $R_1 + R_3$, and $-3R_1 + R_4$ followed by $\frac{1}{2}R_1$. When these operations are performed on H_2 we get

$$H_3 = \begin{bmatrix} \frac{1}{2} & 0 & 0 & 0 \\ 1 & 1 & 0 & 0 \\ 1 & 1 & 1 & 0 \\ \hline -1 & -1 & -1 & 1 \end{bmatrix}.$$

Then

$$M_3 = H_3 M_0 = \begin{bmatrix} & & & & & & \frac{5}{2} \\ & & & & & & 9 \\ & & & & & & 15 \\ \hline 0 & -2 & 0 & -3 & -2 & 0 & f-15 \end{bmatrix}.$$

The last row is nonpositive and a maximum has been attained, namely, $f = 15$, which can be read from the lower right entry as usual. It is important to keep track of the current vectors in the basis after each iteration in order to determine the optimal solution without calculating all of the entries of the last M_i. An efficient way to arrange the tableau for the last example is illustrated next.

			x_1	x_2	x_3	x_4	x_5	x_6	P_0	
basic variables ↓		$M_1 = H_1 M_0$	0	−1	2	1	0	0	5	
		$M_2 = H_2 M_0$	1	2	−2	0	1	0	4	
		$M_3 = H_3 M_0$	−1	1	0	0	0	1	6	M_0
			0	0	0	−2	−1	1	f	
H_1	x_4		1	0	0	0	0			5	
	x_5		0	1	0	0	1*			4	
	x_6		0	0	1	0	−1			6	M_1
			2	1	−1	1	2			$f+8$	
H_2	x_4		1	0	0	0		2*		5	
	x_1		0	1	0	0		−2		4	
	x_6		0	1	1	0		−2		10	M_2
			2	−1	−1	1	0	−5	6	$f-0$	
H_3	x_3		½	0	0	0				5/2	
	x_1		1	1	0	0				9	
	x_6		1	1	1	0				15	M_3
			−1	−1	−1	1	0	−2	0	−3	−2	0	$f-15$	

From M_3 we can read maximum $f = 15$ at $(9, 0, \tfrac{5}{2}, 0, 0, 15)$.

APPLICATIONS

For those who wish to study applications of linear programming in a particular area of interest, a list of references is given on pages 325–337 of [11] (Gass). A more comprehensive list may be found in [30] (Riley and Gass).

EXERCISES

Solve Exercises 1–11 by the revised simplex method.

1. Maximize $2x_1 + x_4$, subject to

$$\begin{cases} x_1 + x_2 + 2x_3 + x_4 = 2, \\ x_1 + 2x_2 + x_3 \le 5, \end{cases} \text{ and } X \ge 0.$$

2. Minimize $3x_1 + x_3$, subject to

$$\begin{cases} x_1 + x_2 + x_3 = 3, \\ x_1 + 2x_2 + x_4 = 7, \end{cases} \text{ and } X \ge 0.$$

3. Maximize $x_1 + 2x_2$, subject to

$$\begin{cases} x_1 + x_2 \le 4, \\ x_1 + 4x_2 \le 7, \end{cases} \text{ and } X \ge 0.$$

4. Maximize $x_1 - 3x_2$, subject to

$$\begin{cases} x_1 - x_2 \ge -1, \\ x_1 + x_2 \le 1, \end{cases} \text{ and } X \ge 0.$$

5. Maximize $2x_1 + x_2 + 6x_3 + x_4$, subject to

$$\begin{cases} x_1 + 3x_2 + x_3 + x_4 \le 4, \\ x_1 + x_3 + 2x_4 \le 5, \\ x_2 + x_3 \le 2, \end{cases} \text{ and } X \ge 0.$$

6. Maximize $4x_1 - 6x_2 + 5x_3$, subject to

$$\begin{cases} -x_1 + x_2 \le 1, \\ x_2 + 2x_3 \le 4, \\ 2x_1 + x_3 \le 6, \\ 2x_2 + x_3 \le 1, \end{cases} \text{ and } X \ge 0.$$

7. Minimize $x_1 + x_2 + 4x_3$, subject to

$$\begin{cases} -x_1 - x_2 + x_3 \ge 2, \\ -x_1 + x_2 + 2x_3 \ge 1, \end{cases} \text{ and } X \ge 0.$$

8. Minimize $2x_1 + x_2$, subject to

$$\begin{cases} x_1 + x_2 \ge 1, \\ x_1 - x_2 \ge -1, \end{cases} \text{ and } X \ge 0.$$

9. Maximize $2x_1 + x_2 + x_4$, subject to

$$\begin{cases} x_1 + x_2 - x_3 \le 10, \\ x_1 - x_2 + x_4 = 1, \end{cases} \text{ and } X \ge 0.$$

10. Maximize $x_1 - x_2 + 2x_3$, subject to

$$\begin{cases} - x_2 + 2x_3 - x_4 \le 2, \\ -x_1 + 2x_2 \le 4, \\ x_1 + x_2 + x_3 + x_4 \le 8, \end{cases} \text{ and } X \ge 0.$$

11. Minimize $2x_1 + 3x_2 - x_3 + x_4$, subject to

$$\begin{cases} x_1 + 4x_2 - x_3 + 2x_4 \le 10, \\ 2x_1 + 2x_2 + 5x_3 - x_4 \le 30, \\ x_1 + x_2 + x_3 - 3x_4 = 8, \end{cases} \text{ and } X \ge 0.$$

NEW VOCABULARY

linear programming problem 11.1
feasible solution 11.1
feasible set 11.1
nonnegativity constraints 11.1
structural constraints 11.1
objective function 11.1
integer programming 11.1
slack variable 11.2
basic variable 11.2
nonbasic variable 11.2
basic solution 11.2
basic feasible solution 11.2
feasible basis 11.2

degenerate basic solution 11.2
nondegenerate basic solution 11.2
optimal solution 11.2
convex combination 11.3
simplex method 11.4
cycling 11.5
evaluator 11.6
pivot 11.8
artificial variable 11.9
augmented linear programming problem 11.9
revised simplex method 11.10

REFERENCES

The following books have been referred to in the applied examples of this book. Those marked with an asterisk have many elementary applications of matrix algebra, and it is recommended that those readers who are interested in applications should at least scan the relevant parts of these books.

[1] Alling, D. W., The After History of Pulmonary TB: A Stochastic Model, *Biometrics*, 14, 1958, pages 527–547.
[2] Ames, W. F., Canonical Forms for Nonlinear Kinetic Differential Equations, *Ind. & Eng. Chem. Fundamentals*, 1, 1962, pages 214–218.
[3] Bernadelli, H., Population Waves, *J. Burma Res. Soc.*, 31, 1941, pages 1–18.
[4] Buerger, M. J., *X-Ray Crystallography*, John Wiley & Sons, New York, 1942.
[5] Cayley, Arthur, *The Principals of Bookkeeping by Double Entry*, The University Press, Cambridge, 1907.
[6] Cole, R. H., *Theory of Ordinary Differential Equations*, Appleton-Century-Crofts, New York, 1968.
[7] Davis, P. J., *The Mathematics of Matrices*, Blaisdell Publishing Co., New York, 1965.
*[8] Elgerd, O. I., *Control Systems Theory*, McGraw-Hill, New York, 1967.
[9] Eves, Howard, *Elementary Matrix Theory*, Allyn & Bacon, Boston, 1966.
[10] Flegg, H. G., *Boolean Algebra and its Application*, John Wiley & Sons, New York, 1964.
[11] Gass, S. I., *Linear Programming*, Third Edition, McGraw-Hill, New York, 1969.
[12] Glicksman, A. M., *An Introduction to Linear Programming and the Theory of Games*, John Wiley & Sons, New York, 1963.

REFERENCES

[13] Hollingsworth, C. A., *Vectors, Matrices, and Group Theory for Scientists and Engineers*, McGraw-Hill, 1967.

*[14] Ijiri, Yuji, *Management Goals and Accounting for Control*, Rand McNally & Company, Chicago, 1965.

[15] Jaaskelainen, Veikko, *Accounting and Mathematical Programming* (Lithographed) Helsinki, Finland, 1969, pages 11–14.

*[16] Johnston, J. B., Price, G. B., Van Vleck, F. S., *Linear Equations and Matrices*, Addison-Wesley Press, Reading, Mass., 1966.

*[17] Kemeny, J. G., Schleifer, A., *Finite Mathematics with Business Applications*, Prentice-Hall, Englewood Cliffs, N.J., 1962.

*[18] Kemeny, J. G., Snell, J. L., Thompson, G. L., *Introduction to Finite Mathematics, Second Edition*, Prentice Hall, Englewood Cliffs, N.J., 1966.

[19] Leslie, P. H., On the Use of Matrices in Certain Population Mathematics, *Biometrika*, 33, 1945, pages 183–212.

[20] Levine, J., Some Elementary Cryptanalysis of Algebraic Cryptography, *American Mathematical Monthly*, Vol. 68, No. 5, 1961, pages 411–418.

[21] Li, C. C., *Population Genetics*, University of Chicago Press, Chicago, 1954.

[22] Nahikian, H. M., *Topics in Modern Mathematics*, Macmillan, New York, 1966.

*[23] Noble, Ben, *Applications of Undergraduate Mathematics in Engineering*, Macmillan, New York, 1967.

[24] Parker, F. D., Boolean Matrices and Logic, *Mathematics Magazine*, Vol. 37, No. 1, pages 33–38.

[25] Perry, Kenneth Jr., An Application of Linear Algebra to Petrologic Problems: Part I Mineral Classification, *Geochimica et Cosmochimica Acta*, 1967, Vol. 31, Pergamon Press Ltd., Northern Ireland, Vol. 31, 1967, pages 1043–1078.

[26] Perry, Kenneth Jr., A Computer Program for Representation of Mineral Chemical Analyses in Terms of End Member Molecules, *Contributions to Geology*, Vol. 7, No. 1, 1968, pages 7–14.

[27] Perry, Kenneth Jr., Representation of Mineral Chemical Analyses in 11-dimensional Space: Part II Amphiboles, *Lithos*, Vol. 1, No. 4, 1968, pages 307–321.

[28] Perry, Kenneth Jr., Representation of Mineral Chemical Analyses in 11-dimensional Space: Part I Feldspars, *Lithos*, Vol. 1, No. 3, 1968, pages 202–218.

[29] Pipes, L. A., *Matrix Methods for Engineering*, Prentice-Hall, Englewood Cliffs, N.J., 1963.

[30] Riley, V., and Gass, S. I., *Bibliography on Linear Programming and Related Techniques*, The Johns Hopkins Press, Baltimore, 1958.

*[31] Searle, S. R., *Matrix Algebra for the Biological Sciences*, John Wiley & Sons, New York, 1966.

[32] Williams, J. D., *The Compleat Strategyst, Revised Ed.*, McGraw-Hill, New York, 1966.

APPENDIX A

Cartesian Product : a pairing operation

Consider two sets A and B. Let a be an element of A, and let b be an element of B; then construct an ordered pair (a, b). The set of all such ordered pairs that can be constructed from A and B is called the *Cartesian product* of A and B, and it is denoted by $A \times B$.

Example. Let $A = \{a_1, a_2, a_3\}$ and $B = \{b_1, b_2\}$; then

$$A \times B = \{(a_1, b_1), (a_1, b_2), (a_2, b_1), (a_2, b_2), (a_3, b_1), (a_3, b_2)\}.$$

Notice that $A \times B$ is the set of *all* possible ordered pairs that can be constructed.

Example. Let R be the set of all real numbers. Then $R \times R$ is the set of all ordered pairs of real numbers; the elements of this set of ordered pairs are used to identify the points on a plane with respect to two perpendicular coordinate axes according to the Cartesian coordinate system, hence the name "Cartesian" product.

APPENDIX A

APPENDIX B

Proofs of Certain Theorems

Theorem A.8.2. *If $b = \{\beta_1, \ldots, \beta_n\}$ is a basis of a vector space $V(F)$, and if $\gamma_1, \gamma_2, \ldots, \gamma_n$ are arbitrary vectors of $V(F)$, then there exists exactly one linear transformation $T: V(F) \to V(F)$ such that*

$$T(\beta_i) = \gamma_i \quad (i = 1, 2, \ldots, n).$$

Proof. The proof is split into three parts. We must first show that there is such a transformation, then we must show that the transformation is linear, and finally we must show that the transformation is unique.

PART I. Form the linear combinations $k_1\gamma_1 + k_2\gamma_2 + \cdots + k_n\gamma_n$ and $k_1\beta_1 + k_2\beta_2 + \cdots + k_n\beta_n$ where the k_i's are arbitrary elements of F. Since b is a basis, every element of $V(F)$ is uniquely expressible in the form

$$k_1\beta_1 + \cdots + k_n\beta_n,$$

and we can define a transformation T over $V(F)$ according to

(1) $$T(k_1\beta_1 + \cdots + k_n\beta_n) = k_1\gamma_1 + \cdots + k_n\gamma_n.$$

If we set $k_1 = 1, k_2 = 0, \ldots, k_n = 0$, then

$$T(\beta_1) = \gamma_1.$$

(The symbols 0 and 1 represent the additive and multiplicative identities of F, and we have used the result of Exercise 11 of Section 6.2.) Likewise, if we set $k_1 = 0, k_2 = 1, k_3 = 0, \ldots, k_n = 0$, then

$$T(\beta_2) = \gamma_2.$$

If this procedure is continued we find that we have a transformation T over $V(F)$ such that

$$T(\beta_i) = \gamma_i \quad (i = 1, 2, \ldots, n).$$

PART II. Let α and δ be arbitrary elements of $V(F)$; then, of course, each can be written as a linear combination of the basis vectors:

$$\alpha = a_1\beta_1 + \cdots + a_n\beta_n,$$

and

$$\delta = d_1\beta_1 + \cdots + d_n\beta_n.$$

We must show that

$$T(\alpha + \delta) = T(\alpha) + T(\delta),$$

and, for k belonging to F,

$$T(k\alpha) = kT(\alpha).$$

STATEMENT	REASON
$T(\alpha + \delta)$	
$= T\{(a_1 + d_1)\beta_1 + \cdots + (a_n + d_n)\beta_n\}$	By coordinate vector addition, since $\{\beta_1, \ldots, \beta_n\}$ is a basis.
$= (a_1 + d_1)\gamma_1 + \cdots + (a_n + d_n)\gamma_n$	By (1) of Part I.
$= (a_1\gamma_1 + \cdots + a_n\gamma_n)$ $+ (d_1\gamma_1 + \cdots + d_n\gamma_n)$	By distributive, associative, and commutative properties. (Definition 10 of Section 6.4.)
$= T(\alpha) + T(\delta).$	By (1).
$T(k\alpha) = T\{ka_1\beta_1 + \cdots + ka_n\beta_n\}$	By multiplication of a vector by a scalar.
$= ka_1\gamma_1 + \cdots + ka_n\gamma_n$	By (1).
$= k(a_1\gamma_1 + \cdots + a_n\gamma_n)$	By multiplication of a vector by a scalar.
$= kT(\alpha).$	By (1).

PART III. Assume a second linear transformation S over $V(F)$ such that

(2) $$S(\beta_i) = \gamma_i \quad (i = 1, 2, \ldots, n).$$

Now

STATEMENT	REASON
$S(\alpha) = a_1 S(\beta_1) + \cdots + a_n S(\beta_n)$	Because S is linear.
$ = a_1 \gamma_1 + \cdots + a_n \gamma_n$	By assumption (2).
$ = a_1 T(\beta_1) + \cdots + a_n T(\beta_n)$	By Part I.
$ = T(a_1 \beta_1 + \cdots + a_n \beta_n)$	Because T is linear.
$ = T(\alpha).$	

Thus T is unique. ∎

Theorem A.10.6. *Any n by n C-matrix A is similar to an upper triangular matrix whose diagonal entries are the characteristic values of A.*

Proof. The theorem is obvious when $n = 1$. Therefore we assume $n > 1$ and show that there exists an invertible matrix P such that $P^{-1}AP$ is an upper triangular matrix with the characteristic values of A on the main diagonal. Since the entries of A are complex numbers, the characteristic equation of A has n complex roots which are not necessarily distinct.

(1) Let λ_1 be a characteristic value of A with corresponding characteristic vector X_1. Hence

$$AX_1 = \lambda_1 X_1 \quad \text{where } X_1 \neq 0.$$

(2) Let Q be an invertible matrix having X_1 as its first column, and form the product AQ or $A[X_1 \mid Q_2 \mid \cdots \mid Q_n]$. The first column of AQ is $\lambda_1 X_1$ by Statement (1).

(3) Now form the product $Q^{-1}(AQ)$ or $Q^{-1}[\lambda_1 X_1 \mid AQ_2 \mid \cdots \mid AQ_n]$. The first column of this product is $Q^{-1}\lambda_1 X_1$ or $\lambda_1 Q^{-1} X_1$. But $Q^{-1} X_1$ is also the first column of $Q^{-1}Q$ which is the first column of I_n, or $\begin{bmatrix} 1 \\ 0 \\ \vdots \\ 0 \end{bmatrix}$. Therefore $Q^{-1}AQ$ has the form $\left[\begin{array}{c|c} \lambda_1 & F \\ \hline 0 & B \end{array}\right]$, where $\begin{bmatrix} F \\ \hline B \end{bmatrix}$ simply represents all but the first column of $Q^{-1}AQ$.

(4) Let λ_2 be a second characteristic value of A; by Theorem 1 of Chapter 10, similar matrices have the same characteristic values, hence λ_2 is a characteristic value of B (regardless of whether λ_1 equals λ_2 or not).

(5) If $n > 2$, then by a repetition of Statements 1–4 we can show that there exists an invertible matrix R such that $R^{-1}BR$ has the form

$$R^{-1}BR = \left[\begin{array}{c|c} \lambda_2 & G \\ \hline 0 & C \end{array}\right],$$

where $\begin{bmatrix} G \\ \hline C \end{bmatrix}$ simply represents all but the first column of $R^{-1}BR$, and where C is an $n-2$ by $n-2$ invertible matrix with characteristic roots $\lambda_3, \ldots, \lambda_n$ which are also characteristic roots of A.

(6) Because R is invertible, the n by n matrix $S = \begin{bmatrix} 1 & 0 \\ \hline 0 & R \end{bmatrix}$ is invertible and consequently QS is invertible. Thus we have

$$(QS)^{-1}A(QS) = S^{-1}(Q^{-1}AQ)S$$

$$= \begin{bmatrix} 1 & 0 \\ \hline 0 & R^{-1} \end{bmatrix} \begin{bmatrix} \lambda_1 & F \\ \hline 0 & B \end{bmatrix} \begin{bmatrix} 1 & 0 \\ \hline 0 & R \end{bmatrix} = \begin{bmatrix} \lambda_1 & FR \\ \hline 0 & R^{-1}BR \end{bmatrix}$$

$$= \begin{bmatrix} \lambda_1 & \alpha & H \\ 0 & \lambda_2 & \\ \hline 0 & & C \end{bmatrix},$$

where $\begin{bmatrix} H \\ \hline C \end{bmatrix}$ simply represents all but the first two columns of $(QS)^{-1}A(QS)$.

(7) The remaining stages may be accomplished in the same manner as outlined above, and we will obtain a product of invertible matrices which we define to be P. Hence there exists an invertible matrix P such that $P^{-1}AP$ is upper triangular with the characteristic values of A on the main diagonal. □

Theorem A.10.11. *Let $V(R)$ be a real inner product space with an orthonormal basis $b = \{\beta_1, \ldots, \beta_n\}$, and let T be a linear transformation defined over $V(R)$. T is an orthogonal linear transformation if and only if $[T]_b^T = [T]_b^{-1}$.*

Proof. PART I. Prove:

$$(T \text{ is an orthogonal operator}) \Rightarrow ([T]_b^T = [T]_b^{-1}).$$

(1) Because T is an orthogonal transformation, and because the basis is an orthonormal set,

$$\langle T(\beta_i), T(\beta_j) \rangle = \langle \beta_i, \beta_j \rangle = \begin{cases} 0 & \text{if } i \neq j, \\ 1 & \text{if } i = j. \end{cases}$$

(2) Expressing $T(\beta_i)$ and $T(\beta_j)$ as linear combinations of the basis we have

$$\langle T(\beta_i), T(\beta_j) \rangle = \langle (t_{i1}\beta_1 + \cdots + t_{in}\beta_n), (t_{j1}\beta_1 + \cdots + t_{jn}\beta_n) \rangle.$$

(3) By repeated use of item (2) of the definition of an inner product (on page 199) and by use of Statement (1) above, we know that the right side of the last equation is equal to

$$t_{i1}t_{j1} + \cdots + t_{in}t_{jn}.$$

(4) By the definition of $[T]_b$, the last expression is the standard inner product of columns i and j of $[T]_b$. Since we have just shown that this *standard inner product* is equal to the *inner product* of Statement (1) and consequently is 0 if $i \neq j$ and 1 if $i = j$, then the columns of $[T]_b$ form an orthonormal set.

(5) By Theorem 13 of Chapter 4, $[T]_b^{-1} = [T]_b^T$.

PART II. Prove:

$$([T]_b^T = [T]_b^{-1}) \Rightarrow (T \text{ is an orthogonal operator}).$$

(1) Let α and γ be arbitrary elements of $V(R)$, and, with respect to the basis b, let their coordinates be $a_1, \ldots, a_n$ and $c_1, \ldots, c_n$, respectively. Also, let the coordinates of $T(\alpha)$ and $T(\gamma)$ be $a'_1, \ldots, a'_n$ and $c'_1, \ldots, c'_n$, respectively. Then, by repeated use of item (2) of the definition of an inner product (Definition 11 of Chapter 7), and because the basis is an orthonormal set, we have

$$\langle T(\alpha), T(\gamma) \rangle = a'_1 c'_1 + \cdots + a'_n c'_n.$$

(2) But $a'_1 c'_1 + \cdots + a'_n c'_n$ is the single entry of

$$[T(\alpha)]_b^T [T(\gamma)]_b.$$

(3) Since $[T]_b$ is orthogonal,

$$[T(\alpha)]_b^T [T(\gamma)]_b = ([T]_b [\alpha]_b)^T ([T]_b [\gamma]_b)$$
$$= [\alpha]_b^T ([T]_b^{-1} [T]_b) [\gamma]_b$$
$$= [\alpha]_b^T [\gamma]_b.$$

(4) But $[\alpha]_b^T [\gamma]_b$ is a 1 by 1 matrix with entry

$$a_1 c_1 + \cdots + a_n c_n,$$

which is equal to $\langle \alpha, \gamma \rangle$ according to the reasoning used in Statement (1).

(5) We have shown that

$$\langle T(\alpha), T(\gamma) \rangle = \langle \alpha, \gamma \rangle,$$

and hence by Definition 3 of Chapter 10, T is an orthogonal operator. □

Theorem A.11.0. *If elementary row operations are applied to the original simplex matrix M_1 so that the pivot column $\begin{bmatrix} A_k \\ \hline c - z \end{bmatrix}$ becomes $\begin{bmatrix} I_{(j)} \\ \hline 0 \end{bmatrix}$, then the last row of the new simplex matrix is*

$$[C - Z' \mid f - f_2].$$

Proof. The matrix Q will denote the transition matrix which replaces $I_{(j)}$ with A_k in the basis of the column space of A. Primes are used to denote matrices with respect to the new basis.

First, we shall show that if the original simplex matrix M_1 is premultiplied by the matrix

$$R = \left[\begin{array}{c|c} Q^{-1} & 0 \\ \hline -\hat{C}'Q^{-1}+\hat{C} & 1 \end{array}\right],$$

then the pivot column of the simplex matrix becomes $\left[\begin{array}{c} I_{(j)} \\ 0 \end{array}\right]$, and hence R is the product of elementary matrices that effects the elementary row operations stated in the hypothesis. Second, we shall show that the last row of the matrix product RM_1 is

$$[C - Z' \mid f - f_2].$$

PART I.

$$\left[\begin{array}{c|c} Q^{-1} & 0 \\ \hline -\hat{C}'Q^{-1}+\hat{C} & 1 \end{array}\right]\left[\begin{array}{c} A_k \\ \hline c_k - z_k \end{array}\right] = \left[\begin{array}{c} Q^{-1}A_k \\ \hline -\hat{C}'Q^{-1}A_k + \hat{C}A_k + c_k - z_k \end{array}\right]$$

$$= \left[\begin{array}{c} I_{(j)} \\ \hline -\hat{C}'I_{(j)} + \hat{C}A_k + c_k - z_k \end{array}\right].$$

Because $\hat{C}'I_{(j)} = c_{(1)} \cdot 0 + c_{(2)} \cdot 0 + \cdots + c_k \cdot 1 + \cdots + c_{(m)} \cdot 0 = c_k$ and because $\hat{C}A_k = z_k$ by definition, then the last matrix is equal to

$$\left[\begin{array}{c} I_{(j)} \\ 0 \end{array}\right].$$

PART II. The last row of the product

$$\left[\begin{array}{c|c} Q^{-1} & 0 \\ \hline -\hat{C}'Q^{-1}+\hat{C} & 1 \end{array}\right]\left[\begin{array}{c|c} A & P_0 \\ \hline C - Z & f - f_1 \end{array}\right]$$

is

$$[-\hat{C}'Q^{-1}A + \hat{C}A + C - Z \mid -\hat{C}'Q^{-1}P_0 + \hat{C}P_0 + f - f_1]$$
$$= [-\hat{C}'A' + \hat{C}A + C - Z \mid -\hat{C}'P_0' + \hat{C}P_0 + f - f_1].$$

Because

$$\hat{C}A = \hat{C}[A_1 \mid \cdots \mid A_n] = [z_1 \mid \cdots \mid z_n] = Z,$$

and likewise

$$\hat{C}'A' = \hat{C}'[A_1' \mid \cdots \mid A_n'] = [z_1' \mid \cdots \mid z_n'] = Z',$$

and because

$$\hat{C}P_0 = f_1 \quad \text{and} \quad \hat{C}'P_0' = f_2,$$

then the last row of the product RM_1 becomes

$$[C - Z' \mid f - f_2]. \quad \square$$

ANSWERS TO SELECTED ODD-NUMBERED EXERCISES

1.1 Page 3

1. (a) Domain is the set $\{a, b, c\}$, Range is the set $\{p, q\}$;
   (b) Images are p and q;
   (c) $T = \{(a, p), (b, q), (c, q)\}$;
   (d)

3. (a) Domain: set of real numbers, Range: set of real numbers;
   (b) $T(x) = x^3 + 1$, where x is a real number;
   (c) $T = \{(x, T(x)) \mid T(x) = x^3 + 1\}$;
   (d)
   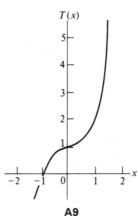

ANSWERS TO SELECTED ODD-NUMBERED EXERCISES

5. (a) Domain: set of nonnegative integers, Range: set of nonnegative odd integers;
(b) $T(k)$ is the set $\{1, 3, 5, 7, 9, 11, 13, \ldots\}$;
(c) $T = \{(k, T(k)) \mid T(k) = 2k + 1,\ k\text{ is a nonnegative integer}\}$;
(d)

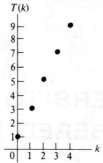

7. Because the image of 9 is not unique.

1.2 Page 9

1. (a) 2 by 3, not an R-matrix; (b) 2 by 1, an R-matrix;
(c) 1 by 3, not an R-matrix; (d) 1 by 1, an R-matrix.

3. (a) $a_{12} = 9$; (b) $a_{23} = 1$; (c) $a_{13} = 4$.

5. $A = \begin{bmatrix} 2 & 1+i & 4 \\ 2 & 1 & i \\ -3 & 0 & 2 \end{bmatrix}$.

7. $\begin{bmatrix} 2 & 3 & 4 & 1 & 0 \\ 9 & 3 & 2 & 0 & 1 \end{bmatrix}$.

9. (a) $x = 5,\ y = 2$; (b) Impossible; (c) $x = 1,\ y = 0$.

11.
$$\begin{array}{c} \\ \#1 \\ \#2 \\ \#3 \\ \#4 \end{array} \begin{array}{cccc} \#1 & \#2 & \#3 & \#4 \\ \begin{bmatrix} 0 & 0 & 0 & 1 \\ 1 & 0 & 1 & 1 \\ 0 & 0 & 0 & 0 \\ 1 & 0 & 0 & 0 \end{bmatrix} \end{array}$$
where $a_{ij} = 1$ if i can speak to j and $a_{ij} = 0$ if i cannot speak to j. No station speaks to itself.

1.3 Page 14

1. $\begin{bmatrix} 6 & 1 \\ 5 & 5 \end{bmatrix}$.

3. Not possible—not conformable.

5. $\begin{bmatrix} 6 & 3 & 2 \\ 4 & 9 & 2 \\ 2 & 2 & 3 \end{bmatrix}$.

7. (a) $\begin{bmatrix} 6 & -3 \\ -9 & -12 \end{bmatrix}$; (b) $\begin{bmatrix} 4 & 0 \\ 2 & -6 \end{bmatrix}$; (c) $\begin{bmatrix} -2 & 1 \\ 3 & 4 \end{bmatrix}$;

(d) $\begin{bmatrix} -4 & -1 \\ -6 & 5 \end{bmatrix}$; (e) $\begin{bmatrix} -5 & 2 \\ \frac{11}{2} & \frac{19}{2} \end{bmatrix}$; (f) $\begin{bmatrix} 4 & -1 \\ -2 & -7 \end{bmatrix}$;

(g) $\begin{bmatrix} 3 & -\frac{1}{2} \\ -\frac{1}{2} & -5 \end{bmatrix}$.

9. No.   **11.** No.

15. Theorem 2; the associative property of addition.

2.1 Page 22

1. (a) $A = \begin{bmatrix} 3 & 1 & 1 \\ 2 & 0 & 1 \\ 1 & 0 & 2 \end{bmatrix}, [A \vdots B] = \begin{bmatrix} 3 & 1 & 1 & 4 \\ 2 & 0 & 1 & 5 \\ 1 & 0 & 2 & 1 \end{bmatrix}$;

(b) $A = \begin{bmatrix} 2 & 1 \\ 1 & -1 \\ 1 & 1 \end{bmatrix}, [A \vdots B] = \begin{bmatrix} 2 & 1 & 5 \\ 1 & -1 & 1 \\ 1 & 1 & 3 \end{bmatrix}$;

(c) $A = \begin{bmatrix} 1 & 1 & 1 & 1 \\ 2 & 1 & 0 & 1 \end{bmatrix}, [A \vdots B] = \begin{bmatrix} 1 & 1 & 1 & 1 & 1 \\ 2 & 1 & 0 & 1 & 2 \end{bmatrix}$;

(d) $A = \begin{bmatrix} 1 & 1 \\ 2 & -1 \end{bmatrix}, [A \vdots B] = \begin{bmatrix} 1 & 1 & -3 \\ 2 & -1 & -6 \end{bmatrix}$.

3. $(-1, 4)$.   **5.** $(3, -2, 1)$.   **7.** $(\frac{7}{2}, \frac{5}{2})$.   **9.** $(2, \frac{1}{2}, -1, 0)$.

13. The graphs of the two linear equations intersect at one point, namely, $(-1, 4)$.

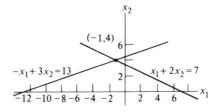

15. The graphs of the three linear equations intersect at the point $(\frac{7}{2}, \frac{5}{2})$.

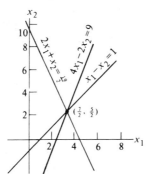

A12 ANSWERS TO SELECTED ODD-NUMBERED EXERCISES

17. No.

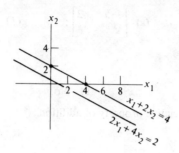

$\begin{bmatrix} 1 & 2 & | & 4 \\ 2 & 4 & | & 2 \end{bmatrix} \overset{-2R_1+R_2}{\sim} \begin{bmatrix} 1 & 2 & | & 4 \\ 0 & 0 & | & -6 \end{bmatrix}$. The last row implies $0x_1 + 0x_2 = -6$; No solution (x_1, x_2) exists that will satisfy this equation.

19. (a) If two systems are equivalent, then one system of linear equations over C is transformed into the other linear system over C by means of one elementary operation on the equations.

(b) No.   **21.** (a) Yes;   (b) Yes.

2.2 Page 27

1. $(b)(d)(f)(g)(h)$ are echelon; $(b)(f)(h)$ are reduced echelon.

3. No because $\begin{bmatrix} 2 & 1 & 4 \\ 0 & 1 & 3 \\ 2 & 0 & 1 \end{bmatrix} \overset{\text{row}}{\sim} \begin{bmatrix} 1 & 0 & 1 \\ 0 & 1 & 3 \\ 0 & 0 & 0 \end{bmatrix}$.

5. $\begin{bmatrix} 1 & 4 & 3 & 2 \\ 1 & 8 & 0 & 2 \\ 2 & 0 & 4 & 2 \end{bmatrix} \overset{\text{row}}{\sim} \begin{bmatrix} 1 & 4 & 3 & 2 \\ 0 & 1 & -\frac{3}{4} & 0 \\ 0 & 0 & 1 & \frac{1}{4} \end{bmatrix}$, this echelon matrix is not unique.

7. $\begin{bmatrix} 2 & 1 & | & 6 \\ 2 & 4 & | & 9 \end{bmatrix} \overset{\text{row}}{\sim} \begin{bmatrix} 1 & 0 & | & \frac{5}{2} \\ 0 & 1 & | & 1 \end{bmatrix}$, $\begin{cases} x = \frac{5}{2}, \\ y = 1. \end{cases}$

9. (a) If B is obtained from A by interchanging two rows, then A can be obtained from B by interchanging the same two rows. If B is obtained from A by multiplying some row by a nonzero scalar, then A can be obtained from B by multiplying the same row by the reciprocal of this scalar. If B is obtained from A by adding to the ith row a scalar multiple of the jth row, then A can be obtained from B by adding to the ith row of B the negative of that scalar multiple of the jth row of B.

(b) Because only elementary row operations are used to obtain B from A, and hence by part (a) only elementary row operations are used to obtain A from B.

11. $A \overset{\text{row}}{\sim} G$, $B \overset{\text{row}}{\sim} E$.

13. $\begin{bmatrix} 1 & 4 & 3 & 2 \\ 0 & 1 & -\frac{3}{4} & 0 \\ 0 & 0 & 1 & \frac{1}{4} \end{bmatrix}, \begin{bmatrix} 1 & 0 & 6 & 2 \\ 0 & 1 & -\frac{3}{4} & 0 \\ 0 & 0 & 1 & \frac{1}{4} \end{bmatrix}$.

ANSWERS TO SELECTED ODD-NUMBERED EXERCISES   A13

2.3 Page 36

1. $\begin{cases} x_1 = \frac{7}{16}, \\ x_2 = \frac{11}{16}. \end{cases}$

3. Rank of A is 1; Rank of $[A \vdots B]$ is 2; thus inconsistent.

5. $\begin{cases} x_1 = 1 + x_3, \\ x_2 = 1 - x_3; \end{cases}$ (2, 0, 1).   7. $\begin{cases} x_1 = 0, \\ x_2 = 0, \\ x_3 = 0. \end{cases}$ 9. $\begin{cases} x_1 = 0, \\ x_2 = 1. \end{cases}$

11. $\begin{cases} x_1 = 0 + x_2, \\ x_3 = 4 - 2x_2; \end{cases}$ (1, 1, 2).   13. $\begin{cases} x_1 = -4 - x_4, \\ x_2 = 0 - x_4, \\ x_3 = 6 + x_4; \end{cases}$ $(-4, 0, 6, 0)$.

15. No; x_2 and x_3 form a set of fundamental variables; x_1 and x_3 form a set of fundamental variables.

17. $\begin{cases} x = 1 - z, \\ y = 2; \end{cases}$ (1, 2, 0); (0, 2, 1); (2, 2, −1).

21. $\begin{cases} 4 \text{ No. 1 Trucks,} \\ 6 \text{ No. 2 Trucks,} \\ 2 \text{ No. 3 Trucks.} \end{cases}$ 23. $\begin{cases} i_1 = \frac{3}{140} E_0, \\ i_2 = \frac{1}{140} E_0, \\ i_3 = \frac{1}{70} E_0. \end{cases}$

25. (a) There is no unique answer; 4 type 1 trucks must be sent, and if k type 3 trucks are sent, then $(8 - k)$ type 2 trucks must be sent.
    (b) There will be exactly 9 solutions because in order to send fully loaded trucks, k must be an integer such that $0 \leq k \leq 8$.

27. (a) Consistent system with unique solution; (b) consistent system with infinite number of solutions; (c) consistent system with infinite number of solutions; (d) inconsistent system; (e) consistent system with unique solution; (f) consistent system with infinite number of solutions; (g) consistent system with unique solution; (h) consistent system with infinite number of solutions.

3.1 Page 49

1. (a) $\begin{bmatrix} 2 & 1 \\ 8 & -1 \end{bmatrix}$;   (b) $\begin{bmatrix} 2 \\ -6 \end{bmatrix}$;   (c) $[8 \quad 2]$;   (d) $\begin{bmatrix} 4 & 2 \\ 3 & 1 \end{bmatrix}$;

   (e) not possible—not conformable;   (f) $\begin{bmatrix} 6 & -2 \\ 9 & -3 \end{bmatrix}$.

3. (a) $n = r$;   (b) m by t;   (c) $t = m$;   (d) r by n;   (e) $m = r = t = n$.

5. $m = p$.   7. $\begin{cases} 2x_1 = 2, \\ x_1 + 3x_2 = 1, \\ 4x_1 + 2x_2 = 3. \end{cases}$

9. No, because in general $ABAB \neq AABB$, since in general $BA \neq AB$.

11. (a) $AB = BA = \begin{bmatrix} 6 & 8 \\ 10 & 12 \end{bmatrix}$;   (b) Yes.

A14 ANSWERS TO SELECTED ODD-NUMBERED EXERCISES

13. No.

15. $\left[\sum_{k=1}^{3} a_{ik}b_{kj}\right]_{(2,1)}$.

19. $AN = \begin{bmatrix} 16 \\ 38 \end{bmatrix}$. The entries of this matrix represent the total number of Gadget R and Gadget S that can be produced in one week by the two factories.

21. (a) $A + A^2 = \begin{bmatrix} 0 & 1 & 1 & 2 \\ 1 & 0 & 2 & 2 \\ 0 & 1 & 0 & 1 \\ 1 & 1 & 1 & 0 \end{bmatrix}$. (b) $\begin{cases} \#2 \text{ has five influence channels,} \\ \#1 \text{ has four influence channels,} \\ \#4 \text{ has three influence channels,} \\ \#3 \text{ has two influence channels.} \end{cases}$

23.

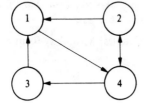

3.2 Page 60

1. $A^{-1} = \begin{bmatrix} \frac{3}{10} & \frac{1}{10} \\ -\frac{4}{10} & \frac{2}{10} \end{bmatrix}$; B is noninvertible;

$C^{-1} = \begin{bmatrix} \frac{2}{7} & -\frac{3}{7} & 0 \\ -\frac{1}{7} & -\frac{2}{7} & 1 \\ \frac{1}{7} & \frac{2}{7} & 0 \end{bmatrix}$; $D^{-1} = \frac{1}{19}\begin{bmatrix} 4 & 4 & -1 \\ -1 & -1 & 5 \\ 9 & -10 & -7 \end{bmatrix}$.

3. $A^{-2} = \frac{1}{9}\begin{bmatrix} 1 & -8 \\ 0 & 9 \end{bmatrix}$; $A^{-3} = \frac{1}{27}\begin{bmatrix} 1 & -26 \\ 0 & 27 \end{bmatrix}$.

7. $(AB)^{-1} = B^{-1}A^{-1} = \begin{bmatrix} 0 & 1 \\ \frac{1}{3} & -\frac{8}{3} \end{bmatrix}$; $(A^{-1})^{-1} = A = \begin{bmatrix} 3 & 2 \\ 0 & 1 \end{bmatrix}$.

9. $\begin{cases} x_1 = \frac{2}{3}, \\ x_2 = -\frac{1}{3}. \end{cases}$ **11.** $\begin{cases} x_1 = -\frac{1}{2}, \\ x_2 = -\frac{1}{2}, \\ x_3 = \frac{3}{2}. \end{cases}$

13. No, because B is noninvertible by Theorem 10.

21. $X = A^{-1}B = \begin{bmatrix} \frac{3}{4} & -\frac{1}{2} \\ -\frac{1}{4} & \frac{1}{2} \end{bmatrix}\begin{bmatrix} 24 \\ 32 \end{bmatrix} = \begin{bmatrix} 2 \\ 10 \end{bmatrix}$.

23. (a) $D = A^{-1}CB^{-1}$; (b) $D = (AC)^{-1}CB^{-1}$; (c) $D = (BA)^{-1}$; (d) $D = B$; (e) $D = (A^{-1} - I_n)B$.

3.3 Page 66

1. $\begin{bmatrix} I_2 & 0 \\ \hline 0 & 0 \end{bmatrix}$. **3.** $[I_2 \mid 0]$. **5.** $\begin{bmatrix} I_2 & 0 \\ \hline 0 & 0 \end{bmatrix}$. **7.** $E = \begin{bmatrix} 0 & 1 \\ 1 & 0 \end{bmatrix}$.

9. $E = \begin{bmatrix} 1 & 0 \\ 0 & 9 \end{bmatrix}$. **11.** $E = \begin{bmatrix} 1 & 0 \\ 4 & 1 \end{bmatrix}$.

ANSWERS TO SELECTED ODD-NUMBERED EXERCISES   A15

13. $E = \begin{bmatrix} 1 & 0 \\ 2 & 1 \end{bmatrix}$, $E^{-1} = \begin{bmatrix} 1 & 0 \\ -2 & 1 \end{bmatrix}$, $A = \begin{bmatrix} -1 & 2 \\ -7 & 4 \\ 6 & 0 \end{bmatrix}$.

19. Any elementary row transformation matrix can also be obtained from the identity matrix by elementary column operations and hence is an elementary column transformation matrix. This should be demonstrated for each of the three elementary operations.

21. If their normal forms are equal, then one can be transformed into the other by elementary operations according to Exercises 17 and 20.

3.4 Page 72

1. $PA = (E_2 E_1)A = \begin{bmatrix} 1 & 0 \\ 0 & 3 \end{bmatrix} \begin{bmatrix} 1 & 0 \\ 1 & 1 \end{bmatrix} A = \begin{bmatrix} 1 & 0 \\ 3 & 3 \end{bmatrix} A = \begin{bmatrix} 1 & 4 \\ 9 & 30 \end{bmatrix}$.

3. $B = PA$ where $P = \begin{bmatrix} 1 & 0 & 0 \\ 0 & 1 & 0 \\ 3 & 0 & 1 \end{bmatrix} \begin{bmatrix} 0 & 1 & 0 \\ 1 & 0 & 0 \\ 0 & 0 & 1 \end{bmatrix} = \begin{bmatrix} 0 & 1 & 0 \\ 1 & 0 & 0 \\ 0 & 3 & 1 \end{bmatrix}$.

5. No.   7. $\begin{bmatrix} 1 & 1 \\ 0 & 1 \end{bmatrix} \begin{bmatrix} 1 & 0 \\ 0 & 2 \end{bmatrix} = \begin{bmatrix} 1 & 2 \\ 0 & 2 \end{bmatrix}$.   9. $B = kA^{-1}$.

17. (a) If $\begin{bmatrix} A \\ \hline I_n \end{bmatrix} \underset{\text{col}}{\sim} \begin{bmatrix} I_n \\ \hline P \end{bmatrix}$ for a given C-matrix A, then P is the inverse of A.

(b) $\begin{bmatrix} 1 & 2 \\ 1 & 3 \\ \hline 1 & 0 \\ 0 & 1 \end{bmatrix} \underset{\text{col}}{\sim} \begin{bmatrix} 1 & 0 \\ 0 & 1 \\ \hline 3 & -2 \\ -1 & 1 \end{bmatrix} \Rightarrow \begin{bmatrix} 1 & 2 \\ 1 & 3 \end{bmatrix}^{-1} = \begin{bmatrix} 3 & -2 \\ -1 & 1 \end{bmatrix}$.

4.1 Page 86

1. (a) symmetric;   (b) neither, because the matrix is not square;
   (c) skew-symmetric;   (d) neither; $A \neq A^T$, $A \neq -A^T$.

3. (a) $(A^T)^T = \begin{bmatrix} 3 & 4 \\ 2 & 0 \end{bmatrix}^T = \begin{bmatrix} 3 & 2 \\ 4 & 0 \end{bmatrix} = A$;   (b) $\begin{bmatrix} 2 & 0 \\ 7 & 4 \\ 6 & 8 \end{bmatrix} = (AB)^T = B^T A^T$.

5. $(A - B)^T = [A + (-B)]^T$ By definition of subtraction of matrices.
   $ = A^T + (-B)^T$ By part (ii) of Theorem 1.
   $ = A^T + (-1)B^T$ By part (iv) of Theorem 1.
   $ = A^T - B^T$.   By definition of subtraction.

7. $(A^T)(A^{-1})^T = (A^{-1}A)^T$ By part (iii) of Theorem 1.
   $\phantom{(A^T)(A^{-1})^T} = I^T$ By definition of inverse.
   $\phantom{(A^T)(A^{-1})^T} = I$.   By definition of transpose.
   Likewise $(A^{-1})^T A^T = I$.   By same reasoning.
   Therefore $(A^{-1})^T = (A^T)^{-1}$.   If A^T has an inverse, then it is unique.

9. $(A + A^T)^T = A^T + (A^T)^T$ By part (ii) of Theorem 1.
   $ = A^T + A$ By part (i) of Theorem 1.
   $ = A + A^T$.   Matrix addition is commutative.

21. $\begin{bmatrix} 3 & \frac{3}{2} & -1 \\ \frac{3}{2} & 3 & 0 \\ -1 & 0 & 2 \end{bmatrix} + \begin{bmatrix} 0 & \frac{3}{2} & 0 \\ -\frac{3}{2} & 0 & -2 \\ 0 & 2 & 0 \end{bmatrix}.$

23. $\begin{array}{c} \\ \\ \\ \end{array} \begin{array}{ccc} \#1 & \#2 & \#3 \\ \begin{bmatrix} 25 & 19 & 23 \\ 19 & 41 & 17 \\ 23 & 17 & 34 \end{bmatrix} & \begin{array}{l} \#1 \\ \#2. \\ \#3 \end{array} \end{array}$

25. (a) $a_{32} = 7$; (b) $a_{32} = -7$.

4.2 Page 92

1. (a) Neither, because matrix is not square;
   (b) Neither, because $A \neq A^*$ and $A \neq -A^*$. Observe that the entries on the main diagonal are not all real (thus not Hermitian), and that the entries on the main diagonal are not all either pure imaginary numbers or zeros (thus not skew-Hermitian);
   (c) Hermitian; (d) Hermitian; (e) Skew-Hermitian.

3. Yes, $\bar{A} = A$ hence $A = A^T \Rightarrow A = A^*$.

5. Yes. $A = (\bar{A})^T \Rightarrow a_{ii} = \bar{a}_{ii} \Rightarrow a_{ii}$ is real. $A = (\bar{A})^T \Rightarrow m = n$.

7. (a) $\frac{1}{2}\begin{bmatrix} 0 & 4+i \\ 4-i & 0 \end{bmatrix} + \frac{1}{2}\begin{bmatrix} 2i & i \\ i & -4i \end{bmatrix}$; (b) $\frac{1}{2}\begin{bmatrix} 0 & 4+i \\ 4-i & 0 \end{bmatrix} + \frac{i}{2}\begin{bmatrix} 2 & 1 \\ 1 & -4 \end{bmatrix}.$

9. (a) $iA = \begin{bmatrix} -1 & -1+i \\ -1-i & 0 \end{bmatrix}$, $-iA = \begin{bmatrix} 1 & 1-i \\ 1+i & 0 \end{bmatrix}$;

   (b) $AA^* = A^*A = \begin{bmatrix} 3 & 1-i \\ 1+i & 2 \end{bmatrix}$;

   (c) $A^2 = \begin{bmatrix} -3 & -1+i \\ -1-i & -2 \end{bmatrix}$; (d) $K + iS = \begin{bmatrix} 0 & 1 \\ -1 & 0 \end{bmatrix} + i\begin{bmatrix} 1 & 1 \\ 1 & 0 \end{bmatrix}.$

4.3 Page 100

1. (a) $\alpha \cdot \beta = 22$; (b) $A \cdot B = -\frac{6}{5}$;
   (c) Not possible—different number of components;
   (d) $A \cdot B = \dfrac{40 + 3\sqrt{6} + 12i}{7}$, $A \cdot B \neq B \cdot A$, because B is not a real coordinate vector;
   (e) $A \cdot B = 6i$, $A \cdot B \neq B \cdot A$, because A is not a real coordinate vector.

3. (a) $\sqrt{37}$—not normal; (b) 3—not normal; (c) $\sqrt{22}$—not normal;
   (d) $\sqrt{47/49}$—not normal; (e) 1—normal.

5. $(12/\sqrt{145}, 1/\sqrt{145}, 0)$. 7. (b) and (f) are orthonormal.

9. $A \cdot B$ not possible; $\|A\| = \sqrt{38}$; $\|B\| = \sqrt{5}$.

11. Every real number is a complex number, and $b = \bar{b}$ if b is real.

13. $(-2, 1)$ is one answer.

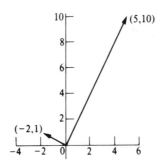

4.4 Page 103

1. (a) Orthogonal; (b) not orthogonal; (c) orthogonal.
3. (a) Orthogonal; (b) not orthogonal; (c) orthogonal.
5. (a) Not unitary; (b) not unitary; (c) unitary; (d) not unitary.
7. (a) and (c) are unitary because an R-matrix is unitary if and only if it is orthogonal.
21. (a) $x = \pm 1$; (b) $x = \pm 1$.

4.5 Page 108

1. A, E, F are upper triangular.
3. E and F are diagonal; F is a scalar matrix.
5. A, B, E are lower triangular block matrices.
7. (a) True; (b) false; (c) true; (d) false;
(e) true; (f) false; (g) true; (h) false; (i) false.
9. The ith row of DA is d_{ii} times the ith row of A.
11. A lower triangular matrix.
13. $A + B = \begin{bmatrix} 2 & 0 & 0 \\ 6 & 3 & 0 \\ 7 & 1 & 5 \end{bmatrix}$; $AB = \begin{bmatrix} 0 & 0 & 0 \\ 12 & 2 & 0 \\ 11 & 1 & 4 \end{bmatrix}$.

5.1 Page 115

1. (a) 20; (b) -2; (c) $-1 + 2i$; (d) $-3t$; (e) 0; (f) -18.
3. $4! = 24; n!$ **5.** t is odd but not unique.

5.2 Page 119

1. (a) $2 \begin{vmatrix} 2 & 2 \\ 4 & 0 \end{vmatrix} - 3 \begin{vmatrix} 3 & 0 \\ 4 & 0 \end{vmatrix} + (-1) \begin{vmatrix} 3 & 0 \\ 2 & 2 \end{vmatrix} = -22$;

ANSWERS TO SELECTED ODD-NUMBERED EXERCISES

(b) $(-1)\begin{vmatrix} 3 & 0 \\ 2 & 2 \end{vmatrix} - 4\begin{vmatrix} 2 & 0 \\ 3 & 2 \end{vmatrix} + 0 = -22;$

(c) $0 - 2\begin{vmatrix} 2 & 3 \\ -1 & 4 \end{vmatrix} + 0 = -22;$

(d) $(-1)^{3+2}\begin{vmatrix} 2 & 0 \\ 3 & 2 \end{vmatrix} = -4;$ (e) $\begin{vmatrix} 3 & 2 \\ -1 & 0 \end{vmatrix} = 2.$

3. 4.   5. (a) 18;   (b) 59.

7. (a) $x = 4, y = 1$;   (b) $x = \frac{11}{2}, y = -5, z = -\frac{1}{2}$.

5.3 Page 125

1. (a) True;   (b) true;   (c) false.
3. -8.
5. 108 by application of Theorem 8 three times.
9. 15.
11. (a) -17;   (b) -5;   (c) 0;   (d) -13;   (e) 39.

5.4 Page 130

9.

STATEMENT	REASON
(1) A can be expressed as a product of elementary row transformation matrices. That is, $A = E_1 E_2 \cdots E_p.$	(1) By Theorem 23 of Chapter 3 since A is invertible.
(2) $\|A\| = \|E_1\|\|E_2\| \cdots \|E_p\|.$	(2) By repeated use of Theorem 10 of this chapter.
(3) The determinant of every elementary row transformation matrix is nonzero.	(3) By Corollaries to Theorems 2, 8, 9.
(4) $\|A\| \neq 0.$	(4) Statements (2) and (3).

5.5 Page 133

1. (a) False. If the determinant of every fourth order submatrix is zero, then the rank is 3 *or less*.
   (b) True; $|A| = |A^T|$, and for every submatrix of A there is a corresponding submatrix of A^T such that the determinants of each are the same number.
   (c) True; if $A_3 = -A_3^T$ then $|A_3| = |-A_3^T| = (-1)^3|A_3^T|$, and $|A_3| = -|A_3^T|$ only if $|A| = 0$.
3. (a) 2;   (b) 3;   (c) 1;   (d) 0;   (e) 1.
5. 2.   7. 2.   9. 2.   11. $\begin{bmatrix} I_2 & 0 \\ \hline 0 & 0 \end{bmatrix}_{(4,6)}.$
13. $x \neq 1$; no, because there exists a 2 by 2 submatrix whose determinant is not zero.

ANSWERS TO SELECTED ODD-NUMBERED EXERCISES A19

15. $\begin{bmatrix} 1 & 2 & 0 & 0 \\ 3 & 4 & 0 & 0 \\ 0 & 0 & 0 & 0 \\ 0 & 0 & 0 & 0 \end{bmatrix}$ is one example.

17. (a) Rank of A is less than 4;   (b) rank of A is 4.

19. (a) The three points are collinear if and only if the slopes of the line segments between two pairs of these three points are equal, that is, $\dfrac{5 - y_1}{4 - x_1} = \dfrac{6 - y_1}{7 - x_1}$.

This expression, however, is identical to $\det \begin{bmatrix} x_1 & y_1 & 1 \\ 4 & 5 & 1 \\ 7 & 6 & 1 \end{bmatrix} = 0$.

5.6 Page 138

1. (a) The system is inconsistent; the rank of A does not equal the rank of $[A \vdots B]$.
   (b) The system is consistent and $r = 2 = n$; therefore there is a unique solution.
   (c) Consistent. $r = 3 = n$, therefore there exists a unique solution.
   (d) Consistent. $r = 2$, $n = 3$, $r < n$. There are an infinite number of solutions.
   (e) The system is homogeneous because $B = 0$; therefore the system is consistent. $r = 2 < n$ hence there are an infinite number of solutions.
   (f) The system is homogeneous because $B = 0$; therefore the system is consistent. $r = 3 = n$ hence there is a unique solution.
   (g) Consistent. $r = 3 < n$ hence there are an infinite number of solutions.
   (h) The system is inconsistent because the rank of A does not equal the rank of $[A \vdots B]$.

3. (a) (Rank of augmented matrix) = (rank of coefficient matrix) = 6. (Theorem 13.)
   (b) There must be at least 6 equations because $r = 6$.

5. Solution is unique; moreover the solution is the trivial solution.

7. r is the order of a submatrix of A, and the order of A is m by n, hence $r \leq n$ and $r \leq m$.

5.7 Page 146

1. (a) $\begin{bmatrix} 1 & -3 \\ 4 & 2 \end{bmatrix}$;   (b) $\begin{bmatrix} -1 & 3 & 0 \\ 2 & -1 & 0 \\ 2 & -1 & -5 \end{bmatrix}$;   (c) $\begin{bmatrix} -4 & 0 & 0 & 2 \\ 0 & 0 & -2 & 0 \\ 0 & -4 & 0 & 0 \\ 0 & 0 & 0 & -2 \end{bmatrix}$.

3. $A^{-1} = \begin{bmatrix} \tfrac{3}{10} & \tfrac{1}{10} \\ -\tfrac{4}{10} & \tfrac{2}{10} \end{bmatrix}$; B^{-1} does not exist;

$C^{-1} = \begin{bmatrix} \tfrac{2}{7} & -\tfrac{3}{7} & 0 \\ -\tfrac{1}{7} & -\tfrac{2}{7} & 1 \\ \tfrac{1}{7} & \tfrac{2}{7} & 0 \end{bmatrix}$; $D^{-1} = \tfrac{1}{19}\begin{bmatrix} 4 & 4 & -1 \\ -1 & -1 & 5 \\ 9 & -10 & -7 \end{bmatrix}$.

5. (a) Rank is 3;   (b) Rank is less than 3.

7. They are equal.   9. $\det G = ad - bc = \det(\operatorname{adj} G)$.

11. (a) $612 = |AB| = |A||B|$;   (b) $14 = |AB| = |A||B|$.

13. $\det\left(\begin{bmatrix} 1 & 2 \\ 3 & 4 \end{bmatrix} + \begin{bmatrix} 1 & -1 \\ -1 & 1 \end{bmatrix}\right) \neq \det\begin{bmatrix} 1 & 2 \\ 3 & 4 \end{bmatrix} + \det\begin{bmatrix} 1 & -1 \\ -1 & 1 \end{bmatrix}.$

(counterexample.)

19. (a) Unique solution;   (b) infinite number of solutions.

6.1   Page 153

1. (a) Not a group, because the set is not closed under $+$; (b) not a group, because of postulate (3); (c) a group; (d) a group; (e) not a group, because of postulates (2) and (3); (f) not a group, because of postulates (1), (2), and (3); (g) a group; (h) not a group, because of postulate (3); (i) a group; (j) not a group, because of postulate (3); the element 0 has no inverse; (k) a group; (l) a group.

3. g and l.   **5.** Yes.

7. (a) No. The set is not closed under addition. For example,

$$\begin{bmatrix} 1 & 1 & 1 \\ 0 & 1 & 1 \\ 0 & 0 & 1 \end{bmatrix} + \begin{bmatrix} 1 & 0 & 0 \\ 1 & 1 & 0 \\ 1 & 1 & 1 \end{bmatrix}$$

is not a triangular matrix.

(b) No, for the same reason as part (a).

9. (a) No, the set is not closed under addition;   (b) Yes.

11. Yes.

6.2   Page 155

1. (a) Not a ring, because the set is not closed under addition;
(b) Not a ring, because the set is not closed under $\odot$;
(c) Not a ring, because the set is not closed under $(\cdot)$; also postulates 3(a) and 4(a) fail;
(d) A ring;   (e) a ring;   (f) a ring.

3. (a) Parts (e) and (f);   (b) none.

5. (a) Because all entries of the tables are elements of the set $\{w, u, v, x\}$.
(b) $u \oplus v = v = v \oplus u$; $\oplus$ is a commutative operation.
(c) To justify postulate 2(a) all possible arrangements of 4 elements taken 3 at a time would have to be verified. There would be 24 of these arrangements. A similar situation would exist in the justification of postulates 2(b) and 5.
(d) Because for any element a in the set, $a \oplus u = u \oplus a = a$.
(e) x is its own inverse for $\oplus$.   (f) Yes.

7. Yes.

6.3   Page 160

1. (a) Not a field, because of postulate 4(b);   (b) a field;
(c) Not a field, because the set is not closed under $(\cdot)$; also because of postulates 3(a) and 4(a);

ANSWERS TO SELECTED ODD-NUMBERED EXERCISES   A21

(d) Not a field, because the set is not closed under the dot product;
(e) A field;
(f) Not a field, because of postulate 1(b) and the set is not closed under addition.

3. (a) Yes; (b) 0 for $\oplus$, and 1 for $\odot$;
(c) If for an element a we let a^{-1} denote the inverse of a with respect to a given binary operation then for $\oplus$, $1^{-1} = 2, 0^{-1} = 0$, and $2^{-1} = 1$; for $\odot$, $1^{-1} = 1$ and $2^{-1} = 2$;
(d) Yes; (e) Yes; yes.

7. (a) Yes; every ring postulate is a field postulate;
(b) No; there are certain field postulates that are not ring postulates.

9. (b) $\frac{5}{23} - \frac{1}{23}\sqrt{2}$.

6.4   Page 165

1. A vector space.   **3.** A vector space.
7. (a) Yes; (b) no; the set of vectors is not closed under scalar multiplication.
9. No, because, among other reasons, the set would not be closed under addition.

6.5   Page 168

1. A linear algebra.
3. Not a linear algebra, because the set is not closed under multiplication.
5. The vector spaces of Exercises 1, 2, and 4 are associative algebras.
7. The vector spaces of Exercises 1, 2, and 4 are linear algebras with unity.
11. No; because the set of vectors is *not* closed under the dot product.
13. Yes.

6.6   Page 174

1. (a) $\begin{bmatrix} 0 & 1 \\ 1 & 1 \end{bmatrix}, \begin{bmatrix} 0 & 0 \\ 1 & 0 \end{bmatrix}$; (b) $\begin{bmatrix} 0 & 0 \\ 0 & 0 \end{bmatrix}, \begin{bmatrix} 1 & 1 \\ 1 & 1 \end{bmatrix}$.

3. (a) $f(x, y) = \{x' \cap (x \cup y)\} \cup \{(y \cup x) \cap (x' \cup y)\}$;   (b) $f(x, y) = y$.

7.1   Page 183

1. (a) Linearly dependent;   (b) linearly independent;
(c) linearly independent;   (d) linearly dependent;
(e) linearly dependent;   (f) linearly independent;
(g) linearly independent;   (h) linearly dependent.

3. $(a, b, c) = a(1, 0, 0) + b(0, 1, 0) + c(0, 0, 1)$.

5. No, any vector that does not lie in the plane of the two given vectors cannot be expressed as a linear combination of the two given vectors. These two vectors do span a subspace of $V(R)$.

A22   ANSWERS TO SELECTED ODD-NUMBERED EXERCISES

7. (a) No;   (b) yes;   (c) yes;   (d) yes.

9. (a) Three-space;   (b) one of the coordinate planes in three-space;
   (c) the plane in three-space in which the two given vectors lie.

11. The set $\{(0, 0)\}$ is linearly dependent. The set $\{(1, 2)\}$ is linearly independent.

15. (a) Set is linearly independent; yes;
    (b) Set is linearly dependent; no; the subspace is the vector space of dimension 2 formed by the set of all linear combinations of the given set of vectors;
    (c) The first set is linearly independent and the second set is linearly dependent. Neither set spans the space; the first subspace is the vector space of dimension 2 formed by the set of all linear combinations of the given set of vectors; the second subspace is the vector space of dimension 1 consisting of the set of all scalar multiples of $2x^2 + 1$.

17. (a) $(2, 3, 5) = \frac{1}{3}(6, 9, 15) + 0(3, -1, 1)$;
    (b) $(3, 5) = -\frac{17}{3}(7, 1) + \frac{16}{3}(8, 2)$;
    (c) $(1, 0, 0) = -\frac{1}{2}(3, 2, 4) + \frac{1}{4}(10, 4, 8)$;
    (d) impossible.

7.2   Page 190

3. The sets of parts (b), (d), and (e) are not a basis; part (b), because the set does not span the space; part (d), because the set is not linearly independent; part (e), because the set does not span the space, nor is the set linearly independent.

5. No, because all three vectors could be collinear or coplanar, and hence not span the space nor be linearly independent.

7. The dimension is three or less.

9. (a) 1;   (b) 2;   (c) 1;   (d) 2;   (e) 3;   (f) 4.

11. The vector space $V(F)$ consisting of the set of all linear combinations of the rows of a matrix A over F is called the *row space* of A.

17. Each set of all geometric vectors that are coplanar forms a two-dimensional subspace; each set of all geometric vectors that are collinear forms a one-dimensional subspace.

7.3   Page 197

1. (a) $\begin{bmatrix} 7 \\ 2 \end{bmatrix}$;   (b) $\begin{bmatrix} 5 \\ 2 \end{bmatrix}$;   (c) $\begin{bmatrix} -4 \\ 3 \end{bmatrix}$.

3. (a) $\begin{bmatrix} 2 \\ 3 \end{bmatrix}$;   (b) $\begin{bmatrix} \frac{12}{5} \\ -\frac{1}{5} \end{bmatrix}$;   (c) $\begin{bmatrix} \frac{23}{5} \\ \frac{1}{5} \end{bmatrix}$.

5. (a) $\begin{bmatrix} 1 \\ 2 \\ 3 \\ 4 \end{bmatrix}$;   (b) $\begin{bmatrix} -\frac{3}{2} \\ 1 \\ \frac{3}{2} \\ 4 \end{bmatrix}$;   (c) $\begin{bmatrix} -1 \\ \frac{1}{2} \\ \frac{3}{2} \\ 3 \end{bmatrix}$.

7. (a) $\begin{bmatrix} 1 \\ i \end{bmatrix}$; (b) $\begin{bmatrix} \frac{1+i}{2} \\ \frac{1+i}{-2} \end{bmatrix}$; (c) $\begin{bmatrix} \frac{3+i}{2} \\ \frac{1+3i}{2} \end{bmatrix}$.

9. One answer is: the subspace of all coordinate vectors from $V_3(R)$ with the first component equal to 0 is isomorphic to $V_2(R)$. There is a 1-to-1 mapping, defined by $(0, a, b) \overset{T}{\mapsto} (a, b)$, from the subspace onto $V_2(R)$ such that the results of addition and scalar multiplication are preserved by the mapping.

7.4 Page 203

1. 97. **3.** 0. **5.** 59. **7.** 18.

7.5 Page 209

1. $\|\alpha\| = \sqrt{5}$, $\|\alpha - \beta\| = \sqrt{29}$, $\cos\theta = \dfrac{7}{\sqrt{65}}$; no.

3. $\|A\| = \sqrt{46}$, $\|A - B\| = \sqrt{62}$, $\cos\theta = \dfrac{-5}{2\sqrt{69}}$; no, no.

5. $\|f\| = \frac{4}{5}\sqrt{10}$, $\|f - g\| = \frac{4}{5}\sqrt{10}$, $\cos\theta = \frac{1}{12}\sqrt{105}$; no.

7.6 Page 215

1. (a) Not orthogonal; (b) orthonormal; (c) not orthogonal.
3. (a) Not orthogonal; (b) orthonormal; (c) orthogonal.
5. $\{(1, 2, 3), (\frac{1}{2}, -1, \frac{1}{2}), (-\frac{20}{7}, -\frac{5}{7}, \frac{10}{7})\}$. **7.** $\{(1, 0), (1, 1)\}$.

8.1 Page 224

1. (a) $(-2, 4)$; (b) $(0, -6)$.

3. A domain: set of real numbers. A codomain: set of coordinate vectors with two real components. Range: set of coordinate vectors with two real components whose second component is always twice the first component. There are other answers to this Exercise.

5. A domain: set of 2 by 1 column R-matrices. A codomain: set of all 2 by 1 column R-matrices. Range: set of 2 by 1 column R-matrices with zero as the first entry. There are other answers to this Exercise.

7. A domain: $\{x \mid -1 \leq x \leq 1,$ and x is a real number$\}$.
A codomain: the set of all real numbers.
Range: $\{T(x) \mid 0 \leq T(x) \leq 1,$ and $T(x)$ is a real number$\}$.
There are other answers to this Exercise.

A24   ANSWERS TO SELECTED ODD-NUMBERED EXERCISES

9. A domain: $\{x \mid x \geq 1 \text{ or } x \leq -1, \text{ and } x \text{ is a real number}\}$.
A codomain: the set of all real numbers.
Range: $\{T(x) \mid 1 < T(x), \text{ and } T(x) \text{ is a real number}\}$.
There are other answers to this Exercise.

11.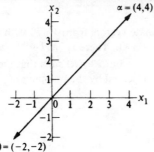

13.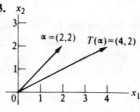

15.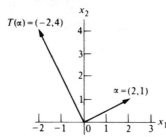

19. $T(\alpha) = \begin{bmatrix} 2a_1 \\ a_2 \end{bmatrix}$.

21. $T(\alpha) = \begin{bmatrix} -3a_2 \\ 3a_1 \end{bmatrix}$.

23. $T(\alpha) = \begin{bmatrix} (a_1 + a_2) \\ (a_1 + a_2) \end{bmatrix}$.

25. Transformation is linear because $T(\alpha + \beta) = T(\alpha) + T(\beta)$ and $T(k\alpha) = kT(\alpha)$.

27. Transformation is linear because $T(\alpha + \beta) = T(\alpha) + T(\beta)$ and $T(k\alpha) = kT(\alpha)$.

29. Transformation is not linear, because $T(\alpha + \beta) \neq T(\alpha) + T(\beta)$ or because $T(k\alpha) \neq kT(\alpha)$.

31. Transformation is not linear, because $T(\alpha + \beta) \neq T(\alpha) + T(\beta)$ or because $T(k\alpha) \neq kT(\alpha)$.

33. Yes, because $D_x^2(f(x) + g(x)) = D_x^2 f(x) + D_x^2 g(x)$ and $D_x^2 k f(x) = k D_x^2 f(x)$.

8.2   Page 231

1. (a) $\begin{bmatrix} 7 \\ 8 \\ 1 \end{bmatrix}$;   (b) $\begin{bmatrix} 8 \\ 7 \\ 1 \end{bmatrix}$;   (c) $\begin{bmatrix} 7 \\ 0 \\ 1 \end{bmatrix}$.

3. $\begin{bmatrix} 1 & 2k \\ 0 & 1 \end{bmatrix}$.

5. $\begin{bmatrix} 4 & 0 \\ 0 & 4 \end{bmatrix}$.

7. $\begin{bmatrix} 1 & 0 \\ 0 & -1 \end{bmatrix}$.

ANSWERS TO SELECTED ODD-NUMBERED EXERCISES   A25

9. $\begin{bmatrix} 1 & 0 \\ 0 & 0 \end{bmatrix}$.   11. $\begin{bmatrix} 1 & 0 & 0 \\ 0 & 1 & 0 \\ 0 & 0 & -1 \end{bmatrix}$.   13. $\begin{bmatrix} 1 & 0 & 0 \\ 0 & 1 & 0 \\ 0 & 0 & 0 \end{bmatrix}$.

15. $\begin{bmatrix} 0 & 0 & 2 & 0 \\ 0 & 0 & 0 & 6 \\ 0 & 0 & 0 & 0 \\ 0 & 0 & 0 & 0 \end{bmatrix}$.   17. $\begin{bmatrix} \cos\theta & -\sin\theta \\ \sin\theta & \cos\theta \end{bmatrix}$.

19. $\begin{bmatrix} 0 & 3 & 0 \\ 0 & 6 & 6 \\ 0 & -6 & -6 \end{bmatrix}$.   21. $\begin{bmatrix} 1 & 2 & 2 \\ 0 & 1 & 0 \\ 0 & 0 & 1 \end{bmatrix}$.

23. $(x, y) \stackrel{T}{\mapsto} (-y, x)$; this transformation is known as the "90° rotation" transformation.

25. $\begin{bmatrix} 0 & 0 & 0 & 0 \\ 0 & 1 & 0 & 2 \\ 0 & 0 & 0 & 0 \\ 0 & 0 & 0 & 1 \end{bmatrix}$.

27. $\begin{bmatrix} a_0 & 0 & a_1 & 0 \\ 0 & (a_2 + a_1 + a_0) & 0 & (2a_2 + a_1) \\ 0 & 0 & a_0 & 0 \\ 0 & 0 & 0 & (a_2 + a_1 + a_0) \end{bmatrix}$.

8.3 Page 237

1. $(6 + 4k, 4)$.   3. $(5, 2)$.   5. $(4a_1, 4a_2)$.
7. $(a_1, -a_2)$.   9. $(a_1, 0)$.   11. $(a_1, a_2, -a_3)$.
13. $(a_1, a_2, 0)$.   15. $30x + 12$.   17. $(\frac{5}{2}\sqrt{3} - 1, \frac{5}{2} + \sqrt{3})$.
19. $6x - 18$.   21. $x^2 - 6x + 7$.
23. $3a_1 + 3a_0 x + (4a_2 + 2a_1)e^x + (2a_2 + 2a_1 + 2a_0)xe^x$.

8.4 Page 248

1. $P = \begin{bmatrix} 0 & 1 \\ 1 & 0 \end{bmatrix}$; $[\alpha]_e = \begin{bmatrix} 4 \\ 3 \end{bmatrix}$.   3. $P = \begin{bmatrix} 1 & 0 \\ 1 & 1 \end{bmatrix}$; $[\alpha]_e = \begin{bmatrix} 8 \\ -2 \end{bmatrix}$.

5. $P = \begin{bmatrix} \frac{1}{2} & \frac{1}{2} \\ \frac{1}{2} & -\frac{1}{2} \end{bmatrix}$; $[\alpha]_e = \begin{bmatrix} 14 \\ 2 \end{bmatrix}$.   7. $P = \begin{bmatrix} 0 & 1 \\ 1 & 0 \end{bmatrix}$; $[\alpha]_e = \begin{bmatrix} 2 \\ 6 \end{bmatrix}$.

9. $P = \begin{bmatrix} -1 & -1 \\ 0 & \frac{1}{2} \end{bmatrix}$; $[\alpha]_e = \begin{bmatrix} -10 \\ 4 \end{bmatrix}$.

11. $P = \begin{bmatrix} 1 & 0 & 0 \\ 0 & 1 & 0 \\ -1 & 0 & 1 \end{bmatrix}$; $[\alpha]_e = \begin{bmatrix} 6 \\ 7 \\ 14 \end{bmatrix}$.

13. The statements of the proof are:
$\alpha = a_1 \varepsilon_1 + a_2 \varepsilon_2 + \cdots + a_n \varepsilon_n$
$= a_1(p_{11}\beta_1 + \cdots + p_{1n}\beta_n) + \cdots + a_n(p_{n1}\beta_1 + \cdots + p_{nn}\beta_n)$
$= (a_1 p_{11} + \cdots + a_n p_{n1})\beta_1 + \cdots + (a_1 p_{1n} + \cdots + a_n p_{nn})\beta_n$. Therefore,

A26 ANSWERS TO SELECTED ODD-NUMBERED EXERCISES

$$[\alpha]_b = \begin{bmatrix} (a_1 p_{11} + \cdots + a_n p_{n1}) \\ \vdots \\ (a_1 p_{1n} + \cdots + a_n p_{nn}) \end{bmatrix} = \begin{bmatrix} p_{11} & \cdots & p_{n1} \\ \vdots & & \vdots \\ p_{1n} & \cdots & p_{nn} \end{bmatrix} \begin{bmatrix} a_1 \\ a_2 \\ \vdots \\ a_n \end{bmatrix} = P[\alpha]_e.$$

15. $e = \left\{ \begin{bmatrix} 0 \\ 1 \end{bmatrix}, \begin{bmatrix} -1 \\ 0 \end{bmatrix} \right\}.$

8.5 Page 254

1. $\begin{bmatrix} 1 & 0 \\ 0 & 6 \end{bmatrix}.$ **3.** $\begin{bmatrix} 2 & 0 \\ 0 & 4 \end{bmatrix}.$ **5.** $\begin{bmatrix} 2 & 0 \\ 1 & 5 \end{bmatrix}.$

7. $[T]_e = \begin{bmatrix} 2 & 0 \\ 0 & 1 \end{bmatrix}.$ **9.** $[T]_e = \begin{bmatrix} 1 & -1 \\ 0 & 2 \end{bmatrix}.$

11. (a) $[\alpha]_b = \begin{bmatrix} 2 \\ 4 \end{bmatrix}; [T]_b = \begin{bmatrix} 1 & 5 \\ 0 & 1 \end{bmatrix};$ (b) $[T(\alpha)]_b = \begin{bmatrix} 22 \\ 4 \end{bmatrix};$

(c) $P = \begin{bmatrix} 0 & -1 \\ 1 & 0 \end{bmatrix};$ (d) $[T]_e = \begin{bmatrix} 1 & 0 \\ -5 & 1 \end{bmatrix};$

(e) (i) $[T(\alpha)]_e = [T]_e[\alpha]_e;$ (ii) $[T(\alpha)]_e = (P^{-1}[T]_b P)[\alpha]_e;$
(iii) $[T(\alpha)]_e = P^{-1}[T(\alpha)]_b = P^{-1}[T]_b[\alpha]_b;$
(iv) $T(2, 4) = (22, 4) = 4(0, 1) - 22(-1, 0) = 4\varepsilon_1 + (-22)\varepsilon_2$, therefore
$[T(\alpha)]_e = \begin{bmatrix} 4 \\ -22 \end{bmatrix}.$

8.6 Page 258

1. $T(\alpha) = (3, 4, 7).$ **3.** $T(\alpha) = (2, 9, 4).$

5. $[T]_b^{b'} = \begin{bmatrix} 1 & 0 \\ 0 & 1 \\ 1 & 1 \end{bmatrix}.$ **7.** $[T]_b^{b'} = \begin{bmatrix} 0 & 1 \\ 1 & 0 \\ 0 & 2 \end{bmatrix}.$

9. $[T]_b^{b'} = \begin{bmatrix} 0 & 0 & 0 \\ 1 & 0 & 0 \\ 0 & \frac{1}{2} & 0 \\ 0 & 0 & \frac{1}{3} \end{bmatrix}; [T(\alpha)]_{b'} = \begin{bmatrix} 0 \\ 1 \\ 1 \\ \frac{5}{3} \end{bmatrix}.$

11. $[T]_b^{b'} = \begin{bmatrix} 0 & -\frac{1}{2} & -\frac{1}{3} \\ 1 & 1 & 0 \\ 0 & \frac{1}{2} & 1 \\ 0 & 0 & \frac{1}{3} \end{bmatrix}; [T(\alpha)]_{b'} = \begin{bmatrix} -\frac{1}{3} \\ 3 \\ 1 \\ \frac{1}{3} \end{bmatrix}.$

8.7 Page 261

1. $[T]_e^{e'} = \begin{bmatrix} 1 & \frac{2}{3} \\ 2 & 8 \\ 5 & 5 \end{bmatrix}.$ **3.** $[T]_e^{e'} = \begin{bmatrix} 32 & 12 \\ 84 & 24 \\ 48 & 12 \end{bmatrix}.$ **5.** $[T(\alpha)]_{e'} = \begin{bmatrix} \frac{1}{3} \\ 4 \\ \frac{5}{2} \end{bmatrix}.$

ANSWERS TO SELECTED ODD-NUMBERED EXERCISES   A27

7. $[T(\alpha)]_{e'} = \begin{bmatrix} 3 \\ 0 \\ 4 \end{bmatrix}$.

9. $[T]_b^{b'} = \begin{bmatrix} 0 & 0 & 0 \\ 1 & 0 & 0 \\ 0 & \frac{1}{2} & 0 \\ 0 & 0 & \frac{1}{3} \end{bmatrix}$; $[T]_e^{e'} = \begin{bmatrix} 0 & 0 & 0 \\ \frac{1}{2} & 0 & 1 \\ 0 & \frac{1}{4} & 0 \\ 0 & 0 & \frac{1}{3} \end{bmatrix}$; $[T(\alpha)]_{e'} = \begin{bmatrix} 0 \\ 1 \\ \frac{1}{4} \\ \frac{1}{2} \end{bmatrix}$.

11. $[T]_b^{b'} = \begin{bmatrix} -1 & -\frac{1}{2} & -\frac{1}{3} \\ 1 & 0 & 0 \\ 0 & \frac{1}{2} & 0 \\ 0 & 0 & \frac{1}{3} \end{bmatrix}$; $[T]_e^{e'} = \begin{bmatrix} 0 & \frac{1}{3} & \frac{1}{3} \\ 0 & \frac{1}{2} & 0 \\ 1 & 0 & 0 \\ -1 & -\frac{5}{6} & -\frac{1}{3} \end{bmatrix}$; $[T(\alpha)]_{e'} = \begin{bmatrix} 1 \\ \frac{1}{2} \\ 2 \\ -\frac{7}{2} \end{bmatrix}$.

13. (a) $Q = \begin{bmatrix} 1 & 0 \\ 2 & 1 \end{bmatrix}$; $Q^{-1} = \begin{bmatrix} 1 & 0 \\ -2 & 1 \end{bmatrix}$;

(b) $[T]_b^{e'} = Q^{-1}A = \begin{bmatrix} 4 & 1 & 1 & 0 \\ -3 & 0 & -2 & 1 \end{bmatrix}$.

15. (a) $Q = \begin{bmatrix} 4 & 0 \\ 5 & 1 \end{bmatrix}$; $Q^{-1} = \begin{bmatrix} \frac{1}{4} & 0 \\ -\frac{5}{4} & 1 \end{bmatrix}$;

(b) $[T]_b^{e'} = Q^{-1}A = \begin{bmatrix} 1 & \frac{1}{4} & \frac{1}{4} & 0 \\ 0 & \frac{3}{4} & -\frac{5}{4} & 1 \end{bmatrix}$.

8.8 Page 266

1. $(x+y, y)$, $(y, y-x)$, $(x+y, -x)$, $(ky, -kx)$, $(kx, kx+ky)$, $(-x, -y)$.

5. $S = \begin{bmatrix} 0 & 1 \\ -1 & 0 \end{bmatrix}$, $T = \begin{bmatrix} 1 & 0 \\ 1 & 1 \end{bmatrix}$, $S+T = \begin{bmatrix} 1 & 1 \\ 0 & 1 \end{bmatrix}$, $ST = \begin{bmatrix} 1 & 1 \\ -1 & 0 \end{bmatrix}$, $3T = \begin{bmatrix} 3 & 0 \\ 3 & 3 \end{bmatrix}$.

7. Identity for addition is defined by $(x,y) \mapsto (0, 0)$.
Identity for multiplication is defined by $(x,y) \mapsto (x, y)$.
The inverse, for addition, of T is defined by $(x, y) \mapsto (-x, -y-x)$.
The inverse, for multiplication, of T is defined by $(x, y) \mapsto (x, y-x)$.

11. $\alpha \xmapsto{T+S} (3a)x^2 + (2b + 6a)x + (2b + c)$;
$\alpha \xmapsto{TS} 6a$;
$\alpha \xmapsto{ST} 6a$;
$\alpha \xmapsto{kS} k(6ax + 2b)$;
$\alpha \xmapsto{kT} k(3ax^2 + 2bx + c)$;
$\alpha \xmapsto{TT} 6ax + 2b$.

13. $T = \begin{bmatrix} 0 & 1 & 0 & 0 \\ 0 & 0 & 2 & 0 \\ 0 & 0 & 0 & 3 \\ 0 & 0 & 0 & 0 \end{bmatrix}$; $S = \begin{bmatrix} 0 & 0 & 2 & 0 \\ 0 & 0 & 0 & 6 \\ 0 & 0 & 0 & 0 \\ 0 & 0 & 0 & 0 \end{bmatrix}$; $S+T = \begin{bmatrix} 0 & 1 & 2 & 0 \\ 0 & 0 & 2 & 6 \\ 0 & 0 & 0 & 3 \\ 0 & 0 & 0 & 0 \end{bmatrix}$;

$ST = \begin{bmatrix} 0 & 0 & 0 & 6 \\ 0 & 0 & 0 & 0 \\ 0 & 0 & 0 & 0 \\ 0 & 0 & 0 & 0 \end{bmatrix}$; $3T = 3\begin{bmatrix} 0 & 1 & 0 & 0 \\ 0 & 0 & 2 & 0 \\ 0 & 0 & 0 & 3 \\ 0 & 0 & 0 & 0 \end{bmatrix}$.

9.1 Page 278

1. $\lambda = 1$ is a characteristic value of T, and any nonzero vector of the form $(a, 0)$ is a characteristic vector of T.

3. $\lambda^2 - 7\lambda + 6 = 0$; $\lambda_1 = 1, \lambda_2 = 6$;
$$X_1 = \pm(\tfrac{1}{5}\sqrt{5})\begin{bmatrix}2\\1\end{bmatrix}, \quad X_2 = \pm(\tfrac{1}{5}\sqrt{5})\begin{bmatrix}1\\-2\end{bmatrix}.$$

5. $\lambda^2 - 6\lambda + 8 = 0$; $\lambda_1 = 2, \lambda_2 = 4$;
$$X_1 = \pm(\tfrac{1}{2}\sqrt{2})\begin{bmatrix}-1\\1\end{bmatrix}, \quad X_2 = \pm(\tfrac{1}{10}\sqrt{10})\begin{bmatrix}-3\\1\end{bmatrix}.$$

7. $(\lambda - 3)(\lambda - 1)\lambda = 0$; $\lambda_1 = 3, \lambda_2 = 1, \lambda_3 = 0$;
$$X_1 = \pm(\tfrac{1}{6}\sqrt{6})\begin{bmatrix}2\\1\\1\end{bmatrix}, \quad X_2 = \pm(\tfrac{1}{2}\sqrt{2})\begin{bmatrix}0\\-1\\1\end{bmatrix}, \quad X_3 = \pm(\tfrac{1}{3}\sqrt{3})\begin{bmatrix}-1\\1\\1\end{bmatrix}.$$

9. $X_1 = \begin{bmatrix}-k_1\\k_1\end{bmatrix}, X_2 = \begin{bmatrix}k_2\\4k_2\end{bmatrix}$, where k_i are arbitrary.

11. $X_1 = \begin{bmatrix}0\\k_1\\0\end{bmatrix}, X_2 = \begin{bmatrix}k_2\\0\\k_2\end{bmatrix}, X_3 = \begin{bmatrix}-k_3\\0\\k_3\end{bmatrix}$, where k_i are arbitrary.

13. $X_1 = \begin{bmatrix}3/\sqrt{13}\\2/\sqrt{13}\end{bmatrix}; \lambda X_1 = \begin{bmatrix}18/\sqrt{13}\\12/\sqrt{13}\end{bmatrix}; X_2 = \begin{bmatrix}-2/\sqrt{13}\\3/\sqrt{13}\end{bmatrix}; \lambda X_2 = \begin{bmatrix}14/\sqrt{13}\\-21/\sqrt{13}\end{bmatrix}.$

15. $X_1 = \begin{bmatrix}-1/\sqrt{5}\\2/\sqrt{5}\\0\end{bmatrix}; \lambda X_1 = \begin{bmatrix}-1/\sqrt{5}\\2/\sqrt{5}\\0\end{bmatrix}; X_2 = \begin{bmatrix}1\\0\\0\end{bmatrix}; \lambda X_2 = \begin{bmatrix}3\\0\\0\end{bmatrix}; X_3 = \begin{bmatrix}2/\sqrt{5}\\0\\1/\sqrt{5}\end{bmatrix};$
$$\lambda X_3 = \begin{bmatrix}8/\sqrt{5}\\0\\4/\sqrt{5}\end{bmatrix}.$$

17. n characteristic values.

19. $\lambda^2 - 7\lambda + 6 = 0 \Rightarrow \begin{bmatrix}2 & -2\\-2 & 5\end{bmatrix}^2 - 7\begin{bmatrix}2 & -2\\-2 & 5\end{bmatrix} + 6\begin{bmatrix}1 & 0\\0 & 1\end{bmatrix} = \begin{bmatrix}0 & 0\\0 & 0\end{bmatrix}.$

21. Let a, a', b, and b' be expressed as column matrices. Show that $Ta = a'$ and $Tb = b'$; $\lambda_1 = \tfrac{3}{2}, \lambda_2 = \tfrac{2}{3}$.

9.2 Page 285

1. $\begin{bmatrix}3 & 5\\-1 & 2\end{bmatrix}$ is one example. 3. $\begin{bmatrix}3 & i\\i & 2\end{bmatrix}$ is one example.

5. Discriminant: $(a+d)^2 - 4(ad - bc) = (a-d)^2 + 4bc$.
   If $bc > 0$, the discriminant is positive. The characteristic values are equal if the discriminant is zero, that is, if b is equal to $\dfrac{(a-d)^2}{-4c}$.

ANSWERS TO SELECTED ODD-NUMBERED EXERCISES   A29

7. Any two characteristic vectors X_i, X_j corresponding to two distinct characteristic values λ_i, λ_j of a symmetric R-matrix are orthogonal. Exercises 3 and 7 are symmetric.

9. $\begin{bmatrix} -1 & 1 \\ 1 & 4 \end{bmatrix}$; $\begin{bmatrix} 0 & 1 & -1 \\ 1 & 0 & 0 \\ 0 & 1 & 1 \end{bmatrix}$ are two possible modal matrices.

11. $M = \begin{bmatrix} 1 & -1 & 3 \\ 0 & 1 & 1 \\ 0 & 0 & 2 \end{bmatrix}$; not orthogonal.   13. $M = \begin{bmatrix} 1 & 1 \\ i & -2i \end{bmatrix}$.

19. The system is stable because $\lambda_1 = \frac{2}{3}$, $\lambda_2 = \frac{1}{2}$, and both are less than one; hence x_n and y_n converge to zero as $n \to \infty$.

9.3   Page 293

1. One answer is $6u^2 + v^2 = 6$.   3. One answer is $3u^2 - v^2 = 1$.
5. One answer is $3u^2 + v^2 = 9$.   7. One answer is $2u^2 + 2v^2 + 4w^2 = 4$.
9. One answer is $\frac{1}{2}u^2 - \frac{1}{2}v^2 = 4$.

11. One answer is $P = \begin{bmatrix} 1/\sqrt{3} & 0 & 2/\sqrt{6} \\ -1/\sqrt{3} & 1/\sqrt{2} & 1/\sqrt{6} \\ 1/\sqrt{3} & 1/\sqrt{2} & -1/\sqrt{6} \end{bmatrix}$ with $P^T A P = \begin{bmatrix} 4 & 0 & 0 \\ 0 & 1 & 0 \\ 0 & 0 & 1 \end{bmatrix}$.

17. One answer is $(\frac{9}{2} - \frac{1}{2}\sqrt{41})u^2 + (\frac{9}{2} + \frac{1}{2}\sqrt{41})v^2 = 6$.

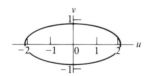

10.2   Page 303

3. One answer is $P = \begin{bmatrix} -3 & 1 \\ 4 & 1 \end{bmatrix}$; $\lambda_1 = -1$, $\lambda_2 = 6$.

5. One answer is $P = \begin{bmatrix} -1 & -4 \\ 1 & -1 \end{bmatrix}$; $\lambda_1 = -1$, $\lambda_2 = 4$.

7. One answer is $P = \begin{bmatrix} 1 & -2 & 1 \\ -3 & 2 & 0 \\ 0 & 1 & 0 \end{bmatrix}$; $\lambda_1 = 0$, $\lambda_2 = 2$, $\lambda_3 = 3$.

9. Modal matrix $P = \begin{bmatrix} a & b \\ -a & -b \end{bmatrix}$ is noninvertible. (a and b are arbitrary.)

11. Modal matrix $P = \begin{bmatrix} 0 & b & c \\ 0 & 0 & 0 \\ a & 0 & 0 \end{bmatrix}$ is noninvertible. Other modal matrices are found by rearranging the columns of P, but they too are noninvertible. (a, b, and c are arbitrary.)

13. Yes.

19. $A^4 = \begin{bmatrix} -49 & 130 \\ -65 & 146 \end{bmatrix}$.

10.3 Page 313

1. $[x \; y \; z] \begin{bmatrix} 1 & \frac{1}{2} & 2 \\ \frac{1}{2} & 1 & 3 \\ 2 & 3 & -3 \end{bmatrix} \begin{bmatrix} x \\ y \\ z \end{bmatrix} = [6]$.

3. $[w \; x \; y \; z] \begin{bmatrix} 1 & 4 & 0 & 0 \\ 4 & 0 & 3 & 0 \\ 0 & 3 & 1 & -2 \\ 0 & 0 & -2 & 1 \end{bmatrix} \begin{bmatrix} w \\ x \\ y \\ z \end{bmatrix} = [7]$.

5. $P = \begin{bmatrix} 1/\sqrt{2} & -1/\sqrt{2} \\ 0 & 1/\sqrt{2} \end{bmatrix}$, $P^T CP = \begin{bmatrix} 1 & 0 \\ 0 & 1 \end{bmatrix}$.

7. $P = \begin{bmatrix} 1 & 0 & -2 \\ 0 & 1 & 0 \\ 0 & 0 & 1 \end{bmatrix}$, $P^T CP = \begin{bmatrix} 1 & 0 & 0 \\ 0 & 1 & 0 \\ 0 & 0 & -1 \end{bmatrix}$.

9. $P = \begin{bmatrix} 1/\sqrt{3} & -3i/\sqrt{29} \\ 0 & 1/\sqrt{29} \end{bmatrix}$, $P^T CP = \begin{bmatrix} 1 & 0 \\ 0 & 1 \end{bmatrix}$.

11. $P = \begin{bmatrix} 1 & 0 & -2 \\ 0 & 0 & 1 \\ 0 & 1/3i & 0 \end{bmatrix}$, $P^T CP = \begin{bmatrix} 1 & 0 & 0 \\ 0 & 1 & 0 \\ 0 & 0 & 0 \end{bmatrix}$.

13.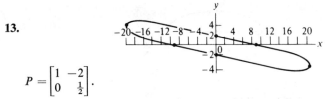

$P = \begin{bmatrix} 1 & -2 \\ 0 & \frac{1}{2} \end{bmatrix}$.

15. $f(x_1, x_2, x_3) = c_{11}x_1^2 + c_{22}x_2^2 + c_{33}x_3^2 + 2c_{12}x_1x_2 + 2c_{13}x_1x_3 + 2c_{23}x_2x_3$; $T: V_3(R) \xrightarrow{T} V_3(R)$ where $X \mapsto P^{-1}X = U$; $g(u_1, u_2, u_3) = d_{11}u_1^2 + d_{22}u_2^2 + d_{33}u_3^2$.

17. *Proof of Theorem 9.* By Theorem 8, A is congruent to a matrix of the form

$$D = \begin{bmatrix} d_{11} & & 0 \\ & \ddots & \\ 0 & & d_{nn} \end{bmatrix}.$$

Because the d_{ii} are complex, we can let the elements of invertible P be $p_{ij} = 0$ if $i \neq j$, otherwise

$$p_{ii} = \begin{cases} \dfrac{1}{\sqrt{d_{ii}}} & \text{if } d_{ii} \neq 0, \\ 1 & \text{if } d_{ii} = 0, \end{cases} \quad (i = 1, \ldots, n).$$

Hence, if we assume that all $d_{ii} = 0$ are listed last, then

$$P^T DP = \left[\begin{array}{c|c} I_r & 0 \\ \hline 0 & 0 \end{array} \right].$$

ANSWERS TO SELECTED ODD-NUMBERED EXERCISES   A31

10.4 Page 317

1. T is orthogonal. Angles, lengths, and distances are preserved.
3. T is not orthogonal. Angles, lengths, and distances are not preserved.
5. T is not orthogonal.   7. T is not orthogonal.
9. T is not orthogonal.   11. T is orthogonal.
13. Even though $[T]_b^{-1} = [T]_b^T$, we cannot conclude from Theorem 11 that T is orthogonal, because the basis is *not* an orthonormal basis.
15. T is orthogonal.   19. T is unitary.

10.5 Page 326

1. One answer is $6u^2 + v^2 = 6$.   3. One answer is $3u^2 - v^2 = 1$.
5. One answer is $3u^2 + v^2 = 9$.   7. One answer is $2u^2 + 2v^2 + 4w^2 = 4$.
9. One answer is $\frac{1}{2}u^2 - \frac{1}{2}v^2 = 4$.
11. One answer is $P = \begin{bmatrix} 1/\sqrt{3} & 0 & 2/\sqrt{6} \\ -1/\sqrt{3} & 1/\sqrt{2} & 1/\sqrt{6} \\ 1/\sqrt{3} & 1/\sqrt{2} & -1/\sqrt{6} \end{bmatrix}$ with $P^T A P = \begin{bmatrix} 4 & 0 & 0 \\ 0 & 1 & 0 \\ 0 & 0 & 1 \end{bmatrix}$.
13. One answer is $(\frac{9}{2} - \frac{1}{2}\sqrt{41})u^2 + (\frac{9}{2} + \frac{1}{2}\sqrt{41})v^2 = 6$.

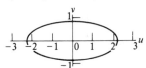

15. One answer is $\begin{bmatrix} 1+2i & 0 \\ 0 & 1-2i \end{bmatrix}$; A is normal but not Hermitian.

10.6 Page 333

1. Indefinite.   3. Positive definite.
5. $P = \begin{bmatrix} \sqrt{2} & \frac{1}{2}\sqrt{2} \\ 0 & \frac{1}{2}\sqrt{6} \end{bmatrix}$; A is positive definite.
7. Positive definite.   9. Indefinite.
11. Positive semidefinite.   13. Positive semidefinite.
21. Minimum $f = -8$ at $(2, 2)$.
23. No maximum nor minimum.

11.1 Page 340

1. Min $x_1 + x_2 = 9$ at $(6, 3)$.   3. Max $2x_1 + x_2 = 8$ at $(4, 0)$.
5. Min $x_1 + 2x_2 = 9$ at any point along the line $x_1 + 2x_2 = 9$ between $(3, 3)$ and $(9, 0)$.

7. Since the lines corresponding to the objective function, $f = 1{,}000x_1 + 1{,}000x_2$, and the structural constraint, $x_1 + x_2 \geq 5$, have the same slope, any point of the set $\{(x_1, x_2) \mid 1 \leq x_1 \leq 3 \text{ and } x_1 + x_2 = 5\}$ will be a minimum point.

11. 100 operations at Refinery 1 and 10 operations at Refinery 2.

13. Two type A trucks and 6 type B trucks.

15. Program A twice a week and Program B four times a week.

17. Minimize CX, subject to $\begin{cases} AX \geq P_0, \\ X \geq 0, \end{cases}$ where $C = [1\ 2]$, $X = \begin{bmatrix} x_1 \\ x_2 \end{bmatrix}$, $A = \begin{bmatrix} 2 & 1 \\ 1 & 1 \\ 1 & 2 \end{bmatrix}$,

and $P_0 = \begin{bmatrix} 8 \\ 6 \\ 9 \end{bmatrix}$.

19. Maximize CX, subject to $\begin{cases} AX \geq P_0, \\ X \geq 0, \end{cases}$ where $C = [30{,}000\ \ 10{,}000]$, $X = \begin{bmatrix} x_1 \\ x_2 \end{bmatrix}$,

$A = \begin{bmatrix} -20 & -10 \\ 1 & 1 \end{bmatrix}$, and $P_0 = \begin{bmatrix} -80 \\ 6 \end{bmatrix}$.

11.2 Page 346

1. $\begin{bmatrix} -1 & -1 & 1 & -1 & 0 \\ -2 & 1 & 1 & 0 & -1 \end{bmatrix} \begin{bmatrix} x_1 \\ x_2 \\ x_3 \\ x_4 \\ x_5 \end{bmatrix} = \begin{bmatrix} 2 \\ 1 \end{bmatrix}$ and $X \geq 0$; $(0, 0, 0, -2, -1)$ is a

basic nonfeasible and nondegenerate solution; $(0, 0, 2, 0, 1)$ is a basic feasible and nondegenerate solution.

3. $\begin{bmatrix} 1 & 2 & 1 & -1 & 0 \\ 1 & 1 & 0 & 0 & 1 \\ 1 & 1 & 1 & 0 & 0 \end{bmatrix} \begin{bmatrix} x_1 \\ x_2 \\ x_3 \\ x_4 \\ x_5 \end{bmatrix} = \begin{bmatrix} 8 \\ 5 \\ 2 \end{bmatrix}$ and $X \geq 0$; $(0, 0, 2, -6, 5)$ is a basic

nonfeasible and nondegenerate solution; $(2, 0, 0, -6, 3)$ is a basic nonfeasible and nondegenerate solution.

5. (b) $(0, 0)$, $(5, 0)$, $(7, 0)$, $(1, 2)$, $(0, \frac{7}{3})$, $(0, \frac{5}{2})$.

(c) $\begin{bmatrix} 1 & 2 & 1 & 0 \\ 1 & 3 & 0 & 1 \end{bmatrix} \begin{bmatrix} x_1 \\ x_2 \\ x_3 \\ x_4 \end{bmatrix} = \begin{bmatrix} 5 \\ 7 \end{bmatrix}$; basic solutions: $(1, 2, 0, 0)$, $(7, 0, -2, 0)$,

$(5, 0, 0, 2)$, $(0, \frac{7}{3}, \frac{1}{3}, 0)$, $(0, \frac{5}{2}, 0, -\frac{1}{2})$, $(0, 0, 5, 7)$.

(d) Basic feasible solutions: $(1, 2, 0, 0)$, $(5, 0, 0, 2)$, $(0, \frac{7}{3}, \frac{1}{3}, 0)$, $(0, 0, 5, 7)$.

(e) No.

ANSWERS TO SELECTED ODD-NUMBERED EXERCISES   A33

7. (a) Maximize CX subject to $AX = P_0$ and $X \geq 0$ where

$$C = [4 \quad -6 \quad 5 \quad 0 \quad 0 \quad 0 \quad 0], \quad X = \begin{bmatrix} x_1 \\ x_2 \\ x_3 \\ x_4 \\ x_5 \\ x_6 \\ x_7 \end{bmatrix},$$

$$A = \begin{bmatrix} -1 & 1 & 0 & 1 & 0 & 0 & 0 \\ 0 & 1 & 2 & 0 & 1 & 0 & 0 \\ 2 & 0 & 1 & 0 & 0 & 1 & 0 \\ 0 & 2 & 1 & 0 & 0 & 0 & 1 \end{bmatrix}, \quad P_0 = \begin{bmatrix} p_{10} \\ p_{20} \\ p_{30} \\ p_{40} \end{bmatrix};$$

(b) $V(R)$ is the vector space of all 7 by 1 column R-matrices, and $U(R)$ is the column space of
$$\begin{bmatrix} -1 & 1 & 0 & 1 & 0 & 0 & 0 \\ 0 & 1 & 2 & 0 & 1 & 0 & 0 \\ 2 & 0 & 1 & 0 & 0 & 1 & 0 \\ 0 & 2 & 1 & 0 & 0 & 0 & 1 \end{bmatrix}$$
or the vector space of all 4 by 1 column R-matrices; $V(R)$ has dimension 7 and $U(R)$ has dimension 4;

(c) $b = \left\{ \begin{bmatrix} -1 \\ 0 \\ 0 \\ 0 \\ 0 \\ 0 \\ 0 \end{bmatrix}, \begin{bmatrix} 0 \\ 1 \\ 0 \\ 0 \\ 0 \\ 0 \\ 0 \end{bmatrix}, \begin{bmatrix} 0 \\ 0 \\ 1 \\ 0 \\ 0 \\ 0 \\ 0 \end{bmatrix}, \begin{bmatrix} 0 \\ 0 \\ 0 \\ 1 \\ 0 \\ 0 \\ 0 \end{bmatrix}, \begin{bmatrix} 0 \\ 0 \\ 0 \\ 0 \\ 1 \\ 0 \\ 0 \end{bmatrix}, \begin{bmatrix} 0 \\ 0 \\ 0 \\ 0 \\ 0 \\ 1 \\ 0 \end{bmatrix}, \begin{bmatrix} 0 \\ 0 \\ 0 \\ 0 \\ 0 \\ 0 \\ 1 \end{bmatrix} \right\};$

(d) $b' = \left\{ \begin{bmatrix} 1 \\ 0 \\ 0 \\ 0 \end{bmatrix}, \begin{bmatrix} 0 \\ 1 \\ 0 \\ 0 \end{bmatrix}, \begin{bmatrix} 0 \\ 0 \\ 1 \\ 0 \end{bmatrix}, \begin{bmatrix} 0 \\ 0 \\ 0 \\ 1 \end{bmatrix} \right\}, e' = \left\{ \begin{bmatrix} 0 \\ 2 \\ 1 \\ 1 \end{bmatrix}, \begin{bmatrix} 1 \\ 0 \\ 0 \\ 0 \end{bmatrix}, \begin{bmatrix} 0 \\ 1 \\ 0 \\ 0 \end{bmatrix}, \begin{bmatrix} 0 \\ 0 \\ 1 \\ 0 \end{bmatrix} \right\};$

(e) $X = \begin{bmatrix} 0 \\ 0 \\ 0 \\ 1 \\ 4 \\ 6 \\ 1 \end{bmatrix}$ is one answer.

11.3   Page 350

1. Minimum is 0 at $(0, 0, 4, 9)$; maximum is $\frac{27}{4}$ at $(\frac{9}{4}, 0, \frac{7}{4}, 0)$.

3. Minimum is 0 at $(0, 0, 4)$; maximum is 8 at $(2, 2, 0)$.

5. Minimum is 36 at $(\frac{30}{7}, \frac{2}{7})$ and at $(0, 6)$, and at all points on the line segment connecting these two points.

A34 ANSWERS TO SELECTED ODD-NUMBERED EXERCISES

7. (a) $(0, 0, 3, 4)$, $(0, 4, 0, -2)$, $(2, 0, 0, 0)$, $(0, 2, 2, 0)$;
   (b) $(0, 0, 3, 4)$, $(2, 0, 0, 0)$, and $(0, 2, 2, 0)$ are feasible and $(2, 0, 0, 0)$ is degenerate;
   (c) basis $\{A_3, A_4\}$ corresponds to $(0, 0, 3, 4)$, basis $\{A_2, A_3\}$ corresponds to $(0, 2, 2, 0)$, basis $\{A_2, A_4\}$ corresponds to $(0, 4, 0, -2)$ and bases $\{A_1, A_2\}$, $\{A_1, A_3\}$, and $\{A_1, A_4\}$ correspond to $(2, 0, 0, 0)$;
   (d) $(0, 2, 2, 0)$, $\{A_2, A_3\}$.

11.4 Page 354

1. (a) Does not meet the restriction because $I_{(2)}$ is not present in the matrix A;
   (b) does not meet the restriction because $I_{(1)}$ is not present in the matrix A;
   (c) does meet the restriction.

3. $A_1 = \begin{bmatrix} 1 \\ 3 \\ 0 \end{bmatrix}$, $A_2 = \begin{bmatrix} 1 \\ 1 \\ 2 \end{bmatrix}$, $A_4 = \begin{bmatrix} 0 \\ 1 \\ 0 \end{bmatrix}$, $I_{(3)} = \begin{bmatrix} 0 \\ 0 \\ 1 \end{bmatrix}$, $P_0 = \begin{bmatrix} 4 \\ 2 \\ 3 \end{bmatrix}$; $\dfrac{p_{20}}{a_{21}} = \dfrac{2}{3}$, $\dfrac{p_{10}}{a_{12}} = \dfrac{4}{1}$.

11.5 Page 361

1. (a) $\begin{bmatrix} 2 & 1 & 1 & 0 \\ 1 & 1 & 0 & 1 \end{bmatrix} \begin{bmatrix} x_1 \\ x_2 \\ x_3 \\ x_4 \end{bmatrix} = \begin{bmatrix} 6 \\ 5 \end{bmatrix}$; $(0, 0, 6, 5)$;
   (b) $(3, 0, 0, 2)$;
   (c) Yes; no; the choice of a minimum ratio (in Theorem 2) requires that if A_1 enters the basis, then A_3 must leave the basis if the new solution is to be basic and feasible;
   (d) $(0, 5, 1, 0)$; $(1, 4, 0, 0)$.

3. (a) $\{I_{(1)}, I_{(2)}, A_1\}$;   (b) $\{I_{(1)}, I_{(2)}, A_2\}$;   (c) $\{A_3, I_{(2)}, I_{(3)}\}$.

5. (a) $(0, 0, 0, 8, 4, 2)$;   (b) $A' = \begin{bmatrix} 0 & -1 & 5 & 1 & 0 & -3 \\ 0 & -3 & 1 & 0 & 1 & -1 \\ 1 & 1 & -1 & 0 & 0 & 1 \end{bmatrix}$;   (c) $\begin{bmatrix} 2 \\ 2 \\ 2 \end{bmatrix}$;

   (d) $(2, 0, 0, 2, 2, 0)$;   (e) $Q^{-1} = \begin{bmatrix} 1 & 0 & -3 \\ 0 & 1 & -1 \\ 0 & 0 & 1 \end{bmatrix}$.

7. (a) $(0, 0, 0, 8, 4, 2)$;   (b) $\begin{bmatrix} 1 & 0 & 4 & 1 & 0 & -2 \\ 3 & 0 & -2 & 0 & 1 & 2 \\ 1 & 1 & -1 & 0 & 0 & 1 \end{bmatrix}$;   (c) $\begin{bmatrix} 4 \\ 8 \\ 2 \end{bmatrix}$;

   (d) $(0, 2, 0, 4, 8, 0)$;   (e) $Q^{-1} = \begin{bmatrix} 1 & 0 & -2 \\ 0 & 1 & 2 \\ 0 & 0 & 1 \end{bmatrix}$.

9. (a) $\{I_{(1)}, A_1, I_{(3)}\}, \{I_{(1)}, I_{(2)}, A_1\}$;
   (b) $(2, 0, 1, 0, 0)$, $(2, 0, 1, 0, 0)$;   (c) degenerate basic feasible solution.

11. (a) $(0, 0, 2, 1, 6)$; (b) $2\begin{bmatrix}1\\0\\0\end{bmatrix} + 1\begin{bmatrix}0\\1\\0\end{bmatrix} + 6\begin{bmatrix}0\\0\\1\end{bmatrix} = \begin{bmatrix}2\\1\\6\end{bmatrix}$;

(c) $a_{12}\begin{bmatrix}1\\0\\0\end{bmatrix} + a_{22}\begin{bmatrix}0\\1\\0\end{bmatrix} + a_{32}\begin{bmatrix}0\\0\\1\end{bmatrix} = \begin{bmatrix}a_{12}\\a_{22}\\a_{32}\end{bmatrix}$;

(d) $P_0 = tA_2 + (2 - ta_{12})I_{(1)} + (1 - ta_{22})I_{(2)} + (6 - ta_{32})I_{(3)}$;

(e) $t = \dfrac{6}{a_{32}}$; (f) $\left(0, \dfrac{6}{a_{32}}, 2 - \dfrac{6a_{12}}{a_{32}}, 1 - \dfrac{6a_{22}}{a_{32}}, 0\right)$;

(g) $t = \dfrac{6}{a_{32}}$ is min $\left\{\dfrac{p_{i0}}{a_{i2}}$ for $i = 1, 2, 3$ and $a_{i2} > 0\right\}$.

13. (a) $b = \{I_{(1)}, \ldots, I_{(n)}\}$; no;
    (b) $b' = \{I_{(1)}, \ldots, I_{(m)}\}$;
    (c) yes;
    (d) yes; $Q^{-1}A = A'$ or $A = QA'$, where Q is an elementary row transformation matrix.

15. No, A_1 cannot replace $I_{(2)}$ in the basis, because Theorem 2 requires $a_{ik} > 0$ if $I_{(i)}$ is to be replaced by A_k, and in this example $a_{21} = 0$.

11.6 Page 369

1. (a) $\hat{C} = [0 \ 0]$; (b) $c_1 - z_1 = 1, c_2 - z_2 = -3$; (c) A_1.
3. (a) $\hat{C} = [2 \ 0 \ 0]$; (b) $c_1 - z_1 = -1, c_2 - z_2 = -3$; (c) A_1 or A_2.
5. (a) $x_{(1)} = 4, x_{(2)} = 4, x_{(3)} = 6, f_1 = 8, \hat{C} = [0 \ 2 \ 0], C = [1 \ 1 \ 0 \ 0 \ 2 \ 0]$,
   $Z = [0 \ 2 \ 4 \ 0 \ 2 \ 0], C - Z = [1 \ -1 \ -4 \ 0 \ 0 \ 0], A_1$;
   (b) A_2 or A_3; (c) A_4; (d) A_6; (e) A_5.

7. (a) Change in f per unit of A_k if A_k is brought into the basis ($t \neq 0$).
   (b) To determine which nonbasis vector should be brought into the basis at each iteration.
   (c) $z_k = \hat{C} \cdot A_k$.
   (d) The coefficients of the basic variables in the objective function.
   (e) c_k is the coefficient of x_k in the objective function.
   (f) Given in the proof of Theorem 3.

9. Maximum $f = 4$ at $(2, 0, 0, 0)$.

11. Consider the linear programming problem with m structural constraints: Minimize CX, subject to $AX = P_0$, and $X \geq 0$, where $P_0 \geq 0$, and all of the columns of I_m are columns of A; if there exists some column of A, say A_k, such that $A_k \not\leq 0$ and $c_k - z_k$ is negative, then CX at the current basic feasible solution is greater than or equal to CX at the basic feasible solution obtained by introducing A_k into the basis. If, however, all $c_k - z_k$ are nonnegative, then minimum CX occurs at the current basic feasible solution.

13. No, because the theorem says that if there is some column of A, say, A_k, such that $c_k - z_k$ is positive, then CX **may** not be maximized at the current basic feasible solution.

11.7 Page 374

1. A_2 must enter the basis but $A_2 = \begin{bmatrix} -2 \\ -1 \end{bmatrix}$; therefore, f is unbounded.

3. (b) $\begin{bmatrix} 1 & -1 & 1 & 0 \\ -3 & 1 & 0 & 1 \end{bmatrix} \begin{bmatrix} x_1 \\ x_2 \\ x_3 \\ x_4 \end{bmatrix} = \begin{bmatrix} 2 \\ 3 \end{bmatrix}$;

   (c) $C = [1\ -2\ 0\ 0]$, $\hat{C} = [0\ 0]$, $Z = [0\ 0\ 0\ 0]$, $C - Z = [1\ -2\ 0\ 0]$;
   (d) The new basis is $\{A'_3, A'_2\}$ and because the coefficients of x_3 and x_2 in the objective function are 0 and -2 respectively, then $\hat{C} = [0\ -2]$. $C - Z' = [-5\ 0\ 0\ 2]$.

5. After one iteration, A'_1 must enter the basis but $A'_1 = \begin{bmatrix} -13 \\ -3 \end{bmatrix}$; therefore, the objective function is unbounded.

9. After one iteration, A'_1 must enter the basis but $A'_1 = \begin{bmatrix} -1 \\ -1 \end{bmatrix}$; therefore, the function $2z_1 + z_2$ is unbounded.

11.8 Page 381

1. $\begin{bmatrix} 1 & 1 & 2 & 1 & 0 & | & 2 \\ 1 & 2 & 1 & 0 & 1 & | & 5 \\ \hline 1 & -1 & -2 & 0 & 0 & | & f-2 \end{bmatrix}$; $f = 4$ at $(2, 0, 0, 0)$.

3. $\begin{bmatrix} 1 & 1 & 1 & 0 & | & 4 \\ 1 & 4 & 0 & 1 & | & 7 \\ \hline 1 & 2 & 0 & 0 & | & f-0 \end{bmatrix}$; $f = 5$ at $(3, 1)$.

5. $\begin{bmatrix} 1 & 3 & 1 & 1 & 0 & 0 & | & 4 \\ 1 & 0 & 1 & 2 & 0 & 1 & 0 & | & 5 \\ 0 & 1 & 1 & 0 & 0 & 0 & 1 & | & 2 \\ \hline 2 & 1 & 6 & 1 & 0 & 0 & 0 & | & f-0 \end{bmatrix}$; $f = 16$ at $(2, 0, 2, 0)$.

7. $\begin{bmatrix} 1 & -2 & 1 & 0 & | & 4 \\ 1 & 1 & 0 & 1 & | & 6 \\ \hline -1 & 4 & 0 & 0 & | & f-8 \end{bmatrix}$; $f = 4$ at $(4, 0, 0)$.

9. $\begin{bmatrix} 2 & 0 & -1 & 0 & 1 & 0 & 0 & | & 4 \\ 1 & 1 & 1 & 1 & 0 & 1 & 0 & | & 8 \\ -1 & 2 & 0 & -1 & 0 & 0 & 1 & | & 2 \\ \hline -1 & 2 & 0 & 1 & 0 & 0 & 0 & | & f-0 \end{bmatrix}$; $f = \frac{34}{3}$ at $(0, \frac{10}{3}, 0, \frac{14}{3})$.

11. $\begin{bmatrix} -1 & 5 & -2 & 1 & 0 & 0 & | & 10 \\ 2 & 1 & -1 & 0 & 1 & 0 & | & 5 \\ 1 & -1 & 2 & 0 & 0 & 1 & | & 4 \\ \hline 2 & -1 & 1 & 0 & 0 & 0 & | & f-0 \end{bmatrix}$; $f = \frac{31}{5}$ at $(\frac{14}{5}, 0, \frac{3}{5})$.

13. $x_1 + x_2$ is unbounded.

11.9 Page 387

1. A_3.

3. A_3.

5. $\begin{bmatrix} -\frac{1}{2} & -\frac{3}{2} & 0 & -1 & \frac{1}{2} & 1 & -\frac{1}{2} & \vdots & \frac{3}{2} \\ -\frac{1}{2} & \frac{1}{2} & 1 & 0 & -\frac{1}{2} & 0 & \frac{1}{2} & \vdots & \frac{1}{2} \\ \hline (3+\frac{1}{2}M) & (-1+\frac{3}{2}M) & 0 & M & (2-\frac{1}{2}M) & 0 & (-2+\frac{3}{2}M) & \vdots & f-\frac{3}{2}M-2 \end{bmatrix}$.

7. $\begin{bmatrix} -\frac{1}{2} & \frac{3}{2} & 0 & -1 & \frac{5}{2} & 0 & 1 & -\frac{5}{2} & \vdots & \frac{11}{2} \\ -\frac{1}{2} & -\frac{1}{2} & 1 & 0 & -\frac{1}{2} & 0 & 0 & \frac{1}{2} & \vdots & \frac{5}{2} \\ -\frac{1}{2} & & \frac{5}{2} & 0 & 0 & \frac{1}{2} & 1 & 0 & -\frac{1}{2} & \vdots & \frac{7}{2} \\ \hline (\frac{5}{2}+\frac{1}{2}M) & (\frac{1}{2}-\frac{3}{2}M) & 0 & M & (-\frac{1}{2}-\frac{5}{2}M) & 0 & 0 & (\frac{1}{2}+\frac{7}{2}M) & \vdots & f-\frac{11}{2}M+\frac{5}{2} \end{bmatrix}$.

9. $f = \frac{7}{2}$ at $(0, 0, \frac{1}{2}, \frac{3}{2})$.   **11.** $f = 5$ at $(1, 3)$.

13. $f = -1$ at $(1, 3, 2)$.

15. (a) If there exists some column of A, say A_k, such that $A_k \leq 0$ and $c_k - z_k$ is positive for maximum problems and $c_k - z_k$ is negative for minimum problems.

(b) If an optimum solution has been found and there is at least one nonbasic $A_j \not< 0$ that is not in the basis for which $c_j - z_j = 0$.

(c) Artificial vectors are in the basis with corresponding nonzero artificial variables and the augmented problem has been optimized. (Contrapositive of Theorem 5.)

17. The objective function, $2x_1$, is unbounded.

11.10 Page 392

1. $f = 4$ at $(2, 0, 0, 0)$.   **3.** $f = 5$ at $(3, 1)$.
5. $f = 16$ at $(2, 0, 2, 0)$.   **7.** $f = 8$ at $(0, 0, 2)$.
9. f is unbounded.   **11.** $f = 2$ at $(\frac{10}{3}, 0, \frac{14}{3}, 0)$.

Index

Abelian group, 150
Addition of coordinate vectors, 95
Addition of matrices, 10
Address of an entry of a matrix, 4
Adjoint matrix, 139
Algebra,
   associative, 167
   binary Boolean, 170
   Boolean, 169
   commutative, 167
   division, 167
   linear, 167
   with unity, 167
Algebra with unity, 167
Algebraic system, 149
Angle between vectors, 205
Artificial variables, 382
Associative Algebra, 167
Associative property,
   for addition of matrices, 11
   for multiplication of matrices, 44
Augmented linear programming problem, 382
Augmented matrix, 17

Basic feasible solution, 345
Basic solution, 345
Basic variables, 345
Basis change, 239, 259
Basis of a vector space, 185
Binary Boolean algebra, 170

Binary operation, 149
   closed set under, 149
Binary relation, 25, 251
Boolean Algebra, 169
Boolean Matrices, 170

Cancellation property,
   for addition of matrices, 11
   for multiplication of matrices, 56, 57
Canonical form for equivalence, 62, 63
Canonical set under equivalence, 63
Cartesian product, 25, 149, A1
Cayley-Hamilton theorem, 273
Cayley product, 40
Change of basis, 239, 259
Characteristic,
   equation, 271
   function, 273
   polynomial, 273
   roots of a linear operator, 269
   roots of a matrix, 271
   value of a linear operator, 269
   value of a matrix, 271
   vector of a linear operator, 269
   vector of a matrix, 271
Closed set, 149
C-matrix, 4
Codomain of a mapping, 1, 217
Coefficient matrix, 17

Cofactor, 117
Cofactor expansion of a determinant, 117
Cofactor matrix, 139
Cofactoral matrix, 175
Column equivalent matrices, 61
Column matrix, 4
Column of a matrix, 4
Column space, 188
Column vector, 4
Combination, linear, 178
Commutative algebra, 167
Commutative group, 150
Commutative property,
 for addition of matrices, 11
Commutative ring, 155
Commutator, 52
Complete solution of a system of linear equations, 32
Complex entries of a matrix, 4
Complex inner product space, 202
Components of a coordinate vector, 94
Composite of transformations, 264
Conformable matrices,
 for addition, 10
 for multiplication, 42
Congruence transformation, 288, 306
Congruent matrices, 288, 306
Conjugate matrix, 88
Conjunctive matrices, 311
Consistent system of equations, 29
Constraints, 335
 nonnegativity, 338
 structural, 338
Contraction, 218
Convex combination of vectors, 350
Coordinate matrix, 195
Coordinates, 94
Coordinates of a vector,
 with respect to a basis, 195
Coordinate vector(s),
 addition of, 95
 component of, 94
 definition of, 94
 dimension of, 94
 dot product of, 95

Coordinate vector(s) (*cont.*)
 equality of, 95
 geometric representation of, 96
 length of, 96
 magnitude of, 96
 multiplication by scalar, 95
 nonzero, 94
 normal, 96
 normalized, 97
 real, 94
 standard inner product of, 95
 unit, 96
 zero, 94
Cramer's Rule, 119
Cross product of matrices, 52
Cryptography, 222
Crystallography, 245
Cycling, 360, 379

Degeneracy, 349, 360, 385, 387
Degenerate basic solution, 345, 349
Degree of a polynomial, 151
Dependence, linear, 180
Determinant, 111
 cofactor expansion of, 117
 definition of, 113
 expansion of, 113
 function, 111
 history of, 111
 line of, 121
 order of, 114
Diagonal block matrix, 105
Diagonal, main, of a matrix, 4
Diagonal matrix, 104, 288
Dimension of a coordinate vector, 94
Dimension of a vector space, 186
Direct sum, 105
Directed line segment, 162, 163
Distance between two vectors, 204
Distributive property, 44, 45
Division Algebra, 167
Division Ring, 155
Domain of a mapping, 1, 217
Dot product of coordinate vectors, 95
Dot product function, 199

Echelon form, 26

INDEX

Echelon matrix, 25
Eigenvalues, 269
Eigenvector, 269
Elementary column operations, 61
Elementary column transformation matrix, 63
Elementary matrices, 64
Elementary operations, 61
Elementary operations on equations, 18
Elementary row operations, 24
Elementary row transformation matrix, 63
Entry of a matrix, 3
Equal matrices, 5
Equivalence relation, 251
Equivalent matrices, 61
Equivalent systems of linear equations, 18
Euclidean space, 202
Evaluator, 365
Expansion of a determinant, 113

Feasible set, 336
Feasible solution, 336
Field, 157
Finite-dimensional vector space, 188
Finite group, 152
Flexibility matrix, 224
Function, 1, 217
  determinant, 111
  objective, 338
  optimum values of, 335, 351
Fundamental variables, 32

Gauss-Jordan method, 28, 33
Generate a vector space, 179
Geometric representation of a vector, 96, 163
Geometric solution of linear programming problem, 336
Geometric Vector, 162
  addition, 162
  length, 163
  multiplication by scalar, 163
  zero, 162
Gram-Schmidt formulas, 211

Gram-Schmidt orthogonalization, 211
Group, 149
  Abelian, 150
  commutative, 150
  finite, 152

Hermitely congruent matrices, 311
Hermitian matrix, 89
Homogeneous system of linear equations, 34

Identity element, 52
Identity matrix, 5, 43
Image, 1
Image set, 2
Inconsistent system of equations, 29
Indefinite real quadratic form, 327
Independence, linear, 181
Index of a matrix, 308
Inequality of real matrices, 337
Inner product function, 199
Inner product space,
  complex, 202
  Euclidean, 202
  real, 202
  unitary, 202
Inner product of vectors, 199
Integer Programming, 339
Interchange, 112, 113
Intersection of matrices, 170
Inverse element, 52
Inverse matrix, 53, 67, 139
Inversion of matrices,
  adjoint method, 140
  by elementary row operations, 54
Invertible matrix, 53
Isomorphic vector spaces, 193
Isomorphism, 193

Jordan product, 52

Latent roots, 269
Left cancellation property, 57
Left distributive property, 44
Length of a coordinate vector, 96
Length of a vector, 203
Lie product, 52

Line of a determinant, 121
Linear algebra,
  definition, 167
Linear combination, 178
Linear dependence, 180
Linear equations, systems of, 16
  consistent, 29
  equivalent, 18
  homogeneous, 34
  inconsistent, 29
  nonhomogeneous, 34
  rank of, 135
  relation to matrices, 17
  solution of, 16
    basic, 345
    by Cramer's Rule, 119
    by Gauss-Jordan elimination method, 28, 33
    by matrix inversion, 57
    complete, 32
    particular, 32
    trivial, 34
    unique, 16
Linear independence, 181
Linear operator, 232
Linear programming, 335
  degeneracy, 349, 360, 385, 387
  history, 338
  methods of solution,
    geometric, 336
    simplex, 352, 375
Linear transformation, 2, 219, 263
Lower triangular block matrix, 107
Lower triangular matrix, 106

Magnitude of a coordinate vector, 96
Magnitude of a vector, 203
Main diagonal of a matrix, 4
Mapping, 1, 217
Matrix,
  adjoint, 139
  augmented, 17
  Boolean, 170
  characteristic value of, 271
  characteristic vector of, 271
  coefficient, 17
  cofactor, 139

Matrix (*cont.*)
  column, 4
  column space of, 188
  conjugate, 88
  coordinate, 195
  definition of, 3
  determinant of, 111, 113
  diagonal, 104
  diagonal block, 105
  echelon, 25
  elementary, 64
  entry of, 3
  Hermitian, 89
  identity, 5, 43
  indefinite, 327
  inverse, 53
  inversion of, 54
  invertible, 53
  lower triangular, 106
  lower triangular block, 107
  methods for solution of system of linear equations, 33, 57
  modal, 281
  negative definite, 327
  negative semidefinite, 327
  noninvertible, 53
  nonsingular, 53
  normal form of, 62
  notation for linear programming problem, 337
  null, 4
  order of, 4
  orthogonal, 101
  partitioned, 5
  positive definite, 327
  positive semidefinite, 327
  rank of, 131
  real, 4
  of real quadratic form, 287, 304
  reduced echelon, 25
  representation of a transformation, 226, 256
  row, 4
  row space of, 191
  scalar, 105
  singular, 53
  skew-Hermitian, 89

Matrix (*cont.*)
  skew-symmetric, 79
  square, 4
  stochastic, 47
  symmetric, 78
  trace of a, 201
  tranjugate, 88
  transformation, 226, 256
  transition, 239
  transpose of, 76
  triangular, 106
  triangular block, 107
  unitary, 102
  upper triangular, 106
  upper triangular block, 107
  zero, 4
Minor of an entry of a matrix, 116
Modal matrix, 281
Multiplication of matrices, 40
Multiplication of a coordinate vector by a scalar, 95
Multiplication of a matrix by a scalar, 11

Negative definite real quadratic form, 327
Negative integral powers, 57
Negative semidefinite real quadratic form, 327
Nonassociative operation, 52
Nonbasic variables, 345
Nondegenerate basic solution, 345
Nonhomogeneous system of equations, 34
Noninvertible matrix, 53
Nonsingular matrix, 53
Norm of a vector, 203
Normal coordinate vector, 96
Normal form, 62
Normalized characteristic vector, 272
Normalized coordinate vector, 97
Normalized vector, 204
Null matrix, 4

Operation,
  binary, 149
  elementary, 61

Operation (*cont.*)
  elementary column, 61
  elementary row, 24
Operator, linear, 232
Optimal solution, 345
Order of a determinant, 114
Order of a matrix, 4
Orthogonal basis, 210
Orthogonal congruence transformation, 319
Orthogonal coordinate vectors, 97
Orthogonal linear operator, 314
Orthogonal matrix, 101
Orthogonal set, 97, 206
Orthogonal similarity transformation, 319
Orthogonal vectors, 206
Orthogonally congruent matrices, 288, 320
Orthogonally similar matrices, 288, 320
Orthonormal basis, 210
Orthonormal coordinate vectors, 97
Orthonormal set, 97, 206
Orthonormal vectors, 206

Parameter, 32
Particular solution of a system of linear equations, 32
Partitioning of a matrix, 5
  in matrix inversion, 107
Pauli spin matrices, 47, 92
Pivot column, 376
Pivot entry, 376
Pivot row, 376
Polynomial(s)
  addition of, 151
  definition of, 150
  degree of, 151
  equality of, 151
  multiplication of by scalar, 161
  multiplication of, 166
  zero, 151
Population matrix, 275
Positive definite real quadratic form, 327

Positive semidefinite real quadratic form, 327
Postmultiplication, 43
Powers of a matrix,
 negative integral, 57
 positive integral, 46
 rational, 301, 302
Premultiplication, 43
Probability vector, 47
Product of coordinate vectors,
 dot, 95
 standard inner, 95
Product of determinants, 142
Product of linear transformations, 264
Product of vectors,
 inner, 199
Projection, of a vector, 219

Quadratic form, 286, 304
Quaternions, 175

Range of a mapping, 2
Rank of a matrix, 131
Rank of a system of linear equations, 135
Real coordinate vector, 94
Real inner product space, 202
Real matrix, 4
Real quadratic form, 286, 305
Reduced echelon matrix, 25
Reflection, 218
Relations,
 binary, 25, 251
 equivalence, 251
Revised simplex method, 389
Right cancellation property, 56
Right distributive property, 45
Ring, 154
 commutative, 155
 division, 155
 with unity, 154
R-matrix, 4
Rotation transformation, 218, 270
Row equivalent matrices, 25
Row of a matrix, 4

Row matrix, 4
Row space, 191
Row vector, 4

Scalar, 11, 162
 multiple of a coordinate vector, 95
 multiple of a matrix, 11
 multiple of a polynomial, 161
Scalar matrix, 105
Schwarz inequality, 205
Set,
 closed, 149
 feasible, 336
 finite, 152
 infinite, 152
 unbounded feasible, 371
Shear, 219
Similar matrices, 251, 297
Similarity transformation, 296
Simplex matrix, 375
Simplex method, 352, 375
Simplex tableau, 375
Singular matrix, 53
Skew-Hermitian matrix, 89
Skew-symmetric matrix, 79
Slack variables, 343
Solution,
 basic, 345
 basic feasible, 345
 complete, 32
 degenerate basic, 345, 349
 feasible, 336
 optimal, 345
 particular, 32
 of system of equations, 16
 trivial, 34
Span a vector space, 179
Square matrix, 4
Standard inner product, 95
Stiffness matrix, 224
Stochastic matrix, 47
Stretching, 218
Submatrix, 5
Subring, 156
Subspace, 180
Subtraction, matrix, 12
Sum of matrices, 10

Summation notation, 41, 114
Symmetric matrix, 78
System of linear equations. *See* linear equations

Trace of a matrix, 201
Tranjugate matrix, 88
Transformation, 1, 217
  congruence, 288
  elementary, 63
  linear, 2, 219, 263
  matrix, 222
  orthogonal, 314
  orthogonal congruence, 288
  orthogonal similarity, 288
  of the plane, 218
  similarity, 296
  unitary, 316
Transition, 245
Transition matrix, 239
Transpose of a matrix, 76
Triangular block matrix, 107
Triangular matrix, 106
Trivial solution, 34

Unbounded feasible set, 371
Unbounded objective function, 371
Underlying field of a vector space, 164
Union of matrices, 170
Unique solution, 16
Unit coordinate vector, 96
Unit vector, 204
Unitarily congruent matrices, 291
Unitarily similar matrices, 291
Unitary linear operator, 316
Unitary matrix, 102
Unitary space, 202
Unitary transformation, 316
Upper triangular block matrix, 107
Upper triangular matrix, 106

Vector(s),
  basis, 355, 363
  column, 4
  coordinate, 94
  coordinates with respect to a basis, 195
  definition of, 162, 164
  geometric, 162
  geometric representation of, 96, 163
  inner product of, 199
  length of, 203
  linear combination of, 178
  linear dependence of, 180
  linear independence of, 181
  linear transformation of, 219
  magnitude of, 203
  matrix transformation of, 222
  norm of, 203
  normalized, 204
  orthogonal, 206
  orthonormal, 206
  probability, 47
  row, 4
  unit, 204
Vector space,
  basis for, 185
  definition of, 162
  dimension of, 186, 188
  finite-dimensional, 188
  geometric examples of, 162-164
  subspace of, 180
  underlying field of, 164

Work, 206, 207

Zero coordinate vector, 94
Zero geometric vector, 162
Zero matrix, 4
Zero polynomial, 151